OCEAN WAVE KINEMATICS, DYNAMICS AND LOADS ON STRUCTURES

PROCEEDINGS OF THE
1998 INTERNATIONAL OTRC SYMPOSIUM

April 30–May 1, 1998
Houston, Texas

SPONSORED BY
Offshore Technology Research Center
American Society of Civil Engineers

EDITED BY
Jun Zhang

1801 ALEXANDER BELL DRIVE
RESTON, VIRGINIA 20191–4400

Abstract: An international symposium on "Wave Kinematics, Dynamics and Loads on Structures" was held in Houston, Texas, from April 30 to May 1, 1998. The proceedings of this symposium provides a collection of papers presenting recent advances in the understanding, measurement and prediction of wave kinematics, wave dynamics, and wave loads acting on offshore and coastal structures. This international exchange of knowledge and data on new theories, models, statistics, and measurements of ocean waves and wave-structure interactions presents the state of the art. It provides a valuable reference for professionals engaged in the fields of ocean science and engineering.

Library of Congress Cataloging-in-Publication Data

Ocean wave kinematics, dynamics and loads on structures: proceedings of the 1998 International OTRC Symposium: Houston, Texas, April 30-May 1, 1998/edited by Jun Zhang.
    p.   cm.
    "1998 International OTRC Symposium on Ocean Wave Kinematics, Dynamics and Loads on Structures, held at the Houston Plaza Hilton Hotel in Houston, Texas"–Pref.
    ISBN 0-7844-0336-8
    1. Ocean waves–Congresses. 2. Kinematics–Congresses. I. Zhang, Jun.
II. International OTRC Symposium on Ocean Wave Kinematics, Dynamics and Loads on Structures (1998: Houston, Texas)
GC205.0253   1998   98-15974
551.47'02–dc21   CIP

Any statements expressed in these materials are those of the individual authors and do not necessarily represent the views of ASCE, which takes no responsibility for any statement made herein. No reference made in this publication to any specific method, product, process or service constitutes or implies an endorsement, recommendation, or warranty thereof by ASCE. The materials are for general information only and do not represent a standard of ASCE, nor are they intended as a reference in purchase specifications, contracts, regulations, statutes, or any other legal document. ASCE makes no representation or warranty of any kind, whether express or implied, concerning the accuracy, completeness, suitability, or utility of any information, apparatus, product, or process discussed in this publication, and assumes no liability therefore. This information should not be used without first securing competent advice with respect to its suitability for any general or specific application. Anyone utilizing this information assumes all liability arising from such use, including but not limited to infringement of any patent or patents.

Photocopies. Authorization to photocopy material for internal or personal use under circumstances not falling within the fair use provisions of the Copyright Act is granted by ASCE to libraries and other users registered with the Copyright Clearance Center (CCC) Transactional Reporting Service, provided that the base fee of $8.00 per chapter plus $.50 per page is paid directly to CCC, 222 Rosewood Drive, Danvers, MA 01923. The identification for ASCE Books is 0-7844-0336-8/98/ $8.00 + $.50 per page. Requests for special permission or bulk copying should be addressed to Permissions & Copyright Dept., ASCE.

Copyright © 1998 by the American Society of Civil Engineers,
All Rights Reserved.
Library of Congress Catalog Card No: 98-15974
ISBN 0-7844-0336-8
Manufactured in the United States of America.

Contents

Preface ix

Keynote Presentation

On the Importance of Understanding Ocean Waves Kinematics for Calculation of Dynamics and Loads on Offshore Structures 1
  Ove T. Gudmestad*

Wave Theory

A Second-Order Solution of the Flap Wavemaker Problem 9
  Jaw-Fang Lee*, Cheng-Chi Liu, and Yuan-Jyh Lan

Steep Transient Waves in a Directionally Spread Sea 17
  T. B. Johannessen* and C. Swan

Hybrid Wave Models and Their Applications 25
  Jun Zhang*, Jun Yang, and Jiangang Wen

Long Waves–Short Waves Resonance in Three Dimensions 34
  Derek W.C. Lai, Andy T. Chan, and Kwok W. Chow*

Wave Kinematics

Some Aspects on Wave Kinematics and Wave Measurements 40
  Igor Prislin* and John Halkyard

Irregular Wave Kinematics from PUV Gauge 48
  Steven A. Hughes* and Rodney J. Sobey

Wave Kinematics Based on a Lagrangian Formulation 56
  Geir Moe*, Øivind A. Arntsen, and Svein H. Gjøsund

Visualization of Wave Kinematics on the World Wide Web: An Interactive Instructional Tool 64
  F.A. Hardjanto, S.A. Kinnas*, H. Hart, and M.H. Kim

The Schrödinger Method for Water Wave Kinematics 76
  Karsten Trulsen*, Jean-Marc Temmos, Kristian B. Dysthe,
  Ove T. Gudmestad, and Manuel G. Velarde

Wave Simulation

Nonlinear Force Calculations by Numerical Wave Tank Simulations 84
  M.H. Kim*, M.S. Celebi, and J.C. Park

Fully Nonlinear Simulation of Ocean Waves in Interaction with Offshore Structures: Project "UNDA" 96
  Even Mehlum*

Simulation of Nonlinear Waves in Shallow-Water Basins 100
  Keh-Han Wang* and Weimin Li

Second Order Wave Forces on 3-D Bodies in a Current 108
    Jesper Skourup*, Kwok Fai Cheung, Harry B. Bingham, and
    Bjarne Büchmann

Numerical Simulation of Wave Effects on a Combined Breakwater and Platform
Configuration 116
    Hamn-Ching Chen* and Woei-Min Lin

Wave and Structure Interaction: Dynamic Response

Nonlinear Inertial Loading in Steep 2-D Water Waves 126
    C. Swan*, T. Bashir, and Ove T. Gudmestad

Nonlinear Wave Loading and Dynamic Response of Drag-Dominated Offshore
Platforms 134
    Daniel Karunakaran*, Nils Spidsøe, and Ove T. Gudmestad

On the Contribution of the Mooring System to the Damping of the Slow Oscillation
of Moored Floating Structure 147
    Ju Fan, Xiaohong Chen, and Xianglu Huang*

Nonlinear Loads and Responses of Ships and Offshore Platforms in Deterministic
and Random Waves 155
    P.D. Sclavounos*, S. Kim, Y. Kim, and D.C. Kring

Spectral Response Surfaces, Designer Waves and the Ringing of Offshore
Structures 171
    Peter S. Tromans* and I. Ketut Suastika

Forced Motions of a Pneumatic Floating Platform 179
    Kwok Fai Cheung*, Ludwig H. Seidl, Amal Phadke, and Sing Kwan Lee

Wave and Structure Interaction: Impact Loads

A Study of Plunging Wave Impact Pressures on Placed Block Revetments 187
    A.L. Gopinath, Eng-Soon Chan*, and Hin-Fatt Cheong

Hurricane Wave Forces on the Decks of Offshore Platforms 195
    Robert Bea*, James Stear, Tao Xu, and Rafael Ramos

Nonlinear Loads on Vertical Columns in Laboratory High Seas 203
    C.H. Kim* and J. Zou

An Investigation of Wave Forces for Design of a Cruise Ship Pier: Bridgetown,
Barbados 215
    Mark Mattila*, Michael Tarbotton, and Andrew Cornett

Forecasting Rig Heave for Drilling Operations in Harsh Environments 225
    Colin Grant*, Stéphane Cornut, Roger Dyer, Martin Holt, and John Mitchell

Wave and Structure Interaction: Load Uncertainty and Reliability

On Wave and Force Uncertainty for Steel Structures 233
    Léon A. Harland* and Paul H. Taylor

Long-Term Response Analysis of Fixed and Floating Structures 240
    Sverre Haver*, Gro Sagli, and Tone M. Gran

Advanced Volterra Models for Fluid-Structure Interactions 249
  Sungbin Im, Edward J. Powers*, and In-Seung Park

Hurricane Wave Conditions for Design and Requalification of Platforms in the Bay
of Campeche, Mexico 257
  Robert Bea*, Joseph Suhayda, Zhaohui Jin, and Rafael Ramos

Reliability and Risk Considerations in Offshore Design and Operations 265
  Cyril E. Arney* and David J. Wisch

Wave Breaking

A Second Order Numerical Algorithm for Simulation of Breaking Wave Impacts
on Structures 273
  Eng-Soon Chan*, Zhi Zong, and Chih-Young Liaw

Breaking Wave Geometry with Emphasis on Steepness and Curvature 281
  Pål F. Lader*, Guttorm Grytøyr, Dag Myrhaug, Bjørnar Pettersen

Numerical Modeling of Breaking Waves over Porous Structures 289
  Pengzhi Lin*, Kuang-An Chang, Philip L.-F. Liu, and Tsutomu Sakakiyama

Breaking Wave Statistics in Nearshore and Surf Zones 297
  Ming-Yang Su*, J.C. Wesson, W.J. Teague, and R.E. Burge

Discriminating Breaking in Deep Water Waves 305
  Richard Seymour*, Charles-Alexandre Zimmermann, and Jun Zhang

Long-Term Wave Evolution

Wind Wave Prediction in Finite Depth Water 313
  I.R. Young*

Numerical Modeling of Storm Waves for Hurricanes Erin and Felix 321
  Lihwa Lin*

WAM4 Model/Data Comparison in the NW Gulf of Mexico 334
  S.F. DiMarco and L.C. Bender, III*

Long-Term and Extreme Waves in the Gulf of Mexico 342
  Chung-Chu Teng*

Wave Statistics

Parametric Characterization of Surface Wave Data 350
  John M. Niedzwecki* and J.W. van de Lindt

A Global Database for Wind Speed and Wave Height 354
  I.R. Young*

Joint Statistics of Waves and Water Levels 362
  David P. Hurdle and Gerbrant Ph. van Vledder*

Wave Crest Distributions: Observations and Second Order Theory 372
  George Z. Forristall*

Wave Climate Parametrization of the Distribution of Maximum Wave Heights
and Crest Elevations 396
  Marc Prevosto and Raymond Nerzic*

Wave Transform from Deep to Shallow Water

Numerical Forecasting of Infra-Gravity Waves near Harbour Entrances 404
   Alexandru Sheremet*

Development and Study of the Green's Function for Water Waves over Variable Bathymetry Domains 412
   Gerassimos A. Athanassoulis* and Konstadinos A. Belibassakis

Focussed Wave Groups on Deep and Shallow Water 420
   Paul H. Taylor* and Erwin Vijfvinkel

Turbulence Measurement in a Towing Tank Using Hot-Film Anemometry 428
   Shukai Wu*

Coastal Process

Turbulent Wave Boundary Layers in the Surf- and Swash-Zone: Analysis of Small Scale Field Measurements 432
   Volker Müller*

Wind Effects on Wave Transformation and Undertow 439
   Joshua Carter, William Hobensack, Douglas Kennedy, and Daniel Cox*

Numerical Simulation of Bed-Load Sediment Transport by Long-Shore Currents 446
   M. Di Natale* and G. Piccirillo

Deposition of Cohesive Sediments Under Waves 454
   S. Abdel-Mawla, A. Matheja*, and C. Zimmermann

Risk Analysis for Wave-Suspended Sediments 462
   Katherine A. Larm and Billy L. Edge*

Coastal Structures

A Risk Analysis of a Submerged Breakwater 470
   Chai-Cheng Huang* and Jia-Shiang Lin

Effects of Onshore Winds on Velocity of Wave Runup 478
   Donald L. Ward*, Jun Zhang, and Christopher G. Wibner

On the Radiation Boundary Conditions for Time-Dependent Parabolic Wave Calculations 486
   Tai-Wen Hsu* and Chih-Chung Wen

Pipelines

A Response-Based Method for Developing Joint Metocean Criteria for Seabed Pipeline Design 494
   Kevin C. Ewans*

Wave Forces on Partially Submerged Pipe Breakwater 505
   J.S. Mani*

On the Average Wave Steepness 513
   Elzbieta M. Bitner-Gregersen, Carlos Guedes Soares, and Alexandra Silvestre

Subject Index	521
Author Index	525

PREFACE

The 1998 International OTRC Symposium on 'Ocean Wave Kinematics, Dynamics Loads on Structures' (WAVE`98) was jointly sponsored by the Offshore Technology Research Center (OTRC) and the American Society of Civil Engineers (ASCE). It was held at the Houston Plaza Hilton Hotel in Houston, Texas from April 30 to May 1, 1998. The first International OTRC Symposium was held in May 1997 in College Station, Texas, under the sole sponsorship of the OTRC, bringing together scientists and engineers with a common interest in the nonlinear dynamics of wave basin model tests.

As evidenced by many journal papers and discussions in related conferences and workshops, the proper understanding of wave kinematics and dynamics of rough seas is crucial to the successful design of offshore and coastal structures and the study of pollutant transport in ocean waters. In recent years, many efforts have been made to develop this understanding and to quantify the kinematics and dynamics of steep ocean waves and their impact on structures. During the meeting of the 'Waves and Wave Forces Committee' of the ASCE at Los Angeles, CA, in May 1996, it was felt an international symposium to exchange and disseminate the knowledge gained from these various efforts was needed. The idea was supported by the executive committee of the Waterway, Port, Coastal and Ocean Division of the ASCE. Encouraged by the success of the first International OTRC Symposium, it was decided that this symposium should further promote the exchange of knowledge between academia and practitioners. The idea was also supported by the OTRC which joined with ASCE in the organization and promotion of the symposium through its strong ties with the offshore industry and provided the needed financial support.

In response to the call for papers, about 70 abstracts were received and reviewed by the organizing committee. Finally, 5x papers were selected for inclusion in the proceedings. Due to time and technical constraints, it was not possible to include in the proceedings all the presentations of the symposium. The collection of papers included in this symposium proceedings reflects recent advances in understanding, measurements, and estimates of wave kinematics, dynamics and their effects on the wave loads acting on offshore and coastal structures. In many aspects, these papers represent the state of the art of wave theories, wave models, statistics and measurements of ocean waves, and wave-structure interactions. Hence, the proceedings provide an important reference for professionals working in the fields of ocean science and engineering.

The valuable assistance and support of the members of the WAVE`98 Organizing Committee, the ASCE Waves and Wave Forces Committee, the Fluid/Structure Group of the OTRC and the Ocean Engineering Program of the Civil Engineering Department at Texas A&M University are gratefully acknowledged. The sponsorship of the Offshore Technology Research Center and the American Society of Civil Engineers is greatly appreciated.

Jun Zhang
Chair of the WAVE`98 Organizing Committee
Associate Professor
Ocean Engineering Program
Department of Civil Engineering
Texas A&M University
College Station, Texas 77843-3136

On the Importance of Understanding Ocean Wave Kinematics for
Calculation of Dynamics and Loads on Offshore Structures.

Ove T. Gudmestad
Statoil
4035 Stavanger, Norway

Abstract

For calculation of hydrodynamic loads on offshore structure, Morison's equation is normally being used in the case of slender or relatively slender structure, and diffraction theory is being used in most other applications. While the Morison equation is empirical, diffraction theory is linked to the basic hydrodynamic principles. None of the methods, however, seem to be capable of explaining the dynamic phenomenon occurring when steep waves interact with offshore structures causing a phenomena like ringing of these structures. Attempts have therefore been made to modify the load calculation methods, generally by including higher order forcing terms in the case of Morison equation, and higher order forcing terms as well as a higher order expansions of the governing hydrodynamic equation and the boundary conditions in the case of diffraction analysis. These aspects are discussed further in the paper, see also Faltinsen, 1990.

The paper emphasis on that the accurate calculation of the loading is contingent upon the accurate knowledge of ocean wave kinematics. Focus will in this respect be put on the numerical values of the kinematics as well as the contribution of the kinematics from linear and higher order components of the wave motion. It's believed that this approach will lead to an improved understanding of the dynamics of offshore structures and some current research topics in this field will be reviewed.

Furthermore, an attempt will be made to point to directions of future research in this field with the ultimate objective of accurate calculation of dynamics and loads on offshore structures.

Introduction

The objective of this paper is to document the link between the needs of the offshore industry and the research related to further understanding of ocean water wave kinematics. It is thought to be important to document this link for those who finance the research, i.e. private companies and government grants, and for the researchers themselves. It is hoped that the paper and the symposium will enhance the interest in resolving current problems related to wave kinematics and loading issues, and that the symposium will serve as inspiration for further work and its financing.

Offshore Engineering, present trends

Current trends in offshore engineering include use of slender and thus dynamic sensitive fixed structures. These are minimum cost monotower structures or jack ups for shallow to medium deep waters, and slender tower structures for deeper waters, like the Draugen monotower concrete platform for Haltenbanken offshore Norway. Furthermore, the trend is clearly towards floating structures like semisubmersibles, tension leg platforms, ships and spar-type structures, even for water depths where fixed jacket structure traditionally have been selected. Although an increased well productivity rate for the deepwater reservoirs has been one of the reasons for this change in Gulf of Mexico waters, other reasons are believed to relate to increased confidence in prediction of loads and response of offshore structure, and to increased awareness of the low costs of removal and the possible reuse of floating structures. The last item is thought to become more and more important as the environmental sensitivity of the offshore platform removal process is coming more and more to the attention of the offshore oil industry.

Current trends in offshore engineering have furthermore included increased reliance on the analytical approach for calculation of load and response of structures, and less emphasis on model testing and in-situ data collection from offshore platforms. This trend was, however, challenged by the detection of the ringing response (transient higher order response of importance for calculation of extreme loads) of the Heidrun tension leg platform and the Draugen monocolumn fixed concrete platform, both installed at the Haltenbanken area offshore mid-Norway in the mid. 1990's. This location is characterised by high waves for considerable parts of the year. By detection of the dynamic higher order response phenomena, the renaissance of the larger testing facilities came about, (OTRC, 1997). At present the Norwegian guidelines for estimating load and load effects on offshore structures (NPD, 1997) are being rewritten to include recommendations for tank-testing wherever there could exist significant uncertainties related to loads on or response of offshore structures.

Through the application of dynamic sensitive fixed structures, and in particular of floating structures, an increased requirement to improved understanding of hydrodynamic phenomena has been seen. Throughout several years analytic tools to assess loads and response of such structures have been improved. Of particular importance for the design of floating structures has been a joint industry programme (WAMIT) managed by the Department of Ocean Engineering at Massachusetts Institute of Technology. (Newman, 1996) The limitation of such computer programmes have also been further understood during the lastest years. It is, for example, realised that the effects of steep and near breaking waves cannot be predicted accurately since the flow around the structures is considerably distorted in such waves and since the convergence of the analytic solutions tends to break down under such extreme loading phenomena. Furthermore, estimation of added mass and in particular damping are of particular uncertainty.

The Shell group of companies has over the last decade spent much efforts on developing a synthetic design wave (a "Newwave") characterised by the statistical properties of the sea state. Reports on excellent comparison between in situ measurements and prediction have been published (First presented by Tromans et al., 1991). The approach has, however, not gained full international confidence at present, as it is thought that a full presentation of basic theory and computer source code is necessary for a full scientific review of theory and results.

Over many years the API recommended practice for design of offshore fixed (jacket) structures (API RP2A) was gradually developed into a de-facto internationally recognised state-of-art design code. The document was, however, developed from the recommended practice for design of shallow water structures into a document that was used for deep water dynamic sensitive structures as well. A particular problem related to an inconsistent description of Morison type of loading, through use of a set of relatively low hydrodynamic coefficients and relatively high values for wave kinematics, found from Stoken V order wave theory or use of Stream Function theory. The latest version of the API recommended practice (API, 1993) has, however, rectified this approach through presenting a more consistent method for calculation of loads and response, utilising state of art knowledge related to both wave kinematics, forcing mechanism and hydrodynamic load coefficients. (see also Gudmestad and Moe, 1996).

Loads and response of offshore structures

In order to calculate the response of offshore structures, state of art has been to design the structures to withstand the loads from a deterministic design wave, characterised by a certain wave height and period, and described by

Stokes V or Stream Function theory. Other descriptions have also been presented, some of them based on the statistical properties of a design storm. An alternative approach, gaining more popularity with the increased availability of computer power, has been to characterise the sea by a wave spectrum, and the significant wave height of a design storm and to generate a synthetic storm sea in the form of a time series which is then run across the structure. The design load and response (action) is thereafter calculated on the basis of the statistics of the extremes. For nonlinear analysis it is, however, necessary to make assumptions related to the composition (and decomposition) of wave components, to the characterisation of individual wave components in the surface zone, to force calculation near the surface zone and to the statistics of extremes.

In order to explore the future of how loads and response of offshore structures will be estimated analytically, we could point to some different possibilities:

- As increasing computer power is becoming available, the basic hydrodynamic equation (the Laplace equation) and the boundary conditions on the surface, the sea bottom and the structure could in principle be solved, whereby the load and response in any defined wave situation could be established. The design parameters could be found through integration of the pressure field along the structure. It is, however, felt that a full time history analysis will be extremely time consuming even with next generation of computers. Estimation of damping values would still be difficult.
- Full numerical analysis, based on Navier Stokes equation, could be conducted to understand vortices and include viscosity. This approach could e.g. be followed for calculation of loads and response of risers, as soon as efficient computer source codes are becoming commercially available.
- The solution of the hydrodynamic equation and the boundary conditions could be developed into higher order series converging more rapidly than present mathematical expansions. This method will involve much analytical development and would require most complicated expansions of surface and wave-structure boundary conditions.
- The load and response could be calculated through a refinement of to-days practice for slender volume structures where the different terms are established separately. These terms are firstly the description of the sea state into linear (or possibly non linearly interacting) components, establishment of wave kinematics (wave velocity and acceleration and wave pressure field) under the waves, calculation of wave loads utilising a fully adequate wave force formalism which incorporates surface effects like run up along platform legs and effects of wave flow plast platform legs. Furthermore, the extrapolation of the calculated wave load and response data into design values must be given appropriate attention as part of the process.

In view of the above discussion, certain problems are listed below, documenting the needs for further work:

- So far it should be noted that it has not been possible to fully explain the "ringing phenomena" and the associated loads on the structures, as no papers so far has been published which fully can reproduce the results of tank testing of offshore structures.
- During the winter of 1995, very high waves caused considerable problems to the Brent B platform in the North Sea (Swan, Taylor and van Langen, 1997) and to the Veslefrikk B platform not far away. Large, very high waves were also reported measured at the Europipe Jacket platform at the Draupner pipeline connection location further south in the North Sea. The explanation of such "freak wave" phenomena and the associated wave loading need further work.
- It is, furthermore, obvious that any new type of offshore structure or the application of a known structural type in new areas with wave and current conditions different from known areas of application (for example in the case of strong currents, steep waves or long periodic waves) could lead to unknown load and response phenomena. Wave tank testing is recommended for all such incidents. Decay tests are necessary to determine damping.

For calculations of loads on and response of "large volume structures", defined as structures where wave reflection plays an important role, diffraction loading effects are determined by analysing both incoming and outgoing (reflected) waves. Main emphasis has been placed on finding higher order analytical solutions to the hydrodynamical equation and the boundary equations. Finding solutions to cases with complicated geometrical forms (e.g. platforms with several columns) and to solve wave-structure interaction problems near the surface are yet to be fully resolved. Of particular concern is the wave surface form on the leeward side of platform columns, i.e. how the wave moves around and behind the columns. Higher order force components would emerge as waves are passing behind the columns in steep waves. These force components are of particular importance, and are thought to contribute to the ringing response of offshore structures (Faltinsen, Newman and Vinje, 1995). These phenomena are clearly seen in wave tanks when large and steep waves pass by slender structures. As these phenomena are occurring under specific wave conditions only, the response are of transient dynamic character. Further uncertainties relate to the interaction of reflected waves from the different columns, sometimes leading to enhancement of the water level in between platform columns. (Swan, Taylor and van Langen, 1997.)

For more slender offshore structures, the wave force has so far almost exclusively been determined by the two term Morison equation, composed by the addition of an inertia term proportional to the acceleration of the water wave particles and a drag term proportional to the square of the water particle velocity. The inertia term could analytically be determined in the case of a uniform wave particle acceleration field, while the drag term is found to be present in the case when water with constant velocity is passing behind the structure. In both cases, however, the surface effects are not fully understood. To add these terms in the case of waves passing behind a structure, represents an assumption and the coefficients of the inertia and drag terms are different from cases of accelerated flow and constant velocity, respectively (Moe and Gudmestad, 1997).

To fully account for surface effects and other effects not explained by the use of the Morison equation, it has been proposed to add more terms to the Morison equation (see e.g. Sarpkaya and Isaacson, 1981). A thorough review of the wave loading mechanism on slender structures was also carried out by Bendat and Piersol (1982). Despite these efforts, no adjustment to the two-term Morison equation has been documented to represent a real improvement. Therefore, the two-term Morison equation represents state of art for calculation of loads on slender offshore structures. Improved understanding of near surface phenomena are, however, expected to yield introduction of additional terms in the Morison equation.

Water wave kinematics

In order to calculate wave forces, it is evident from the previous discussion that wave particle acceleration and velocity values must be found. Considering the physics of the phenomenon, it is also obvious that it is the movement of the water particles that causes the forces and the response of offshore structures.

The spectrum of a sea state is composed of waves with different frequencies and amplitudes. The wave-wave interaction is also included in the spectral form. However, all phase information is lost when generating the spectrum. When de-composing a spectrum into linear waves, the nonlinear wave-wave information in the real sea is not re-produced. This should be rectified by allowing for such nonlinear wave-wave interaction phenomena, in analysis of extreme waves and their kinematics (Zhang et al., 1996, Spell et al., 1996).

Regarding water wave kinematics, main emphasis has been on water wave velocity, as the drag term in the Morison equation is proportional to the square of the particle velocity. It has also been possible to measure water wave kinematics under waves in the laboratory. With the introduction of laser doppler velocimetry equipment, the measurements have become particularly accurate (Skjelbreia et al., 1990). On the other hand, it has been much more difficult to obtain accurate wave particle acceleracton values. In principle, measurements cannot be carried out with present equipment, except that the values could be found numerically by finding the change in measured velocities. Lack of accurate data has been striking but new information will be published by Swan, Bashir and Gudmestad (1997). It is thought that this information could contribute to the increased understanding of the behaviour of inertia dominated structures in steep wave conditions.

Due to the combination of problems related to wave decomposition and to the need for linearizing (or presenting series expansion) the surface boundary conditions, the wave kinematics have been particularly difficult to determine in the near surface zone. Main needs for further improvements are thus in this zone.

Further work

In order to improve the understanding of dynamics and loads on offshore structures, there is need to further improve the understanding of wave-wave interaction phenomena and of water wave kinematics, in particular in the surface zone. Furthermore, the accurate estimate of wave load near the surface needs further investigation and finally it is thought necessary to further analyze statistical models for how to extrapolate the data to design values.

It is considered prudent to increase the refinement of this research until theoretical results agree fully with measurements, although for new conditions and when involving new structural types, tank testing would be necessary to ensure that unforeseen response characteristics are not overlooked and that relevant damping values are selected.

List of references

- American Petroleum Institute (API); "Recommended Practice for Planning, Designing and Constructing Offshore Fixed Platforms - API RP2A". API, USA, 1993.
- Bendat, J.J. and Piersol, A.G.; "Spectral analysis of non-linear systems involving square-law operations". Journal of Sound and Vibration, Vol. 81 (2), pp 199-213, 1982.
- Faltinsen, O.M.; "Sea loads on Ships and Offshore Structures". Cambridge University Press, 1990.
- Faltinsen, O.M., Newman J.N. and Vinje, T.; "Non-linear wave loads on a slender vertical cylinder." Journal of Fluid Mechanics Vol. 289, pp. 179-98, 1995.
- Gudmestad, O.T. and Moe, G.; "Hydrodynamic Coefficients for Calculation of Hydrodynamic Loads on Offshore Truss Structures". Marine Structures, Vol. 9, No 8 pp 745-758, 1996.

- Moe, G. and Gudmestad, O.T.; " Prediction of Morison type Forces in Irregular Thigh Reynolds Number Waves". Proc. ISOPE Conference, Honolulu, 1997.
- Newman, J.N.; "Non-linear scattering of long waves by a vertical cylinder." In: "Waves and non-linear processes in hydrodynamics." (Eds: J. Grude et al.), Kluwer Academic Publishers, 1996.
- Norwegian Petroleum Directorate (NPD); "Guidelines for determination of load and load effects". NPD, Stavanger, Norway, 1997.
- Offshore Technology Research Centre (OTRC); International Conference on "Nonlinear Design Aspects of Physical Model Tests". OTRC, College Station, USA, May 1997.
- Sarpkaya and Isaacson, M.; "Mechanics of Wave Forces on Offshore Structures". Van Nostrand Reinhold Company, New York, 1981.
- Skjelbreia, J., Berek, E., Bolen, Z., Gudmestad, O.T., Heideman, J., Ohmart, R.D., Spidsoe, N. and Tørum, A., "Wave Kinematics in Irregular Waves." Proc. OMAE, Vol. 1A pp 223-228, Stavanger 1991.
- Spell, C.A., Zhang. J. and Randall, R.E.; "Hybrid wave model for unidirectional irregular waves - part II. Comparison with laboratory measurements". Applied Ocean Research, Vol. 18, pp. 93-110, 1996.
- Swan, C., Taylor, P.H. and van Langen, H.; "Observation of wave-structure Interaction for a multi-legged Concrete Platform." Submitted to Journal of Fluid Mechanics, 1997.
- Swan , C., Bashir T. and Gudmestad, O.T.; "Accelerations in steep two-dimensional water waves with implications for non-linear wave loading". Submitted for publication, Aug. 1997.
- Tromans, P.S., Aaturk, A. and Hagermeijer, P.; "A new model for the kinematics of large ocean waves - applications as a design wave." Proc. 1st Int. Conf. on Offshore and Polar Engineering, Edinburgh Vol. 3, pp. 64-71, 1991.
- Zhang, J., Chen, L., Ye, M. and Randall, R.; "Hybrid wave model for unidirectional irregular waves - part I. Theory and numerical scheme." Applied Ocean Research Vol. 18, pp. 77-92 (1996).

A Second-Order Solution of the Flap Wavemaker Problem

Jaw-Fang Lee[1], Cheng-Chi Liu[2], and Yuan-Jyh Lan[3]

Abstract

A second-order solution of the flap wavemaker problem is presented in this paper. The present solution is derived based on complex variables, which is different from previous solutions derived using real variables by Hudspeth and Sulisz(1991), and Sulisz and Hudspeth(1993b). The perturbation method is used to solve the nonlinear boundary value problem. In the second-order solution, unwanted real parts produced by multiplication of imaginary parts of the linear complex solutions are subtracted from the formulation so that correct solutions can be obtained. Complete solutions up to the second order are presented. The present analytic solutions are compared favorably with experimental results by Wu(1987). Using the present theory, the characteristics of the second-order waves generated by the flap wavemaker are studied. The waves generated in the wave channel consist of the Stokes waves and the wavemaker free waves. Analysis of the mass transport shows that the free waves generate negative value of the Stokes drift. Therefore, wavemaker generated waves satisfy conservation of mass, and have zero mass transport.

Introduction

The generation of waves in the wave channel provides useful information such as wave deformation, wave characteristics et al., for ocean waves in the

---

[1] Professor, Dept. Hydraulics & Ocean Engineering, National Cheng Kung University, Tainan, Taiwan, ROC.
[2] PhD student, Dept. Hydraulics & Ocean Engineering, National Cheng Kung University, Tainan, Taiwan, ROC.
[3] PhD student, Dept. Hydraulics & Ocean Engineering, National Cheng Kung University, Tainan, Taiwan, ROC.

coastal region, and therefore is very important in ocean engineering (Madsen, 1971; Flick and Guza, 1980; Wu, 1987). The wavemaker theory provides a useful tool to simulate waves in the wavemaker channel. The difficulties of analyzing the wavemaker problem are mainly on the moving boundaries at the free surface and the wavemaker boundary, and the inherent kinetic and dynamic non-linearity of the problem (Kim, 1984). Not until recently, a complete second-order solution for a generic-type wavemaker problem was proposed by Sulisz and Hudspeth(1993a). The perturbation method was used to analyze the problem, and Taylor series expansions was used for the moving free surface and the wavemaker boundaries. Real variables were used in deriving the second-order solution. The Stokes drift in the two-dimensional wave flume was calculated. The solution method was extended to calculate wave loading on the wavemaking structures (Sulisz and Hudspeth, 1993b), on a horizontal rectangular cylinder (Sulisz and Johnasson, 1992), and on submerged structures (Sulisz, 1993). The method similar to Vantorre (1986) was used to obtain correct the second-order solutions.

Problem description

Consider a flap wavemaker located at one end of a semi-infinite channel of constant water depth as shown in Figure 1. A Cartesian coordinate system is used

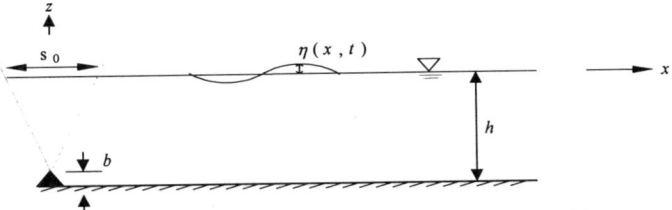

Figure 1 Definition sketch of the flap wavemaker problem.

with origin located at still water level, x-axis pointed to the right, and z-axis pointed upward. The water depth is h and the free surface elevation η. The wavemaker is given a periodic motion with frequency σ, and the displacement function $\xi(z,t)$ defined by

$$\xi(z,t) = \begin{cases} i\dfrac{s_0}{2(h-b)}(z+h-b)e^{-i\sigma t} &, \ -(h-b) < z < 0 \\ 0 &, \ -h < z < -(h-b) \end{cases} \qquad (1)$$

where s_0 is the stroke, b is the height of the hinge above the bottom. The generated waves can be described by potential function $\Phi(x,z,t)$, and the boundary value problem solving for wave potential Φ can be written as

$$\nabla^2 \Phi = 0 \tag{2}$$

$$-\Phi_{,z} = 0 \quad \text{on } z = -h \tag{3}$$

$$-\Phi_{,t} + \frac{1}{2}|\vec{V}|^2 + g\eta + B(t) = 0 \quad \text{on } z = \eta(x,t) \tag{4}$$

$$-\Phi_{,z} = \eta_{,t} - \Phi_{,x}\eta_{,x} \quad \text{on } z = \eta(x,t) \tag{5}$$

$$-\Phi_{,x} = \xi_{,t} - \xi_{,z}\Phi_{,z} \quad \text{on } x = \xi(z,t) \tag{6}$$

in which g is gravitational constant, $B(t)$ is Bernoulli constant, the subscript comma indicates differentiation. The dynamic condition Equations (4) and kinematic condition Equation (5) can be combined as

$$-\Phi_{,tt} - g\Phi_{,z} + (\vec{V}^2)_{,t} + \vec{V}\cdot\nabla(\frac{1}{2}\vec{V}^2) + B'(t) = 0 \tag{7}$$

At infinite x the waves propagating outward satisfy radiation condition, and are of finite value. The perturbation method is used to obtain boundary value problems for each orders. The first order problem and the corresponding solution are quite common, and can be found in textbook, e.g., Dean and Dalrymple (1984).

The second order problem

The second order problem can be written as

$$\nabla^2 \Phi_2 = 0 \quad , 0 \le x < \infty \ ; \ -h \le z \le 0 \tag{8}$$

$$-\Phi_{2,z} = 0 \quad , z = -h \tag{9}$$

$$\Phi_{2,tt} + g\Phi_{2,z} = 2\Phi_{1,x}\Phi_{1,xt} + 2\Phi_{1,z}\Phi_{z,t} - \eta_1\Phi_{1,ttz} \\ - g\eta_1\Phi_{1,zz} + dB_2/dt \quad , \quad z = 0 \tag{10}$$

$$-\Phi_{2,x} = \xi\Phi_{1,xx} - \xi_{,z}\Phi_{1,z} \quad , x = 0 \tag{11}$$

The free surface elevation of the second order η_2 has the following form,

$$\eta_2 = \frac{1}{g}\left\{\Phi_{2,t} - \frac{1}{2}\left(\Phi_{1,x}^2 + \Phi_{1,z}^2\right) + \eta_1\Phi_{1,tz} + B_2\right\} \quad , z = 0 \tag{12}$$

Since the second order problem contains two non-homogeneous boundaries, therefore, in order for the problem to be solvable it is first divided into two parts, $\Phi_2 = \Phi_2^s + \Phi_2^f$, Φ_2^s is to satisfy the problem with non-homogeneous free surface boundary condition, Φ_2^f satisfies the problem with non-homogeneous wavemaker boundary condition. And correspondingly free surface elevation be expressed as

$$\eta_2^s = \frac{1}{g}\left\{\Phi_{2,t}^s - \tfrac{1}{2}\left(\Phi_{1,x}^2 + \Phi_{1,z}^2\right) + \eta_1 \Phi_{1,tz} - B_2\right\} \tag{13}$$

$$\eta_2^f = \frac{1}{g}\Phi_{2,t}^f \tag{14}$$

Since complex variables are used in the derivation, multiplication of the first order solution can produce unwanted real parts and need to be discarded. The method used by Vantorre (1986) to avoid this problem can be written as

$$\mathrm{Re}(\alpha e^{-i\sigma t})\cdot\mathrm{Re}(\beta e^{-i\sigma t}) = \mathrm{Re}\left[\tfrac{1}{2}\left(\alpha\cdot\beta e^{-i2\sigma t}\right) + \tfrac{1}{2}\alpha\cdot\overline{\beta}\right] \tag{15}$$

where α and β are complex, and the over bar indicates complex conjugate. Equation (15) means real parts of the two sides of the equation are equal. Note that from Equation (15) a time-dependent term and a time-independent term are produced, and similarly results in Equations (10)-(12). By applying wave periodicity the second order Bernoulli constant B_2 in Equation (12) can be shown to be $-T_0^2/4$. The solution of Φ_2^s and Φ_2^f take the forms of $\Phi_2^{*s} + \varphi_2^s$ and $\Phi_2^{*f} + \varphi_2^f$ where Φ_2^{*s} and Φ_2^{*f} are time-dependent functions, and φ_2^s and φ_2^f and are time-independent. The solutions can be summarized as followings.

$$\Phi_2^{*s} = \sum_{n=0}^{\infty}\sum_{m=0}^{\infty} C_{2nm}^s \cos\left[(k_{1n}+k_{1m})(z+h)\right]e^{-(k_{1n}+k_{1m})x} e^{-i2\sigma t} \tag{16}$$

$$\varphi_2^s = \sum_{n=1}^{\infty} D_{2n0}^s \cos\left[(k_{10}-k_{1n})(z+h)\right]e^{(k_{10}-k_{1n})x}$$

$$+ \sum_{n=0}^{\infty}\sum_{m=1}^{\infty} D_{2nm}^s \cos\left[(k_{1n}+k_{1m})(z+h)\right]e^{-(k_{1n}+k_{1m})x} \tag{17}$$

where the coefficients C_{2nm}^s, D_{2n0}^s and D_{2nm}^s are integration coefficients. The solution of Φ_2^{*f} can be written as

$$\Phi_2^{*f} = \sum_{\ell=0} C_{2\ell}^f \cos\left[k_{2\ell}(z+h)\right]e^{-k_{2\ell}x}e^{-i2\sigma t} \tag{18}$$

where $k_{20} = -iK_2$, $k_{2\ell}$ satisfies the second order dispersion equation

$$4\sigma^2 = -gk_{2\ell}\tan k_{2\ell}h \quad ,\ell = 0,1,2,3,\cdots \tag{19}$$

The coefficient $C_{2\ell}^f$ in Equation (18) can be obtained by utilizing the orthogonal function. The solution of φ_2^f can be written as

$$\varphi_2^f = D_{20}^f x + \sum_{\ell=1}^{\infty} D_{2\ell}^f \cos\left[\mu_\ell(z+h)\right]e^{-\mu_\ell x} \tag{20}$$

where D_{20}^f and $D_{2\ell}^f$ are integration coefficients.

$$\mu_\ell = \frac{\ell\pi}{h}, \quad \ell = 1,2,3,\cdots \tag{21}$$

The free surface elevation

The free surface elevation η calculated from wave potential contains the first order η_1 and the second order η_2. The second order quantity equals $\eta_2^s + \eta_2^f$. η_2^s is induced from the free surface boundary condition, and η_2^f is duced by the wavemaker condition. The propagation waves away from the wavemaker can be written as

$$\eta^p = a_{1p}\cos(K_1 x - \sigma t) + a_{2p}^{*s}\cos[2(K_1 x - \sigma t)] + \left|a_{2p}^f\right|\cos(K_2 x - 2\sigma t + \theta_2) \tag{22}$$

where θ_2 is the phase angle.

The mass transport

The mass momentum flux or mass transport can be expressed as

$$M = \left\langle \int_{-h}^{\eta} \rho U dz \right\rangle \tag{23}$$

where U is the horizontal fluid velocity, $\langle \cdot \rangle$ indicates time average of wave period. Equation (23) is calculated by first applying Taylor series expansion about z=0, then substituted into velocities up to the second order. Equation (23) can be rewritten as $M = M_\Phi + M_\varphi$, M_Φ represents result of time-dependent functions, M_φ represents result of time-independent functions. The time-dependent result can be further formulated as

$$M_\Phi = M_\Phi^e(x) + M_\Phi^\infty \tag{24}$$

where

$$M_\Phi^e(x) = \rho \frac{a_{1p}}{2}\left[\sum_{n=1}^{\infty}\frac{T_n}{k_{1n}}\cos(k_{1n}h)e^{-k_{1n}x}\left[k_{1n}\cos(K_1 x) - K_1\sin(K_1 x)\right]\right] \tag{25}$$

$$M_\Phi^\infty = \frac{\rho g K_1 a_{1p}^2}{2\sigma} \tag{26}$$

Equation (26) is the so called Stokes drift. On the other hand, the time-independent result can be expressed as

$$M_\varphi = M_\varphi^e(x) + M_\varphi^\infty \tag{27}$$

where

$$M_\varphi^e(x) = -\rho\left(\frac{a_{1p}}{2}\right)\sum_{n=1}^{\infty}\frac{T_n}{k_{1n}}\cos(k_{1n}h)e^{-k_{1n}x}\left[k_{1n}\cos(K_1 x) - K_1\sin(K_1 x)\right] \quad (28)$$

$$M_\varphi^\infty = -\rho D_{20}^f h = -\frac{\rho g K_1 a_{1p}^2}{2\sigma} \quad (29)$$

Equation (29) indicates that there is return flow toward the wavemaker. Notice that in the wavemaker theory the Stokes drift M_Φ^∞ is balanced by the mean return flow M_φ^∞. This is true because in the wavemaker problem the Laplace equation is solved which indicates the conservation of mass.

Results and Discussion

To show accuracy of the present theory, analytic solution up to the second order is compared to experimental results by Wu (1987). The hinged height of the wavemaker is 0.035m, the period and stroke of the wavemaker motion are 1.23sec and 0.12m, respectively. The constant water depth is 0.4m. The data acquisition sampling rate was 74 times per second. In Figure 2 the real line is analytic solution and circles represent experimental results, Figure 2(a) is water elevation at four times water depth, and Figure 2(b) at five times water depth. It can be seen that the comparisons show very good agreement.

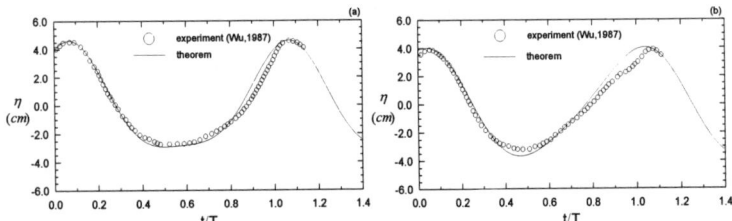

Figure 2 Comparison of present analytic solution with experiment by Wu(1987). T=1.23sec, s_0=0.12m, h=0.4m. (a)x/h=4 (b)x/h=5

Using the present analytic solution, wave forms generated by the wavemaker under different conditions are studied. As shown in Figure 3 non-dimensional wave height versus distance from the wavemaker, Figure 3(a), 3(b), and 3(c) are for $K_1 h$=0.5, 1.5, and 3.0, respectively. Sinusoidal wave forms are generated for relative shallow depth, whereas the deeper depth increases the irregular wave forms. In figure 3(c) second wave peaks are shown in the wave troughs.

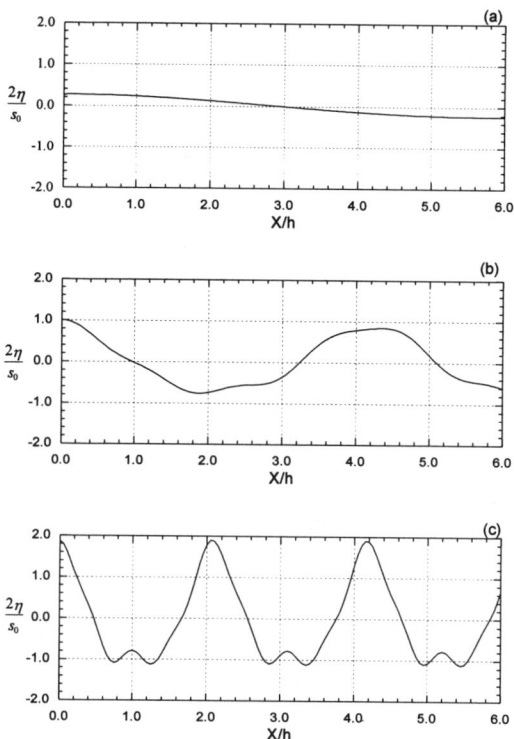

Figure 3 Wave forms generated by the wavemaker $s_0/h=0.3$, b/h=0. (a) $K_1h=0.5$, (b) $K_1h=1.5$, (c) $K_1h=3.0$.

The analysis of the wavemaker theory shows that the second order component resulted from the second order non-homogeneous free surface condition and the second order wavemaker boundary condition. The free surface condition induced the well known Stokes second order wave, while the wavemaker condition generated free waves. The analysis of mass transport showed that the Stokes wave induced Stokes drift, and the free wave generated an amount negative to the Stokes drift. In other words, in the wave channel the generated waves carry zero mass transport, that is, the fluid mass remains conserved. One should be aware that in this analysis a semi-infinite channel is considered, therefore, no effects of end wall reflection induced.

Conclusion

In this paper detailed derivation of the second order wavemaker theory using complex variables is presented. The flap type wavemaker is considered. The analytic solution is verified by the solution derived using real variables. Comparison with experimental results also shows favorable agreement. From the analysis the second order wavemaker solution consists of the Stokes waves and the wavemaker free waves. The analysis of mass transport also indicates that the former induces Stokes drift, and the later generated an equal but negative amount. Therefore, in the wave channel the generated waves up to the second order carry no mass transport, which implies the solution satisfies mass conservation.

References

Dean, R. G. and Dalrymple, R.A., Water Wave Mechanics for Engineers and Scientists, Prentice-Hall, Inc., Englewood Cliffs, New Jersey 07632, U.S.A., 1984.

Flick, R.E. and Guza, R.T., Paddle Generated Waves in Laboratory Channels, J. Waterway, Port, Coastal, and Ocean Engineering, ASCE, Vol.106, No.1, pp.79-97, 1980.

Hudspeth, R.T. and Sulisz, W., Stokes Drift in Two Dimensional Wave Flumes, J. Fluid Mech., Vol.230, pp.209-229, 1991.

Kim, T.I., Mass Transport in Laboratory Water Wave Flumes, Ph.D. Thesis, Oregon State University, Corvallis, OR, USA, 1984.

Madsen, O.S., On the Generation of Long Waves, J. of Geophys. Res., Vol.36, pp.8672-8683, 1971.

Sulisz, W., Diffraction of Second-Order Surface Waves by Semi-submerged Horizontal Rectangular Cylinder, J. WPCOE, ASCE, Vol.119, No.2, pp.160-171, 1993.

Sulisz, W. and Hudspeth, R.T., Complete Second-Order Solution for Water Waves Generated in Wave Flumes, J. Fluids and Structures, Vol.7, pp.253-268, 1993a.

Sulisz, W. and Hudspeth, R.T., Second-Order Wave Loads on Planar Wavemaker, J. Waterway, Port, Coastal, and Ocean Engineering, ASCE, Vol.119, No.5, pp.521-536, 1993b.

Sulisz, W. and Johnasson, M., Second-Order Wave Loading on a Horizontal Rectangular Cylinder of Substantial Draught, Applied Ocean Research, 14, pp.333-340, 1992.

Vantorre, M., Third-Order Theory for Determining the Hydrodynamic Forces on Axi-symmetric Floating or Submerged Bodies in Oscillatory Heaving Motion, Ocean Engng, Vol.13, No.4, pp.339-371, 1986.

Wu, Y-C., Constant Wave Form Generated by A Hinged Wavemaker of Finite Draft in Water of Constant Depth, Proc. 9th Ocean Engineering Conference in ROC, pp.552-569, 1987.

Steep Transient Waves in a Directionally Spread Sea

T.B. Johannessen[*] & C. Swan[*]

Abstract

The present paper is concerned with the nonlinear dynamics of focused wave groups in which the wave energy is spread in both the frequency and the directional domains. Experimental observations and numerical calculations are presented in which the evolution of a number of wave forms, with varying frequency and directional spreads, are considered. In each case the intensity of the wave, measured in terms of a linear amplitude sum, is varied from a near-linear condition to the limit of incipient wave breaking. These results show that for a given linear amplitude sum an increase in the directional spread leads to a reduction in the nonlinear wave-wave interactions. Conversely, the maximum nonlinear crest elevation, observed just prior to the onset of wave breaking, increases with the directional spread assuming the frequency bandwidth remains constant. These results clearly suggest that if the water particle kinematics are to be accurately predicted, an appropriate wave model must incorporate the frequency distribution, the nonlinearity and the directional spread.

Introduction

Field data suggests that ocean waves have a distribution of energy in both the frequency and the directional domains (Jonathan et al., 1994). Evidence suggests that the frequency spectrum is relatively broad-banded, implying that large changes in a wave group envelope will take place over less than a typical wave period or wave length. As a result, any regular wave representation will inevitably be inaccurate. Furthermore, the distribution of energy in the directional domain is also relatively broad. For example, Jonathan et al. (1994) show that in the northern North sea wind-waves arising in large storms have a typical directional spread corresponding to a normal distribution with a standard deviation of 30°.

---

[*] Department of Civil Engineering, Imperial College, London. SW7 2BU.

Although linear theory can describe many of the properties of ocean waves, nonlinearity is undoubtedly significant. This is particularly true in the vicinity of extreme events which are typically used as 'design' waves. Indeed, recent laboratory observations of extreme 2-D wave groups, produced by the focusing of energy due to frequency dispersion (Baldock et al, 1996), have shown that the nonlinear wave-wave interactions may increase the maximum linearly predicted crest elevation by as much as 30%. However, although detailed analysis of field observations clearly identify nonlinear effects, they are of a reduced magnitude. An obvious explanation for this difference lies in the directionality (or three-dimensional character) of real ocean waves.

The present paper will address this point, and will present the results of an experimental study concerning the nonlinear evolution of large isolated wave groups with a spread of energy in both the frequency and the directional domains. A new, fully nonlinear, directional wave model is also introduced, and the results shown to be in good agreement with the measured data. Comparisons between these results will highlight the importance of incorporating the directional spread when seeking to model an extreme wave form.

Experimental Study

The experimental work was undertaken in the wide wave basin at the University of Edinburgh (figure 1). This facility has a plan area of 25m×11m, a uniform depth of 1.2m, and is equipped with 75 numerically controlled wave paddles, each 0.3m wide. To limit the occurrence of reflected waves, large passive absorbers were located at the downstream end of the wave basin and along one of the side walls. Within this facility preliminary tests showed that waves could be generated within a directional range of $\theta=\pm 45°$ and a frequency range of $0.6Hz \leq f \leq 1.7Hz$.

For each of the wave fields generated, the surface elevation was sampled at 45 spatial positions in the vicinity of the focus position ($x\approx 0$, $y\approx 0$) using standard surface piercing wave gauges. Furthermore, for a number of selected wave cases, the x-component of the horizontal wave induced velocity was measured at closely spaced vertical positions underneath the measured position of the maximum crest elevation using Laser Doppler Anemometry (LDA).

As part of our preliminary study, considerable time was spent calibrating and validating the wave basin. The usual problems associated with wave reflections, which are notoriously strong in many wave basins, did not pose a significant problem in the present study. This is because the nature of a focused wave group is such that they are almost completely dispersed when they reach the downstream wave absorbers. Any energy reflected or scattered from the wave absorbers will thus be negligible compared with the energy density in the vicinity of the extreme event. The success of the calibration process is clearly demonstrated in figure 2. This describes the time-history of the water surface elevation, $\eta(t)$, relating to a small amplitude wave group measured at the linearly predicted focus position. Near-perfect agreement is observed between the measured data and linear theory. In contrast, figure 3 concerns a highly nonlinear, near-breaking, wave event and provides a sequence of time-histories measured at 100mm intervals along the x-axis (or mean wave direction). In this, and indeed all other nonlinear wave cases, the position of the maximum crest elevation is shifted downstream in both space and time. Nevertheless, one maximum crest elevation is clearly observed, indicated by the thicker solid line. Furthermore, the largest crest elevation occurring at t≈-1sec includes several overlying profiles recorded from a single pilot gauge. These demonstrate the repeatability of the test conditions. Further details of the experimental study are given in Johannessen and Swan (1998).

Numerical Modelling

The numerical model which has been adopted is an extension of the two dimensional time-stepping formulation originally proposed by Fenton & Rienecker (1980). Assuming that the flow is irrotational and that the required continuity condition is represented by Laplace's equation, a velocity potential Φ defined by $\underline{u}=\nabla\Phi$, where $\underline{u}$ is the velocity vector, can be expressed as:

$$\Phi(x,y,z;t) = \sum_{m=0}^{\infty} Cos(mk_y y) \sum_{n=0}^{\infty} \left(\left(A_{nm}Cos(nk_x x) + B_{nm}Sin(nk_x x)\right) \frac{\cosh(k_{nm}[z+d])}{\sinh(k_{nm}d)} \right) \quad (1)$$

where $k_{nm}=\sqrt{(nk_x^2 + mk_y^2)}$, and (k_x, k_y) define the large fundamental length scales, $\lambda_x=2\pi/k_x$ and $\lambda_y=2\pi/k_y$, over which the solution is assumed to be periodic in the x and y directions respectively. Likewise, the free surface elevation, which must be similarly periodic in space, is given by:

$$\eta(x,y;t) = \sum_{m=0}^{\infty} Cos(mk_y y) \sum_{n=0}^{\infty} \left(\left(a_{nm}Cos(nk_x x) + b_{nm}Sin(nk_x x)\right) \right) \quad (2)$$

Both equations (1) and (2) utilise the fact that the experimental wave fields are symmetric in y, although this is not a formal necessity. Furthermore, the series coefficients A_{nm}, B_{nm}, a_{nm}, b_{nm}, are assumed to be functions of time only. In this form the velocity potential satisfies both the governing field equation ($\nabla^2\Phi=0$) and the bottom boundary condition ($\Phi_z=0$ on $z=-d$). The remaining constraints represent the nonlinear free surface boundary conditions (both kinematic and dynamic) evaluated on the water surface ($z=\eta$). After some re-arrangement these can be written as:

$$\Phi_t = -\left(g\eta + \tfrac{1}{2}(\Phi_x^2 + \Phi_y^2 + \Phi_z^2)\right) \quad (3)$$

$$\eta_t = \Phi_z - \left(\Phi_x \eta_x + \Phi_y \eta_y\right) \quad (4)$$

Longuet-Higgins and Cokelet (1976) first noted that in this form the right side of equations (3) and (4) involve no time derivatives. As a result, if an initial spatial description of the water surface elevation and its associated velocity potential are known, it is possible to time-march the solution such that η and Φ can be defined at all times. In the present cases the initial conditions, at some early time prior to the focus event, can be calculated from linear or second-order theory on the basis that the wave group is fully dispersed. To apply this procedure it must be assumed that no significant energy lies above the truncation wave numbers Nk_x and Mk_y. Provided this is indeed the case, equations (3) and (4) can be solved at $2N(M+1)$ spatial locations (corresponding to a grid spacing equal to half the wavelength of the shortest wave component) in order to define the time derivatives of the coefficients (A_{nm_t}, B_{nm_t}, a_{nm_t} and b_{nm_t}). Once these have been determined the solution can be time-stepped using the Adams, Bashford, Moulton formulation (Gear, 1971) and the solution procedure repeated. Within the present scheme the time derivatives of the free surface coefficients may be evaluated using a Fast Fourier Transform (FFT). Unfortunately, the time derivatives of the velocity potential are functions of η, and are thus evaluated by solving a set of linear simultaneous equations using an lower-upper (LU) matrix decomposition. As a result, the numerical formulation is time consuming and requires parallel computing power for the steepest directional wave cases. The present numerical formulation has been implemented on a Fujitsu AP1000 parallel computer, and it has been found that the most computationally demanding wave groups may be evaluated accurately with run times of up to 16 hours. Further details of the numerical procedure are given in Johannessen (1997).

Discussion of Results

Figure 4 describes the time-history of the water surface elevation measured at the focal location. This case corresponds to a short-crested or strongly directional wave event (s=4, where s is the Mitsuyasu spreading parameter) generated within a broad-banded frequency spectrum ($0.71Hz \geq f \geq 1.67Hz$). The numerical calculations are shown to be in good agreement with the measured data, and show a marked improvement over both the linear and the second-order solution (Sharma and Dean, 1981). Figure 5 concerns the same wave case but provides a spatial description of the water surface profile. Once again, the numerical model is in good agreement with the laboratory data, and in particular accurately reproduces the 'spatial' steepness of the wave (ie. $d\eta/dx$). This is clearly very important in terms of predicting the underlying water particle accelerations, and also has implications for both wave slamming and wave run-up. The maximum near-surface velocities arising beneath this wave are considered on figure 6. In this case the measured data is also compared with a typical design solution (Wheeler, 1970) based on the measured water surface elevation, $\eta(t)$. It is clear from this comparison that the numerical model provides the best description of the underlying water particle kinematics.

To further examine the effect of a directional spread, two separate investigations were undertaken. In the first, the linear sum of the component wave amplitudes generated at the wave paddles was held constant (A=55mm), and the global maximum crest elevation measured for a range of directional spreads. This data is presented on figure 7, where cases B, C, and D respectively correspond to a broad-banded frequency spectrum ($1.67Hz \geq f \geq 0.71Hz$), an intermediate frequency spectrum ($1.43Hz \geq f \geq 0.77Hz$) and a narrow-banded frequency spectrum ($1.25Hz \geq f \geq 0.83Hz$). It is clear that the effect of introducing a directional spread is to dramatically reduce the maximum crest elevation. Furthermore, the bulk of the reduction occurs between a unidirectional wave group ($1/s=0$, where s is again the Mitsuyasu spreading parameter) and a long-crested group with small directional spread. This result represents a real weakening of the nonlinear wave-wave interaction due to the underlying directionality. It is thought that this is caused by a reduction in absolute wave front steepness and curvature due to the fact that in a directional wave, this steepness can be distributed around the perimeter of the wave (ie. it is not constrained in one plane).

In the second study the amplitude of the wave components generated at the wave paddles was progressively increased until there was evidence of incipient wave breaking in the vicinity of the focal point. This allowed the variation in the maximum possible crest elevation, for a given underlying frequency bandwidth, to be recorded as a function of the directional spread. This data is shown on figure 8 and clearly suggests that the limiting crest elevation increases with the directional spread. Indeed, figure 8 suggests that the difference between a unidirectional wave group ($1/s=0$) and a shorted-crested wave group (with $1/s \approx 0.25$) can lead to an increase in the limiting crest elevation by as much as 25%.

Conclusions

The nonlinear evolution of a large number of focused directional wave groups has been investigated experimentally, and a new numerical model shown to be in good agreement with the laboratory data. The principle advantage of this numerical model is that it includes nonlinearity, unsteadiness and directionality. Furthermore, it does not require a detailed description of a nonlinear water surface profile, since it is based upon a linear description of the underlying wave spectrum. The results presented suggest that for a given linear crest height and frequency bandwidth, the effect of increasing directionality is to significantly reduce the nonlinearity of a wave group. However, the maximum nonlinear crest elevation, which may be obtained before further growth is limited by wave breaking, increases with the directionality of the wave field, provided the frequency bandwidth is held constant. This has important implications for the specification of a 'design' wave.

Acknowledgements

The present study has formed part of the research programme 'Uncertainties in Loads on Offshore Structures' sponsored by EPSRC through MTD Ltd and jointly funded with: Amoco (UK) Exploration Company, BP Exploration Operating Co Ltd, Brown & Root Ltd, Exxon Production Research Company, Health and Safety Executive, Norwegian Contractors a.s., Shell UK Exploration and Production, Den norske Stats Oljeselskap a.s. and Texaco Britain Ltd. EPSRC Grant Reference GR/J23976. The authors are also grateful for further funding from the Norwegian Research Council (norges Forskningsråd).

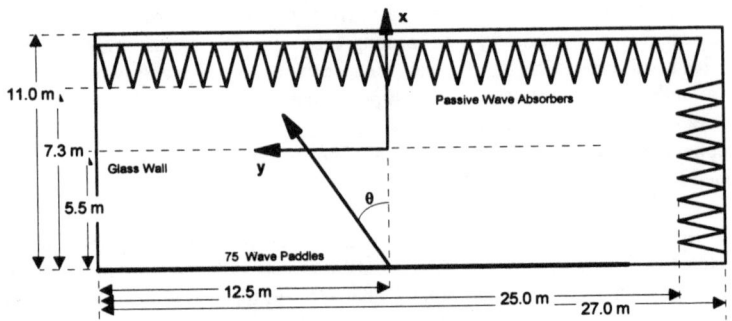

Figure 1. Experimental facility.

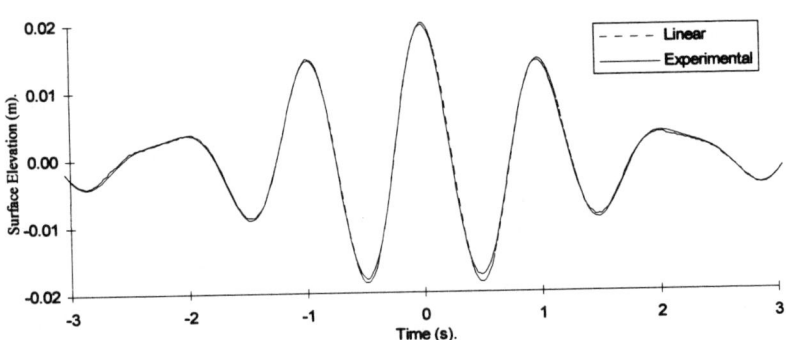

Figure 2. Small amplitude wave group. A comparison with linear theory.

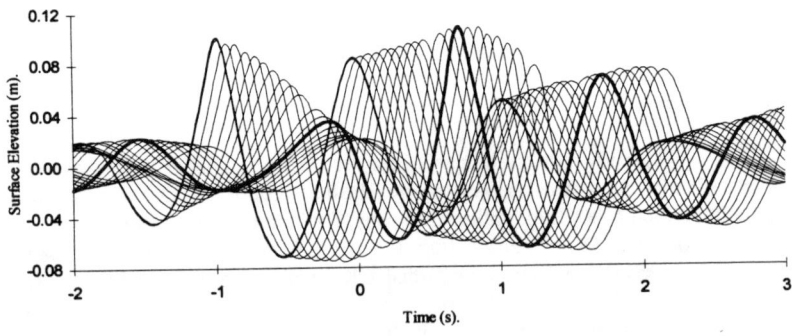

Figure 3. Nonlinear wave group. $\eta(t)$ at equally spaced intervals along the x-axis.

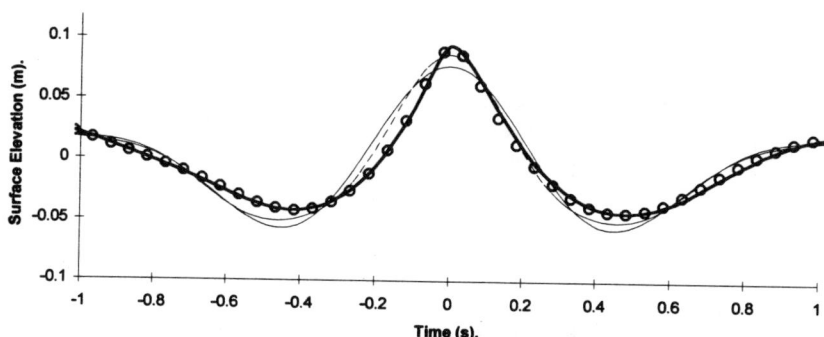

Figure 4. Time-history of the water surface elevation, $\eta(t)$.

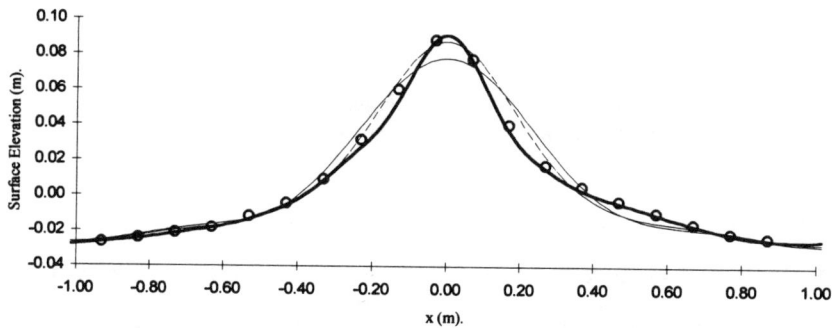

Figure 5. Spatial description of the water surface elevation, $\eta(x)$.

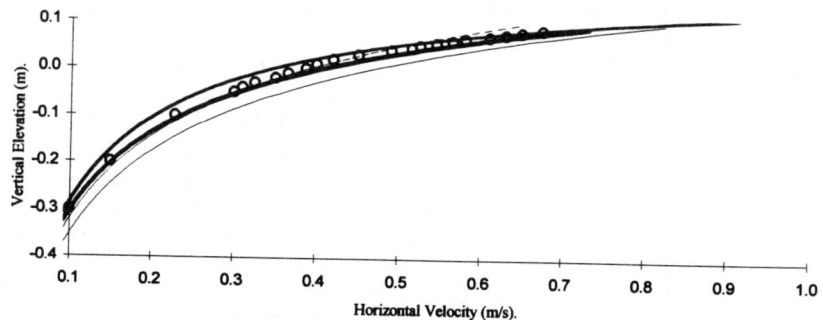

Figure 6. Horizontal velocity profile, $u(z)$.
° measured data, ——— linear theory, ---- second-order theory, ——— numerical model, and ——— Wheeler theory (only applicable to figure 6).

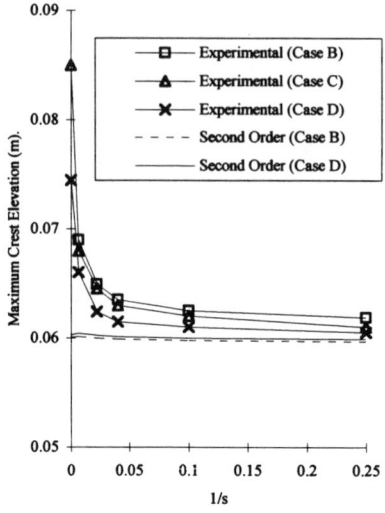
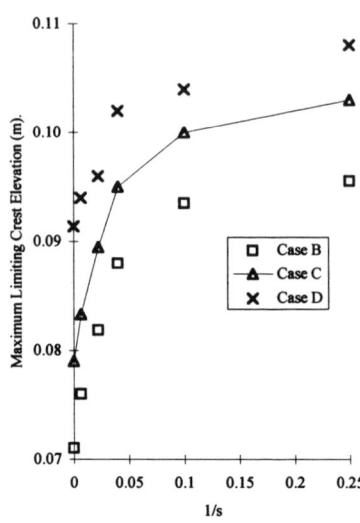

Figure 7. Crest elevation vs. directional spread with a linear amplitude sum of A=0.055m

Figure 8. Maximum limiting crest elevation vs. directional spread.

References

Baldock,T.E., Swan,C. & Taylor,P.H., (1996), A laboratory study of non-linear surface waves on water, Phil. Trans. Roy. Soc. London, Ser. A, 354, pp 649-676.
Fenton,J.D. & Rienecker,M.M., Accurate numerical solutions for nonlinear waves, 7th Int. Conf. on Coastal Engng., Sydney, 1, pp 50-69.
Gear,S.W., (1971), Numerical initial value problems in ordinary differential equations, Englewood Cliffs Inc., Prentice Hall.
Johannessen,T.B., (1997), The Effect of Directionality on the Nonlinear Behaviour of Extreme Ocean Waves, PhD Thesis, Univ. of London.
Johannessen,T.B. & Swan,C. (1998), A laboratory study of directionally spread surface waves on water. Submitted to Phil. Trans. Roy. Soc. Ser. A.
Jonathan, P.,Taylor,P.H. & Tromans,P.S., (1994), Storm Waves in the Northern North Sea, Proc. 7th Int. Conf. on the Behaviour of Offsh. Str., Mass., 2, pp 481-494.
Longuet-Higgins,M.S. & Cokelet,E.D.,(1976), The deformation of steep surface waves on water. A numerical method of computation., Proc. Roy. Soc., Ser. A, 350, pp 1-26.
Sharma,J.N. & Dean,R.G., (1981), Second-Order Directional Seas and Associated Wave Forces, Society of Petroleum Engineers J., 4, pp 129-140.
Wheeler,J.D., (1970), Method for calculating forces produced by irregular waves, Proc. 1st Annual offsh. Tech. Conf., Houston, 1, pp 71-82.

Hybrid Wave Models and Their Applications

Jun Zhang,[1] Jun Yang [2] and Jiangang Wen[3]

Abstract

Hybrid Wave Models (HWM) use both Mode Coupling Method (MCM) and Phase Modulation Method (PMM) to address the nonlinear interactions between free-wave components in an ocean wave field. Different from most existing methods for nonlinear wave simulation, HWMs consider nonlinear wave interaction in the decomposition of an irregular wave field. They are able to provide deterministic analysis of ocean waves based on their measurements at fixed points and do not require initial and boundary conditions of the wave field. Two examples of their applications to the prediction of wave properties are presented, which demonstrate their usefulness to the analysis and measurements of ocean waves.

Introduction

Ocean surface gravity waves are irregular and short-crested. They are modeled by a superposition of many monochromatic wave components of different frequencies, amplitudes and advancing in different directions. The basic wave components, known as free-wave (or linear) components, obey the dispersion relation. Due to the nonlinear nature of surface waves, these free-wave components interact among themselves. The interactions can be classified into "strong" or "weak" interactions. Strong interactions are observable as "bound-wave" components or modulation of short-wavelength components by long-wavelength components in a duration of about one wave period after the free-wave components start to interact. Bound-wave components in general do not obey the dispersion relation. The effects of strong interactions on the properties of interacting wave components, such as wave amplitudes, wavelengths and frequencies, do not last if the free-wave components no longer advance together. Weak interactions, also known as resonance interactions, may occur when the frequencies and wavelengths of interacting wave components satisfy the resonance condition. They may result in energy transfer among free-wave components of different frequencies and can be substantial only after a duration of hundreds of wave periods (Phillips 1960; Hasselmann 1962). Strong interactions in a wave field may occur at second order in wave steepness and are important

---

[1] Asso. Prof., Ocean Engr. Program, Dept. of Civil Engr., Texas A&M Univ. College Station, TX, USA 77843-3136
[2] Graduate Student, Ocean Engr. Program, Dept. of Civil Engr., Texas A&M Univ. College Station, TX, USA 77843-3136
[3] Research Associate, Ocean Engr. Program, Dept. of Civil Engr., Texas A&M Univ. College Station, TX, USA 77843-3136

to wave propagation in a short duration or short distance and the prediction of wave characteristics in the vicinity of measurements. Weak interactions may occur at third order for waves propagating in deep or intermediate-depth water.

For the study of wave-wave interactions in ocean waves and their effects on wave properties, Hybrid Wave Models (HWM) were recently developed for long-crested and short-crested seas (Zhang et al. 1996; Prislin et al. 1997; Zhang et al. 1998). Current HWMs are accurate up to second order in wave steepness. Hence, they are limited to the study of the wave propagation and wave properties in a short distance from the locations where a wave field is measured or described. However, the HWMs can be extended to include high-order nonlinear effects and can be used to study long-term wave evolution, specially those interactions in the conjuncted area of strong and weak wave interactions.

In comparison with other nonlinear wave models or numerical schemes, the HWMs are unique and innovative in three respects. First, the nonlinear effects are computed using two different perturbation approaches, namely, Mode-Coupling Method (MCM) and Phase Modulation Method (PMM). The truncated analytic solutions for bound-wave components given by the MCM are in the frequency and wavenumber domains and easy to be handled in computation. However, they may not converge if the wavelengths of two interacting wave components are quite different. The corresponding solutions given by the PMM are convergent but their numerical computation is slightly more complicated than the MCM. An ocean wave field may have its energy distributed over a relatively broad frequency range, the combination of the MCM and PPM can provide convergent solutions for the interactions involving wave components over a relatively broad frequency range (Zhang et al. 1996; Zhang et al. 1998). Secondly, most existing nonlinear numerical schemes are able to compute the nonlinear effects based on the prescribed or assumed free-wave components which constitute an ocean wave field. Ocean waves are often described by measurements at fixed points. The measurements actually record the resultant wave properties which are the superposition of those of the free-wave components and their nonlinear interactions. Hence, the results of the Fast Fourier Transfer (FFT) of the measurements are not equal to the free-wave components, specially in the frequency ranges above and below the spectral peak frequency. Based on the measurements, the HWMs decouple the nonlinear wave effects from the measurements of an ocean wave field and then decompose the wave field into a set of free-wave components. Based on the free-wave components, the HWMs compute the wave properties in the vicinity of the measurements by superposing the contributions from the free-wave and bound-wave components, which is similar to the other nonlinear numerical schemes. The third respect is specially related to short-crested or directional ocean waves. The existing methods for analyzing directional ocean waves only provide spectral or statistical information even within the scope of linear wave theory (Borgman 1990). The predictions of related wave properties in the time domain are made with a random-phase assumption. Based on as few as three independent measurements, the Directional HWM (DHWM) is able to decompose a directional ocean wave field into free-wave components without the assumptions of random phases and prior directional spreading function. The wave properties in the vicinity of the measuring location can be deterministically predicted up to second order in wave steepness. Because of these unique features, the HWMs are essential to the study of ocean waves, specially the analysis of wave measurements.

Principles of HWMs

The numerical scheme of the HWMs mainly consists of two steps. The first step is to decompose an wave field into the free-wave components based on its measurements. The second one is to predict wave properties based on the free-wave components obtained in the decomposition. Nonlinear interactions among free-wave components need to be calculated in both steps. The calculation in the first step is for decoupling the nonlinear effects from the measurements and that in the second step for quantifying the nonlinear effects on the resultant wave properties. The formulation for calculating wave interactions and numerical scheme of the first step are briefed below.

The nonlinear solutions for wave-wave interactions in an ocean wave field are truncated at second order. While an ocean wave field consists of many free-wave components and a variety of interactions may occur among these components, at second order, each wave-wave interaction involves only two free-wave components. Although there are many pairs of free-wave components interacting in a wave field they can be calculated following a basic solution derived for the interaction between two free-wave components. The solution for the interaction of two unidirectional free-wave components was derived up to third order by Zhang et al. (1993) in deep water using both MCM and PMM. It was later extended to allow for intermediate depth water by Chen & Zhang (1997) and for the components of different directions by Zhang et al. (1998). Their studies showed that the solutions obtained using the two different methods are identical if the wavelengths of the two interacting free-wave components are close. If their wavelengths are quite different, the solution by the MCM may diverge while that by the PMM still converges. The basic solution for two interacting free-wave components and its extension to an ocean wave field consisting multiple free-wave components were given by Zhang et al. (1996 & 1998). For brevity, they are omitted here.

An amplitude-frequency spectrum in general can be divided into three parts: a very-low frequency range, a "powerful" range which includes all relatively large-amplitude wave components, and a high-frequency tail range, as sketched in Figure 1. Wave components located in both very-low frequency and tail ranges have relatively small amplitudes, and any interactions involving one of these components are insignificant and hence ignored in the computation. The powerful range can be further divided into several bands. The frequencies of the components within the same band are close and their interactions are computed using the MCM. The frequencies of wave components located in different bands are usually quite different and their interactions are modeled by the PMM.

The computation of nonlinear effects is based on the free-wave components while the decomposition of a wave field into the free-wave components needs to decouple the nonlinear effects from the measurements. Therefore, the decomposition is accomplished through iterative processes. Because the nonlinear effects are relatively significant in a high frequency band comparing to those in a low frequency band, the decouple of nonlinear effects from the resultant wave properties proceeds from the low-frequency

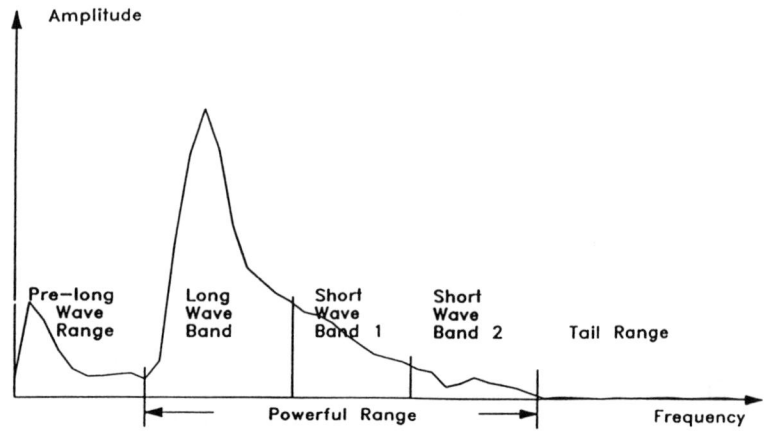

Figure 1: Sketch of frequency band division.

to high-frequency bands. The numerical scheme has been used for analyzing extremely steep ocean waves, such as waves induced by Hurricane Frederic and Andrew in the Gulf of Mexico (Couch & Conte 1996; Zhang et al. 1997) and the storm waves measured by the FULWACK test in the North sea (Forristall 1986). In all cases, the numerical results are convergent and unique.

Applications

Typical ocean surface waves consist of free-wave components of significant amplitudes whose frequencies range from about 0.05 to 0.18 Hz. The nonlinear interactions among these free-wave components may result in bound-wave components of a much wider frequency band ranging from hundredth to one Hz. These bound-wave components are in general much smaller in amplitude than the free-wave components near the spectral peak frequency, but may be comparable to or greater than those of high frequencies (0.15–0.4 Hz) as well as small frequencies (less than 0.05 Hz). In general, the free-wave components near the spectral peak dictate the resultant wave characteristics. That is why linear wave theory is a good approximation in most cases . However, in certain special cases, the contributions from wave components in either very lower or relatively high frequency bands can be crucial to the resultant wave properties and structure responses to waves. Under these circumstances, distinguishing the bound-wave components from free-wave components in the wave decomposition is necessary, and the use of the HWMs can provide more accurate prediction than linear theory or its improved modifications. Demonstrated below are two of these cases.

Kinematics near Steep Wave Crests

It is known that the predicted particle velocities near the steep wave crest based

OCEAN WAVE KINEMATICS, DYNAMICS AND LOADS ON STRUCTURES

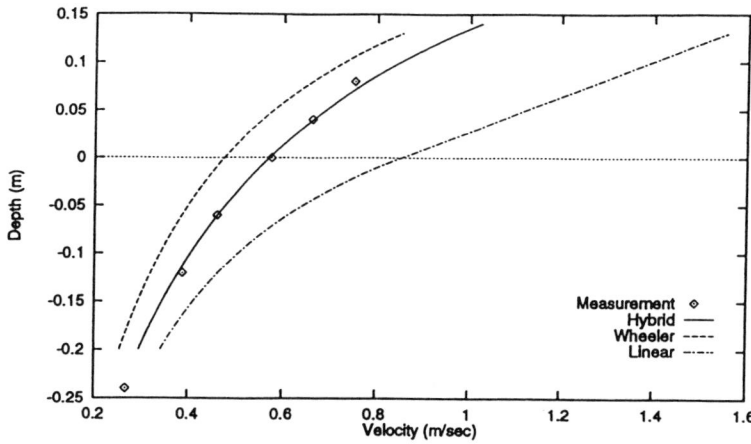

Figure 2: Comparison of predicted (— HWM, - - - Wheeler stretching, and - · - Linear extrapolation) and measured (◇) horizontal velocities at time 16.58 sec.

on linear wave theory are not accurate (Wheeler 1970). To improve the accuracy, modifications known as 'stretching' and 'extrapolation' techniques were made to linear wave theory. Many modification methods have been proposed based on either stretching or extrapolation techniques, or their combination. Two of the most widely used ones are the Wheeler stretching and linear extrapolation methods (Rodenbusch & Forristall 1986). Because they are not based on sound hydrodynamic principles, it is found that: 1) in the case of extreme steep waves, based on the same measured wave elevation the predicted wave kinematics obtained using the two different methods can differ by more than 50 %, which greatly concerns the users in computing wave loads on offshore structures (Steele et al. 1988); 2) the predictions made by both modifications show very large discrepancies with respect to the corresponding measurements (see Figure 2). In general, the Wheeler stretching under-predicts horizontal velocities under wave crests while the linear extrapolation over-predicts them. However, relatively away from the free surface, both predictions are very close to the measurements (Randall et al. 1993; Spell et al. 1996).

Poor predictions of wave kinematics near steep crests made by the modifications of linear wave theory are not by accident. Although the contribution to the surface elevation from the free-wave components near the spectral peak are much greater than that from wave components of relatively high frequencies, the differences in their contributions to wave kinematics near the free surface are much smaller because the particle velocity induced by a wave component is roughly proportional to its frequency and the acceleration to the frequency square. As mentioned before, the bound-wave components are significant in the high-frequency band and their contribution to wave kinematics is quite different from that of the free-wave components of the same frequencies. The HWM distinguishes the bound-wave from the free-wave components in an irregular

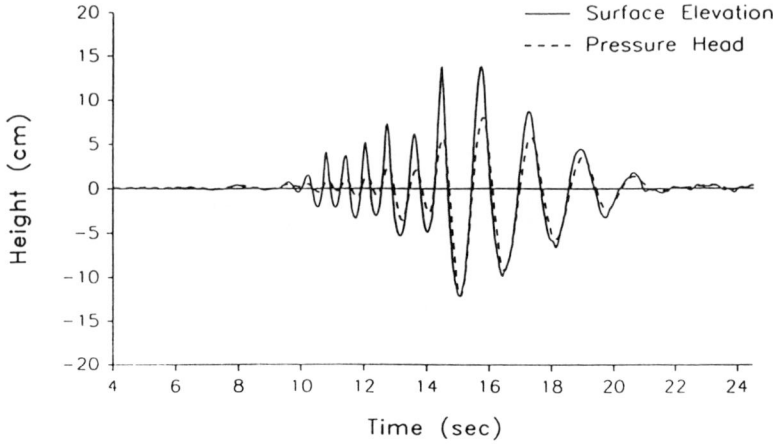

Figure 3: Pair of simultaneously measured surface elevation and pressure head time series. Pressure transducer is at 16 cm below the still water level.

wave field and computes their contributions to wave kinematics accordingly. Therefore, the HWM is able to predict the kinematics more accurately than the modified methods, as witnessed in Figure 2.

Wave Elevation Predicted Based on Pressure

Sea surface elevation is often indirectly measured by pressure transducers located at certain distance below the surface. Wave elevation is determined based on the measured dynamic pressure using a linear transfer function. Nonlinear statistical analysis has shown that nonlinear effects, specially due to second-order strong interactions, are significant at very low frequencies and frequencies about twice of the spectral peak (Herbers & Guza 1991 ; 1992). Yet, the nonlinear effects on measured pressure in the time domain, and specially on the predicted surface elevation using linear theory have not been investigated.

Figure 3 shows a measured surface elevation and its corresponding dynamic pressure recorded about 16 cm below the still water level. Both were measured at the same horizontal coordinate and in a very steep but non-breaking transient wave train generated in the wave flume located at the Hydromechanics Laboratory of Texas A&M Univer-

OCEAN WAVE KINEMATICS, DYNAMICS AND LOADS ON STRUCTURES 31

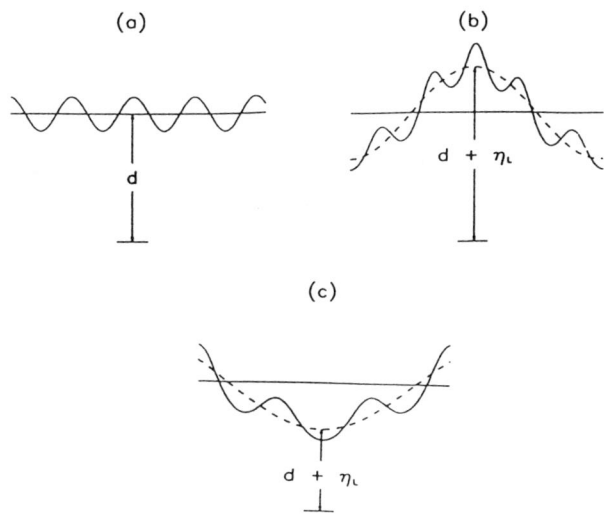

Figure 4: Change of a short wave still water level due to the presence of a long wave: (a) no long wave; (b) a short wave at the crest of a long wave; (c) a short wave at the trough of a long wave.

sity. The measured surface elevation shows larger amplitudes at the crests than at the trough, while the corresponding pressure shows larger amplitudes at the trough. At the first glance, the two records seem to indicate inconsistent measurements. The differences between the two measurements, however, can be qualitatively explained by the strong interaction between a high-frequency and a dominant low-frequency free-wave component.

In an irregular wave train, large crests and troughs are essentially produced when wave components of different frequencies constructively interfere. Short-wave (or high-frequency) components superposed on the surface of long-wave (or low-frequency) components may behave quite differently from their traveling on otherwise calm water. The wavelength of a short-wave component decreases at the crests of the long-wave component and increases in the troughs as a result of the modulation by the long-wave components (Longuet-Higgins & Stewart 1960; Phillips 1981). More importantly, as sketched in Figure 4, the vertical distance of a short-wave component to the pressure transducer becomes greater at the crest of long-wave components and smaller at the trough. It is well known that the wave-induced dynamic pressure decays exponentially with the depth. Thus, when the height of long-wave components is of order of the wavelength of a short-wave component, the measured dynamic pressure induced by the short-wave component at the pressure transducer will be much greater under the long-wave trough than that under its crest.

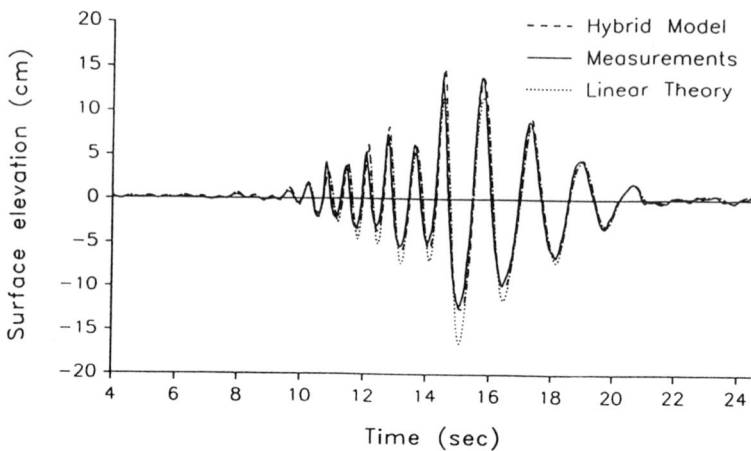

Figure 5: Surface elevation predictions from pressure transducer measurements at 16 cm below the still water level.

Linear wave theory ignores the interactions among the free-wave components. In using a linear transfer function, the surface elevation of each individual wave component is computed as if it travels on otherwise calm water. As explained above, due to the presence of long-wave components, the dynamic pressure induced by the high-frequency wave components heavily depends on the phase of long-wave components. Therefore, the neglect of wave interactions in linear wave theory results in over-prediction in trough heights and under-prediction in crest heights. The comparison between the measured and predicted elevations obtained using linear wave theory clearly shows the over-prediction of trough height and under-prediction of crest height (Figure 5). The total wave height is also over-predicted but not as great as the trough height. The HWM models the strong interactions among free-wave components and hence provides accurate predicted elevation.

Conclusion and Future Work

In the cases of steep irregular waves, nonlinear wave effects on waves properties can be more pronounced than in a periodic wave train. The use of HWMs can greatly improve accuracy of predicted wave properties near the locations of measurements. Two of these effects and their applications to wave prediction and measurements are demonstrated.

The development of the HWMs, especially DHWM, is in its infant stage. There is a plenty of room for improvement. The model is currently truncated at second order in wave steepness. For more accurate prediction, the extension to third order may be necessary. In the future, we hope that more applications will be made to ocean and coastal engineering problems.

Acknowledgment

This research was supported by the Offshore Technology Research Center which is sponsored in part by National Science Foundation Engineering Research Centers Program Grant Number CDR-8721512.

References

Borgman, L. E. 1990. *The Seas: Vol 9, Ocean Engineering Science, Part A*. pp. 121-168. J. Wiley and Sons.

Chen, L. & Zhang, J. 1997. *J. Ocean Engrg.* (in press).

Couch, A.T. & Conte J.P 1996. *Proc. OMAE'96*, Vol I, Part A, 1-13

Forristall, G. Z. 1981. *Proc. 20th Coastal Engineering Conference* ASCE, 208-222, Taipei, Taiwan.

Hasselmann, K. 1962. *J. Fluid Mech.* **12**, 481-500.

Herbers, T. H. C. & Guza, R. T. 1991. *J. Phys. Oceanogr.* **21**(12), 1740-1761.

Herbers, T. H. C. & Guza, R. T. 1992. *J. Phys. Oceanogr.* **22**(5), 489-504.

Longuet-Higgins, M. S. & Stewart, R. W. 1960. *J. Fluid Mech.* **8**, 565-583.

Philips, O. M. 1960. *J. Fluid Mech.* **9**, 193-217.

Philips, O. M. 1981. *J. Fluid Mech.* **107**, 465-485.

Prislin, I., Zhang, J. & Seymour, R.J. 1997. *J. Geophys. Res.* **102**(C6), 12,677-12,688.

Randall, R. E., Zhang, J. & Longridge, J. K. 1993. *J. Ocean Engrg.* **20**, 541-554.

Rodenbusch, G. & Forristall, G. Z. 1986. *Offshore Technology Conference* **OTC 5097**.

Spell, C. A. Zhang, J. & Randall, R. E. 1996. *Applied Ocean Res.* **18**, 93-110.

Steele, K. M., Finn, L. D. & Lambrakos, K. F. 1988. *Proc. 20th Annual Offshore Technology Conf.* **OTC 5783**, Houston, TX.

Wheeler, J. D. 1970. *J. Petroleum Tech.* **249**, 359-367.

Zhang, J., Hong, K. & Yue, D. K. P. 1993. *J. Fluid Mech.* **248**, 107-127.

Zhang, J., Chen, L., Ye, M. & Randall, R. E. 1996. *Applied Ocean Res.* **18**, 77-92.

Zhang, J., Meza Conde, E., Dimarco, S.F. & Heideman J.C. 1997. *Proc. Wave'97* ASCE (in press).

Zhang, J., Yang, J. & Wen, J. and Prislin, I. 1998. *Offshore Technology Research Center* **Rep. No. 98A9475**.

Long Waves - Short Waves Resonance in Three Dimensions

Derek W. C. Lai, Andy T. Chan and Kwok W. Chow [1]

Abstract

Three-dimensional long waves - short waves interactions are studied by a model system of coupled nonlinear Schrödinger and Kadomtsev - Petviashvili equations. Such long - short resonance is relevant in the interactions of surface and internal waves in a two-layer model of the ocean. A one soliton solution is obtained by the bilinear transform. The modulational instability of the plane wave is investigated by Fourier analysis.

Introduction

Interactions of surface and internal waves are of considerable interest in the dynamics of the upper ocean. Resonance of such waves is especially important since energy transfer will occur on a more rapid time scale. In a two-layer model of the ocean two short surface waves and one long interfacial (internal) wave can participate in a degenerate case of triad resonance. Such long wave - short wave resonance occurs when the group velocity of the short surface wave matches the phase velocity of long internal wave (Benney, 1976; Craik, 1985). The governing equations can then be derived by the method of multiple scales :

$$L_t = (SS^\star)_x, \qquad (1)$$

$$iS_t + S_{xx} + LS = 0. \qquad (2)$$

L, S are the long wave and short wave envelope respectively. The superscript $\star$ denotes the complex-conjugate. Special exact solutions, including solitary waves, are known for this system (Ma and Redekopp, 1979).

Due to the special scaling balance the nonlinearity of the long wave L does not enter into the dynamics at this order. It is conceivable that such nonlinearity

---
[1]Department of Mechanical Engineering, The University of Hong Kong, Pokfulam Road, Hong Kong

might become a factor in the higher order terms of the asymptotic expansion. In fact earlier works (Kawahara, Sugimoto and Kakutani, 1975) showed that the effects of surface tension might result in a Korteweg - de Vries type operator for L :

$$L_t + LL_x + L_{xxx} = (SS^*)_x, \tag{3}$$

$$iS_t + S_{xx} + LS = 0. \tag{4}$$

The nonlinearity of the long wave follows the Korteweg - de Vries equation :

$$u_t + 6uu_x + u_{xxx} = 0. \tag{5}$$

The treatment so far is strictly two dimensional as no spanwise dependence is allowed. The present goal is to study the combined effects of these higher order nonlinear terms in L and the three dimensionality of the interaction process. It is well known that a natural three dimensional extension of the Korteweg - de Vries operator is the Kadomtsev - Petviashvili equation :

$$(u_t + 6uu_x + u_{xxx})_x + \sigma u_{yy} = 0 \tag{6}$$

We shall propose a model that incorporates this feature.

The interactions of long waves and short wave packets in nonlinear dispersive media are relevant in many disciplines besides surface - internal waves resonance, e.g., solid state and plasma physics. Many coupled systems of evolution equations have been proposed in the literature to serve as models for the dynamics. One example is the Mel'nikov system (1986) :

$$(u_t + 6uu_x + u_{xxx})_x + \sigma u_{yy} = (SS^*)_{xx}, \tag{7}$$

$$iS_y + S_{xx} + uS = 0, \tag{8}$$

where u is the long-wave amplitude. Hirota, Ohta and Satsuma (1988) have shown that the above equations possess a special structure in terms of the Wronskian.

A remark regarding the introduction of three dimensional effects is in order. Due to the special requirements of the group and phase velocities in the long short resonance, the surface and internal waves might propagate at an angle of each other (Craik, 1985). Hence the introduction of spanwise dependence is more than an academic curiosity. Comparing (2) and (8) one now observes that the asymptotic expansion of fluid dynamics leads to a derivative in t and not y in the short wave equation. Hence we propose to study a coupled system of the nonlinear Schrödinger and the Kadomtsev-Petviashvili equations,

$$(u_t + 6uu_x + u_{xxx})_x + \sigma u_{yy} = (SS^*)_{xx}, \tag{9}$$

$$iS_t + S_{xx} + uS = 0. \tag{10}$$

The factor 6 is introduced in anticipation of the bilinear transformation. Some, but not all, coefficients of the equations can be adjusted by rescaling. The sign of the constant σ will represent the effect of spanwise dispersion. Hydrodynamic waves in ordinary oceanic conditions will have a positive value for σ. This coupled system will now be shown to possess a one-soliton solution using Hirota's (1973) bilinear transformation. We shall also investigate the modulational stability of plane waves of this system.

One-soliton Solution

The equations (9) and (10) can be transformed by using the Hirota bilinear transformation to

$$(D_x D_t + D_x^4 + \sigma D_y^2) f \cdot f = g g^*, \quad (i D_t + D_x^2) g \cdot f = 0, \qquad (11)$$

where f (real) and g (complex) are related to u and S by

$$S = \frac{g}{f}, \quad u = 2(\log f)_{xx}. \qquad (12)$$

The operators D_x, D_y and D_t are the Hirota's transform operators:

$$D_x^m D_y^n D_t^p g \cdot f = (\frac{\partial}{\partial x} - \frac{\partial}{\partial x'})^m (\frac{\partial}{\partial y} - \frac{\partial}{\partial y'})^n (\frac{\partial}{\partial t} - \frac{\partial}{\partial t'})^p g(x,y,t) f(x',y',t')|_{x=x',y=y',t=t'}. \qquad (13)$$

We can now obtain a one-soliton solution,

$$f = 1 + m[1, 1^*] e^{\eta_1 + \eta_1^*}, \quad g = e^{\eta_1}, \qquad (14)$$

where

$$m[1, 1^*] = \frac{1}{2[(p_1 + p_1^*)^4 + i(p_1 + p_1^*)^2 (p_1 - p_1^*) + \sigma(q_1 + q_1^*)^2]}, \qquad (15)$$

and

$$\eta_1 = p_1 x + q_1 y + i p_1^2 t + \eta_0, \qquad (16)$$

where p_1, q_1 and η_0 are free (complex) parameters. The propagation of such a 1-soliton is illustrated in Figures 1, 2, 3.

Modulational Stability

A simple plane wave solution for (9) and (10) is

$$S = S_0 \exp(i k S_0^2 t), \text{ and } u = k S_0^2, \qquad (17)$$

where S_0 is the wave amplitude of the short wave and k is the corresponding wavenumber. To investigate the modulational stability of this wave we impose a perturbation onto (17). Thus,

$$S = S_0 \exp(i k S_0^2 t)(1 + \epsilon B), \text{ and } u = k S_0^2 (1 + \epsilon v), \qquad (18)$$

where ϵ is a small parameter. Substituting solutions (18) into (9) and (10) and collecting terms up to the order of ϵ, we obtain two linearized partial differential equations:

$$iB_t + B_{xx} + kS_0^2 v = 0, \tag{19}$$

$$k(v_{xt} + 6kv_{xx} + v_{xxxx} + \sigma v_{yy}) = B_{xx} + B_{xx}^*. \tag{20}$$

We now write $B = P + iQ$ and $v = R$. The linearized stability equations reduce to

$$P_t + Q_{xx} = 0, \tag{21}$$

$$-Q_t + P_{xx} + kS_0^2 R = 0, \tag{22}$$

$$k(R_{xt} + 6kR_{xx} + R_{xxxx} + \sigma R_{yy}) = 2P_{xx}. \tag{23}$$

One now employs normal mode analysis for P, Q and R and considers

$$P = P_0 e^{i(\alpha x + \beta y - \omega t)}, \quad Q = Q_0 e^{i(\alpha x + \beta y - \omega t)}, \quad \text{and } R = R_0 e^{i(\alpha x + \beta y - \omega t)}. \tag{24}$$

α and β are the wavenumbers in the x and y directions respectively. ω is the angular frequency of the wave. Complex exponential is preferred over the real sine and cosine functions as the former is easier to manipulate. At first sight there appears to be an inconsistency as P, Q, R are assumed to be real, but a little thought reveals that this procedure is sound. Complex ω will imply instability. A cubic equation in ω results when one eliminates the constants P_0, Q_0 and R_0. The stability condition for the interacting waves is now

$$\frac{\omega^3}{\alpha^3} + \left(1 - \frac{\beta^2}{\alpha^4} - \frac{6k}{\alpha^2}\right)\omega^2 - \alpha\omega - (\alpha^4 - 6k\alpha^2 - \sigma\beta^2 + S_0^2) = 0. \tag{25}$$

We now rewrite (25) in terms of the constants A_n, $n = 0, 1, 2, 3$,

$$A_0 \omega^3 + A_1 \omega^2 + A_2 \omega + A_3 = 0. \tag{26}$$

Instability commences when imaginary roots occur. Elementary algebra now gives the condition as

$$-27A_0^2 A_3^2 - 18A_0 A_1 A_2 A_3 + 12A_0 A_2^3 - A_1^2 A_2^2 + 27A_1^3 A_3 < 0. \tag{27}$$

Discussions

A coupled system of nonlinear evolution equations is proposed as a model for long wave - short wave interaction in three dimensions. Modulational instability of plane waves is studied by Fourier analysis. Although a 1-soliton solution can be obtained, preliminary calculations show that the 2-soliton solution does not exist. Hence the system (9, 10) probably is not integrable. The real challenge is to derive (9, 10) from a formal asymptotic expansion of the equations of motion. This task is being actively pursued. When the coefficients are expressed in terms of the physical properties of the two-layer fluid, e.g., the density ratio and the depth ratio, the region(s) of instability of the plane waves can be elucidated completely. Relevant and useful information on the surface and internal wave fields can then be extracted.

References

Benney, D.J. (1976), *Studies in Applied Mathematics*, **55**, 93.

Craik, A.D.D. (1985), *Wave Interactions and Fluid Flows*, Cambridge University Press.

Hirota, R. (1973), *Journal of Mathematical Physics*, **14**, 805.

Hirota, R., Ohta, Y., and Satsuma J. (1988), *Journal of the Physical Society of Japan*, **57**, 1901.

Kawahara, T., Sugimoto, N., Kakutani, T. (1975), *Journal of the Physical Society of Japan*, **39**, 1379.

Ma, Y.C., Redekopp, L.G. (1979), *Physics of Fluids*, **22**, 1872.

Mel'nikov V.K. (1986), *Physics Letters A*, **118**, 22.

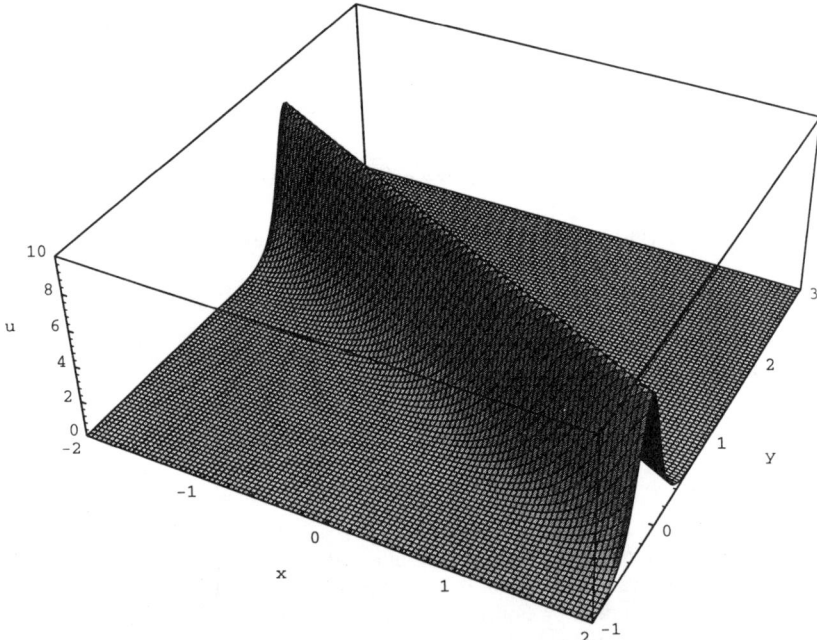

Figure 1 : Nonlinear Schrödinger equation coupled with Kadomtsev-Petviashvili equation: u against x, y for $p_1 = 2 + i$, $q_1 = 4 - 3i$ at $t = 0$.

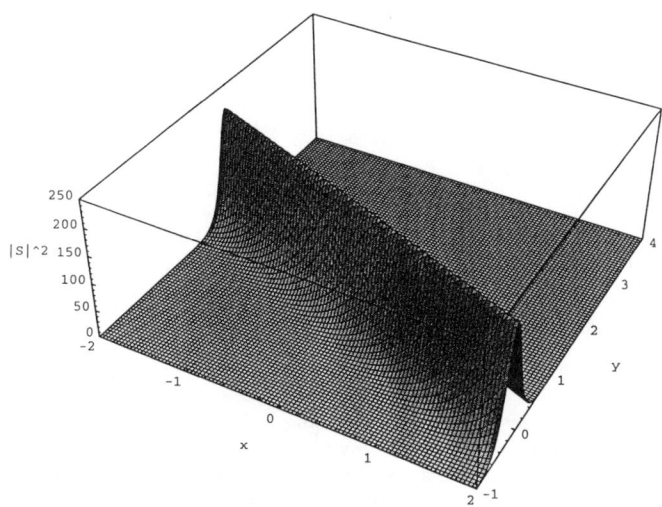

Figure 2 : Nonlinear Schrödinger equation coupled with Kadomtsev-Petviashvili equation: $|S|^2$ against x, y for $p_1 = 2 + i$, $q_1 = 4 - 3i$ at $t = 0$.

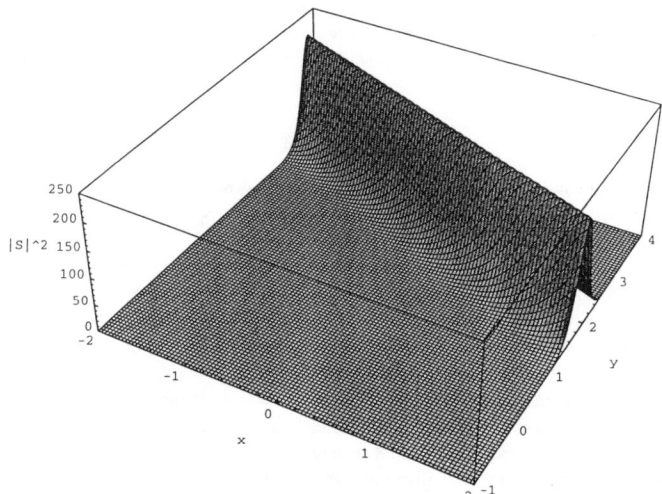

Figure 3 : Nonlinear Schrödinger equation coupled with Kadomtsev-Petviashvili equation: $|S|^2$ against x, y for $p_1 = 2 + i$, $q_1 = 4 - 3i$ at $t = 2$.

Some Aspects on Wave Kinematics and Wave Measurements

Igor Prislin and John Halkyard[1]

Abstract

Wave measurements are important in two aspects. First, they are used to determine the initial and boundary conditions for a variety of contemporary numerical algorithms, known as numerical wave tank. Second, in verification of wave theories, wave measurements obtained at different locations or at different times are used as a criterion against which theory predicted wave properties are compared.

The paper examines the relationship among different wave kinematics measurements and points out potential pitfalls in experimental data. Wave profiles, hydrodynamic pressure, and velocity components measured in a directional wave tank and in the ocean are analyzed by a deterministic nonlinear short-crested wave model up to second order in wave steepness and by a conventional spectral method of the first order. The paper discusses laboratory and in situ instrumentation, data analyses, and physical wave modeling. Finally, some practical solutions and recommendations are indicated.

Introduction

Short-crested irregular ocean surface waves are intuitively modeled by the superposition of free-wave components with different frequencies and propagating in different directions. The interaction among these free-wave components may change their characteristics (amplitude, frequency, wavelength and propagating direction) and the changes are usually described as bound-waves in conventional perturbation methods [*Donelan et al.* 1985; *Zhang et al.* 1993]. To predict the wave characteristics other than the measured or the measured values somewhere near the location where the measurements are taken, the decomposition of an irregular short-crested wave field into its free-wave components is necessary. Whereas the bound-wave components can be calculated straightforwardly if the free-wave components are known, conversely, the decomposition of resultant wave measurements into their free-wave components is much more complicated because the measurements record the resultant properties of the free-wave and bound-wave components [*Long*, 1986].

Decomposition of measured wave field into free-wave components is crucial to a variety of offshore and coastal engineering and environmental science applications. Based on

---

[1] Deep Oil Technology, Inc., 11757 Katy Freeway, #500, Houston, Texas 77079-1709

its free-wave components, the characteristics of a related wave field, such as elevation, pressure, velocity, acceleration, and slope can be predicted. For example, to predict the wave loads on slender elements of a truss spar (Fig. 1) using the Morison equation, the wave kinematics should be known.

This paper presents an application of Directional Hybrid Wave Model (DHWM), a new method for the deterministic decomposition of short-crested ocean waves, to wave kinematic measurements [Prislin ,1996; Prislin et al., 1996, 1997a]. DHWM integrates three fundamental features of ocean waves: wave directionality, initial wave phases, and nonlinear wave-wave interactions. For the first time, all these features have been considered jointly to decompose a measured ocean wave field into an ensemble of free-wave components. The purpose of wave decomposition is to predict wave properties other than those measured if, and only if, the free-wave (linear) components are known, wave properties other than those measured can be predicted. The difficulties of wave decomposition stem from the complicated structure of wave fields and inherently intertwined directional nonlinear waves. The additional challenge is that a wave field is defined by only a limited number of measurements.

DHWM makes possible computations of directional, nonlinear, and initial phase characteristics of linear wave components based on as few as three measurements of short-crested irregular ocean waves. Thus, the DHWM seems to be flexible and robust in analyzing wave measurements. It unifies conventional and contemporary nonlinear wave theories based on the

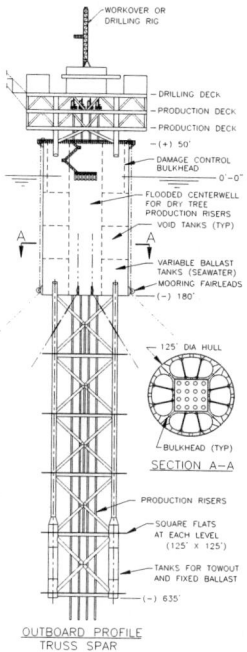

Figure 1: Truss Spar

wave phase modulation, which is attributed to the unidirectional nonlinear hybrid wave theory [Zhang et al. 1993, 1996]. The DHWM is materialized in three major steps. The extended maximum likelihood method [Isobe et al., 1984] is used for the computation of directional energy spreading and phase fitting method is used for initial phases for the computation of free-wave components. The third step accounts for the computation of nonlinear effects up to second order in wave steepness and the subtraction from the measurements [Prislin, 1996; Zhang et al. 1996].

The accuracy of DHWM for deterministic decomposition of short-crested ocean waves was examined by means of numerical simulations and field wave data analyses. The results are highly congruent with both simulated and measured waves [Prislin ,1996; Prislin et al., 1997b].

In this paper the relationship among different wave kinematics measurements and their predicted values are compared and discussed.

Experiments

The wave kinematic measurements were performed at Offshore Model Basin (OTRC) in College Station, Texas [*Prislin*, 1996]. Three types of measurements are considered: surface elevation, hydrodynamic pressure and orthogonal velocity components. All measurements are performed with high accuracy to preserve the relative phases among signals measured by sensors at different locations.

To simulate an ocean wave field, the model scale is determined based on the Froude similarity criterion. For all tests the length ratio is 1:25.

Results
Wave Gauges

Capacitance-type wave gauges were used to measure surface elevations. The wave gauges were calibrated daily before the tests by the static immersion procedure. The calibrations showed that the curves were virtually linear within the calibration range with a standard deviation of 0.6 mm (0.1% of the calibration range).

Pressure Sensors

The SENSOTEC Model GW-100 Series Pressure Transducers of the range 0 to 1 psi were used. The sensors belong to the strain gage type, and are suitable for the underwater measurements. The accuracy of transducers expressed as standard deviation and specified by the manufacturer is 0.2% over the nominal pressure range with continuous resolution. The overload safety factor is 10. The sensor has a cylindrical shape of diameter around 25 mm. Its nominal weight is 0.1 kg. The sensors were calibrated daily before the tests. The calibration was accomplished statically by immersing all sensors simultaneously to a specified depths of up to 0.55 m below the calm water level. Each reading was related to the hydrostatic pressure at a particular depth. Fresh water density, used for hydrostatic pressure computation, was determined based on the measured water temperature. The calibration results show that the curves were virtually linear in the range of interest with an average standard deviation of 0.7 mm in terms of the pressure head (0.2% of the calibration range). If this value is attributed to the uncorrelated noise in pressure measurements, then the average noise (defined here as the standard deviation of calibrated values) can be visualized as an average offset in pressure signal. An exponential growth of amplitude spectral value with frequency (see Fig. 2) does not really represent wave amplitudes but rather the noise in the measured pressure signal. In such a case the hydrodynamic pressure is less than the ambiental noise. To show the effect of noise in pressure to surface elevation conversion, the

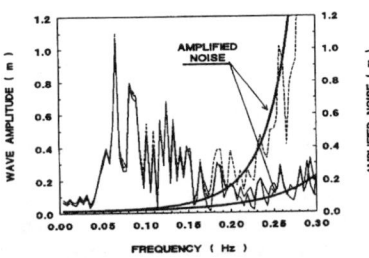

Figure 2: Effect of noise amplification on the prediction of wave amplitude from pressure measurements.

average noise amplitude obtained from the pressure sensor calibration are converted to the wave amplitude and compared to the converted measured pressure at two different depths (see Fig. 2). It is interesting that there are good agreements between converted average noise and converted pressure for frequencies higher than 0.20 Hz. This demonstrates that the pressure signals above certain frequency are indeed insignificant in comparison with the estimated average noise level. In addition, similar fluctuations of converted amplitudes based on two different pressure sensors at two different depths indicate that some wave effects due to the nonlinear wave-wave interactions on the pressure might be present. The higher order pressure components are, however, of at most the same or of a smaller order of magnitude than the recorded noise. If that was not the case, the curves of converted pressure and noise would differ notable. To conclude, if pressure measurements are used for decomposition of the ocean waves, a high pressure signal to noise ratio is crucial to get accurate results.

Velocitymeters

Three types of flow intrusive velocitymeters were used to measure water particle velocities in waves. Two of them were acoustic: an Acoustic Doppler Velocitymeter (ADV) and an Acoustic Velocitymeter (AV). The third velocitymeter was an Electro-Magnetic Velocitymeter (EMV). Unfortunately, none of them were suitable for the short-crested wave measurements in a multidirectional wave basin. However, this was not known before the tests were conducted.

The ADV (Sensotec type) is based on a Doppler frequency shift between the emitted and reflected acoustic beams from small inhomogeneities in water (air bubbles or microscopic floating particles). The water in the OTRC wave tank is continuously filtered; consequently, the reflected acoustic beam is highly dispersed. This results in a very low signal to noise ratio, thus preventing the velocity measurement. The recommended synthetic seeding material for this type of velocitymeter was not used because of its unknown long-term effects on humans, even in a large body of water (the OTRC divers have to perform a variety of underwater activities daily). To avoid this problem, dissolved flour in water was used as a seeding material. This method produced excellent results when the seeding material was hit by the acoustic beams. On the other hand, it was very hard to keep the proper concentration of the seeding material around the probes at all times. A small deviation from the proper concentration of the seedings resulted in significant noisy signal. Approximately one gallon of seeding emulsion was used at each test (for approximately a two-minute run) to seed the water around the ADV. An additional problem was that flour deposits built up on the wave gauges during a test, making the surface elevation measurements useless. Hence, the ADV is not suitable for wave induced velocity measurements in clean water.

The second acoustic velocitymeter (AV Minilab type) is based on differential travel-time between two acoustic beams emitted in the opposite directions. The travel-time is proportional to the component of fluid velocity along the path. This method is more reliable in detecting fluid velocity than the method used in ADV because the travel-time difference is not sensitive to water quality and additional seeding is not required.

The errors that may corrupt velocity measurements arise from the flow disturbance by the sensor itself. *Trivett et al.* [1991] discussed several primary sources of self-induced error. Their error analysis of an acoustic current meter is based on an uniform flow. In the OTRC wave basin measurements, the flow is oscillating and it is found that vortex shedding dominates over other possible sources of error.

The generated vortices are present around the AV probes all the time and they enhanced the super harmonics of velocity components (see Figures 3 and 4). The orientation of the AV in short-crested waves does not have a noticeable influence on vortex shedding, and vortex shedding effects cannot be avoided. The frequency of vortex shedding is roughly the nearest multiple of the flow oscillation frequency expressed as a product of the Strouhal number S and the Keulegan-Carpenter number Kc [Blevins, 1990]. The Strouhal and Keulegan-Carpenter numbers are defined as $S=(f_s D)/U_m$ and $Kc=U_m/(fD)$, respectively. U_m, D, f, and f_s are the mean flow velocity, the characteristic diameter of a bluff structure, flow frequency, and shedding frequency, respectively. If the measured velocity amplitude is 0.2 m/s (model scale) in a wave train of frequency 0.5 Hz (model scale), then the estimated Keulegan-Carpenter number for a structure of the characteristic diameter of 0.025 m (model scale) is about 16. Using the characteristic Strouhal number of 0.2 [Blevins, 1990; Trivett et al., 1991] the vortex shedding frequency could occur around 1.6 Hz (model scale), which is about three times the wave frequency (the third harmonic). This estimation is very close to what can be observed in the velocity measurements (see the velocity component at frequency around 0.3 Hz - full scale in Figure 4). The estimated effective diameter in the above example seems to be correct, although the AV is not a completely bluff body. The subharmonics of the shedding frequency are also present [Blevins, 1990], but in this case they are most likely covered by the first and second wave harmonics. To conclude, it is indisputable that such a strong harmonic (*e.g.* at 0.3 Hz in Fig. 4) in the velocity measurements is not mainly due to wave-wave interactions but rather due to induced velocities from vortex shedding.

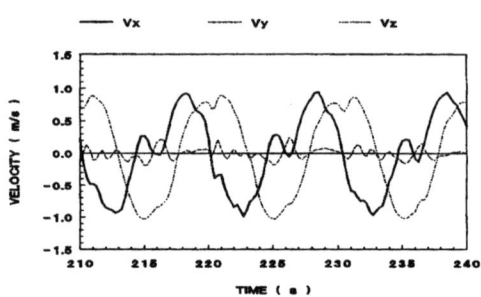

Figure 3: Wave particle velocities measured by an AV in a regular wave train (f=0.1 Hz).

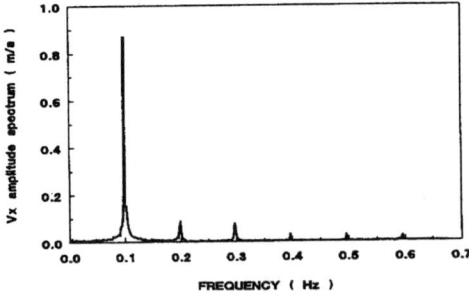

Figure 4: Vx velocity amplitude spectrum measured by an AV in a regular wave train (f=0.1 Hz).

The third velocitymeter, EMV is used to avoid the difficulties encountered by the ADV and AV. The electromagnetic velocitymeter is the Marsh-McBirney Model 511 with a spherical probe with diameter of 38.1 mm (1.5 inch). The water velocity measurement is based on the Faraday principle of electromagnetic induction. The principle states that a conductor (the water) moving in a magnetic field generated from within the sensor, produces a voltage that is proportional to the velocity of the water. Two pairs of electrodes are used to resolve the velocity vector into its two orthogonal horizontal components. The induced voltages across the electrodes are of the same carrier frequency (30 Hz) as the magnet driver frequency. The voltage signals are linearly proportional to the flow velocity components detected by the electrodes. The signals are amplified, synchronized, and filtered to yield two analog voltages, which can be digitized with an analog-digital converter.

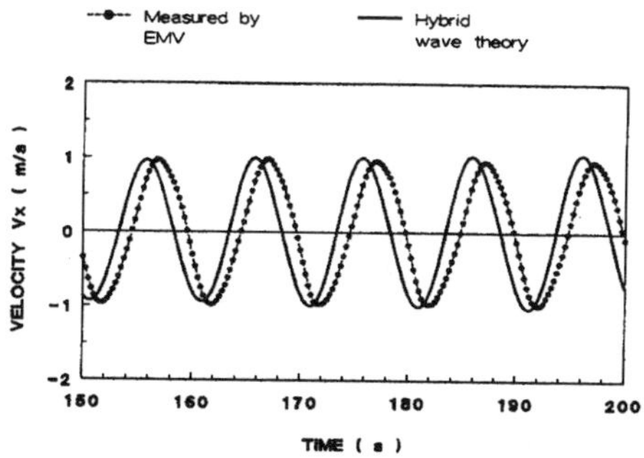

Figure 5: Measured (by EMV) and predicted Vx velocity component in a regular wave train (f=0.1 Hz).

The obtained measurements are smooth and appeared to be well behaved (without spurious peaks as observed in the AV signals). However, later analysis by the proposed wave decomposition methodology revealed that the velocity signals had a phase delay because of the embedded low-frequency filtering of the measured signals (the time constant for the EVM is one second). The measured and predicted horizontal velocity in a monochromatic wave train are shown in Figure 5.

At the same time and at approximately the same location, the surface elevation measurement did not show noticeable delay with respect to the predicted surface elevation signal (see Figure 6). Because of imperfection in velocity measurements, they cannot be used in wave model data analyses by DHWM.

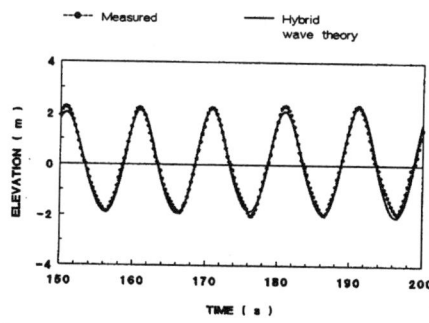

Figure 6: Measured and predicted surface elevation in a regular wave train (f=0.1 Hz).

In addition to the laboratory environment, the DHWM is tested by using the data from the FULWACK experiment in the North Sea on November 24, 1982 [*Forristall*, 1986]. The surface elevation and velocity measurements (EVM-Marsh-McBirney Model 524), 8.53 m below free surface are used to decompose the sea and then the velocity components are predicted and compared with the measured values at 4.88 m above free surface. The results are

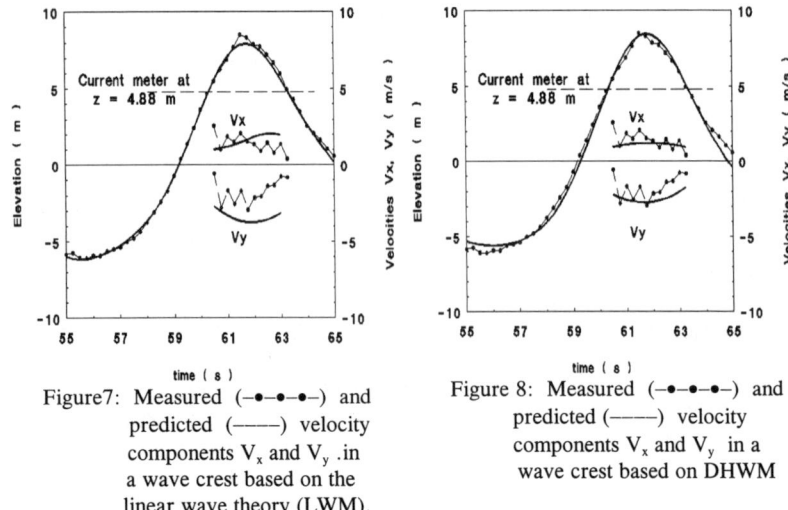

Figure 7: Measured (–•–•–•–) and predicted (——) velocity components V_x and V_y in a wave crest based on the linear wave theory (LWM).

Figure 8: Measured (–•–•–•–) and predicted (——) velocity components V_x and V_y in a wave crest based on DHWM

presented in Figures 7 and 8 using the linear wave model (LWM) and DHWM, respectively[*Prislin et al.*, 1997b].

The predicted velocities agree much better using DHWM than using LWM (see Figs. 7 and 8), which proves that wave nonlinearities cannot be neglected [Foristall *et al*, 1978]. Also, it seems that there are no difficulties in *in situ* velocity measurements, which are present in the wave basin.

Conclusions

The detection of instrument-related variations in wave properties is important in model test conditions when several measured values are scaled. Therefore, all imprecisions in measurements caused by a typical instrument are proportionally augmented. When used as a basis for designing offshore structures, these imprecisions from model tests may translate into costly errors. The DHWM is a valuable tool for the detection of instrument imprecisions, (*e.g.*, the time delay in velocity measurements) which then can be corrected, thus saving resources.

DHWM can predict wave properties other than the measured which can serve as inputs for other wave-structure related programs.

DHWM can be used for predicting wave properties which could be very difficult to measure (*e.g.;* wave induced accelerations, or the velocity measurements in wave crests, such as in the FULWACK experiment).

Further studies should explore the limiting sea state conditions which can be modeled and analyzed by the DHWM in a wave basin. In addition, they should examine the furthest distance from the primary instrument location at which the DHWM predicted wave properties are still in good agreement with the measurements.

Acknowledgments

This research was supported by the Offshore Technology Research Center, which is sponsored in part by National Science Foundation Engineering Research Centers Program Grant Number CDR-8721512 and the Joint Industry Program by Chevron, Exxon, Mobil, Shell, and Texaco. Without this support the research would not be possible. The FULWACK Experiment data are made available to us by Shell. The equipment and support from Deep Oil Technology, Inc. in preparing this paper is gratefully acknowledged.

References

Blevins, R. D., *Flow-induced Vibration,* 2nd ed., 451 pp., Van Nostrand Reinhold, New York, 1990.

Donelan, M. A., J. Hamilton, and W. H. Hui, Directional spectra of wind-generated waves, *Phil. Trans. R. Soc. Lond., A 315,* 509-562, 1985.

Forristall, G. Z., E. G. Ward, V. J. Cardone, and L. E. Borgmann, The directional spectra and kinematics of surface gravity waves in tropical storm Delia, *J. Phys. Ocean. 8,* 888-909, 1978.

Forristall, G. Z., Kinematics in the crests of storm waves, in *Proc. 20th Coastal Engineering Conference,* pp. 208-222, ASCE, Taipei, Taiwan, 1986.

Isobe, M., K. Kondo, and K. Horikawa, Extension of MLM for estimating directional wave spectrum, in *Proc. Symposium on Description and Modeling of Directional Seas, Paper No. A-6,* pp. 1-15, Technical University Denmark, Lyngby, Denmark, 1984.

Long, R. B., Inverse modeling in ocean wave studies, in *Wave Dynamics and Radio Probing of the Ocean Surface,* edited by O.M. Phillips and K. Hasselmann, pp. 571-593, Plenum Press, 1986.

Prislin, I., J. Zhang, and P. Johnson, Deterministic decomposition of irregular short-crested surface gravity waves, Proc. of the Sixth Intern. Offshore and Polar Engr. Conference, Los Angeles, 3, 57-64, 1996.

Prislin, I., Nonlinear deterministic decomposition of short-crested ocean waves, A Dissertation. Texas A&M University, College Station, Texas, 1996.

Prislin, I., J. Zhang, and R. J. Seymour, Deterministic decomposition of deep water short-crested irregular gravity waves, J. of Geophysical Research Vol. 102 No. C6, 12677-12688, 1997a.

Prislin, I. and J. Zhang, Prediction of wave kinematics under the wave crest in short-crested irregular gravity waves. 27th IAHR Congress - Seminar Proc., ed. E. Mansard,, 357-370, 1997b.

Trivett, D. A., E. A. Terray, and A. J. Williams, III, Error analysis of an acoustic current meter, *IEEE J. Oceanic Engineering, 16(4),* 329-337, 1991.

Zhang, J., K. Hong, and D. K. P. Yue, Effects of wavelength ratio on wave modeling, *J. Fluid Mech., 248,* 107-127, 1993.

Zhang, J., L. Chen, M. Ye, and R. E. Randall, Hybrid wave model for unidirectional irregular waves, Part I, Theory and numerical scheme, *J. Applied Ocean Research* Vol. 18, 77-92, 1996.

Irregular Wave Kinematics From PUV Gauge

Steven A. Hughes[1] and Rodney J. Sobey[2]

Abstract: A nonlinear theory is presented for time-domain interpretation of PUV time-series records measured beneath multidirectional, irregular waves in the presence of a uniform current of known magnitude and direction. The theory imposes a local Fourier approximation in a narrow moving local window while satisfying the complete nonlinear kinematic and dynamic free surface boundary conditions. Therefore, predictions of sea surface elevation and directional kinematics throughout the water column accurately portray the actual nonlinear character of the waves. The analysis method can be applied using PUV time series as short as an individual wave.

Introduction

The PUV gauge is a robust field instrument for measuring directional wave conditions in shallow water. The instrument provides simultaneous time-history traces of pressure and two orthogonal components of the horizontal velocity at a submerged location. Directional wave analyses of PUV data have usually relied on linear wave theory to interpret and transform these traces to a directional spectrum of the local sea surface elevations.

The linear analysis methods are based on the Gaussian random wave model where the sea surface is assumed to be the superposition of many linear waves of different frequencies, amplitudes, and direction. Linear transformations between measured kinematic quantities and sea surface elevations or kinematics at other depths are performed in the frequency domain. Inverse Fourier transforms can be used to estimate wave kinematics time series provided the phase information has been retained. However, time-series kinematics cannot be estimated if the directional wave condition is represented by an amplitude spectrum with random phases. Additionally, most linear analyses assume stationary conditions over a duration sufficient to achieve adequate

---
[1]Research Hydraulic Engineer, US Army Engineer Waterways Experiment Station, Coastal and Hydraulics Laboratory, 3909 Halls Ferry Road, Vicksburg, Mississippi 39180-6199 USA.
[2]Department of Civil and Environmental Engineering 412 O'Brien Hall, University of California, Berkeley, CA 94720 USA.

spectral resolution, and depth-uniform currents present in the measurements are usually removed prior to analysis.

It is well known that nonlinear wave influences are maximized in shallow water. The application of linear wave theory to interpret PUV measurements may compromise the value of the observations and distort important detail in the record, particularly if depth-uniform currents are neglected. This could have substantial consequences for project design in shallow water (Dean 1990). This paper describes a nonlinear theory for time-domain interpretation of a PUV time series record in the presence of a uniform current of known magnitude and direction.

Theoretical Development

PUV gauges provide simultaneous traces of submerged pressure $p(t; x_\alpha, z_P)$ and submerged horizontal velocity components $u_\alpha(t; x_\alpha, z_{UV})$. In horizontal tensor notation the position coordinates are (x_α, z) and the velocity coordinates are (u_α, w), with $\alpha = 1$ and $\alpha = 2$ corresponding to the x and y directions respectively of a cartesian coordinate system as shown in Figure 1. The z axis is directed vertically upwards from the horizontal plane of the local mean water level (MWL), in opposition to the gravity vector $\vec{g}$. Both instruments are at the same fixed horizontal position x_α but at different (though fixed and adjacent) vertical elevations z_P and z_{UV}, respectively.

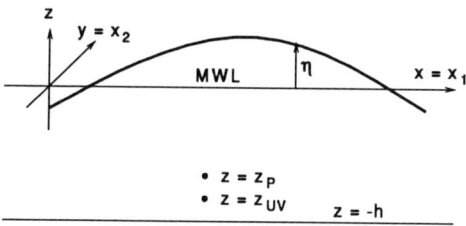

Figure 1: Coordinate system and PUV gauge location.

For incompressible, irrotational flow the field equation describing nonlinear and irregular waves is the Laplace equation, given in horizontal tensor form as

$$\frac{\partial^2 \phi}{\partial x_\alpha \partial x_\alpha} + \frac{\partial^2 \phi}{\partial z^2} = 0 \tag{1}$$

At the free surface the kinematic and dynamic boundary conditions are given as

$$f'_K = w - \frac{\partial \eta}{\partial t} - u_\alpha \frac{\partial \eta}{\partial x_\alpha} = 0 \quad \text{at} \quad z = \eta \tag{2}$$

$$f_D = \frac{\partial \phi}{\partial t} + \frac{1}{2}\left(u_\beta u_\beta + w^2\right) + g\eta = 0 \quad \text{at} \quad z = \eta \tag{3}$$

Assuming a locally horizontal bottom, the bottom boundary condition is simply

$$w = 0 \quad \text{at} \quad z = -h \tag{4}$$

Note that the primitive form of the kinematic free surface boundary condition (Eqn. 2) requires estimates of the gradients $\partial \eta/\partial t$ and $\partial \eta/\partial x_\alpha$, which is somewhat of a problem. This difficulty is avoided without compromise by redefinition (Longuet-Higgins 1962) of the kinematic free surface boundary condition as

$$f_K = f'_K + \frac{1}{g}\frac{Df_D}{Dt}$$
$$= w + \frac{1}{g}\frac{D}{Dt}\left(\frac{\partial \phi}{\partial t}\right) + \frac{u_\beta}{g}\frac{Du_\beta}{Dt} + \frac{w}{g}\frac{Dw}{Dt} \quad \text{at} \quad z = \eta \quad (5)$$

This modified form of the kinematic free surface boundary condition excludes both temporal and spatial gradients of η.

The field equation (Eqn. 1) and the bottom boundary condition (Eqn. 4) are exactly satisfied by nonlinear wave solutions given in terms of the velocity potential function

$$\phi(x_\alpha, z, t) = U_\alpha x_\alpha + \sum_j A_j \frac{\cosh jk(h+z)}{\cosh jkh} \sin j(k_\alpha x_\alpha - \omega t) \quad (6)$$

with velocity components

$$u_\alpha(x_\alpha, z, t) = \frac{\partial \phi}{\partial x_\alpha} = U_\alpha + \sum_j jk_\alpha A_j \frac{\cosh jk(h+z)}{\cosh jkh} \cos j(k_\alpha x_\alpha - \omega t) \quad (7)$$

$$w(x_\alpha, z, t) = \frac{\partial \phi}{\partial z} = \sum_j jkA_j \frac{\sinh jk(h+z)}{\cosh jkh} \sin j(k_\alpha x_\alpha - \omega t) \quad (8)$$

From the irrotational Bernoulli equation, the dynamic pressure is

$$p_d(x_\alpha, z, t) = \bar{B} - \frac{\partial \phi}{\partial t} - \frac{1}{2}\left(u_\alpha u_\alpha + w^2\right) \quad (9)$$

in which

$$\frac{\partial \phi}{\partial t} = -\sum_j j\omega A_j \frac{\sinh jk(h+z)}{\cosh jkh} \cos j(k_\alpha x_\alpha - \omega t) \quad (10)$$

and the Bernoulli constant is (Sobey 1992)

$$\bar{B} = \frac{1}{2}U_\alpha U_\alpha + \frac{1}{4}\sum_j \left(\frac{jkA_j}{\cosh jkh}\right)^2 \quad (11)$$

A prediction of the local kinematics is sought throughout the water column in the immediate neighborhood of the PUV gauge at horizontal position x_α. The pressure sensor is at known elevation z_P and the directional current meter at known elevation z_{UV}. Measured dynamic pressure traces $p_d^{obs} = p_d(t_i; x_\alpha, z_P)$ and measured velocity component traces $u_\alpha^{obs} = u_\alpha(t_i; x_\alpha, z_{UV})$ are available at discrete times t_i. The local water depth h and depth-uniform current U_α are also known.

The unknowns in this local solution are radian frequency ω, the wave number components k_α, the spatial phase $k_\alpha x_\alpha$, the Fourier coefficients A_j and the local water surface elevations $\eta_n = \eta(t_n; x_\alpha)$ in the neighborhood of time t_i. With J Fourier coefficients, N local water surface elevations and treating the spatial phase as a single unknown, there are $4 + J + N$ unknowns in a narrow local window.

At any discrete time t_i there are two active theoretical equations provided by the free surface boundary conditions and three PUV observational equations, i.e.,

$$\begin{aligned}
f_K(\omega, k_\alpha, k_\alpha x_\alpha, A_j, \eta) &= w + \frac{1}{g}\frac{D}{Dt}\left(\frac{\partial \phi}{\partial t}\right) + \frac{u_\beta}{g}\frac{Du_\beta}{Dt} + \frac{w}{g}\frac{Dw}{Dt} = 0 \quad \text{at} \quad z = \eta \\
f_D(\omega, k_\alpha, k_\alpha x_\alpha, A_j, \eta) &= \frac{\partial \phi}{\partial t} + \frac{1}{2}\left(u_\beta^2 + w^2\right) + g\eta = 0 \quad \text{at} \quad z = \eta \\
f_P(\omega, k_\alpha, k_\alpha x_\alpha, A_j, \eta) &= p_d - p_d^{obs} = 0 \quad \text{at} \quad z = z_P \\
f_{u_\alpha}(\omega, k_\alpha, k_\alpha x_\alpha, A_j, \eta) &= u_\alpha - u_\alpha^{obs} = 0 \quad \text{at} \quad z = z_{UV}
\end{aligned} \quad (12)$$

All equations are implicit and algebraic. Additional closure equations are provided by observational and free surface boundary equations at neighboring times within a narrow local window of duration τ_0 that is centered on t_i. For mathematical closure $4 + J + N$ independent equations must be provided; however, in practice additional equations must be specified because the P, U, and V observational equations have measurement error bands that preclude an exact solution. Therefore, the solution within each local window is accomplished by significant overspecification of equations and adoption of a least squares solutions.

The local LFI-PUV theory has two free parameters, the truncation order (J) and the number of η points (N) in each local window. The truncation order has much the same authority as order in an analytical wave theory (e.g., Stokes) or truncation order in Fourier wave theory.

Numerical Implementation

Solutions for the unknowns at each time step are sought in narrow local windows having a target width of $\tau_0 = 0.1T_z$ (single window) or $\tau_0 = 0.2T_z$ (double window), where T_z is the zero-up-crossing period of the record segment. Only the single window is discussed here. For typical applications a very satisfactory representation of the wave nonlinearities is accomplished with $J = 3$ and $N = 3$ which gives 10 unknowns in the local window, i.e., $\omega, k_\alpha, k_\alpha x_\alpha, A_1, A_2, A_3, \eta_1, \eta_2, \eta_3$. The unknown η_n are located within the local window as indicated on Figure 2.

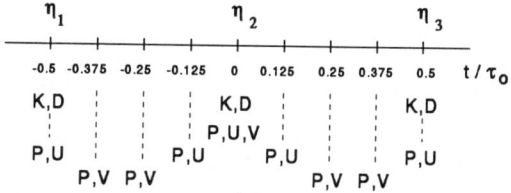

Figure 2: Distribution of equations within a narrow local window.

For a single width window, kinematic (K) and dynamic (D) free surface boundary condition equations are applied at the same locations as the unknown values of η_n. Nineteen observational equations of P, U, and V are also specified within the window as shown on Figure 2. This gives an overspecification of 25 equations to be solved for 10 unknowns in the window. Experience suggested that this level of overspecification is generally necessary to accommodate the observational error bands. For theoretical wave traces with no observational error bands only a closed set of KD and PUV equations is necessary for credible solutions.

The problem formulation requires the least-squares solution of a system of simultaneous, nonlinear, implicit algebraic equations. Numerical solutions routinely used the Dennis, et al. (1981b) NL2SOL code. The algorithm (Dennis, et al. 1981a) is a variation on Newton's method in which part of the Hessian matrix is computed exactly and part is approximated by a secant updating method.

In any high order nonlinear optimization problem with a large number of unknowns, the crucial elements of successful solutions are accurate estimates of the Jacobian and good initial solution estimates. Accurate estimates of the Jacobian required analytical estimates of the partial derivatives of all f equations with respect to all the unknown parameters. These are the quantities $\partial f_K/\partial \omega, \partial f_K/\partial k_1, \partial f_K/\partial k_2, \ldots$ through $\ldots, \partial f_{u_2}/\partial \eta_N$. The analytical derivatives were evaluated and confirmed against finite difference approximations.

Initial solution estimates are provided by first identifying by zero-crossing analysis a leading wave trough, a crest, and a following trough. Each half-wave is treated as an extra-wide window of width about $0.5T_z$. A similar nonlinear optimization approach as discussed above is applied to the extra-wide window. For example, a wave record sampled at 1 Hz has 14 unknowns and 35 equations, whereas a 4-Hz record has 35 unknowns and 140 equations. Using initial estimates provided by a global Airy theory approximation, reasonable approximations are estimated for initial input into the much narrower local windows. The algorithm had most difficulty in finding credible local solutions around the zero-crossings of the P trace. At such regions, profile curvature is small and a narrow local window has poor resolution. In these cases a double window was used to improve resolution of the local profile curvature.

Methodology Evaluation

An initial validation of the LFI-PUV theory and coding was accomplished in various water depths using theoretical PUV traces for uniform, long-crested waves propagating at an oblique angle on a uniform current. Fourier wave theory (Sobey 1989) at truncation order 18 with 25 water surface nodes from crest to trough, provided near exact P, U, and V time series at gauge depths z_P and z_{UV} for use as input to the code. Fourier wave theory also provided traces of the sea surface elevation and nonlinear kinematics at other depths for comparison with the LFI-PUV theory.

Figure 3 presents LFI-PUV predictions for a Fourier wave in 20-m depth having 10-m wave height and 10-s period. The wave was propagated at an angle of -162 degrees relative to the x_1 axis, and the uniform current had components of $U_1 = -0.95$ m/s and $U_2 = -0.31$ m/s. The PUV observational traces elevations were specified at a time interval of 0.5 s in a depth of $z_P = z_{UV} = 10$ m.

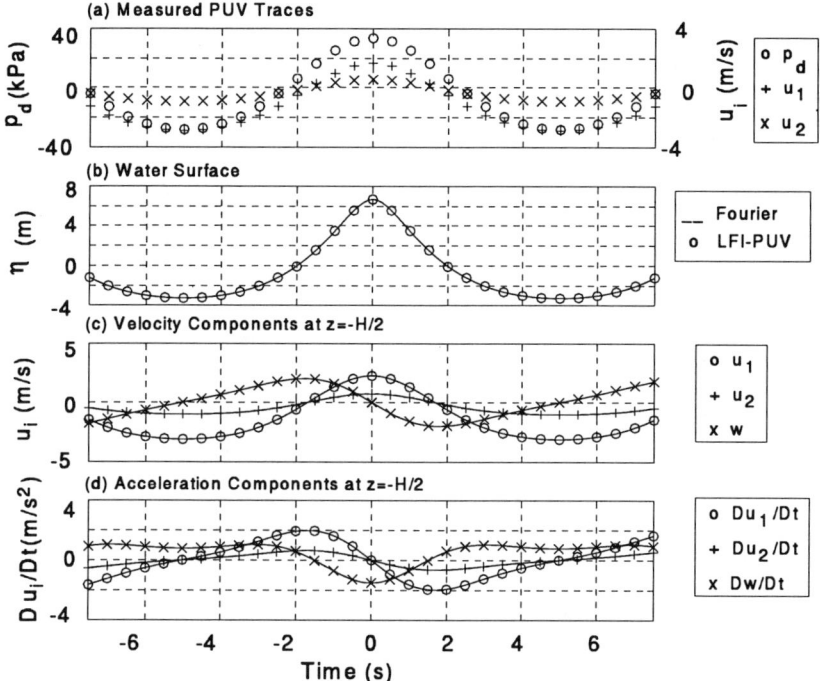

Figure 3: LFI-PUV Predictions for Fourier wave in 20-m depth.

Plot (a) shows the "measured" $p_d(t; z_P)$ trace along with the "measured" $u_1(t; z_{UV})$ and $u_2(t; z_{UV})$ traces. Plot (b) shows the η traces predicted by LFI-PUV (markers) and by the near-exact Fourier wave theory (continuous line). Plot (c) compares all three velocity components at elevation $z = -H/2$ just below the trough, and plot (d) compares all three acceleration components also at the same near-trough elevation. Agreement to near-exact Fourier wave theory is almost perfect. The excellent agreement is achieved at a relatively low order $(J = 3)$, and this is a consistent strength of the LFI-PUV methodology. Similar excellent agreement was achieved for highly nonlinear waves in 5-m and 100-m depths, illustrating the suitability of the LFI-PUV algorithm from shallow to deep water. Application of a "locally linear" analysis method to the same test case resulted in significant underprediction of the wave crest elevation and its associated kinematics (not shown in this paper).

A much more challenging test is application of the LFI-PUV method to actual field PUV records with their inherent measurement error bands. Preliminary results are very encouraging. Plot (a) of Figure 4 presents a section of PUV data measured in 46 m of water at Platform Edith, offshore of Huntington Beach in southern California.

The current meter and the pressure gauge were both at a depth of -7.4 m relative to mean water level, and data were sampled at a 1-Hz rate. The magnitude of the Eulerian current was about 8 cm/s. Plots (b), (c), and (d) of Figure 4 show LFI-PUV predictions of sea surface elevation, velocity components at depth -2 m, and accelerations at depth -2 m, respectively. These predictions are credible, but there was no independent field data for confirmation.

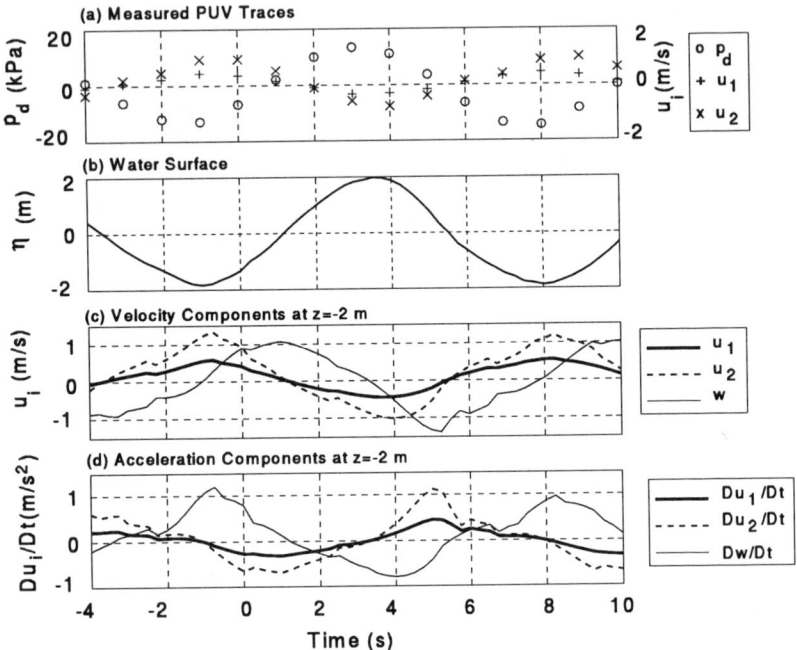

Figure 4: LFI-PUV Predictions for Platform Edith data.

Conclusions

Traditional directional analysis of PUV data results in <u>linearized</u> estimates of directional wave and velocity spectra which assume random phases and typically do not include steady currents. Even when phases are retained, attempts to reconstitute velocity time series at any other location in the water column result in linear estimates with decreasing veracity approaching the free surface.

A new method has been presented for interpreting traces recorded by conventional PUV instruments under irregular multidirectional waves. This time-domain method provides estimates of sea surface elevation and wave kinematics throughout the water column without compromising the essential nonlinear character of the waves. The key to the formulation is retention of the fully nonlinear free surface boundary conditions which are used in seeking a nonlinear least-squares solution in a narrow time window. Also included is the nonlinear interaction between the waves and a steady current. The analysis method provides nearly exact results for theoretical nonlinear steady waves, but application to field data requires overspecification in the solution to allow for instrument error bands.

Results from the locally nonlinear analysis differ substantially from traditional directional spectra. The solution provides time series of sea surface elevation η, radial frequency ω, directional wave numbers k_1 and k_2, phase $k_\alpha x_\alpha$, and Fourier coefficients, A_i. These solved values can be used to calculate corresponding time series for any kinematic parameter at any position in the water column. So essentially a complete field solution is available at each time interval.

The locally nonlinear LFI-PUV method provides a distinct improvement for analyzing practical problems in which individual wave characteristics are critical, e.g., wave loads on offshore structures. Furthermore, because the method does not rely on the entire time series, it is possible to perform selective analysis of individual waves or large wave groups within the time series that may be of particular interest. Also, the local nature of the nonlinear analysis method may find application providing near real-time wave estimates from PUV gauges.

Acknowledgements

The research described and the results presented herein, unless otherwise noted, were obtained from research funded through the *Scour Holes at Inlet Structures* work unit in the *Coastal Inlets Research Program* at the US Army Engineer Waterways Experiment Station (WES). Permission to publish this information was granted by the Chief of Engineers. Mr. Patrick McKinney of WES provided the Platform Edith PUV data.

References

Dean, R. G. (1990). "Water Wave Kinematics: State of the Art and Future Research Needs," in *Water Wave Kinematics* edited by Tørum and Gudmestad, Kluwer Academic Publishers, The Netherlands, pp 743-756.

Dennis, J.E., Gay, D.M., and Welsh, R.E. (1981a). "An Adaptive Nonlinear Least-Squares Algorithm," *ACM Transactions on Mathematical Software*, Vol 7, pp 348-368.

Dennis, J.E., Gay, D.M., and Welsh, R.E. (1981b). "ALGORITHM 573 NL2SOL - An Adaptive Nonlinear Least-Squares Algorithm," *ACM Transactions on Mathematical Software*, Vol 7, pp 369-383.

Longuet-Higgins, M.S. (1962). "Resonant Interactions Between Two Trains of Gravity Waves," *Journal of Fluid Mechanics*, Vol 12, pp 321-332.

Sobey, R.J. (1989). "Variations on Fourier Wave Theory," *International Journal for Numerical Methods in Fluids*, Vol 9, pp 1453-1467.

Sobey, R. J. (1992). "A Local Fourier Approximation Method for Irregular Wave Kinematics," *Applied Ocean Research*, Elsevier Science Publishers, Ltd., The Netherlands, Vol 14, pp 93-105.

WAVE KINEMATICS BASED ON A LAGRANGIAN FORMULATION

Geir Moe[1], Øivind A. Arntsen[1], Svein H. Gjøsund[1]

Abstract

In the present paper the Lagrangian equations for regular water waves are developed to second order for finite water depth in a manner similar to Miche (1944). Under deep water conditions the first order solution is the well known Gerstner wave which, if supplemented by an adjustment of the vertical coordinate, is a Lagrangian wave equation correct to second order. The second order terms become increasingly important as the water depth to wavelenght ratio decreases. The present theory has been compared to experimental results in a wave flume for regular water waves. Here the comparisons include measurements of the water surface profile and Laser measurements of particle velocities. The results compare quite well.

Introduction: the Lagrangian formulation of the wave problem

In the theory of water waves almost all analytical work has been based on Eulerian formulations, however some efforts based on Lagrangian formulation do exist. Then the idea is to formulate equations for the position (x, z) of a physical water particle as a function of two generalized coordinates or 'tags' (a, c) which together uniquely identifies any particle, and of time t, i.e.

$$x = x(a,c,t) \quad , \quad z = z(a,c,t) \tag{1}$$

One important point is to chose the generalized coordinates so that the free surface may be specified as c=constant. For progressive waves the position of the particles at some 'initial' time $t=t_0$ will represent a sinusoidal function, and can therefore not be used as generalized coordinates. Instead the position of the water particles in a still water situation (x_0, z_0) in the same body of water may be used, viz. $a=x_0$ and $c=z_0$. It is also common in the Lagrangian formulation to base the development on

---

[1] Faculty of Civil and Environmental Engineering, The Norwegian University of Science and Technology (NTNU), N-7034 Trondheim, Norway.

the momentum equations with zero viscosity, instead of invoking the assumption of irrotational flow. Starting from the momentum equations in Eulerian formulation

$$\frac{Du}{Dt} = -\frac{1}{\rho}\frac{\partial p}{\partial x}, \qquad \frac{Dw}{Dt} = -\frac{1}{\rho}\frac{\partial p}{\partial z} - g \qquad (2)$$

where Du/Dt and Dw/Dt are the (nonlinear) material accelerations, the equivalent Lagrangian formulation will be derived. In a Lagrangian formulation the material accelerations are quite simply $\partial^2 x/\partial t^2$ and $\partial^2 z/\partial t^2$, i.e. linear, because the partial derivative in accordance with (1) is the rate of change with time when a and c are kept constant, which is another way of saying that one follows the water particle. The pressure gradients in (2) must now be expressed through $\partial p/\partial a$, (i.e. by differentiation with respect to a, while c and t are kept constant,) and $\partial p/\partial c$. Changing now for convenience to a subscript notation, so that $\partial p/\partial a \equiv p_a$ and $\partial p/\partial c = p_c$ etc. it is useful to remember that

$$p_a = p_x x_a + p_z z_a \qquad (3)$$
$$p_c = p_x x_c + p_z z_c$$

From (3) p_x and p_z could be solved for and thus expressed by the other symbols which all represent Lagrangian quantities. Actually here p_a is formed by multiplying the former of equations (2) with x_a and the latter with z_a and adding, yielding (4). Similarly p_c is formed by multiplication with x_c and z_c respectively, yielding (5)

$$x_{tt} x_a + (z_{tt} + g) z_a = -p_a/\rho \qquad (4)$$
$$x_{tt} x_c + (z_{tt} + g) z_c = -p_c/\rho \qquad (5)$$

These are the two momentum equations for 2-D flow in Lagrangian coordinates.

The continuity equation will provide the third equation for the three unknowns x, z and p. In incompressible flow it prescribes that the areas of all infinitesimal material fluid elements must stay constant as time varies, viz.

$$J = \frac{\partial(x,z)}{\partial(a,c)} = x_a z_c - x_c z_a = const \qquad (6)$$

J represents the area of an infinitesimal element in waves identified by the generalized coordinates a and c, divided by the reference area ($da\ dc$). This quantity must be time independent, i.e. $J_t = 0$. This requirement was used by Pierson (1961) and Lamb (1932) who presented a first order solution to the Lagrangian equations (the Gerstner wave) and proved that the accompanying J was time independent. Miche (1944) chose (a, c) to represent the position of the particle in still water. Then the size of a physical element in still water and the same element in waves must be equal, i.e. $J=1$. Quantities such as the position of the crest relative to the still water level can now easily be found, and the momentum equations can also be presented on an attractive form by solving for $-(z_{tt}+g)$ in (4) and (5), and equating the resulting expressions. Then on multiplying these two terms by $z_a z_c$ and ordering one obtains equation (7)

$$x_{tt} = -(p_a z_c - p_c z_a)/\rho \qquad (7)$$

$$z_{tt} + g = -(p_c z_a - p_a z_c)/\rho \qquad (8)$$

and (8) follows in an analogous manner. Equations (7) and (8) are valid only if the continuity equation is on the form $J=1$. Equations (7) and (8) may not necessarily be simpler to work with than equations (4) and (5), but they show more directly how the accelerations depend on the pressure gradient in the Lagrangian coordinate system.

The boundary conditions at the air-water interface is $p=0$ at $c=0$ for the pressure, relative to the atmospheric pressure (neglecting dynamic wind pressures). At a horizontal bottom $z=-d$ the boundary condition is $w=z_c=0$, since the flow adjacent to the bottom is parallel to it. Summarizing

$$\begin{aligned} p &= 0, \quad \text{at } c = 0 \\ w &= z_c = 0, \quad \text{at } c = -d \end{aligned} \qquad (9)$$

Solution by a perturbation scheme

The zeroth order solution is the hydrostatic case, viz.

$$x_0 = a, \; z_0 = c, \; p_0 = p_{atm} - \rho g c \qquad (10)$$

The perturbation expansion in a small parameter ε to 2nd order is obtained by substitution of

$$\left. \begin{aligned} x &= a + \varepsilon x_1 + \varepsilon^2 x_2 \\ z &= c + \varepsilon z_1 + \varepsilon^2 z_2 \\ p &= p_{atm} - \rho g c + \varepsilon p_1 + \varepsilon^2 p_2 \end{aligned} \right\} \qquad (11)$$

In (10) and (11) the subscripts do not imply differentiation, and it can be taken as a general rule that numerical subscripts will be 'straight' indices typically identifying zeroth, first or second order solutions while the letter subscripts (a, c, t) denotes differentiation. Substitution into the continuity equation (6), and the two momentum equations (7) and (8) yields

$$\varepsilon(x_{1a}+z_{1c})+\varepsilon^{2}(x_{1a}z_{1c}+x_{2a}+z_{2c}-x_{1c}z_{1a})=0 \tag{12}$$

$$\varepsilon x_{1tt}+\varepsilon^{2}x_{2tt}=-[\varepsilon(p_{1a}+\rho g z_{1a})+\varepsilon^{2}(p_{1a}z_{1c}+p_{2a}+\rho g z_{2a}-p_{1c}z_{1a})]/\rho \tag{13}$$

$$\varepsilon z_{1tt}+\varepsilon^{2}z_{2tt}=-[\varepsilon(-\rho g x_{1a}+p_{1c})+\varepsilon^{2}(-\rho g x_{2a}+p_{1c}x_{1a}+p_{2c}-p_{1a}x_{1c})]/\rho \tag{14}$$

The first order solution

The terms to first order in ε in (12), (13) and (14) are solved for first. This solution may be found in Miche (1944)

$$\varepsilon x_{1} = -A\frac{\cosh k(c+d)}{\sinh kd}\cos(\omega t - ka) \tag{15}$$

$$\varepsilon z_{1} = A\frac{\sinh k(c+d)}{\sinh kd}\sin(\omega t - ka) \tag{16}$$

$$\frac{\varepsilon p_{1}}{\rho} = -Ag\frac{\sinh kc}{\sinh kd \cosh kd}\sin(\omega t - ka) \tag{17}$$

in which A represents the wave amplitude, and as usual $k = 2\pi/L$ is the wave number with L being the wave length, and $\omega = 2\pi/T$ is the radian frequency with T being the wave period. Here the phase of the wave is shifted by $-\pi/2$ and the vertical coordinate is positive in the opposite direction relative to Miche. A dispersion relation identical to the Euler version also follows

$$\omega^{2} = gk \tanh kd \tag{18}$$

It is seen that the pressure p_1 at the surface (i.e. at $c=0$) is zero, and that the vertical velocity at the bottom (i.e. at $c=-d$) is zero, as required by the two boundary conditions (9). The shape of the surface is found by setting $c=0$ in (16)

$$\eta = z_{1}(a,0,t) = A\sin(\omega t - ka) \tag{19}$$

which represents a trochoid. At the bottom, $c = -d$, the pressure is

$$p_{bottom} = Ag\frac{1}{\cosh kd}\sin(\omega t - ka) \tag{20}$$

If a is replaced by x in (20) this is identical to the expression for the bottom pressure for the Airy wave, but the horizontal water motion of the particles will distort the bottom pressure profile somewhat, relative to the Airy solution. At other depths the expression for the pressure will deviate more from the Airy solution, but this is because in the Lagrange solution the pressure increase is given at the *displaced* position of the particle, while the Euler solution gives the pressure increase at the mean position of the particle. Thus at $z=0$ the Eulerian dynamic pressure at a crest is equal to $\rho g A \sin(\omega t - kx)$, while at $c=0$ the Lagrangian pressure is $p_1 \equiv 0$. In view of (19) these two results agree well, however.

The second order solution

The terms in ε^2 in (12), (13) and (14) are then solved for, treating the first order solutionas known quantities. Then from Miche (1944)

$$\varepsilon^2 x_2 = A^2 k \frac{\sin 2(\omega t - ka)}{4 \sinh^2 kd} \left(1 - \frac{3}{2} \frac{\cosh 2k(c+d)}{\sinh^2 kd}\right) + A^2 U(c) \qquad (21)$$

$$\varepsilon^2 z_2 = A^2 k \frac{\sinh 2k(c+d)}{4 \sinh^2 kd} \left(1 - \frac{3}{2} \frac{\cos 2(\omega t - ka)}{\sinh^2 kd}\right) \qquad (22)$$

$$\frac{\varepsilon^2 p_2}{\rho} = \frac{A^2 kg \sinh kc}{4 \sinh^2 kd} \left[3 \cos 2(\omega t - ka) \left(\frac{\cosh kc}{\sinh^2 kd} - \frac{2 \cosh k(c+d)}{\cosh kd}\right) - \frac{2 \cosh k(c+d)}{\cosh kd}\right] \qquad (23)$$

This can be verified by direct substitution. When comparing to the Miche solution the reversal of the direction of the vertical axis and the phase shift of $-\pi/2$ in $(\omega t - ka)$ must of course be included.

It may be seen from (22) that the effect of the second order terms is to elevate and sharpen the crests, and to elevate and flatten the troughs. The relative magnitude of the second order terms depends on wave steepness (A/L) and on water depth (d/L). Miche (1944) has determined the limit at which a secondary crest develops at the trough position and found that for nonbreaking waves this limit is exceeded for a length to depth ratio above 6.5, i.e. only in quite shallow water.

The last term in (21) represents a steady horizontal current of arbitrary profile. Since all viscous stresses have been neglected in the momentum equation, such a current is permissible, and must be determined on the basis of boundary conditions at vertical sections. Thus in a closed basin within the confines of inviscid flow $U(c)$ must be zero to match boundary conditions at the wave maker and at the wave front. If the flow is to be irrotational, on the other hand, $U(c)$ becomes a "Stokes-like" current

$$A^2 U(c) = A^2 k \omega \frac{1}{2 \sinh kd} \left(\cosh 2k(c+d) - \frac{\sinh 2kd}{2kd}\right) \qquad (24)$$

Deep water case

The above expressions simplify considerably in deep water. Thus the only second order terms that prevails is a steady term

$$\varepsilon^2 z_2 = \tfrac{1}{2} A^2 k e^{2kc} \tag{25}$$

while the deep water limit of the first order terms (15), (16) and (17) with $U(c)=0$ is the well known Gerstner solution, see e.g. Lamb (1932):

$$x = a + \varepsilon x_1 = a - A\, e^{kc} \cos(\omega t - ka) \tag{26}$$
$$z^{(1)} = c + \varepsilon z_1 = c - A\, e^{kc} \sin(\omega t - ka) \tag{27}$$

The second order term (25) represents the fact that a profile with steeper crests than troughs makes the particles rotate about a point that is lifted up the distance z_2. The same result may be obtained by direct integration along the free surface to determine the average water elevation which by assumption must be equal to the still water depth, see e.g. Moe & Arntsen (1996). The same result also follows by starting from the Gerstner solution (26) and (27) and determining the Jacobian in (6) to determine the volume distortions of the fluid elements. Then integrating from the bottom to the actual position, again the expression in (25) is found. This means that when $z=0$ is located at the still water surface then the Gerstner solution for the z-direction is

$$z = z^{(1)} + \varepsilon^2 z_2 = c - A e^{kc} \sin(\omega t - ka) + \tfrac{1}{2} A^2 k e^{2kc} \tag{28}$$

Comparison with experiments

Expressions for the surface elevation and fluid velocity in an Eulerian frame of reference may be derived on the basis of a Lagrangian formulation of particle motions (Moe & Arntsen, 1996). Comparisons of results based on the present equations (here called Miche's equations) and experimental results from Skjelbreia (1991) is shown in figure 1. The second order horizontal current term solution has been set equal to zero, to satisfy the boundary conditions in the wave basin. This means that the motion is not irrotational. Skjelbreia (1991) used a two-component, non-conventional reference beam laser doppler velocimeter (LDV) for the measurements of the fluid velocity in the splash-zone, with an arrangement that allowed measurements to within 1 mm from a horizontal surface. The measurements were considered to be accurate to ± 0.005 m/s with better than 95% confidence. The vertical position was considered to be accurate to ± 0.1 mm. More information on the experimental equipment and procedures can be found in Skjelbreia (1991). The experiments (run R15b) covered the following case: water depth d = 1.3 m, wave-period T = 1.5 sec., wave-height H = 2A = 0.26 m. The results based on Miche's solution are plotted with dotted lines directly on the timeseries plots taken

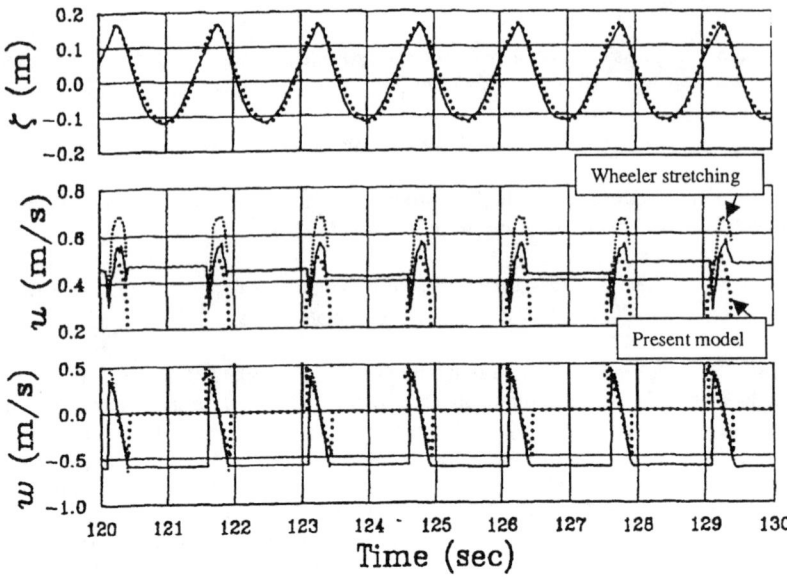

Fig. 1. Surface elevation ζ, horizontal velocity $u(z,t)$, vertical velocity $w(z,t)$. $z = 0.10$ m.

from Skjelbreia (1991). This is done in a relatively crude manner, so that some inaccuracies may have been introduced. The solid lines show the experimental results. In addition, results for u and w based on Wheeler-stretching (Wheeler, 1970) as done by Skjelbreia (1991) has been included These are plotted with another type of dotted lines. By direct measurement on the plots, the total wave-height appears to be greater than the value H = 0.26 m given in the text of the report. We have therefore chosen to perform our calculations with a total waveheight of H = 0.286 m (wave-amplitude a = 0.143 m), i.e. a 10% increase in the H-value. The figure represents a time segment where the experimental results have reached a "steady" situation which is reported to persist for the duration of the run. Earlier in the timeseries there was a transition zone, where the peak values of the horizontal velocity decreased from values corresponding to Wheeler-stretching to the "steady-state" value in the above figure.

The recorded velocities are at a position 0.10 m above still water level, at a location on which there will be fluid only during part of the wave-cycle. In the experimental results the last value is "frozen" when the point comes out of water, while in the theoretical results the curves "jump" to zero when the point comes out of water. The phase of the theoretical curves are chosen so that the instants when the measuring point enters the water for the first wave coincide for the theoretical and experimental case. From the surface elevation plots, it is seen that the experimental results show slightly steeper crests and wider troughs than predicted. Also, for the last two waves, there is a clear asymmetry about a vertical axis through the crest, accompanied by a phase shift between the experimental and theoretical crests. This phase difference can also be recognized in the velocities for the same two waves. The curves representing the vertical velocity are very close to each other. The most interesting difference can be seen in the plots of the horizontal velocity. The peak values from Wheeler-stretching are approximately 20% higher than the experimental peak values, while Miche's (and Gerstner's) theory predicts results that are approximately 10% less than the experimental peak values. If a wave amplitude $A = H/2 = 0.13$ m is used in these calculations, a difference of approximately 15% is found.

Conclusions

The Lagrangian formulation of the wave problem is based on the inviscid momentum equations and continuity. It results in managable algebra up to second order and appears to predict the behaviour of water waves in wave tanks quite well. A tentative conclusion is that such flows are indeed rotational. Improvements of this solution might require modelling of viscous behaviour.

Acknowledgement

The authors gratefully acknowledge the assistance from Dr. Skjelbreia on interpretation and access to the laboratory data.

References

1. Kyozuka, Y. (1995): "Mass transport in two-dimensional tank", preliminary version of paper, private communication.
2. Neumann, G., Pierson, W. J.: *Principles of Physical Oceanography,* Prentice-Hall inc. Englewood Cliffs, N.J., 1966.
3. Lamb, H.: *Hydrodynamics,* Sixth edition, Cambridge University Press, Art.:251, 1932
4. Miche, R. (1944): "Mouvements ondulatoires de la mer en profondeur constante ou décroissante form limte de la houle lors de son déferlement. Application aux digues marine." Ann. Pontes Chaussées 114: 25-78, 131-164, 270-292, 369-406.
5. Moe, G., Arntsen, Ø. (1996): "Particle velocity distribution - Emergence effects", Proceedings of the 25th Conference on Coastal Engineering (ICCE'96), Orlando, Florida.
6. Skjelbreia, J. E. (1991): "Kinematics in Irregular and regular waves", Marintek Draft report, March 1991.
7. Wheeler, J. D. (1970): "Method for calculating forces produced by irregular waves", J. Petroleum Tech., March 1970, pp. 119-137.

VISUALIZATION OF WAVE KINEMATICS ON THE WORLD WIDE WEB: AN INTERACTIVE INSTRUCTIONAL TOOL [1]

F.A. Hardjanto[2] , S.A. Kinnas[3] , H. Hart[4] , M.H. Kim[5]

Abstract

This paper presents an instructional tool for wave kinematics being developed at the University of Texas at Austin and Texas A&M University. The interactive tool is intended to offer visualizations of various conditions of water waves and their effects on the surroundings. The pilot project of this development has been successful in establishing the framework of an interactive instructional tool on the World Wide Web for usage in and out of class. It is anticipated that the final outcome of this work will enhance the learning process on the subject of ocean wave theory by developing an *electronic classroom* or *electronic textbook*.

Introduction

Knowledge of wave theory is essential to understanding the interaction of the ocean with its natural or engineered environment. At the same time, wave theory is also one of the most complex applications of fluid mechanics and thus one of the most challenging to teach. The theory of water waves is governed by highly mathematical concepts, which frequently hamper the instruction of wave theory in class. The students, particularly undergraduates, often encounter difficulties in understanding the physical meaning of the mathematical concepts because they cannot visualize various motions of the waves and their effects on the surroundings. Therefore, an interactive tool for the visualization of wave

---

[1] The corresponding web site address is: *http://www.otrc.utexas.edu/~waves/home.html*
[2] Graduate Research Assistant, Department of Civil Engineering, The University of Texas at Austin, Austin, TX 78712
[3] Assistant Professor, Department of Civil Engineering, The University of Texas at Austin, Austin, TX 78712
[4] Senior Lecturer, Department of Civil Engineering, The University of Texas at Austin, Austin, TX 78712
[5] Associate Professor, Department of Civil Engineering, Texas A&M University, College Station, TX 77843

kinematics is needed to aid the students, especially those who are exposed to the subject for the first time.

The rapid growth of the Internet and the World Wide Web (the Web) provides a convenient way to deliver an interactive instructional tool. The flexibility of the Web in combining text, graphics, animation, even video and sound, offers an interesting and compelling presentation for usage in and out of class. The capability of the Web is continuously growing and it allows the delivery of interactive presentations, such as those written in Java scripts / applets and Common Gateway Interface (CGI) scripts. These modes of delivery enable us to create an *Electronic Classroom*.

This paper describes the development of this interactive, instructional tool on the Web. The web site is being developed and tested for use in CE358: Introductory Ocean Engineering, an upper level undergraduate course at the Civil Engineering Department, The University of Texas at Austin. This site is actually a combination of two: an informational site serving as an additional source of information for students, and an instructional site offering the interactive visualization tool. A brief description of the goals and strategies in developing computer-based instructional media, as well as some issues in the implementation of a class web site, are presented. We will focus on the methodology of creating an interactive instructional site. Finally, the anticipated issues in the development of this class web site in the future are addressed. The CE358 web site is located at *http://www.otrc.utexas.edu/~waves/home.html*.

Goals and Strategies

Our long term goal is to develop computer-based instructional media that will enhance the education of future engineers on the broad subject of ocean wave theory, so that they are capable of coping with the *quantitative* aspects underlying the design of engineered systems and the environmental issues involved. This education should not only provide sufficient understanding of the basics of wave theory but also expose the targeted audience to the *current* experimental and computational techniques for assessing or predicting the effects of the ocean on its surroundings.

To reach these goals, this project is developing the *Electronic Classroom on Ocean Wave Theory* in order to accomplish the following objectives:

- Provide multimedia tools (on the World-Wide Web and CD-ROM) for the *active learning* of the fundamentals of ocean wave theory.

- Disseminate the *state-of-the-art* in experimental and computational techniques - in particular, some of the formulations developed in the research conducted at the Offshore Technology Research Center (OTRC).

In addition, the development of a computer-based instructional tool-set should include learning-process objectives. The tools should be designed to: 1) deliver the information in a more concise and better presentation, 2) create a highly interactive system, 3) promote a self-paced learning process and 4) incorporate formative and summative evaluation strategies (Kinnas and Hart, 1997).

In order to achieve these objectives, certain strategies in developing computer-based instructional media need to be adopted. Many people learn better if they can get a mental picture of an abstract idea, either by seeing or listening, and if they are actively involved in solving problems. Therefore, the ideal content of electronic-based instructional media should incorporate 1) graphics, animation or movies to illustrate and establish the context of the concept, 2) audio combined with graphics to provide multi-sensory input, 3) online tutorial and examples, and 4) online interactive tests for self evaluation (Schmidt, 1997).

The current technology allows us to deliver all of the above components via the Internet. The Web provides an economical delivery of text, graphics, and audio and has the capability to carry an interactive presentation. However, this mode of delivery also demands extensive preparation of media, e. g. creating graphics and animation, scanning pictures, producing audio and video, scripting programs, etc. In addition, a web site also requires regular maintenance of the infrastructure and still has the problem of platform dependence and bandwidth limitations for certain applications, mainly the graphical animation, movies, and audio. Many of them require additional plug-ins on the web browser on the user side.

This expenditure of time and resources means that educators should be sure, up-front, that multimedia will enhance student learning of the particular material. Objectives for student learning should be established at the outset, as should be strategies for evaluating the usability of the web site. Our long term goal, listed above, translated into these objectives for student learning:

- enable students to better understand the math involved in ocean wave theory
- enable students to assess and predict the effects of the ocean on its surroundings

The success of both of these objectives can be evaluated by testing students before and after using the Web tool-set, or by using the same test for students who are learning only in a traditional classroom setting and those using the web tool as a supplement. Measuring the ease of use of the site can be accomplished through electronic tracking devices, student note-pads, and user-feedback forms. These evaluation strategies are further explained in Hart and Kinnas (1998).

Structure of the Web Site

Because learning occurs in different modes at different times, the web site is meant to deliver two types of information: passive and interactive. To address these needs, a class web site generally consists of two types: an informational site and an instructional site. The informational site delivers passive information, which is intended to supplement the information given in class by the instructor. Some examples of passive information include general information on the course, class-notes, things to remember, solutions of homework or examinations, references, links to other sites, etc. On the other hand, the instructional site is designed to allow the user to absorb the information via a "see and do" method. It should allow the user to learn by doing "what-if" simulations; i.e., how certain parameters affect the results. Additionally, the user should be allowed to give feedback to the instructor or the web-master of the site. The instructional site could include tutorials, workbooks, evaluation forms and animated demonstrations. The two most popular methods of building an interactive instructional web site are CGI scripts (combined with Forms or Imagemaps) and Java scripts / applets. Nonetheless, more versatile methods are already available, such as ActiveX controls and VRML (Virtual Reality Modeling Language). For reasons of simplicity, only CGI scripts combined with Forms are used in the current CE358 web site.

General Technical Considerations

Among the issues taken into consideration in the development of this class web site, the following points are worth mentioning:

1. Use of graphics
   In order to serve the students who do not have high-speed Internet connections, the use of graphics is kept to a minimum. Large scale graphics are used only for animation in the visualization tools.

2. Site map
   The ease of use and the user-friendliness of a web site depends heavily on its site map. A site map depicts the organization of the web pages and the links between them. It dictates how the user moves from one page to the other. Site maps can be organized by any of the following schemes: Linear, Hierarchical, Web, or a combination (Schmidt, 1997). A scheme of Hierarchical with Alternatives has been selected for this class web site. This scheme offers a structured directory for incorporating more materials in the future while providing ease of navigation at the same time. Every page has links that allow the user either to jump back to the previous page or to go to the directory above the corresponding page.

3. CGI scripts vs. Java scripts / applets
   The usage of these two items is the central issue in bringing interactivity to the web site. A fully interactive web site should have quick reaction time to user input, capability to reject inadmissible input, and the proper algorithms and hardware to serve multiple users at the same time.

In developing the interactive visualization tools for the water-wave mechanics, the authors were faced with problems related to the numerical evaluation computer codes and the animation generator. Quite extensive number crunching is required to create even a simple animation, and this process must be repeated every time the user submits new input. The development options taken into consideration were:

(a) Java applets containing both the numerical algorithm and the animation generator to run on the user side.

(b) Combination of CGI script and Java applets. This allows the use of existing computer codes, which reside on the server and are written in any standard programming language, while the animation generator is sent to the user via Java applets or any plug-in applications.

(c) CGI script controlling both the numerical evaluation computer codes and the animation generator that resides on the server. The input data and the animation are transferred to and from the server.

The first option is feasible only for very simple cases involving simple numerical calculation and animation. This option is not very appealing considering the possibility of including more involved numerical algorithms in the future. It is not economical to rewrite existing programs in Java.

As for the second option, it was found to be acceptable only for animation involving a small number of frames, unless an additional plug-in application, such as Shockwave, is utilized. When the number of frames increases, the standard Java applets consume a lot of CPU time on the user side to generate the animation.

The third option seems to be the simplest one for the class web site. It does not require a transcription of any existing numerical algorithm into Java, and it does not depend on the speed of the user's CPU to create the animation. The downside of this approach is that a dedicated server is required to handle the requests from many users, which may cause congestion in the server. Moreover, the transmission of large animation files may be bothersome to users with low-speed internet connections.

Technical Specifics of the Site

The interactive instructional site developed for CE358 contains several visualization tools of the flow field under the linear wave theory and Stokes 2nd order wave theory. The tools for certain topics already developed include:

- Particle trajectories and streamlines from linear wave theory

- Particle trajectories from Stokes 2nd order wave theory

- Wave group
- Wave reflection and standing wave
- Combination of waves in linear theory.

These tools are in the form of animated graphics. Currently, only the visualization of the particle trajectories and the streamlines from linear wave theory are fully interactive, in that they allow the user to supply the parameters: water depth, wave height and period through the Web. The elements building up the animation involve Forms in HTML (HyperText Markup Language), CGI script in C language, MATLAB/numerical program, and an Animation Processor (IMAGEMAGICK©[6] by John Cristy was used). Figure 1 shows how these elements are combined to build up an interactive animated presentation.

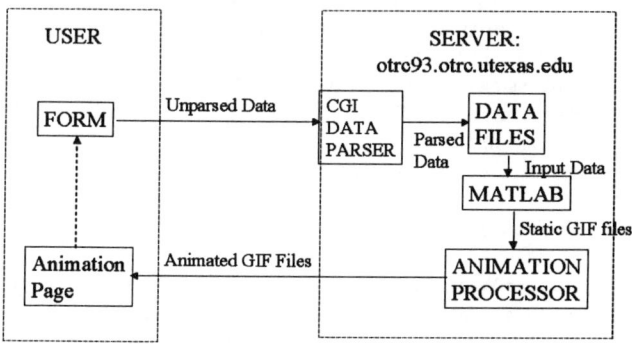

Figure 1: Scheme of the CGI Script for the Interactive Site

In this pilot project, MATLAB was used to perform the numerical computation as well as to create the graphics. In a more involved case, the numerical program and the visualization / graphic software may be separate programs. The animation processor is used to create a single animated GIF file from several static GIF files produced by MATLAB. The whole process of receiving input data, invoking MATLAB and the animation processor, and sending the animation files back to the users is controlled by the CGI script residing on the server.

Some examples of the animation graphics are presented in Figures 2-6. Traveling wave and particle trajectory animations, snapshots of which are shown in Figures 2 and 3, allow the user to make a comparison between the shape of the free surface and the particle trajectories according to linear or Stokes 2nd order wave theories. Moreover, the user can also capture the physical meaning of the kinematic boundary condition (a very basic idea in wave theory)

---

[6]http://www.wizards.dupont.com/cristy/ImageMagick.html

by observing that the particles follow the wave surface. Figure 4 shows a snapshot from an animation of the streamlines of the fluid flow from linear wave theory. The user can observe the differences between pathlines and streamlines in the case of unsteady wave flow. The concept of a wave group is visualized via an animation; a snapshot of which is shown in Figure 5. This animation helps the user understand the difference between Group Velocity and Wave Celerity by observing on the screen the speeds with which the individual wave and the wave group propagate. Figure 6 illustrates the concepts of wave reflection and standing waves. These graphics present the resulting fluid particle trajectories with increasing reflection coefficient, including the case of standing waves (fully reflected waves).

While this procedure for creating an interactive tool on the Web has been quite encouraging, additional issues need to be addressed. A more involved procedure in the CGI script would be needed if the numerical program and the graphics software were separated. Depending on the graphics software, compatibility of the graphic file format may pose an additional problem. All the sequences of converting the graphic file format and generating the animation take a substantial amount of CPU time and resources. Finally, the CGI script opens a security hole in the server. It is well known that many breakages into a computer system are accomplished by attacking the CGI script.

Concluding Remarks

The use of the World Wide Web as a mode of delivery for computer-based instructional media has good potential. The Web provides an economical and convenient way of delivering the components required in instructional media, such as text, graphics, video and audio. A certain level of interactivity can be achieved as long as the proper infrastructure is available.

It is worth mentioning again that the preparation of interactive instructional media requires a substantial amount of time and resources, regular maintenance of the hardware and software, and a careful design of the presentation itself. The technology of the Web is constantly changing, and the author of the web site should keep up with the pace of this revolution. The benefits of using the Web as an interactive instructional media should be continuously evaluated to justify the development and maintenance.

Future Work

The pilot project presented in this paper has shown the enormous potential of an electronic textbook and its interactive learning tools to complement the teaching of the complex and broad subject of ocean wave theory. This electronic textbook will be substantially enhanced to include more topics on an introductory, as well as advanced level. The development will be carried out at The

OCEAN WAVE KINEMATICS, DYNAMICS AND LOADS ON STRUCTURES 71

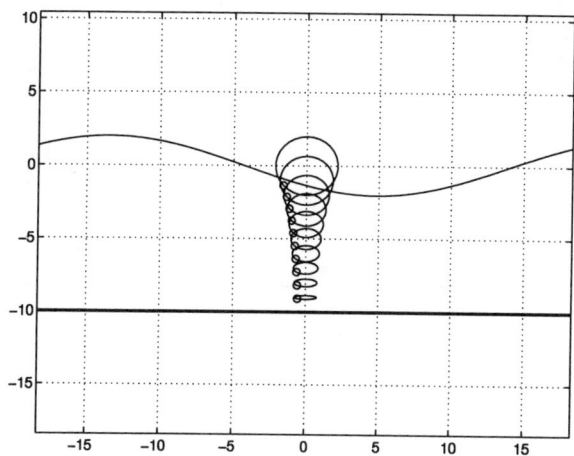

Figure 2: Particle trajectories from linear wave theory, Depth=10.0m, Height=4.0m, Period=5.0s, A snapshot from an animation presentation at *http://www.otrc.utexas.edu/~waves/wow_sto2dee.html*

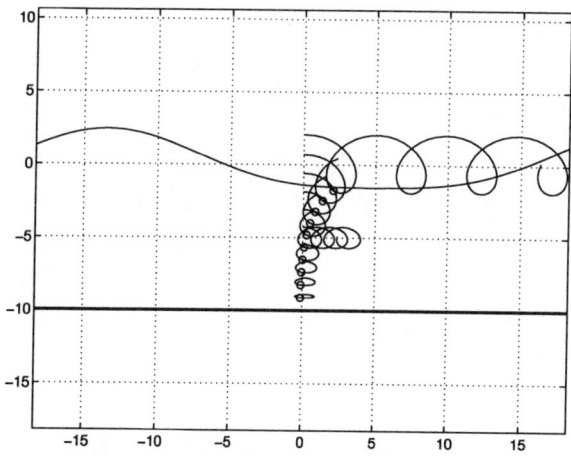

Figure 3: Particle trajectories from Stokes 2nd order wave theory, Depth=10.0m, Height=4.0m, Period=5.0s, A snapshot from an animation presentation at *http://www.otrc.utexas.edu/~waves/wow_sto2dee.html*

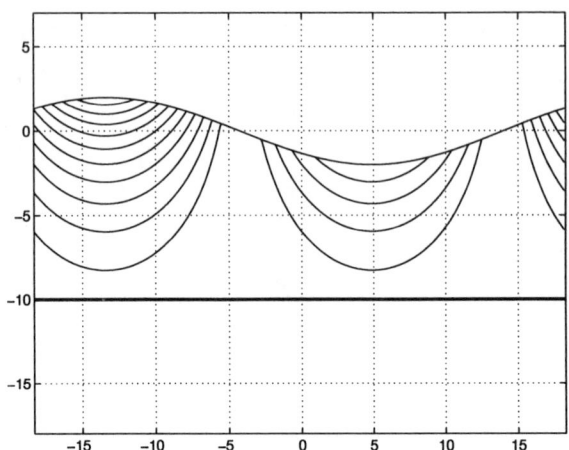

Figure 4: Flow streamlines from linear wave theory, Depth=10.0m, Height=4.0m, Period=5.0s. A snapshot from an animation created after the corresponding input form has been submitted in
http://www.otrc.utexas.edu/~waves/wowform.html

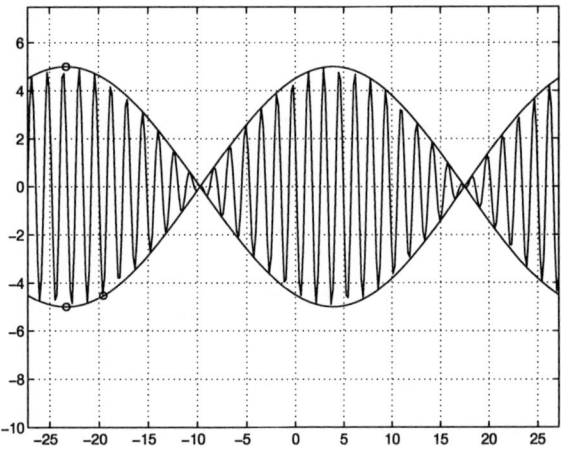

Figure 5: Combined wave in a wave group. A snapshot from an animation presentation at *http://www.otrc.utexas.edu/~waves/wowwvgr.html*

OCEAN WAVE KINEMATICS, DYNAMICS AND LOADS ON STRUCTURES 73

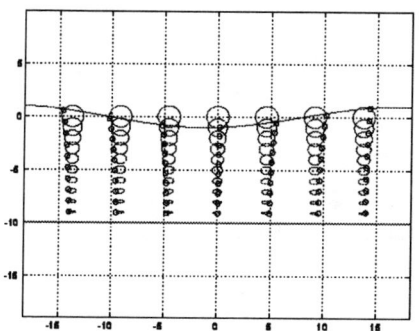
(a) Pure progressive wave, reflection coefficient = 0.0

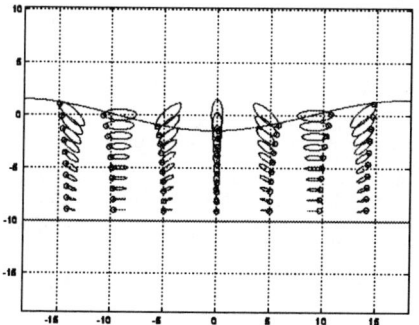
(b) Reflected wave, reflection coefficient = 0.5

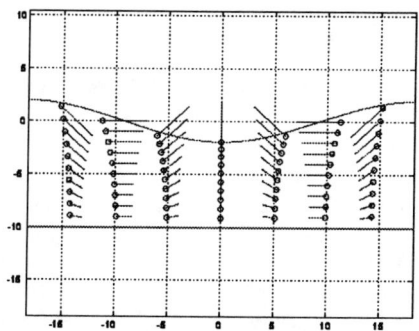
(c) Pure standing wave, reflection coefficient = 1.0

Figure 6: Snapshots from animations showing the effect of reflection coefficient on particle trajectories

University of Texas at Austin and Texas A&M University under the guidance of the Offshore Technology Research Center. Support of this effort will be provided by a three-year Grant sponsored by the National Science Foundation Engineering Research Centers Program (Grant No. 9711733).

Acknowledgments

The support for this project has been provided by Offshore Technology Research Center. Computing resources have been provided by the Offshore Technology Research Center, the Department of Civil Engineering, and the Ocean Engineering Group at the University of Texas at Austin. The authors wish to thank Professor José M. Röesset for his continuous inspiration and support in the creation of the electronic classroom on ocean wave theory.

References

Dean, R.A., and Dalrymple, R.G. (1991). *Water Wave Mechanics for Engineers and Scientists*, World Scientific, Singapore.

Hart, H., and Kinnas, S.A. (1998). "Developing Web-Based Tools for Environmental Courses," *Proceedings*, American Society for Engineering Education, Washington, D.C.

Kinnas, S.A., and Hart, H. (1997). "Developing and Using Textbooks on the Web," Brown Bag Seminars, College of Engineering, The University of Texas at Austin.

Lemay, L. (1996). *Teach Yourself Web Publishing with HTML 3.2*, 3rd. ed., Sams Net Publishers.

Sarpkaya, T., and Isaacson, M. (1981). *Mechanics of Wave Forces on Offshore Structures*, Van Nostrand Reinhold Co., New York.

Schmidt, P.S. (1997). Lecture notes on Electronic-based Instructional Media for Engineering, The University of Texas at Austin.

Appendix: Formulas Used in the Visualization Tools

The graphics in the visualization tools are generated using closed-form formulas whenever possible. These formulas are based on the main assumption of inviscid and irrotational flow. For completeness, a summary is given below. The complete derivation can be found in Sarpkaya and Isaacson (1981) or Dean and Dalrymple (1991). In the following formulas, H is the wave height, L the wave length, d the water depth, k the wave number, ω the wave frequency (in rad/sec), t the time, z the vertical distance (positive above the free surface), x the horizontal distance along the wave direction, and η the wave elevation.

- Particle trajectories from linear wave theory

Horizontal particle displacement:
$$\xi = -\frac{H}{2}\frac{\cosh(k(z+d))}{\sinh(kd)}\cos(kx_0 - \omega t) \quad (1)$$

Vertical particle displacement:
$$\zeta = \frac{H}{2}\frac{\sinh(k(z+d))}{\sinh(kd)}\cos(kx_0 - \omega t) \quad (2)$$

- Stream function from linear wave theory
$$\psi = -\frac{H}{2}\frac{g}{\omega}\frac{\sinh(k(z+d))}{\cosh(kd)}\cos(kx_0 - \omega t) \quad (3)$$

- Particle trajectories from Stokes second order wave theory
  Horizontal particle displacement:
$$\xi = -\frac{H}{2}\frac{\cosh(k(z+d))}{\sinh(kd)}\sin(kx - \omega t)$$
$$+ \frac{H}{8}\left(\frac{\pi H}{L}\right)\frac{1}{\sinh^2(kd)}\left(1 - \frac{3\cosh(2k(z+d))}{2\sinh^2(kd)}\right)\sin(2(kx - \omega t))$$
$$+ \frac{H}{8}\left(\frac{\pi H}{L}\right)\frac{\cosh(2k(z+d))}{\sinh^2(kd)}(\omega t) \quad (4)$$

Vertical particle displacement:
$$\zeta = \frac{H}{2}\frac{\sinh(k(z+d))}{\sinh(kd)}\cos(kx - \omega t)$$
$$+ \frac{3H}{16}\left(\frac{\pi H}{L}\right)\frac{3\sinh(2k(z+d))}{\sinh^4(kd)}\cos(2(kx - \omega t)) \quad (5)$$

- Wave Group
  Wave 1: $\eta_1 = \frac{H}{2}\cos(k_1 x - \omega_1 t)$, Wave 2: $\eta_2 = \frac{H}{2}\cos(k_2 x - \omega_2 t)$

Combined Wave:
$$\eta = \eta_1 + \eta_2 = H\cos(k_p x - \omega_p t)\cos(k_m x - \omega_m t)$$
$$H\cos(k_p x - \omega_p t)\cos\left[k_m\left(x - \frac{\omega_m}{k_m}t\right)\right] \quad (6)$$

where: $k_p = \dfrac{k_1 + k_2}{2}$, $\omega_p = \dfrac{\omega_1 + \omega_2}{2}$, $k_m = \dfrac{k_1 - k_2}{2}$, $\omega_m = \dfrac{\omega_1 - \omega_2}{2}$

As $\omega_m \to 0$ (or $k_m \to 0$), the group velocity is defined as:
$$C_g = \frac{d\omega}{dk} = \frac{1}{2}\left(1 + \frac{2kd}{\sinh(2kd)}\right) \quad (7)$$

The Schrödinger method for water wave kinematics

KARSTEN TRULSEN[1], JEAN-MARC TEMMOS[2], KRISTIAN B. DYSTHE[3], OVE T. GUDMESTAD[4] & MANUEL G. VELARDE[1]

Abstract

The Schrödinger method is developed for computation of water wave kinematics on finite depth, and is applied to laboratory experimental data for irregular waves.

Introduction

Conventional methods for computation of the velocity field under water waves typically fail near the wave crest due to uncertainty in how to treat the high-frequency portion of the spectrum. The failure of the linear Airy method is due to high-frequency contamination: The high-frequency tail of the spectrum is treated as representing linear free-wave oscillations around the mean water level, and is hence assigned too much weight.

The Wheeler stretching method was designed to compensate for the poor performance of the linear Airy method near the crest by a stretching modification that has little foundation in wave hydrodynamics. There is a large number of other similar empirical modifications of the linear Airy method to yield improved results.

Due to limited accuracy of field measurements, the high-frequency tail is the least reliable part of the measured spectrum. The energy carried by the high-frequency tail is in any case a small portion of the total energy. If only the most reliable portion of the measured spectrum is retained, the bandwidth is no longer wide. The high-frequency portion of the spectrum will only be important to the extent that it gives contributions coherent with the spectral peak.

The *nonlinear* Schrödinger method accounts for linear free waves near the spectral peak, in addition to nonlinear waves that are forced by the free waves near the peak. Hence the high-frequency tail is accounted for to the extent that it

---
[1]Instituto Pluridisciplinar, Universidad Complutense de Madrid, Paseo Juan XXIII 1, E-28040 Madrid, Spain
[2]ISITV, University of Toulon, France
[3]Department of Mathematics, University of Bergen, Bergen, Norway
[4]Statoil, Stavanger, Norway

consists of nonlinear waves coherent with, and forced by, the linear free waves near the spectral peak. The *linear* Schrödinger method only accounts for linear free waves near the spectral peak. The philosophy behind the Schrödinger method is to take advantage of the weakly nonlinear narrow banded nature of typical ocean waves, in order to predict the velocity and acceleration fields by an asymptotic perturbation method with acceptable accuracy. In this paper we show that even a leading order *linear* Schrödinger method performs remarkably well in this respect.

Recently, the nonlinear Schrödinger method has been described for computation of the velocity field under *deep-water* waves (Trulsen & Dysthe 1997). Further analysis of synthetic irregular wave data on deep water has demonstrated that the nonlinear Schrödinger method predicts a horizontal velocity under a crest between the predictions of the linear Airy method and the Wheeler stretching method. The nonlinear Schrödinger method generally predicts a horizontal velocity 12–25% larger than the prediction of the Wheeler stretching method near the mean water level, in good agreement with the recommendation of Gudmestad & Haver (1993) and Gudmestad & Karunakaran (1994).

For many applications of practical interest, the waves under consideration are *finite-depth* waves on the continental shelf. In this paper, we extend the Schrödinger method to finite depth. The present theory uses the same scaling assumptions for broader bandwidth weakly nonlinear waves that where first introduced by Trulsen & Dysthe (1996). We believe that these scaling assumptions are representative for typical ocean waves.

Governing equations

The inviscid hydrodynamic equations for the velocity potential $\phi(x, z, t)$ and surface displacement $\zeta(x, t)$ of an incompressible fluid with uniform depth h are

$$\nabla^2 \phi = 0 \quad \text{for} \quad -h < z < \zeta, \tag{1}$$

$$\frac{\partial \zeta}{\partial t} + \frac{\partial \phi}{\partial x}\frac{\partial \zeta}{\partial x} = \frac{\partial \phi}{\partial z} \quad \text{at} \quad z = \zeta, \tag{2}$$

$$\frac{\partial \phi}{\partial t} + g\zeta + \frac{1}{2}(\nabla \phi)^2 = 0 \quad \text{at} \quad z = \zeta, \tag{3}$$

$$\frac{\partial \phi}{\partial z} = 0 \quad \text{at} \quad z = -h. \tag{4}$$

Here g is the acceleration of gravity, the horizontal coordinate is x, the vertical coordinate is z and $\nabla = (\partial/\partial x, \partial/\partial z)$.

In comparison with Trulsen & Dysthe (1996), the assumption on the depth is relaxed to order unity:

$$k_c a = \mathcal{O}(\epsilon), \quad |\Delta k|/k_c = \mathcal{O}(\epsilon^{1/2}), \quad (k_c h)^{-1} = \mathcal{O}(1). \tag{5}$$

Here k_c, a and Δk are characteristic scales for the wavenumber, amplitude and modulation wavenumber, respectively, and $\epsilon \ll 1$ is a small perturbation parameter. In practical terms, the steepness $k_c a$ is typically in the range 0.1–0.12, and the bandwidth $|\Delta k|/k_c$ is typically in the rage 0.35–0.45.

The linearized equations (1)–(4) yield the dispersion relation for gravity waves on finite depth

$$\omega^2 = gk \tanh kh. \tag{6}$$

All of the following results have been made dimensionless by using the wavenumber and frequency of the central wave, k_c and ω_c, related by the linear dispersion relation (6).

The following harmonic expansions for the velocity potential and surface displacement are employed:

$$\phi = \epsilon^{3/2} \bar{\phi} + \frac{1}{2} \left(\epsilon A'_1 e^{i\theta} + \epsilon^2 A'_2 e^{2i\theta} + \cdots + \text{c.c.} \right), \tag{7}$$

$$\zeta = \epsilon^2 \bar{\zeta} + \frac{1}{2} \left(\epsilon B e^{i\theta} + \epsilon^2 B_2 e^{2i\theta} + \cdots + \text{c.c.} \right). \tag{8}$$

Here c.c. denotes the complex conjugate, and the rapid phase is $\theta = x - t$. The slow drift $\bar{\phi}$ and set-down $\bar{\zeta}$ as well as the complex harmonic amplitudes $A'_1, A'_2, \ldots, B, B_2, \ldots$ are functions of the slow modulation variables $\epsilon^{1/2} x$ and $\epsilon^{1/2} t$. For finite depth, it is convenient to assume that the variables $\bar{\phi}, A'_1, A'_2, \ldots$ depend on the basic vertical coordinate z.

The vertical dependence of the harmonic amplitudes of the velocity potential is found from the Laplace equation and the bottom boundary condition:

$$\frac{\partial^2 A'_n}{\partial z^2} - n^2 A'_n + 2ni\epsilon^{\frac{1}{2}} \frac{\partial A'_n}{\partial x} + \epsilon \frac{\partial^2 A'_n}{\partial x^2} = 0 \quad \text{for} \quad -h < z \tag{9}$$

$$\frac{\partial A'_n}{\partial z} = 0 \quad \text{at} \quad z = -h \tag{10}$$

The vertical dependence can then be found by the perturbation expansion

$$A'_n = A_{n,0} + \epsilon^{\frac{1}{2}} A_{n,1} + \ldots \tag{11}$$

The leading-order solution is

$$A_{1,0} = \frac{\cosh(z+h)}{\cosh h} A \tag{12}$$

which evaluates to A at $z = 0$. A is not a function of z. At the next order, the particular solution that vanishes at $z = 0$ can be written as

$$A_{1,1} = -i \frac{(z+h)\sinh(z+h) - h\tanh h \cosh(z+h)}{\cosh h} \frac{\partial A}{\partial x}. \tag{13}$$

Substituting the expansions (7), (8) and (11) into the two surface conditions (2) and (3), we get

$$A = -\frac{i}{\sigma} B - \frac{\epsilon^{\frac{1}{2}}}{\sigma} \frac{\partial B}{\partial t} \tag{14}$$

and the evolution equation

$$\frac{\partial B}{\partial x} + \frac{1}{c_g}\frac{\partial B}{\partial t} + \frac{i\epsilon^{\frac{1}{2}}}{8c_g^3}\left(1 + 2(\sigma^2 - 1)\frac{h}{\sigma}\left(1 - (3\sigma^2 + 1)\frac{h}{\sigma}\right)\right)\frac{\partial^2 B}{\partial t^2} = 0 \qquad (15)$$

where the group velocity is

$$c_g = \frac{1}{2}\left(1 - (\sigma^2 - 1)\frac{h}{\sigma}\right) \qquad (16)$$

and $\sigma = \tanh h$.

The velocity and acceleration fields

The velocity field is found explicitly by taking the gradient of the velocity potential, $(u, w) = \nabla\phi$, and is given by

$$u = \mathrm{Re}\left\{\left(\mathcal{C}_1 B + \frac{i\epsilon^{\frac{1}{2}}}{c_g}\left[\frac{1}{2}\left(1 - (\sigma^2 + 1)\frac{h}{\sigma}\right)\mathcal{C}_1 + (z+h)\mathcal{S}_1\right]\frac{\partial B}{\partial t}\right)e^{i\theta}\right\} \qquad (17)$$

and

$$w = \mathrm{Re}\left\{\left(-i\mathcal{S}_1 B + \frac{\epsilon^{\frac{1}{2}}}{c_g}\left[\frac{1}{2}\left(1 - (\sigma^2 + 1)\frac{h}{\sigma}\right)\mathcal{S}_1 + (z+h)\mathcal{C}_1\right]\frac{\partial B}{\partial t}\right)e^{i\theta}\right\}, \qquad (18)$$

where we have used the notation

$$\mathcal{C}_1 = \frac{\cosh(z+h)}{\sinh h} \qquad \mathcal{S}_1 = \frac{\sinh(z+h)}{\sinh h}.$$

Within the present level of approximation, the Eulerian and Lagrangian accelaration fields are identical and are given by

$$\frac{\partial u}{\partial t} = \mathrm{Re}\left\{\left(-i\mathcal{C}_1 B + \frac{\epsilon^{\frac{1}{2}}}{c_g}[(1 - h\sigma)\mathcal{C}_1 + (z+h)\mathcal{S}_1]\frac{\partial B}{\partial t}\right)e^{i\theta}\right\} \qquad (19)$$

and

$$\frac{\partial w}{\partial t} = \mathrm{Re}\left\{\left(-\mathcal{S}_1 B - \frac{i\epsilon^{\frac{1}{2}}}{c_g}[(1 - h\sigma)\mathcal{S}_1 + (z+h)\mathcal{C}_1]\frac{\partial B}{\partial t}\right)e^{i\theta}\right\}. \qquad (20)$$

Application to laboratory measurements

We use wave data from an experiment at the Norwegian Hydrotechnical Laboratory in Trondheim, see Skjelbreia et al. (1991). The experiment was carried out in a 33 m long and 1 m wide wave tank with a water depth of 1.302 m. Letting x and z be the horizontal and vertical coordinates in the tank, the wave generator was located at $x = -19.05$ m, while simultaneous surface elevation and velocity field measurements were made at $x = 0$. The velocity field could only be measured at one vertical height at a time. Hence the same wave generator signal was used several times in order to obtain a vertical profile of the velocity field under the same wave train. Velocity measurements were carried out at 15 different vertical positions. There are altogether 41 time series denoted by i24_#, where # is a number between 1 and 41. Each time series has 32768 scans with sample period $\tau = 0.025$ s.

While the peak frequency is often used to characterize the frequency spectrum, it is here more relevant to characterize the power spectrum by its mean. For the presented cases, the mean frequency is approximately 25% larger than the peak frequency. The characteristic frequency ω_c is hence determined from the Fourier transform of the surface displacement, $\hat{\zeta}$, by

$$\omega_c = \frac{2\pi M_0}{T} \approx \frac{\sum_j |\omega_j||\hat{\zeta}_j|^2}{\sum_j |\hat{\zeta}_j|^2}, \qquad (21)$$

such that M_0 is an integral number of central wave oscillations in the periodic computational domain of length T. The corresponding characteristic wavenumber k_c is computed from the linear dispersion relation (6). The non-dimensional depth $k_c h$ is in all cases between 1.4 and 1.6, and hence the waves are on strictly finite depth.

We have computed the velocity field under the highest wave in i24_15, which occurred at scan 3903. The velocity field was measured at the vertical position $z = -0.20$ m. The first 3000 scans are skipped, and the next 2048 scans are used for a periodic domain of length $T = 2048\tau = 51.2$ s. The number of central wave oscillations is set to be $M_0 = 25$, and the first harmonic B of the harmonic expansion (8) is defined by bandpassing $M_{bp} = 33$ spectral components of $\hat{\zeta}$ around the central wave. This gives a normalized bandpass width of the approximation $M_{bp}/(2M_0) = 0.66$. The exact choice of the bandpass width should not be critical provided the power spectrum decays sufficiently fast away from the central frequency. The bandpass width should be chosen appropriately to accommodate the scaling assumption for the bandwidth of the power spectrum (5b).

The computations of the Airy and Wheeler methods have been performed with a lowpass cutoff frequency such that 99% of the total energy in the frequency power spectrum is retained. We employ linear extrapolation for the Airy velocity profile above the mean water level.

The measured surface displacement is shown in figure 1 for the periodic time domain. The frequency power spectrum is shown in figure 2, where we have indicated the bandpass filter employed for the first harmonic of the linear Schrödinger method at frequencies $(M_0 \pm M_{bp}/2)/T$ and the lowpass filter frequency employed for the Airy and Wheeler methods.

In figure 3 we show the horizontal velocity profile under the highest wave. The prediction of the Airy and Wheeler methods, as well as the Schrödinger method at relative orders 1, $\epsilon^{1/2}$ and ϵ (we intend to describe the relative order ϵ nonlinear

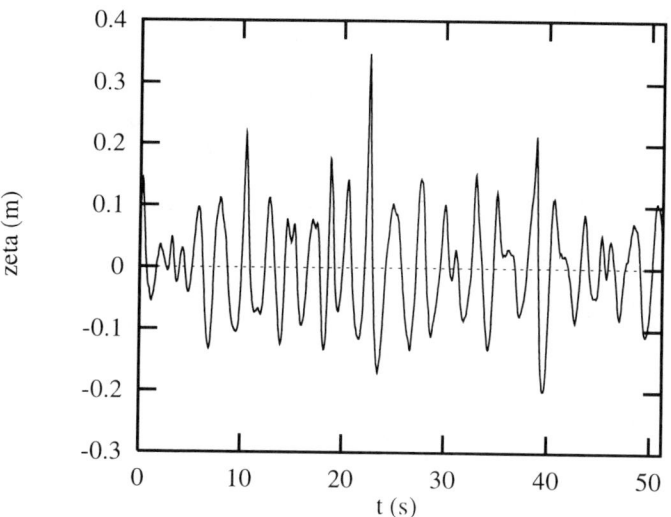

Figure 1: Measured surface displacement within the periodic time domain for computation.

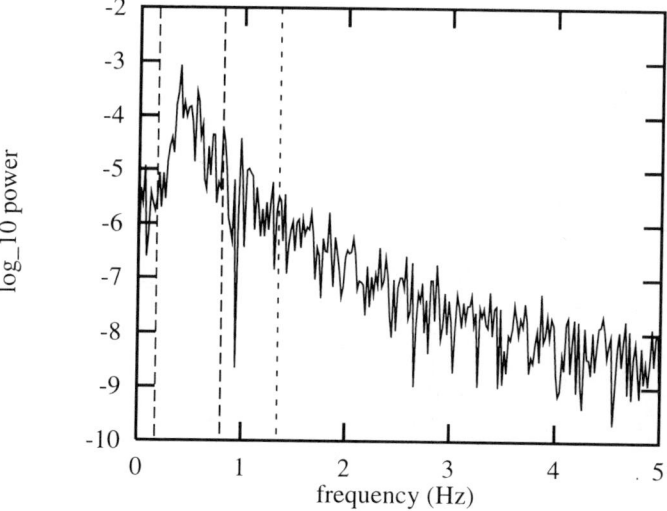

Figure 2: Frequency power spectrum (solid line) with Schrödinger bandpass filter (long dashed lines) and Airy lowpass filter (short dashed line).

Schrödinger method elsewhere). The measured velocity field is also indicated for the actual experiment i24_15 and all the other experiments.

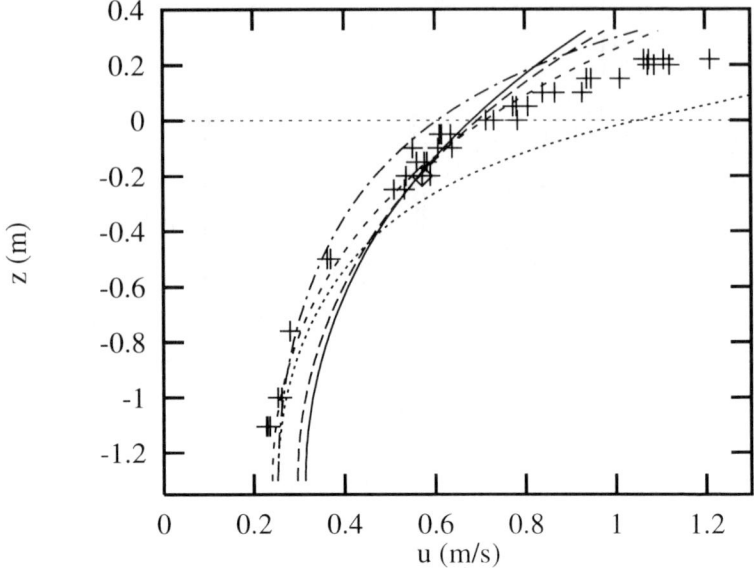

Figure 3: Horizontal velocity field under highest crest: Schrödinger order 1 (solid line), order $\epsilon^{1/2}$ (long dashed line), order ϵ (short dashed line); Airy (dotted line); Wheeler (dashed-dotted line); velocity measured in this experiment (diamond); velocities measured in the other experiments (cross).

Due to the problem of repeating the experiment exactly, while measuring the velocity field at only one location at a time, comparison of the velocity measurements from various experiments is subject to uncertainty. This is evidenced by the spread of the velocity measurements in the figure.

The correction to the Schrödinger method at relative order $\epsilon^{1/2}$ is seen not to give a significant contribution to the predicted velocity field. This is due to the fact that we are considering the velocity field near the global maximum of the time series, and hence $\partial B/\partial t$ should by definition be small. In general, the contribution from the term of order $\epsilon^{1/2}$ in (17) is larger than suggested by the figure.

The correction to the Schrödinger method at relative order ϵ, which we intend to describe elsewhere, includes the leading nonlinear contributions to the velocity field from the zeroth and second harmonics. The nonlinear contributions are most significant near the crest and the bottom. It is seen that for each consecutive order of the asymptotic theory, the approximation becomes slightly better.

Near the equilibrium water level, where the Wheeler method typically suffers from its greatest underprediction of the velocity field, even a leading order linear Schrödinger method gives remarkably good agreement with laboratory data.

Further developments

We have extended the Schrödinger method for water wave kinematics in two dimensions to finite depth, as is required for realistic applications on the continental shelf. Only the leading two linear orders are summarized here, yet we already achieve good agreement with laboratory data.

At the next order, there are nonlinear contributions to the velocity field from the zeroth and second harmonics, as well as additional linear contributions from the first harmonic. The nonlinear contributions are most significant near the crest and the bottom.

We have found that when the leading nonlinear contributions are taken into account, the velocity profile predicted by the nonlinear Schrödinger method corresponds remarkably well with the empirical correction to the Wheeler method suggested by Gudmestad & Haver (1993) (their equation 2). The nonlinear Schrödinger method for finite depth in two and three dimensions is currently being developed and will be presented in the future.

Acknowledgment

This research has been supported by fellowships from the European Union (MAS3-CT96-5016) and the Norwegian Research Council (NFR 109328/410), and by a grant from STATOIL (ANS025701).

References

GUDMESTAD, O. T. & HAVER, S. 1993 Uncertainties in prediction of wave kinematics in irregular waves. *Wave Kinematics and Environmental Forces* **29**, 75–99.

GUDMESTAD, O. T. & KARUNAKARAN, D. 1994 Wave kinematics models for calculation of wave loads on truss structures. In *Proceedings of the 26th annual Offshore Technology Conference*, 413–424.

SKJELBREIA, J. E., BEREK, G., BOLEN, Z. K., GUDMESTAD, O. T., HEIDEMAN, J. C., OHMART, R. D., SPIDSØE, N., & TØRUM, A. 1991 Wave kinematics in irregular waves. In *Proceedings of the 10th International Conference on Offshore Mechanics and Arctic Engineering*, 223–228. ASME.

TRULSEN, K. & DYSTHE, K. B. 1996 A modified nonlinear Schrödinger equation for broader bandwidth gravity waves on deep water. *Wave Motion* **24**, 281–289.

TRULSEN, K. & DYSTHE, K. B. 1997 Water wave kinematics computed with nonlinear Schrödinger theory. In *Proceedings of the 16th International Conference on Offshore Mechanics and Arctic Engineering* **1B**, 7–18. ASME.

Nonlinear Force Calculations by Numerical Wave Tank Simulations

M.H.Kim[1], M.S.Celebi[2] and J.C.Park[3]

Abstract

Fully nonlinear wave interactions with a three-dimensional body in the presence of steady uniform currents are studied using two independent Numerical Wave Tanks (NWT). The potential-NWT simulations are compared with the viscous-NWT simulations. The potential NWT used an indirect Desingularized Boundary Integral Equation Method (DBIEM) and a Mixed Eulerian-Lagrangian (MEL) time marching scheme. The Laplace equation is solved at each time step and the fully nonlinear free surface boundary conditions are integrated with time. A regridding algorithm is devised to eliminate the possible saw-tooth instabilities. The incident waves are generated by a piston-type wavemaker. The outgoing waves are dissipated inside a damping zone by using spatially varying artificial damping on the free surface. The viscous NWT solves a Navier-Stokes (NS) equation by using a finite-difference scheme and a modified marker-and cell (MAC) method in the frame of rectangular-coordinate system. The fully-nonlinear kinematic free-surface condition is satisfied by the density-function technique developed for two fluid layers. Computations are performed for the nonlinear diffractions of steep monochromatic waves by a truncated vertical cylinder with/without uniform currents. The NWT simulations compared favorably with available experimental results.

1. Introduction

Accurate predictions of wave loads and run-up on large offshore or coastal structures in combined wave and currents are of practical importance in the design and operation of these structures. When waves and currents coexist, the overall diffraction pattern and the resultant loading can be significantly different from the wave-only case. Most recent publications on this topic have been based on the perturbation-based potential theory. The perturbation approach in general neglects the short ship-wave systems generated by currents, and thus may not accurately predict the run-up around a structure as current speed increases. Until now, the study of fully-nonlinear wave-current-body interactions in the time domain is very scarce (e.g. *Ferrant, 1997*).

The major difficulties associated with the nonlinear free-surface simulations are (i)

---
1 Associate Professor, Texas A&M University
2 Research Assistant, Texas A&M University
3 Post-doctoral Associate, Texas A&M University, College Station, TX, 77843

such that the $z = 0$ plane corresponds to the calm water level and z is positive upwards. The total velocity potential Φ including steady uniform current U can then be expressed as $\Phi = Ux + \phi(x,y,z,t)$.

The velocity potentials Φ and ϕ describe fluid motions and they satisfy the Laplace equation in the computational domain:

$$\nabla^2 \Phi = \nabla^2 \phi = 0 \qquad (1)$$

On the instantaneous free surface $\eta(x,y,t)$, both the kinematic and dynamic boundary conditions must be satisfied. In addition, the zero-normal-flux condition at the bottom, wall, and body boundaries as well as the proper radiation condition at far field has to be satisfied.

In the MEL approach, a time stepping procedure is used in which a mixed boundary value problem is solved at each time step. At each step, the value of the potential is given on the free surface (Dirichlet boundary condition) and the value of the normal derivative of the velocity potential (Neumann boundary condition) is known on the body, wall, wavemaker, and bottom surfaces. After the mixed boundary value problem is solved, the free surface elevation, potential and its normal derivative are updated. For the time integration, a Runge-Kutta -Fehlberg (fourth-fifth order) method was used and proved to be stable and accurate.

The most common approach to time march the free surface boundary conditions in fully-nonlinear simulations is the material node approach in which the nodes or collocation points follow the individual fluid particles i.e.

$$\frac{D\mathbf{X}_F(t)}{Dt} = U\mathbf{i} + \nabla \phi \qquad (2)$$

which results in

$$\frac{D\eta}{Dt} = \frac{\partial \phi}{\partial z}. \qquad (3)$$

Assuming that the ambient pressure $P_a = 0$, one obtains

$$\frac{D\phi}{Dt} = -g\eta + \frac{1}{2}\nabla \phi \cdot \nabla \phi \qquad (4)$$

where g=gravitational acceleration, $\mathbf{X}_F(t) = (x_F(t), y_F(t), z_F(t))$ is the position vector of a fluid particle on the free surface, and D/Dt is the usual material derivative.

The use of the above kinematic and dynamic free-surface conditions allows the value of the elevation and potential to be stepped forward in time. The spatial derivative of the velocity potential $\nabla \phi$ can then be determined analytically after solving the boundary value problem for ϕ.

2.2 Indirect Desingularized Boundary Integral Equation Method

The mixed boundary value problem must be solved at each time step for the unknown velocity potential ϕ. With the given potential on the free surface and the known normal velocity on the body, wall, wavemaker, and bottom surfaces, the potential at any point in the fluid domain is given by the distribution of Rankine sources:

the complicated nonlinear free-surface boundary conditions that have to be satisfied on the instantaneous free-surface not known *a priori*, (ii) various types of numerical instabilities, (iii) appropriate numerical open-boundary conditions which can simulate open-sea conditions, and (iv) substantial CPU time unless some approximations are used. The use of fully-nonlinear time-dependent free-surface boundary conditions for two-dimensional potential-based surface waves was first introduced by *Longuet-Higgins and Cokelet (1976)*, in which a Mixed Eulerian-Lagrangian (MEL) time stepping technique was adopted. The two-dimensional MEL method was subsequently extended by *Dommermuth and Yue (1987)* to study the motion-induced nonlinear waves by axisymmetric bodies. Later, *Cao, Schultz and Beck (1990, 1991)* and *Cao, Lee and Beck (1992)* used similar time-stepping techniques and the so-called Desingularized Boundary Integral Equation Method (DBIEM) to study various nonlinear free-surface problems. Recently, *Celebi et al. (1997, 1998)* investigated fully-nonlinear wave interactions with a stationary vertical cylinder in a three-dimensional NWT with side and end beaches.

In this paper, the MEL time stepping technique and the DBIEM are used to investigate the fully-nonlinear diffraction of stationary three-dimensional bodies by steep regular waves and steady uniform currents inside a numerical wave tank equipped with a wavemaker and end beach. The numerical wavemaker and end beach are designed to allow the passage of currents and maintain the volume of water constant inside the NWT. The fully-nonlinear simulations with currents are validated through comparison with the perturbation approach of *Kim & Kim (1997)*. It is seen that the fully nonlinear simulations become close to the linear results when wave steepness decreases. When proper discretizations are used, the developed numerical-wave-tank computer program is robust, accurate, and free of numerical instabilities except when incident waves are too steep or current velocities are too large. Its accuracy was verified through convergence test, mass and energy conservations, and comparisons with the experiments of *Mercier & Niedzwecki (1994)* and *Moe (1993)*. The potential NWT results were also reasonably compared with the viscous NWT results. It is seen that the present potential NWT can reliably produce nonlinear free-surface elevation, run-up, pressure, and forces including a spectrum of higher harmonics unless current speed is large.

As current speed increases, the flow separation will be developed around a body and the potential theory becomes less meaningful. As an indication of the applicability of the potential theory, flow separation effects may be neglected when the relevant Keulegan-Carpenter (KC) number is small (say less than 2), which is also partly verified by comparing the potential NWT with an independently developed viscous-flow-based NWT (*Park & Kim, 1998*).

2. Fully Nonlinear Simulations By Potential NWT

2.1 Governing Equation and Boundary Conditions

An ideal, irrotational fluid is assumed so that a velocity potential $\Phi(x, y, z, t)$ exists and the fluid velocity is given by its gradient. It is also assumed that the surface tension on the free surface can be neglected. A Cartesian coordinate system (see Fig.1) is chosen

$$\phi(\mathbf{x}) = \iint_\Omega \sigma(\mathbf{x}_s)(\frac{1}{|\mathbf{x}-\mathbf{x}_s|})d\Omega \tag{5}$$

where Ω is the integration surface outside the fluid domain and σ is the source strength to be determined. Applying the relevant boundary conditions, the desingularized indirect boundary integral equations that must be solved to determine the unknown source strengths are

$$\iint_\Omega \sigma(\mathbf{x}_s)(\frac{1}{|\mathbf{x}_c-\mathbf{x}_s|})d\Omega = \phi_o(\mathbf{x}_c) \qquad (\mathbf{x}_c \in \Gamma_d) \tag{6}$$

and

$$\iint_\Omega \sigma(\mathbf{x}_s)\frac{\partial}{\partial \mathbf{n}}(\frac{1}{|\mathbf{x}_c-\mathbf{x}_s|})d\Omega = \chi(\mathbf{x}_c) \qquad (\mathbf{x}_c \in \Gamma_n) \tag{7}$$

where
$\mathbf{x}_s = (x', y', z')$: a source point on the integration surface,
$\mathbf{x}_c = (x, y, z)$: a field point on the real boundary,
ϕ_o = the given potential value at $\mathbf{x}_c$,
Γ_d = surface on which ϕ_o is given,
χ = the given normal velocity at $\mathbf{x}_c$,
Γ_n = surface on which χ is given.

The integration domain includes the free surface, the bottom surface, the body surface, the wavemaker, the side wall, and the truncation surfaces. The discretized version of the above integral equation is solved for σ by an iterative solver. Subsequently, ϕ can be evaluated by (5). Its spatial derivative $\nabla\phi$ can be obtained from the derivative of (5), where the differentiation of the Rankine source can be done analytically. In case of the flat bottom, image sources can be used to eliminate the integration over bottom surface.

In the desingularized method, the source distribution is outside the fluid domain so that the source points never correspond to the field (or collocation) points, and the resulting integrals are nonsingular and can be straightforwardly integrated by numerical quadratures. The isolated sources are distributed a small distance above each of the nodes. The desingularized distance is in general given by $L_d = l_d(D_m)^{\beta'}$, where D_m represents the local mesh size and l_d and β' are the constant parameters to be selected. If the singularity is located too far from the boundary, the resulting influence-coefficient matrix is likely to be poorly conditioned, which results in slow convergence in solving the matrix iteratively. If the singularity is too close to the boundary, the numerical integration of the kernel may not be robust. Therefore, appropriate l_d and β' values need to be determined. In this paper, β'=0.5 and l_d=0.5~1 (depending on wave length) are chosen after numerical testing. A detailed study with regard to the performance of DBIEM with the desingularization parameters is reported in *Cao et al.* (1991).

2.3 Saw-tooth Instability and Regriding

Longuet-Higgins and Cokelet (1976) first found the so-called *sawtooth* instability during their two-dimensional breaking wave simulations. They employed a smoothing technique to suppress the development of the sawtooth instability. Later researches pointed out that the possible cause of the instability was the concentration of Lagrangian points in the region of higher gradients and the minimum grid size cannot be effectively controlled for given time step. The time evolving deformation of the Largrangian grid system in general results in the numerical degradation and eventual breakdown. In this regard, a regriding algorithm similar to that of *Dommermuth and Yue (1987)* is used to eliminate such instabilities. Thus, a new set of uniformly spaced Lagrangian points are created on the free surface and body surface after every time step. A cubic-spline interpolation technique is then employed to redistribute the boundary values on the new set of nodal points. Then, the time stepping is continued with the fourth/fifth-order Runge-Kutta-Fehlberg time-integration scheme. The same regriding scheme was also applied at the body-free surface intersection line. Using this algorithm, the instabilities are effectively removed and no artificial smoothing is required.

The main disadvantage of regriding is the potential loss of resolution which is usually provided by more closely spaced Lagrangian points in the areas of large gradients. The advantages of regriding over artificial smoothing particularly in the present context are, however, substantial in that the smoothing cannot be straightforwardly applied at the body-free surface intersection line and the crossing of Lagrangian points can be more easily controlled.

2.4 Wave Generation and Absorption

The incoming waves are generated by the prescribed motion of a piston-type wave maker. In the beginning, the wavemaker velocity is gradually increased by using a linear ramp function. The uniform current is then introduced in the entire fluid field to generate the wave-current environment. The end beach is designed to allow the passage of currents.

The fluid particles (the nodes) on the wavemaker are forced to move with the wavemaker in the normal direction, while the particles are allowed to move freely in the other directions. Fully nonlinear free surface boundary conditions are applied along the free surface and wavemaker intersection line. During the simulation, the intersection line is updated and regrided like the body waterline, which is shown to be effective in preventing possible numerical instability there.

Inside the downstream damping zone, the artificial damping is applied only to the dynamic free-surface condition so that waves are damped out as much as possible before they reach the truncation boundary. On the truncation boundary, only uniform currents are passed through. Both ϕ- and ϕ_n-type beaches have been tested, and the following ϕ_n-type beach was shown to be the most efficacious and subsequently used:

$$\frac{D\phi}{Dt} = -g\eta + \frac{1}{2}\nabla\phi \cdot \nabla\phi - \nu\frac{\partial \phi}{\partial n} \tag{8}$$

where ν is a damping coefficient. The use of the ϕ_n-type beach ensures the energy absorption rate to be always positive.

Our numerical experiments show that the choice of proper damping coefficient $\nu(x,y)$ is important to achieve maximum wave dissipation. Inside the damping zone, the damping is chosen to be smooth and gradually increasing. On the transition boundary between the damping zone and the inner solution domain, C^2 or higher continuity have to be satisfied to guarantee high performance. The performance of the artificial beach depends on the ratio of the beach length to wave length. After a series of tests, it is found that the beach length has to be at least one wave length to guarantee minimal reflection. In view of this, an adaptive grid generation technique is used in this paper to optimize the length of the damping zone depending on the pertinent wave length.

3. Viscous NWT

A finite-difference scheme and a modified marker-and-cell (MAC) method are used for alternative NWT simulations with viscous effects to investigate nonlinear wave-body interactions and the role of fluid viscosity. The Navier-Stokes (NS) equation is solved for two fluid layers and the boundary values updated at each time step by a finite-difference time marching scheme in the frame of rectangular coordinate system. The fully nonlinear kinematic free surface condition is satisfied by the density-function technique developed for two fluid layers:

$$\frac{\partial M_\rho}{\partial t} + u\frac{\partial M_\rho}{\partial x} + v\frac{\partial M_\rho}{\partial y} + w\frac{\partial M_\rho}{\partial z} = 0, \tag{9}$$

where the density function M_ρ takes the value between ρ_1 and ρ_2 all over the computational domain and this scalar value has the meaning of porosity in each cell. In case of water-air flow, this value means the volume fraction of water in a cell. The free-surface location is determined to be a point where the marker-density function takes the mean value of ρ_1 and ρ_2.

In the time-marching procedure, the density-function distribution is calculated from the density-function equation and the velocity field is updated through the Navier-Stokes equation. For the time and space differencing of the density-function equation, Adams-Bashforth and flux-split methods are used, respectively. The dynamic free-surface condition is implemented by the so-called irregular-star technique in the solution process of the Poisson equation for the pressure. The incident waves are generated from the inflow boundary by prescribing the particle-velocity profile of linear waves. The fully-nonlinear free-surface condition is applied immediately after the generation of incident waves. The outgoing waves are numerically dissipated inside an artificial damping zone located at the end of the tank. Both physical and numerical dampings are used for this purpose. The uniform currents are gradually introduced in the entire fluid region to minimize transient free-surface responses. The numerical wavemaker and end beach are designed to allow the passage of currents and maintain the volume of water constant inside the NWT. Through various numerical tests, the free-surface, inflow- and open-boundary treatments are found to be robust. The effects of turbulence are not modeled in this NWT. The details of the viscous NWT used here are given in *Park & Kim*. (1998).

4. Numerical Results and Discussions

To demonstrate the usefulness and accuracy of the numerical wave tanks for wave-current-body interactions, we present results here for nonlinear diffractions by a truncated vertical cylinder inside a rectangular tank with side walls (see Figure 1) in the presence of monochromatic waves and uniform coplanar or adverse currents. The fully nonlinear computations are compared with the perturbation results of *Kim & Kim* (1997) (or experimental results, if available) for various wave steepness and current speeds.

For the perturbation method, 288 nine-node quadratic boundary elements are used in the half computational domain. On each element, quadratic interpolation functions are used to describe the variation of velocity potentials.

For the potential-based fully nonlinear free surface calculations, the time step $\Delta t = 0.1$s was selected based on the test for the shortest wave. To check the convergence with the number of nodes, three different free-surface discretizations, N_F=1246, 2420, and 3840, were tested. For the ensuing examples, 2420 nodes were used to discretize the free surface, 186 nodes on the body, 175 nodes on the wavemaker, 175 nodes on the end wall, and 330 nodes on the side wall in a half domain. For the selected spatial and temporal discretizations and the open-boundary scheme used, it is shown that the mass and energy are conserved within 2% error (*Celebi et al, 1998*) for typical wave periods and wave slopes used here.

For the viscous NWT, rectangular grid cells of 4-cm length, 4-cm width, and $H/20$ (H=wave height) height were used. The time increment used was $T/2000$ and the total number of grids in the half fluid domain was 682,500 including the damping zone.

In the following, the potential and viscous NWT simulations are carried out for a stationary vertical truncated cylinder (radius a =0.23m, draft=1.33m in model scale), and the results for $F_n = U/\sqrt{ga} = 0$ are compared with *Mercier and Niedzwecki's* (1994) experiments conducted in the OTRC wave tank at Texas A&M University. For $F_n \neq 0$, the NWT simulations are compared with the perturbation results.

For the potential-based NWT simulations, the incident waves were generated by a harmonic motion of the piston-type wave maker. The wave condition used in *Mercier and Niedzwecki's* (1994) experiments was $\varepsilon = H/gT^2$=0.005 and 0.0075, which corresponds to wave slopes H/λ=0.031, and 0.047. In both NWT simulations and the experiments, the wave steepness was maintained to be constant for all wave frequencies.

To quantitatively observe the nonlinear contributions, a series of higher harmonics were generated by a FFT algorithm. The length of time series used for FFT was $18T$. In Figures 2-4, the first three harmonics of nondimensional wave run-up (where $A = H/2$) at the weather side (θ= 180°) are presented for two different wave steepness $\varepsilon = 0.005$ and 0.0075 (F_n=0). The experimental results were read from *Mercier and Niedzwecki* (1994). The NWT computations generally agree well with experiments for the first-harmonic (wave-frequency) component. As for the second- and third-harmonic components, which are very small compared to the first-harmonic component, the trend looks similar but there exists some discrepancy in magnitude between the predicted and measured values. It is pointed out in *Mercier and Niedzwecki* (1994) that the actual wave heights used for the high-frequency cases are very small, and thus the corresponding measurements of higher harmonics are expected to be less accurate. The discrepancy between the NWT simulations and experiments can also be attributed to the difference in incident waves and viscous effects. It is interesting to see in both experiments and NWT simulations that the dimensionless second- and third-order run-ups are smaller in steeper waves, which cannot be predicted by the linearized theory.

Figures 5 show the first-, second-, and third-order wave run-up at the weather side obtained from the viscous NWT simulations for $F_n = 0$ and $\varepsilon = 0.0075$. The viscous-NWT results are in good agreement with the potential-NWT results. The range of K-C number for this case is 0.4-2. For this kind of small K-C number, the inertia effects are more important and the potential-based NWT simulations produce reasonable results.

To further confirm the accuracy of higher harmonics, the simulated third harmonic horizontal forces by the potential-based NWT are compared in Figure 6 with the theoretical and experimental results presented in *Malenica & Molin (1995)*, where the theoretical results (third-order semi-analytic solutions) are for draft/radius=10 and the experimental results of *Moe (1993)* are for draft/radius=3. The agreement between them is generally good.

Next, the effects of uniform currents on the weather-side run-up are investigated for ε=0.0075 and ω=5.1 rad/s (in model scale or $\omega^2 a/g = 0.61$). In the following, positive Froude number corresponds to coplanar currents. Figure 7 shows perturbation results and Figure 8 fully-nonlinear simulations by the potential-based NWT. From these long time histories, we can see that the open boundary condition works well and wave relection is very small. The shift of wave periods (or wave frequencies) in currents is due to the Doppler effect. It is seen in both figures that the run-up amplitude is increased in coplanar currents and decreased in adverse currents, which is also shown by *Ferrant (1997)*. The rates of increase and decrease are greater in fully-nonlinear simulations.

In Figures 9 and 10, the horizontal forces on the vertical cylinder for ε=0.0075 and ω=5.1 rad/s are plotted for three different current speeds, F_n=-0.02, 0, and 0.02. The trend of NWT and perturbation results is in general similar and the NWT results show greater influence by currents.

Finally Figures 11 and 12 show the change of horizontal forces and weather-side run-up due to the presence of steady uniform coplanar currents. The results are computed by the viscous NWT. The wave period used for this example is 0.87s and ε=0.0075. The general trend looks similar to that of Figures 9 and 10.

5. Concluding Remarks

The three-dimensional fully nonlinear wave-body interactions in steady uniform currents are studied using two different numerical wave tanks. The potential-based NWT is based on a desingularized boundary integral equation method (DBIEM) and mixed Eulerian and Lagrangian (MEL) time marching scheme. The fully nonlinear kinematic and dynamic free-surface conditions were integrated with time by the 4/5th-order Runge-Kutta-Fehlberg method. The incident waves were generated by a piston-type wave maker and the waves downstream were dissipated by the adaptive ϕ_n-type damping layer on the free surface inside the damping zone. The viscous NWT solves a Navier-Stokes (NS) equation by using a finite-difference scheme and a modified marker-and cell (MAC) method in the frame of rectangular-coordinate system. The fully-nonlinear kinematic free-surface condition is satisfied by the density-function technique developed for two fluid layers. In both cases, uniform currents are gradually introduced in the entire fluid and the boundary conditions at the wavemaker, truncation boundary downstream, and the body are modified accordingly.

The developed NWTs were used to study fully nonlinear interactions of monochromatic waves with a truncated vertical cylinder in uniform coplanar or adverse currents. The potential-based NWT simulations were compared with the perturbation-based numerical results calculated by a Time-domain High-Order Boundary Element Method (THOBEM). The NWT simulations were conducted for several Froude numbers and wave slopes and the simulation results become close to the perturbation results as both wave slopes and currents are decreased. The NWT results also compared favorably with the experimental results of *Mercier & Niedzwecki (1994)* and *Moe (1993)* as well as the third-order diffraction theory (without currents) of *Malenica & Molin (1995)*. The effects of currents are shown to be important to the prediction of maximum run-up around an ocean structure. It is also shown that the potential NWT simulations compare well with viscous NWT simulations when current speed is not large.

6. Acknowledgment

This research was supported by the Offshore Technology Research Center through the National Science Foundation Engineering Research Centers Program, Grant Number CDR 8721512.

7. References

Cao, Y., Schultz, W.W. and Beck, R.F., 1990 *Three Dimensional Unsteady Computations of Nonlinear Waves Caused by Underwater Disturbance*, Proc. 18th Symposium on Naval Hydrodynamics, Ann Arbor, MI, USA, 417-427

Cao, Y., Schultz, W.W. and Beck, R.F., 1991 *Three Dimensional Desingularized Boundary Integral Methods for Potential Problems*, Int. Journal of Num. Meth. Fluids, Vol 12, 785-803

Celebi, M.S. and Kim, M.H., 1997 *Nonlinear Wave Body Interactions in a Numerical Wave Tank*, 12th Int. Workshop on Water Waves and Floating Bodies, France

Celebi, M.S., Kim, M.H., and Beck, R.F. 1998 *Fully Nonlinear 3D Numerical Wave Tank Simulation*, Journal of Ship Research, Vol.42, No.1, 33-45

Dommermuth, D.G. and Yue, D.K.P., 1987 *Numerical Simulations of Nonlinear Axisymmetric Flows With a Free Surface*, Journal of Fluid Mechanics, Vol 178, 195-219

Ferrant, P. 1997 *Nonlinear Wave-current Interactions in The Vicinity of a Vertical Cylinder*, 12th Int. Workshop on Water Waves and Floating Bodies, France

Kim, D.J. and Kim, M.H., 1997 *Wave-current Interactions with a Large Three-dimensional Body by THOBEM*, Journal of Ship Research, Vol.41, No.4, 273-285

Longuet-Higgins, M.S. and Cokelet, C.D., 1976 *The Deformation of Steep Surface Waves on Water: I. A Numerical Method of Computation*, Proc. R. Soc., London, A350, 1-26

Malenica, S. and Molin, B., 1995 *Third Harmonic Wave Diffraction by a Vertical Cylinder*, J. of Fluid Mechanics, Vol.302, 203-229

Mercier, R.S. and Niedzwecki, J.M., 1994 *Experimental Measurement of Second Order Diffraction by a Truncated Vertical Cylinder in Monochoromatic Waves,* Proc. 7th Intl. Conf. Behavior of Offshore Structures, MIT, Vol.2, 265-287

Moe, G., 1993 *Vertical Resonant Motions of TLPs,* Final Report NTH Rep. R-1-93

Park, J.C. and Kim, M.H., 1998 *Fully-nonlinear wave-current-body interactions by a 3D viscous NWT,* Conf. ISOPE 98, Montreal

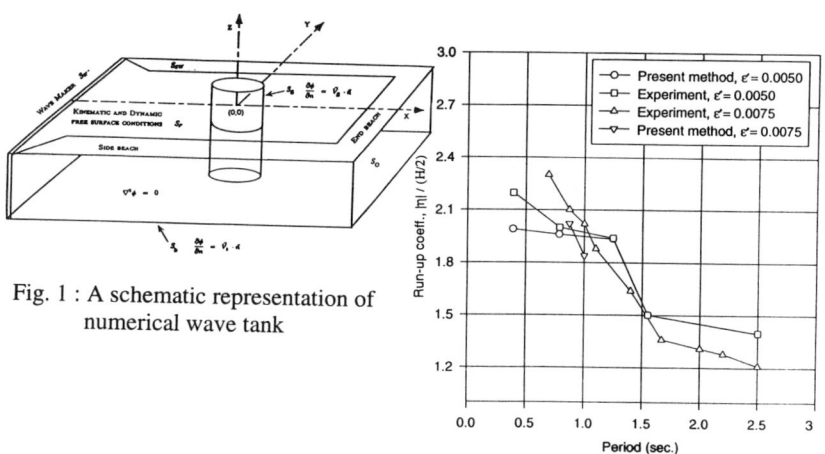

Fig. 1 : A schematic representation of numerical wave tank

Fig. 2 : First-order weather($\theta = \pi$) side run-up on a uniform vertical truncated cylinder for $F_n = 0$

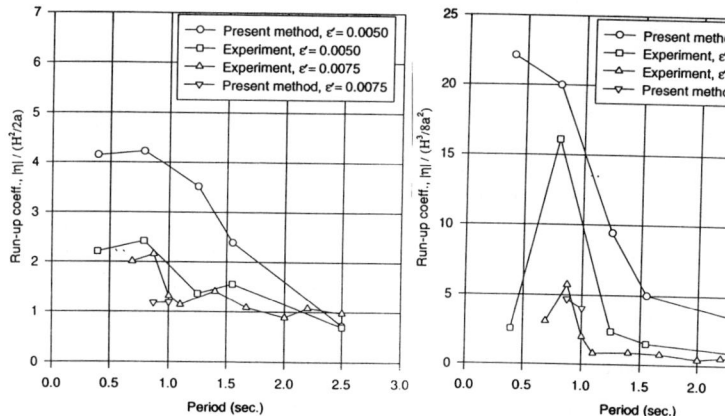

Fig. 3 : Second-order weather side run-up on a uniform vertical truncated cylinder for $F_n = 0$

Fig. 4 : Third-order weather side run-up on a uniform vertical truncated cylinder for $F_n = 0$

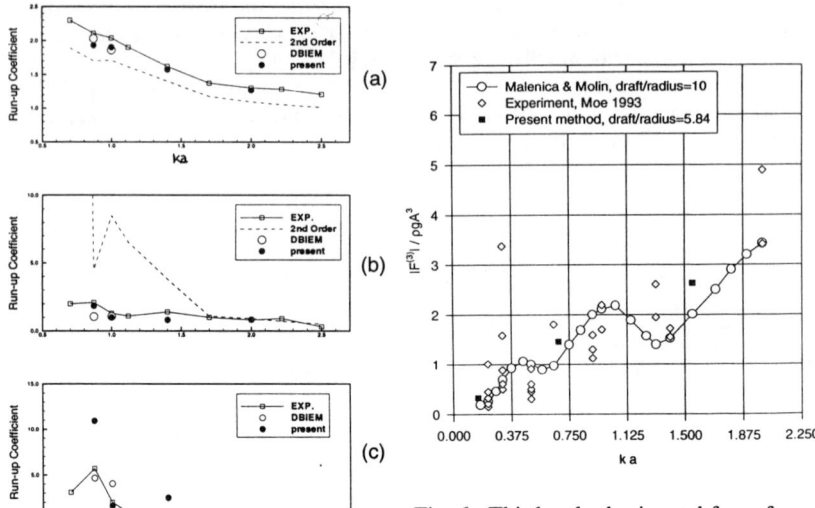

Fig. 5 : Wave run-up at the weather side ; (a) first-order, (b) second-order and (c) third-order

Fig. 6 : Third-order horizontal force for a vertical truncated cylinder

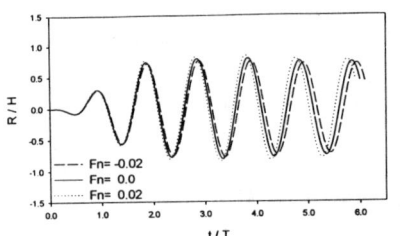

Fig. 7 : Weather side run-up in adverse and co-planar currents for a truncated vertical cylinder by perturbation theory

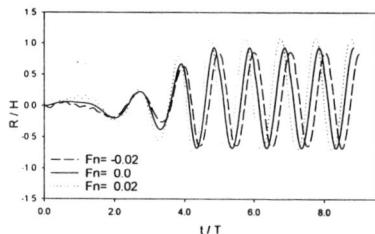

Fig. 8 : Weather side run-up in adverse and co-planar currents for a truncated vertical cylinder by potential NWT

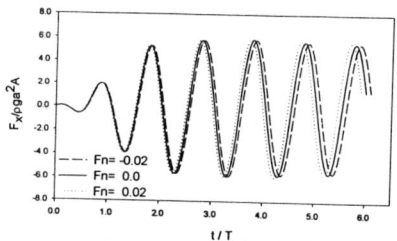

Fig. 9 : Horizontal force in adverse and co-planar currents for a truncated vertical cylinder by perturbation theory

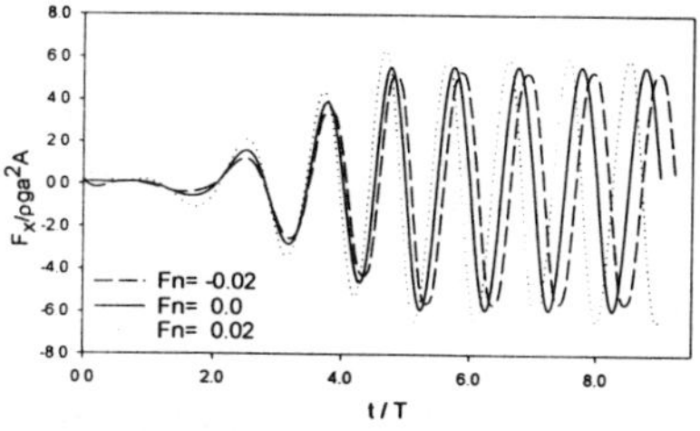

Fig. 10 : Horizontal force in adverse and co-planar currents for a truncated vertical cylinder by potential NWT

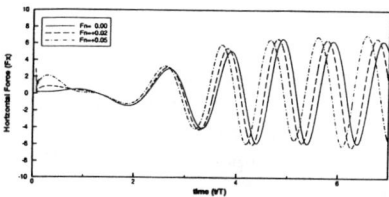
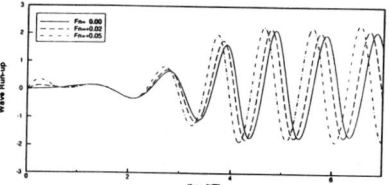

Fig. 11 : Horizontal force in co-planar currents for a truncated vertical cylinder by viscous NWT

Fig. 12 : Weather side run-up in co-planar currents for a truncated vertical cylinder by viscous NWT

Fully Nonlinear Simulation of Ocean Waves in interaction with Offshore Structures – Project "UNDA"

Even Mehlum[1]

1. Introduction

Within the framework of the so called "Potential Theory" of ocean waves, it is necessary to perform fully nonlinear simulations in the time domain. Some kind of time stepping method has to be devised and the Laplace equation must be solved at each time step. However, the time dependence of the whole problem does not involve the Laplace equation itself, therefore it is possible to devise a kind of *eigenfunction expansion* (i.e. a spectral method, see e.g. [2] or [4]) to solve the Laplace equation. In this paper the set of eigenfunctions is determined once and for all, i.e., outside the time stepping. The time dependence enters into the coefficients of this expansion, i.e., into the time dependent individual weightings of the eigenfunctions.

The intention of this paper is to introduce two fragments of the simulation method used in the UNDA-project which, we believe, are genuinely novel. The UNDA-project (WAVE=UNDA in Latin) was initiated in 1993 by SINTEF's Institute for Applied Mathematics in Oslo and later sponsored by 5 industrial partners viz. Norsk Hydro, Statoil, Saga Petroleum, Conoco and Shell. The goal is to compute load data on offshore structures as well as dynamics and kinematics.

The first fragment is discussed in Section 1 where some properties of the so called "Spline Functions" are reviewed. In the UNDA-project spline functions are employed both as approximants in the solution process of Laplace's equation and to represent the free surface elevation.

As the second fragment, a method for introducing fixed bodies in the water volume is presented in Section 2. The proposed method is based on the intro-

---
[1]Department of Applied Mathematics, SINTEF, P.O.Box 124 Blindern, N-0314 Oslo, Norway

duction of the triharmonic operator together with tensor product splines. It can be regarded as a variety of the familiar singularity method. One of its merits is the possibility of maintaining a continuous description of the potential everywhere in the fluid and also inside the bodies.

In Section 3 the two fragments are connected to the solution process for the fully nonlinear water wave system, and we conclude by some further discussion and a brief mention of results. [1] provides a much more detailed presentation than the present extended abstract.

2. Spline functions and differential equations

Spline functions (or spline curves) are piecewise analytic functions, of which any two consecutive pieces are joined together at so called "knots" with continuity requirements on the function value and the first few derivatives.

We shall here limit our discussion to cubic splines. These splines consist of cubic polynomials between knots. The function value and the first two derivatives are continuous in passing a knot from one cubic polynomial to the next. There are many different notations, names and algorithms in use regarding cubic splines, most notably "B-splines". We have preferred to use explicit cubic polynomials.

The one-dimensional cubic spline is, in a certain sense, the smoothest possible interpolating function between a sequence of data points. In particular, we are given a set of I points $[x_i, f_i]$ ($x_i > x_{i-1}$) and want to determine a function $f(x)$ which satisfies the following two criteria:

i) The integral $\int [f''(x)]^2 dx$ is minimized (the smoothness criterion);

ii) $f(x_i) = f_i$, $i = 1, \ldots, I$ (the interpolation criterion).

The function $f(x)$ which satisfies this constrained variational problem uniquely, is the cubic spline with knots coinciding with the data points x_i (see [5]). More information on spline functions is readily available in [6].

3. Introduction of bodies in the water

In three-dimensional simulations of water waves we introduce a potential ϕ that must satisfy the Laplace equation $\nabla^2 \phi = 0$ in the fluid domain as well as the boundary conditions on the free surface. The introduction of three-dimensional fixed bodies, fully submerged or surface-piercing, in the water volume by the proposed method results in internal boundary conditions inside the computational domain. These conditions have the physical interpretation that water is not allowed to flow through the body surfaces.

A traditional way to achieve this is to introduce singularities inside the body and adjust, via equation solving, the strength of these singularities so that the boundary conditions are met. The singularities are the familiar sources, dipoles etc. An alternative method is furnished by the "Boundary Panel" method which is an application of Green's Theorem. However, as a variety of the singularity method, we have found it advantageous to replace the discrete singularities with continuous and, in a certain sense smooth, trivariate functions inside the volumes occupied by the bodies. To be specific, we require

$$\nabla^6 \phi = 0$$

inside the bodies in contrast to the Laplace equation $\nabla^2 \phi = 0$ in the water volume. (Strictly speaking, ϕ inside the bodies is no longer a potential.) The very different differential equation satisfied by the "potential" inside the bodies makes the them "stand out". There are two motivations behind choosing the triharmonic operator inside the bodies:

1. It satisfies a variational criterion for smoothness (minimum variation of $\int [\nabla(\nabla^2 \phi)]^2$) and provides a generalization into 3D of quintic splines. A similar criterion was applied in [5]. Today such criteria are common practice in scattered data approximation, see e.g. [3].

2. It is the lowest order operator that can satisfy *three* requirements on the boundary, i.e., function value, normal derivative and the Laplace equation.

This method of including bodies makes it possible to maintain a continuous description of the potential everywhere in the fluid and also inside the bodies. Furthermore, it is possible to work with separated variables in a Cartesian coordinate system.

4. Time stepping and and Discussion

The time stepping of the whole problem is realized by an explicit scheme, in which the surface elevation and the potential at the surface are updated by the following rewritten forms of the kinematic and dynamic boundary conditions at the free surface (see [7]):

$$\eta_t = (1 + \eta_x^2 + \eta_y^2)G - \eta_x F_x - \eta_y F_y,$$
$$F_t = -g\eta - \frac{1}{2}F_x^2 - \frac{1}{2}F_y^2 + \frac{1}{2}(1 + \eta_x^2 + \eta_y^2)G^2,$$

where $F(x,y,t) = \phi(x,y,\eta(x,y,t),t)$ and $G(x,y,t) = \phi_z(x,y,\eta(x,y,t),t)$.

There would seem to be a rather deep reason why the application of splines, which are non-analytic, is advantageous in ocean wave modelling. The key point can be described intuitively as follows: The discontinuities in the higher order derivatives of the spline functions, which arise from the smoothness criterion, help to restrict the non- (or almost non-) analytic behaviour of the solution to a local area. Such behaviour occurs close to bodies and in areas of peaking waves. (Breaking is beyond the scope of this work.)

The results from UNDA-simulations have undergone extensive checking against experiment. The main conclusion is that the simulations compare very well with experiment and that important physical phenomena (e.g. "ringing") can be explored via such simulations.

References

[1] Cai, X. and Mehlum, E. (1996), Two fragments of a method for fully nonlinear simulation of water waves. *Waves and Nonlinear Processes in Hydrodynamics.* Edited by Grue, J. et al. Kluwer Academic Publishers, pp. 37-50.

[2] Canuto, C., Hussaini, M. Y., Quarteroni, A. and Zang, T. A. (1988), Spectral methods in fluid dynamics. Springer-Verlag.

[3] Greiner, G. (1994), Variational design and fairing of spline surfaces. EUROGRAPHICS '94. Dæhlen, M. and Kjelldahl, L. (Guest Editors), Blackwell Publishers, **13**, pp. 143-154.

[4] Liu, Y., Dommermuth, D. G. and Yue, D. K. P. (1992), A high-order spectral method for nonlinear wave-body interactions. *J. Fluid Mech.*, **245**, pp. 115-136.

[5] Mehlum, E. (1969), Curve and surface fitting based on variational criteriae for smoothness. *PhD thesis*, University of Oslo, Norway and SINTEF report.

[6] Press, W. P., Teukolsky, S. A., Vetterling, W. T. and Flannery, B. P. (1992), Numerical Recipes in C. *The Art of Scientific Computing.* 2nd Edition. Cambridge University Press.

[7] Zakharov, V. E. (1968), Stability of periodic waves of finite amplitude on the surface of a deep fluid. *J. Appl. Mech. Tech. Phys.*, **9**, pp. 190-194 (English tansl.)

Simulation of Nonlinear Waves in Shallow-Water Basins

Keh-Han Wang[1] and Weimin Li[2]

Abstract

The propagation and transformation of nonlinear shallow-water waves in irregular basins were simulated using a generalized Boussinesq model. The hydrodynamic interaction between nonlinear waves and cylinder arrays was also studied. The Boussinesq equations were solved numerically in a curvilinear coordinate system. An Euler's predictor-corrector finite-difference algorithm was applied for numerical computation. A second-order cnoidal wave solution was used as an incident wave condition. A set of open boundary conditions was also developed to effectively transmit waves out of the computational domain. The simulated free-surface elevations for waves propagating in a harbor with different layout of inner and outer breakwaters are presented. The evolution of nonlinear waves and their interactions with large cylinder arrays are also presented and discussed.

Introduction

Study of wave propagation and transformation in a shallow-water basin is of great importance to coastal engineering practice. With an increase of the application of large cylindrical structures in the sea, the hydrodynamic interaction between water waves and cylinder arrays has also received extensive attention. The phenomenon of wave oscillation in a harbor has been studied numerically or analytically by many researchers (e.g., Ippen and Goda (1963), Lee (1971), Lepelletier and Raichlen (1987, 1988), Chou and Han (1994)). However, the overall evolution and transformation of nonlinear waves in a more realistic, irregular shallow-water basin have not been studied extensively.
Starting from the study by MacCamy and Fuchs (1954), many efforts have been directed toward understanding the wave diffraction and wave forces on multiple bodies (e.g., McIver and Evans (1984); Linton and Evans (1990), Masuda et. al. (1986), Abul-Azm and Williams (1989) and Isaacson and Cheung (1992)). In an effort to model the three-dimensional nonlinear waves in shallow water, Wu (1981) presented a set of equations of Boussinesq class, which was called the generalized Boussinesq (gB) model. The linear and nonlinear analysis of the hydrodynamic

---

[1] Associate Professor [2] Graduate student, Department of Civil and Environmental Engineering, University of Houston, Houston, TX, 77204-4791

interaction between cnoidal waves (or solitary waves) and a vertical cylinder were investigated by Isaacson (1983), Wang et. al. (1992), Jiang and Wang (1995) and other researchers. It has been concluded that the linear diffraction theories for force computation are not sufficient to describe the nonlinear interaction between waves and structures.

The simulation of wave propagation in a harbor basin with variable depth and the obstruction of internal coastal structures may become very complicated due to the irregularity of basin boundary and the nonlinear characteristics of incident waves. It is intended in this study to develop a generalized wave prediction model which is capable of simulating spatial and temporal variation of propagation and transformation of large-amplitude waves by obstacles in shallow water of variable depth. The Boussinesq equations (Wu, 1981) was used as the model equations. The curvilinear coordinate transformation was incorporated into the model equations to simplify the computational domain and to facilitate the application of boundary condition on complicated basin and structural boundaries. A set of radiation conditions was applied along the open boundaries to effectively propagate waves out of the computational domain. Simulations were conducted for cnoidal waves propagating past a convex ramp bottom topography. The model was also extended to study the evolution of wave propagation in a harbor with combined inner and outer breakwaters. The results showing nonlinear wave interaction with an array of four cylinders are also presented.

Boussinesq Model for Nonlinear Shallow-Water Waves

To describe weakly nonlinear and weakly dispersive nonlinear shallow water waves propagating in a variable water depth, we adopt the Boussinesq equations developed by Wu (1981) for numerical simulations. The equations in terms of layer-mean horizontal velocities and free-surface elevation are given as

$$\zeta_t + \nabla \cdot [(h+\zeta)\bar{\mathbf{u}}] = -h_t \tag{1}$$

$$\bar{\mathbf{u}}_t + \bar{\mathbf{u}} \cdot \nabla \bar{\mathbf{u}} + \nabla \zeta = \frac{h}{2}\frac{\partial}{\partial t}\nabla[h_t + \nabla \cdot (h\bar{\mathbf{u}})] - \frac{h^2}{6}\frac{\partial}{\partial t}\nabla^2 \bar{\mathbf{u}} \tag{2}$$

Here, ζ is the wave elevation and h denotes the arbitrary water depth. $\bar{\mathbf{u}}$ represents the depth-averaged velocity vector. The basic equations (1) to (2) are dimensionless in form with (x,y,z,h,ζ) all referenced to h_o, $\bar{\mathbf{u}}$ to c_o, and t to h_o/c_o. Here, $c_o = (gh_o)^{1/2}$ and h_o is a representative constant water depth. A two-equation model described by the layer-mean velocity potential and free-surface elevation can be expressed as

$$\zeta_t + \nabla \cdot [(h+\zeta)\nabla\bar{\phi}] = -h_t + \nabla \cdot \{[\frac{h}{2}\{h_t + \nabla \cdot (h\nabla\bar{\phi})\} - \frac{h^2}{3}\nabla^2\bar{\phi}]\nabla h\} \tag{3}$$

$$\bar{\phi}_t + \frac{1}{2}(\nabla\bar{\phi})^2 + \zeta = \frac{h}{2}\frac{\partial}{\partial t}[h_t + \nabla \cdot (h\nabla\bar{\phi})] - \frac{h^2}{6}\nabla^2\bar{\phi}_t \tag{4}$$

where $\bar{\phi}$ is the layer-mean velocity potential.

Numerical Method and Boundary Conditions

The curvilinear coordinate transformation is applied to simplify the computational domain and to facilitate the applications of the boundary conditions.

Equations (1) and (2) (or (3) and (4)) are transformed into the curvilinear coordinate system for numerical computation. The numerical method applied to computation of the basic equations is based on the method established by Wang et al. (1992) using a modified Euler's predictor-corrector algorithm for time advancing and a central difference representation for the space derivatives. The free-surface elevation is calculated explicitly, whereas the depth-averaged velocity component (or velocity potential) is obtained through an implicit-iteration procedure. The detailed numerical procedure and the applications of the initial and boundary conditions can be found in Jiang and Wang (1995)

Results and Comments

The propagation of cnoidal waves through a channel with uneven bottom is studied by solving Eqs. (1) and (2). Here, we simulate cnoidal waves propagating past a convex bottom topography. The convex ramp configuration is shown in Fig. 1. The dimensionless water depth changes from 1 to 1/2.

A sequence of three-dimensional perspective view plots of the wave profiles is presented in Fig. 1. We note that the primary wave crest is amplified at the central region due to wave focusing as waves propagate over the convex ramp. Initially, the focusing waves, although having higher amplitude, lag behind the wave front outside of the central region. A convex wave crestline is formed. Upon entering the region of shallower water depth, the higher amplitude waves along the centerline propagate faster than the relatively smaller amplitude waves along the side walls, and eventually catch up the lower amplitude portion of the wave crestline. Then, the wave energy is redistributed along the crestline, causing the non-uniform wave amplitude sloshes back and forth across the channel. A series of secondary wave crest is emerged and propagates behind the main incident wave as shown in Fig. 1. The results present interesting phenomena of wave focusing and the coherent structure of wave propagating due to the well-balanced interplay of both nonlinear and dispersive effects.

The present model was extended to simulate wave propagation in a harbor with the layout of internal and external breakwaters. This coupled harbor geometry is shown in Fig. 2. The inner basin is protected by two perpendicular breakwaters with an opening G2. The outer basin is formed by two inclined breakwaters at an angle of θ_B. The entrance opening is defined as G1.

The evolution of a normal incident wave after it enters a coupled harbor is simulated. We select G1=0.5L and G2=L, where L denotes wave length. A time sequence of three-dimensional perspective plots of free-surface elevation is presented in Fig. 3. The wave diffraction pattern after waves encountering the breakwaters is clearly modeled.

Selected cases were simulated to compare the model results with laboratory measurements conducted by Lee (1990). Here, the amplification factor Kd at selected points in the harbor is presented for comparison. The amplification factor is defined as Kd = H/H_I, where H_I is the incident wave height. Under the condition of G1 = 0.5L, G2 = L and H_I = 0.066 h_0 (L =2m, h_0=0.5m), the simulated Kd and measured data are plotted against the longitudinal direction (x coordinate) for different transverse coordinates (e.g. y = 0, y = 0.25L and y = 0.5L) in Fig. 4. The results indicate that the simulated wave heights are in fairly good agreement with the laboratory measured data.

The numerical simulations for cnoidal waves propagating past an array of four cylinders were conducted in a domain of 0≤x≤60 and 0≤y≤30. Here, Eqs. (3) and (4) were solved to obtain the free-surface elevation. The wave parameters were taken to be

$H_I / h_O = 0.5$, $L/ h_O = 12$ and the separation distance between two adjacent cylinders is 5.07R/ h_O. R is the radius of each individual cylinder.

The free-surface contour plots for t=40 and t=90 are presented in Fig. 5. These show the diffracted wave pattern and the strong wave-wave interaction occurring for a four-cylinder configuration. The interference between scattered waves is revealed in the contour plots. We also note that a crosswise amplitude modulation appears in the main wave train. The results indicate the diffracted waves may be trapped in the region between cylinders, giving rise to a large transverse force acting on each of the transverse cylinders. The time variation of the in-line wave force on the frontal and rear cylinders for $H_I / h_O = 0.3$ is shown in Fig. 6. The force on the rear cylinder is smaller than that on the frontal cylinder due to the shielding effect by the frontal cylinder and the two transversely-oriented cylinders. The forces on the rear cylinder are affected by the complex wave interference and differ considerably from those for a single cylinder or an array of two tandem cylinders.

Conclusions

Based on the generalized Bousssinesq equations, a numerical investigation for cnoidal waves propagating in an irregular basin was conducted. The evolution of cnoidal waves and their interactions with large cylinder arrays were also studied. Numerical examples were simulated to show the stability and accuracy of the wave prediction model. The simulated wave heights are in fairly good agreement with the laboratory measured data.

References

Abul-Azm A. G. & Williams A. N. (1989). J. of Fluids and Structures, **3,** 17-36.
Chou, C.R. and Han,W.Y. (1994). Proc.,24th Int. Coast. Engrg. Conf., ASCE, New York, N. Y., 2987-3001.
Ippen, A.T. and Goda,Y.(1963). Report No.59, Hydrodynamics Lab. ,M.I.T.
Isaacson, M. and Cheung, K.F. (1992). J. Wtrwy, Port, Coast. and Oc. Engrg., ASCE, **118**(5), 496-516.
Jiang, L. & Wang, K.H. (1995). Applied Ocean Research. 17(1995),277-289.
Lee, H.S. (1990). Proc. 4th Pacific Cong. Marine Sci. & Tech. (PACON'90), University of Hawaii at Manoa, **2**,257-264.
Lee, J. J. (1971). J. Fluid Mech., **45**,375-394.
Lepelletier, T.G. and Raichlen, F. (1987). J. of Waterway, Port, Coastal and Ocean Engineering, **113**, pp. 381-400.
Lepelletier, T.G. and Raichlen, F. (1988). J. of Waterway, Port, Coastal and Ocean Engineering, **114**, pp. 1-23.
Linton, C.M. and Evans, D.V. (1990). J. Fluid Mech., **215**, 549-569.
MacCamy, R. and Fuchs, R.A. (1954). Beach Erosion Board Technical Memo 69.
Masuda K., Kato, W. & Ishizuka H. (1986). Proc. 5th Int. OMAE Conf, Tokyo, Japan, Ed. Jin S. Chung et al. , ASME, New York , 345-352.
Mciver P. & Evans D. V. (1984). Applied Ocean Research **6**, 101-107.
Wu, J. K. and Liu, Philip L.-F.(1990). J. Fluid Mech. **217**, 595-613.
Wang, K. H., Wu, T. Y. & Yates, G. T. (1992). J. of Waterway, Port, Coastal and Ocean Eng., ASCE. **118**, 551-566.
Wu, T.Y. (1981). J. Eng. Mech. Div. (ASCE), **107,** 501-522.

104 OCEAN WAVE KINEMATICS, DYNAMICS AND LOADS ON STRUCTURES

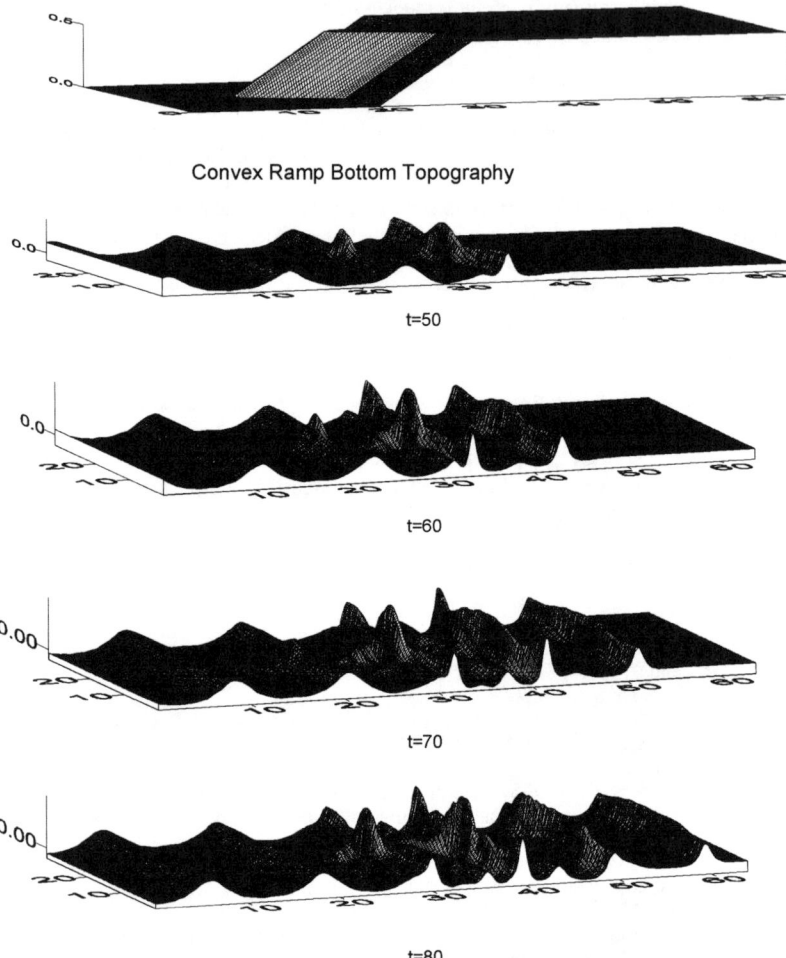

Fig.1. A time sequence of three-dimensional perspective view plots of cnoidal waves propagating past a convex ramp bottom topography.

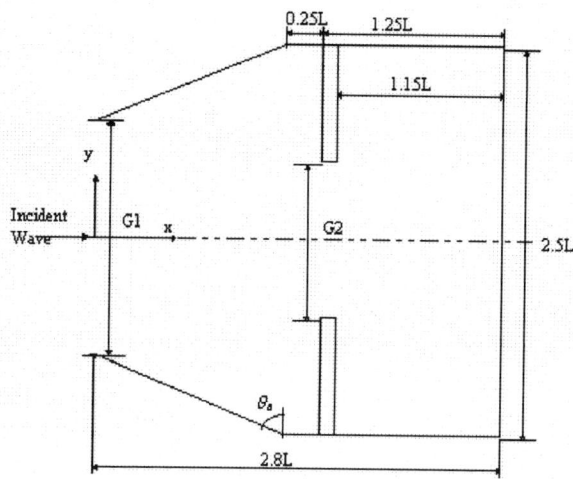

Fig.2. A coupled harbor geometry with inner and outer breakwaters.

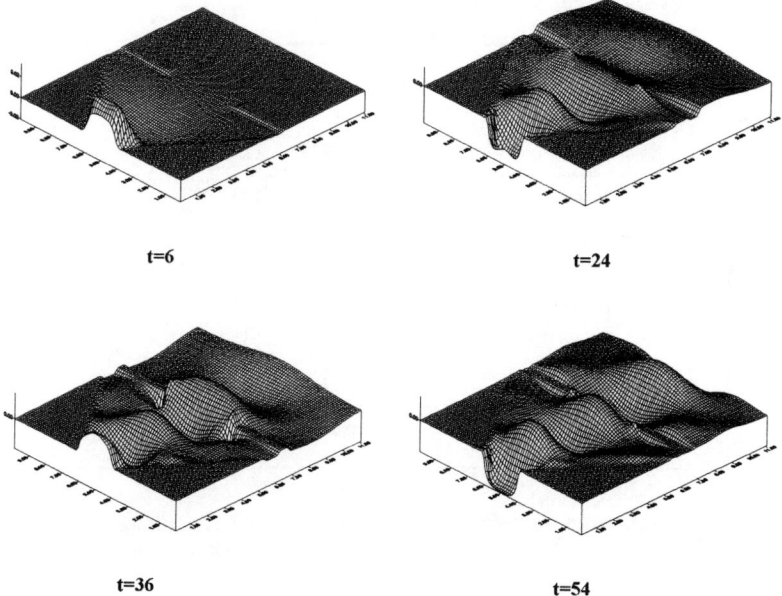

Fig.3. A time sequence of three-dimensional perspective view plots of the free-surface elevation in a coupled harbor with G1/L=0.5, G2/L=1.0

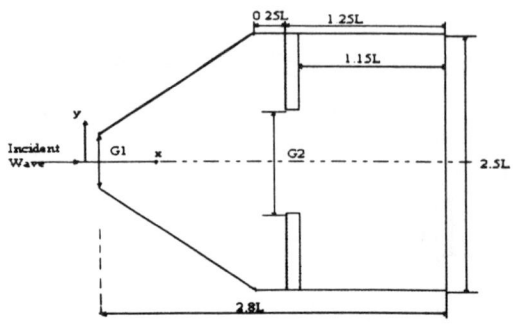

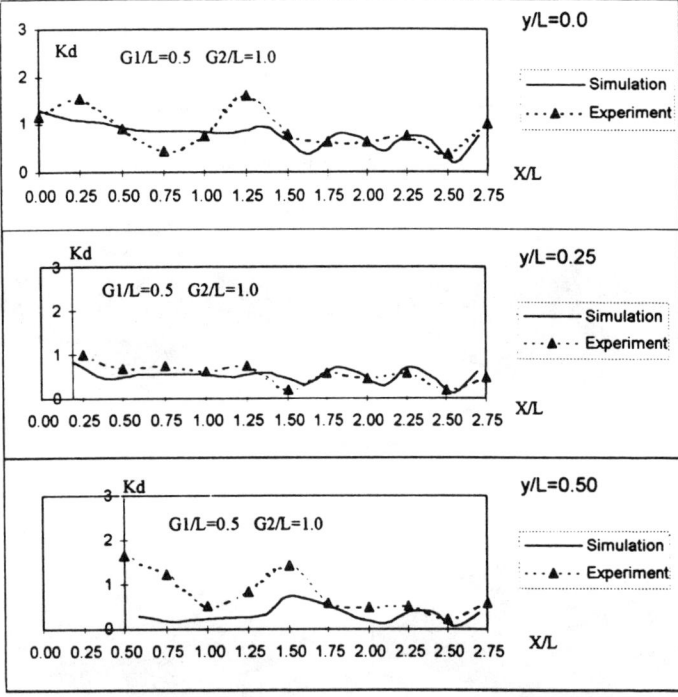

Fig. 4. Comparison of wave amplification factor obtained from present model and laboratory measurements for the case of G1/L=0.5 and G2/L=1.

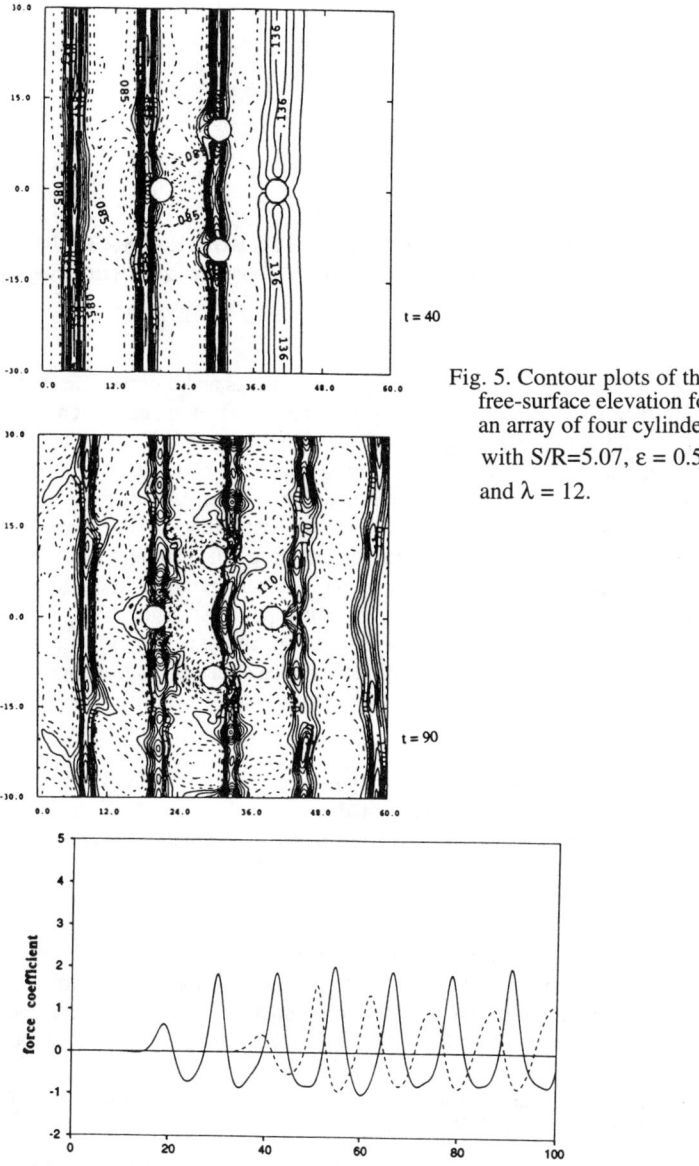

Fig. 5. Contour plots of the free-surface elevation for an array of four cylinders with S/R=5.07, $\varepsilon = 0.5$ and $\lambda = 12$.

Fig. 6. Time variations of wave forces on the two tandem cylinders of a four-cylinder configuration. The solid line represents forces on the frontal cylinder and the dashed line shows the forces on the rear cylinder.

SECOND ORDER WAVE FORCES ON 3D BODIES IN A CURRENT
Jesper Skourup[1], Kwok Fai Cheung[2], Harry B. Bingham[1] and Bjarne Büchmann[3]

Abstract

Non-linear loads on a fixed body due to waves and a current are investigated. Potential theory is assumed to be valid, and a 3D Boundary Element Method, combined with a time stepping procedure, is used to solve the problem. The exact free-surface boundary conditions are Taylor expanded about the still water level so that the solution is obtained on a time invariant geometry. A formulation correct to second order in the wave steepness and to first order in the current speed is developed. Numerical results are given for the wave drift damping coefficient and for the second order oscillatory forces on a fixed vertical circular cylinder in waves and a current. The second order oscillatory force components on the body in waves and current are new results. It is shown that the current can have a significant influence on the amplitude of the second order oscillatory forces in the diffraction regime.

Introduction

The hydrodynamic load on a body exposed to steep waves has been subject to intense study during the last couple of decades. A numerical solution to the fully nonlinear wave-structure interaction problem is very demanding, both with respect to computing time and data storage requirements. A simplification of the fully non-linear problem is generally required in order to obtain practical solutions. Many studies have used Stokes' perturbation technique (with wave steepness as the small parameter) to reduce the non-linear problem to a sequence of linear problems over a time-invariant geometry. The first order model is today standard practice in engineering design, while inclusion of second-order effects without a current has been presented by a number of authors (see e.g. Molin, 1979, Eatock Taylor and Hung,

---

[1] International Research Centre for Computational Hydrodynamics (ICCH), Danish Hydraulic Institute, DK-2970 Hørsholm, DENMARK

[2] Department of Ocean Engineering, University of Hawaii at Manoa, Honolulu, HI 96822, USA

[3] Institute of Hydrodynamics and Water Resources, Technical University of Denmark, DK-2800 Lyngby, DENMARK

1987, Kim and Yue, 1989, Lee and Newman, 1994, and Newman, 1996). Some of these authors have also applied the second order theory to the special geometry of a bottom mounted circular cylinder, which is a very common geometry in marine structures and for which a semi-analytic solution exists (Kriebel, 1990). When the diameter of the cylinder is greater than about 1/5 of the length of the waves impinging on it, flow separation and viscous effects may reasonably be omitted from the analysis. The numerical models by Eatock Taylor and Hung (1987), Kriebel (1990) and Newman (1996) are all formulated for a cylinder, while the model by Kim and Yue (1989) is applicable to general axi-symmetric bodies and the second order WAMIT model (cf. Lee and Newman, 1994) is applied to arbitrary body shapes.

Including the effects of a steady current into the analysis of the interaction between a fixed body and waves is analogous to considering a structure moving with steady forward speed in waves. Zhao and Faltinsen (1988), Nossen et al. (1991), Grue and Biberg (1993) and Teng and Eatock Taylor (1995) have developed two-dimensional and three-dimensional boundary-element methods for studying linear wave diffraction and radiation by arbitrary bodies with a small forward speed.

The typical Stokes perturbation problem can be formulated either in the time-domain or in the frequency-domain, and the above mentioned authors have all presented solutions in the frequency-domain (i.e. canonical solutions with time-harmonic forcing at one frequency, and combinations of two frequencies, which are then summed up to obtain the complete particular solution). When considering non-linear effects it can be advantageous to solve directly for the particular solution in the time-domain, as was done by Isaacson and Cheung (1992) using a boundary-element method. Calculations for the wave runup and forces on a bottom mounted circular cylinder from this method compared favourably with the frequency domain results by Kim and Yue (1989). Cheung et al. (1996) included a current in their 3D model and presented results of the wave runup and the forces on a vertical circular cylinder, correct to first order in the wave steepness and to first order in the current speed. For the cases considered they found that the current had a significant effect on the runup and on the second order steady force on the structure, while the first order oscillatory force was affected to a lesser extent by the current. These results were later confirmed by the numerical calculations by Kim and Kim (1997). An extension of the formulation by Cheung et al. (1996) was made by Büchmann et al. (1997) who included the effects at second order in wave steepness. The wave runup on a fixed vertical circular cylinder was considered, and it was found that the second order part of the solution could give a significant contribution to the maximum runup on the structure, and that the magnitude of the current speed played a major role for the runup on the cylinder. The forces on the cylinder were not considered in Büchmann et al. (1997) and this is the subject of the present paper. First and second order oscillatory forces and second order steady forces on a fixed structure due to waves and a current are computed. The first and second order force components in waves and the first order and mean second order forces in waves and current compare well with numerical results by others. The second order oscillatory forces on the structure in combined waves and current are new results.

Mathematical Formulation

A potential flow is assumed, with boundary conditions expanded up to second order and applied on the mean positions of the free surface and body boundaries (see Büchmann et al, 1997). The total velocity potential is separated into a known incident potential, ϕ_I, and a scattering potential, ϕ_S, representing the effects due to the body:

$$\phi(\vec{x},t) = \phi^{(0)}(\vec{x}) + \varepsilon \left[\phi_I^{(1)}(\vec{x},t) + \phi_S^{(1)}(\vec{x},t) \right] + \varepsilon^2 \left[\phi_I^{(2)}(\vec{x},t) + \phi_S^{(2)}(\vec{x},t) \right] + ... \quad (1)$$

In (1) ε is the perturbation parameter, $\vec{x}$ is an observation point, t is the time, the superscripts denote the order of the expansion and the zeroth order term is due to the current. The boundary value problem is formulated for the scattered field alone. All waves in the domain then become outgoing and the lateral boundaries can be defined as absorbing boundaries. The active wave absorption method used here is similar to the one used at the Danish Hydraulic Institute for wave absorption in physical flumes. The motion of a wave absorber is a function of the time history of the wave absorber position and of the free surface elevation at the wave absorber. These are transformed to an updated wave absorber position by use of a digital filter designed to match a theoretically determined transfer function. The same technique may also be used in a 3D model by considering a finite number of 2D wave absorbers placed next to each other and working independently. Each absorber is then governed by the same digital recursive filter and by the local time history of the position of the absorber and the elevation there. An extension to a fully 3D active absorption method is given in Schäffer and Skourup (1996), but it has not yet been implemented into the present version of the program. In the numerical simulations the absorbers all work in the piston mode, but digital filters are also available for hinged flap wave absorbers. The wave absorber boundary condition is of the Neumann type.

To compute the scattered potential, the boundary value problem is re-cast as a boundary integral equation via Green's 2nd identity

$$\alpha(\vec{x})\phi_s^{(k)}(\vec{x},t) = \int_\Gamma G(\vec{x},\vec{\xi}) \frac{\partial \phi_s^{(k)}(\vec{\xi},t)}{\partial n} - \phi_s^{(k)}(\vec{\xi},t) \frac{\partial G(\vec{x},\vec{\xi})}{\partial n} d\Gamma \quad (2)$$

where $\vec{\xi} = (\xi_1, \xi_2, \xi_3)$ is the position vector of an integration point situated at the boundary Γ of the domain, subscript n indicates differentiation along the outwards normal vector at $\vec{\xi}$, and the factor $\alpha(\vec{x})$ depends on the position of the observation point ($\alpha(\vec{x}) = 2\pi$ for $\vec{x}$ situated at a smooth part of the boundary). Equation (2) is discretized using a panel method with the kernel function $G(\vec{x},\vec{\xi}) = 1/|\vec{x} - \vec{\xi}|$, and the variation over a panel of both the potential and the geometry is taken to be linear. Collocation is performed at the corners of each panel, and the resulting linear system of equations is solved by LU factorisation at the first time level and then by back-substitution at each subsequent time step. The free surface boundary conditions are integrated using 4th order Adams-Bashforth and Adams-Moulton schemes. Further details concerning the mathematical formulation and the numerical solution can be found in Büchmann et al. (1997).

Forces on a Structure

The pressure, p, at the surface of the body is given by the Bernoulli equation

$$p = -\rho\left(\frac{\partial \phi}{\partial t} - \tfrac{1}{2}|\nabla\phi|^2 - gz\right) \tag{3}$$

where ρ is the density of water. The force vector components on the body are found by integrating the pressure over the instantaneously wetted body surface.

$$\vec{F} = \int_{\Gamma_c} p\vec{n}'\, d\Gamma \tag{4}$$

In (4) $\vec{n}'$ is the generalised normal vector on the body surface given as $\vec{n}' = (n_x, n_y, n_z, yn_z - zn_y, zn_x - xn_z, xn_y - yn_x)$. The vector $\vec{F} = (F_x, F_y, F_z, M_x, M_y, M_z)$ represents the components of wave forces and moments in the three translational and rotational directions, respectively. Taylor expansion of the force vector about the still water level gives

$$\vec{F} = \int_{\Gamma_c'} p\vec{n}'\, d\Gamma + \frac{1}{2}\int_{w_0} \eta p\vec{n}'\, dw + \ldots \tag{5}$$

The last integral in (5) is the water line integral around the contour of the body, Γ_c' is the body surface below still water level and w_0 denotes distance along the still waterline.

The Bernoulli equation (3) is substituted into (5) and perturbation expansions of ϕ and η are introduced. Collecting terms at each order the force vectors on the body due to waves and current at first and second order becomes:

$$\vec{F}^{(1)} = -\rho \int_{\Gamma_c'} \left(\frac{\partial \phi^{(1)}}{\partial t} + \nabla\phi^{(0)}\nabla\phi^{(1)}\right) \vec{n}'\, d\Gamma \tag{6}$$

$$\vec{F}^{(2)} = -\rho \int_{\Gamma_c'} \left(\frac{\partial \phi^{(2)}}{\partial t} + \nabla\phi^{(0)}\nabla\phi^{(2)} + \tfrac{1}{2}\nabla\phi^{(1)}\nabla\phi^{(1)}\right) \vec{n}'\, d\Gamma + \frac{\rho g}{2}\int_{w_0} \left(\eta^{(1)}\right)^2 \vec{n}'\, dw \tag{7}$$

The first and second terms in the first integral in (7) are due to the second order potential (and current) while the third term is the square of the gradient of the first order potential. The second integral accounts for the wave force due to the pressure close to the water surface and associated with the fluctuation of the free surface around the body. When the incident wave is mono-chromatic, the second order force can be separated into a steady (drift) force (i.e. the time averaged second order force) and a second order oscillatory force at twice the wave frequency.

Numerical Results

A bottom mounted surface-piercing vertical circular cylinder in finite water depth is chosen for the numerical simulations. The radius of the cylinder is a and the water depth is h. The interaction between waves and this structure has been treated frequently in the literature (both numerically and analytically) and reference results for the wave forces and the wave runup on the cylinder can easily be obtained. However, for combined waves and a current, non-linear results are scarce. The computational results presented in this paper are all obtained with a discretisation of half the domain (we exploit the symmetry of the problem) into about 3000 nodes, and the computing time for one simulation with about 800 time steps is less than 20 CPU minutes on an IBM RS6000 work station.

The forces on the cylinder are computed as function of ka for $a=h$ at Froude numbers $Fn = U/\sqrt{ga}$ between –0.1 and 0.1 (the wave number is k and the current speed is U). The first order forces are in good agreement with the results by Cheung et al. (1996). Note that the latter results were previously shown to agree well with those of Nossen et al. (1991) for the special case of a hemisphere at the free surface.

The second order steady drift force on the cylinder is calculated as the mean value from the time series of the second order force determined from (7). The second order drift force can of course also be determined on the basis of the first order results since all second order terms in (7) vanish when the time average is taken. A further check of the present model has been made by verifying that this indeed is the case. The second order drift forces also compare well with the numerical results by Chéung et al. On the basis of the steady drift force in following and opposing current (Froude numbers of –0.05 and 0.05) the wave drift damping coefficient $B_x /\rho g A^2 a$ is determined (where A is the wave amplitude). A good agreement with the frequency domain result by Grue and Biberg (1993) is seen as shown in Fig. 1.

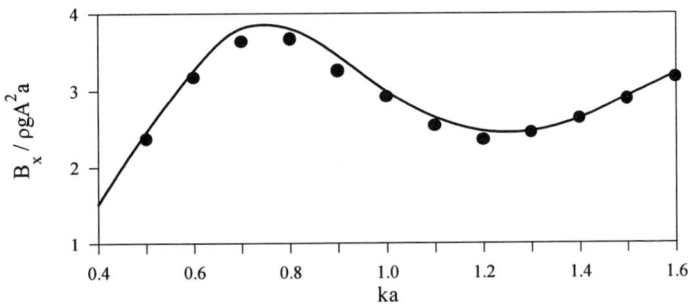

Figure 1. Wave drift damping coefficient for vertical circular cylinder as function of ka and for $h=a$. ● : present results, ——— : Grue and Biberg (1993).

The amplitudes of the second order oscillatory in-line force and of the second order overturning moment on the cylinder as functions of ka are shown in Figs. 2a, 2b, 3a and 3b for the cases of no current ($Fr=0.0$), following current ($Fr=0.05$ and 0.1) and opposing current ($Fr=-0.05$ and -0.1). A comparison with the analytical second order zero-current solution by Eatock Taylor and Hung (1987) shows good agreement.

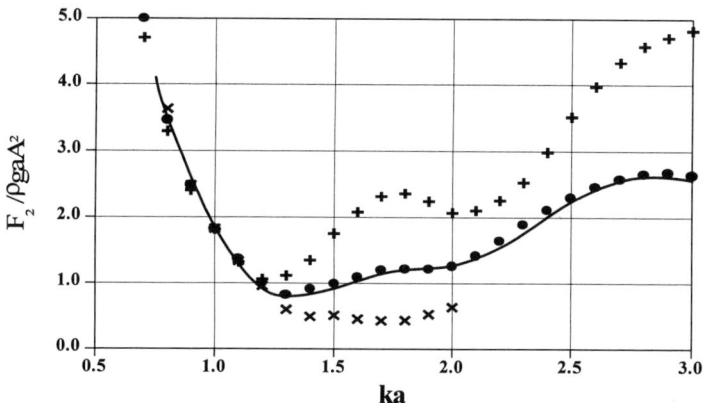

Figure 2a. Amplitude of second order oscillatory force on a vertical circular cylinder as function of ka and for $h=a$. ———— : Eatock Taylor and Hung (1987), ✕ Fr=-0.05, ● Fr=0.00 (no current), + Fr=0.05.

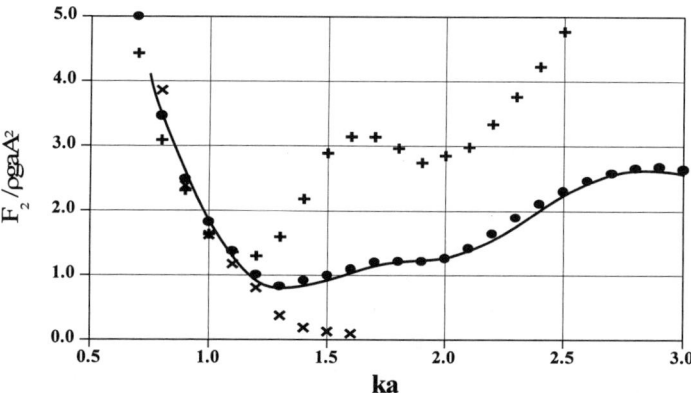

Figure 2b. Amplitude of second order oscillatory force on a vertical circular cylinder as function of ka and for $h=a$. ———— : Eatock Taylor and Hung (1987), ✕ Fr= -0.10, ● Fr=0.00 (no current), + Fr=0.10.

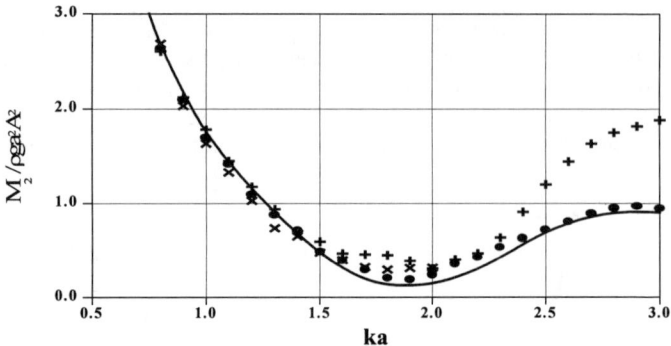

Figure 3a. Amplitude of second order oscillatory overturning moment on vertical circular cylinder as function of ka and for $h=a$. ——— : Eatock Taylor and Hung (1987), × Fr=-0.05, ● Fr=0.00 (no current), + Fr=0.05.

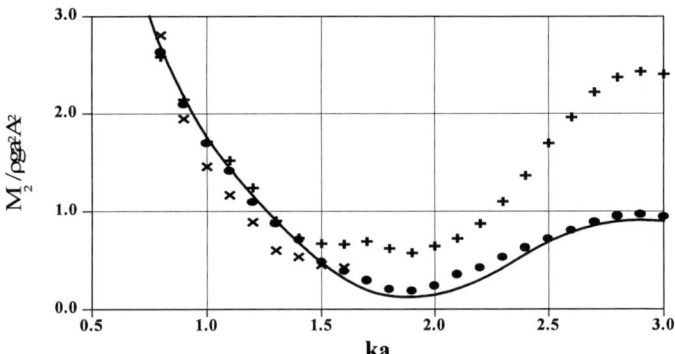

Figure 3b. Amplitude of second order oscillatory overturning moment on vertical circular cylinder as function of ka and for $h=a$. ——— : Eatock Taylor and Hung (1987), × Fr=-0.10, ● Fr=0.00 (no current), + Fr=0.10.

The amplitude of the second order oscillatory force on the cylinder in an opposing current with $Fr=-0.1$ cannot be evaluated when ka exceeds about 1.6. This is explained by the "$\tau=1/4$ effect". This is the point where the linear group velocity of the scattered waves is equal to the current speed, or $\tau=\omega_0 U/g=\tau_c$. The critical oscillation frequency is a function of the water depth, but in deep water $\tau_c=1/4$. For a Froude number of -0.1, it is found that $ka=1.56$ at $\tau=1/4$ for the free second harmonic of the incident wave, limiting the range of ka for this case. In the results with following current the amplitude of the second order oscillatory force is seen to deviate from the zero-current case when ka exceeds about 1.2. For higher values of ka the influence on the force amplitude due to the current is seen to become substantial.

Conclusions

A time-domain 3D boundary element method is developed to study the interaction between waves and a current with a fixed structure. The problem is formulated over a time invariant geometry using Taylor expansions of the free surface conditions and perturbation techniques for the variables of the problem. A formulation correct to second order in the wave steepness and to first order in the current speed is given. The first and second order oscillatory forces and the second order mean force on a fixed vertical circular cylinder in waves and a current are determined. The results for the amplitude of the second order oscillatory forces and moments in waves and current are new. It is shown that the current can have a significant influence on the amplitude of the second order oscillatory forces. These are found to deviate significantly from the second order forces in waves alone when the ratio of structure dimensions to the length of the incident waves falls within the diffraction regime. A comprehensive parameter study will be published in the future.

Acknowledgements

This work was funded by the Danish National Research Foundation.

References

Büchmann, B, Skourup, J and Cheung, KF (1997). Runup on a structure due to waves and current. Proc.7th Int. Offshore and Polar Engineering Conf., ISOPE-97, Honolulu, USA,vol. 3, 48-55.

Cheung, KF, Isaacson, M and Lee, JW (1996). Wave diffraction around a three-dimensional body in a current. ASME *Journal of Offshore Mechanics and Arctic Engineering*, 118(4), 247-252.

Eatock Taylor, R and Hung, SM (1987). Second order diffraction forces on a vertical cylinder in regular waves. *Applied Ocean Research*, 9(1), 19-30.

Grue, J and Biberg, D (1993). Wave forces on marine structures with small speed in water of restricted depth. *Applied Ocean Research*, 15(3), 121-135.

Isaacson, M and Cheung, KF (1992). Time domain second-order wave diffraction in three dimensions. *Journal of Waterway, Port, Coastal and Ocean Engineering*, 118(5), 496-516.

Kim, DJ and Kim, MH (1997). Wave-current-body interaction by a time-domain high-order boundary element method. Proc. 7th Int. Offshore and Polar Eng. Conf., ISOPE-97, Honolulu, USA, vol. 3, 107-115.

Kim, MH and Yue, DKP (1989). The complete second-order diffraction solution for an axisymmetric body. Part 1. Monochromatic incident waves. *Journal of Fluid Mechanics*, 200, 235-264.

Kriebel, DL (1990). Nonlinear wave interaction with a vertical circular cylinder. Part I: Diffraction theory. *Ocean Engineering*, 17(4), 345-377.

Lee, C-H and Newman, JN (1994). Second-order wave effects on offshore structures. Proc. 7nd Int. Conf. Behaviour of Offshore Structures, BOSS-94, MIT, Boston, USA, vol 1. 133-146.

Molin, B (1979). Second order diffraction loads upon three-dimensional bodies. *Applied Ocean Research*, 1(4), 197-202.

Newman, JN (1996). The second order wave force on a vertical cylinder. *Journal of Fluid Mechanics*, 320, 417-443.

Nossen, J, Grue, J and Palm, E (1991). Wave forces on three-dimensional floating bodies with forward speed. *Journal of Fluid Mechanics*, 227, 135-160.

Schäffer, HA and Skourup, J (1996). Active absorption of multidirectional waves. Proc. of the 25th Int. Conf. on Coastal Engineering, Orlando, Florida, USA, vol.1, 55-66, ASCE, New York, 1997.

Teng, B and Eatock Taylor, R (1995). Application of a higher order BEM in the calculation of wave run-up on bodies in a weak current. *Int. Journal of Offshore and Polar Engineering*, 5(3), 219-224.

Zhao, R and Faltinsen, O (1988). Interaction between waves and current on a two-dimensional body in the free surface. *Applied Ocean Research*, 10(2), 87-99.

Numerical Simulation of Wave Effects on A Combined Breakwater and Platform Configuration

Hamn-Ching Chen,[1] Member, ASCE, and Woei-Min Lin[2]

Abstract

A new 3-D potential flow numerical method has been developed to study nonlinear wave diffraction around complex geometries. This new method is based on a multi-block finite-analytic scheme. A chimera domain decomposition technique is used to model complex geometry and to connect overlapped grids by interpolating information across block boundaries. Three numerical examples, including waves around a combined breakwater and large floating platform, are presented in this paper to illustrate the effectiveness of this new method.

Introduction

Very large floating structures in the coastal area have recently been considered for use as an airport. The platform can be on the order of several thousand meters long. This type of platform typically requires the protection of a breakwater from severe weather conditions. For the design of a large platform, the wave field behind the breakwater is one the most important design considerations. The breakwater must be designed to achieve the specified design waves for the platform. Therefore, the breakwater efficiency has to be established carefully.

Computational methods are available for computing the propagation of a typhoon wind/wave from the open ocean to the shore and to compute the relationship between the wave field outside and inside the breakwater. However, these methods typically ignore the interaction between the breakwater and the platform. Furthermore, large-domain computations for this type of platform are still not practical in actual designs due to geometry complexity and prohibitively high computation cost.

In this study, an efficient new 3-D unsteady numerical method, CHAMPS (CHimera finite Analytic Method Potential-flow Solver), was developed to com-

---

[1] Associate Professor, Department of Civil Engineering, Texas A & M University, College Station, TX 77843
[2] Senior Scientist, Ship Technology Division, Science Applications International Corporation, 134 Holiday Court, Suite 318, Annapolis, MD 21401

pute the wave mitigation behind a breakwater and the wave field around a combined breakwater and platform configuration. This new method solves the Laplace equation on structured multi-block grids using the finite-analytic method of Chen, Patel, and Ju (1990). Within each computational block, velocity potential was solved on a general curvilinear, body-fitted coordinate system. This method is an extension of Chen and Lee (1996, 1998) and Lee and Chen (1996) for calculations of fully nonlinear free surface flows generated by bodies moving in calm water.

For body-wave interaction problems, the incident ambient wave field is generated by a numerical wave-maker. For fully nonlinear wave cases, the nonlinear free surface boundary conditions can be enforced on the exact free surface for accurate resolution. For weakly nonlinear wave cases, the nonlinear free surface boundary conditions may be specified on the mean free surface to reduce the simulation time. Wave effects on the structure can be calculated by the unsteady pressure field. This information can be used for further structural analysis.

For modeling geometries, a chimera domain decomposition technique has been incorporated to connect embedded, patched, or overlapped grids by interpolating information across block boundaries. The chimera scheme provides a very effective and flexible treatment of complex geometries and flow conditions in multiply-connected flow domains, since each component grid can be generated separately and linked in arbitrary combinations. Moreover, selective grid refinements can be efficiently performed in areas of high gradients without a significant increase in total CPU time. The method is very general and can model both bottom topography and shorelines in addition to physical structures. More detail on the numerical method is given in the next two sections of this paper.

Three examples are given in this paper to demonstrate the effectiveness of this new method. The first example is a classical diffraction problem of waves through a single gap in a breakwater. The second example is the wave diffraction behind a finite width breakwater. The third example is a combined breakwater and floating platform configuration. The third example clearly illustrates seen the necessity of using a chimera scheme for design assessment of different breakwater and platform arrangements.

Mathematical Formulation

For the simple breakwaters and combined breakwater and platform configurations considered here, the associated boundary value problem can be constructed using a Cartesian coordinate system $x^i = (x, y, z)$. The regular waves generated by the wavemaker are traveling in the negative-x direction, and the z-axis is pointing upward with $z = 0$ representing the mean water surface. It is assumed that the fluid is incompressible and inviscid and that the flow is irrotational. This implies the existence of a velocity potential function $\phi(x, y, z, t)$ such that the velocity can be described by $\vec{V} = \nabla \phi$. Due to the conservation of fluid mass, the velocity potential ϕ satisfies the Laplace equation:

$$\nabla^2 \phi = 0 \tag{1}$$

Also, the momentum equations reduce to Bernoulli's equation

$$p + \phi_t + \frac{1}{2}\nabla\phi \cdot \nabla\phi = 0 \tag{2}$$

where p is the dynamic pressure.

On the free surface, the normal velocity of a point on the wave surface should be equal to the normal velocity of a fluid particle at that point. This kinematic free surface boundary condition can be expressed as

$$\eta_t + \phi_x \eta_x + \phi_y \eta_y - \phi_z = 0 \quad on \ z = \eta \tag{3}$$

where η is the wave elevation. For the inviscid fluid, the dynamic free surface condition requires that the normal pressure on the free surface be equal to the atmospheric pressure. When both the viscosity and surface tension are neglected, the dynamic condition on the exact free surface can be written as

$$\phi_t + \frac{1}{2}\nabla\phi \cdot \nabla\phi + \frac{\eta}{Fr^2} = 0 \quad on \ z = \eta \tag{4}$$

where Fr is the Froude number $U_o/\sqrt{gL}$ based on a reference velocity U_o, a reference length L, and the gravitational acceleration g.

On the wetted body surface, the normal velocity of a point on the body must be equal to the normal velocity of the adjacent fluid, $\vec{V} \cdot \vec{n} = \vec{V_s} \cdot \vec{n}$, where $\vec{V}$ and $\vec{V_s}$ are the fluid and body velocities, respectively, and $\vec{n}$ is the unit normal of the body surface. The same Neumann boundary conditions has also been used for the bottom and side wall boundaries. On the upstream boundary, the velocity potential for the incident wave is specified:

$$\phi = \sum_{n=1}^{\infty} \frac{A_n}{\omega_n} \frac{cosh k_n(h+z)}{cosh k_n h} \sin\left[k_n(x\cos\theta_n + y\sin\theta_n) - \omega_n t + \varepsilon_n\right] \tag{5}$$

where A_n is the wave amplitude, h is the water depth, k_n is the wave number, ω_n is the wave frequency, θ_n is the wave heading, and ε_n is the phase shift.

In the far field, a radiation condition or absorbing beach must be employed to avoid wave reflections from the downstream boundary. Open boundaries enclosing the fluid domain are artificial and essentially arbitrary. The fluid domain is truncated at some distance from the fluid region of interest. Physically, the aim of the radiation boundary condition is to eliminate unphysical wave reflections from the open boundaries; otherwise, those reflected waves would interact with the waves inside the solution domain and contaminate the solution. Several methods, such as the Rayleigh viscosity method, the sponge layer method (Chan,

1977), and Sommerfeld's radiation condition (Orlanski, 1976; Chan, 1977), have been considered for the handling of radiation conditions on open boundaries. In this study, Sommerfeld's radiation condition has been used to eliminate wave reflection from the downstream boundary.

Chimera Laplace Solver

In the present chimera Laplace method, the solution domain is first decomposed into a number of computational blocks. The body-fitted numerical grids for different part of the geometry are generated separately. The PEGSUS program of Suhs and Tramel (1991) was employed to determine the interpolation information for linking grids. It outputs a complete set of information corresponding to the interpolation performed. The most important information is included in a "blanking" array, the interpolation and boundary lists, the interpolation stencils, and the maps showing boundary and interpolation node locations. The "blanking" array and interpolation stencils are incorporated into the present flow solver to remove hole points and update boundary conditions from the linking grids.

Within each computational block, the finite-analytic method of Chen, Patel, and Ju (1990) was employed to solve the Laplace equation for velocity potential on a general curvilinear, body-fitted coordinates system. For fully nonlinear free surface flows, the wave elevation was updated for each time step using the kinematic free surface boundary condition (3) while the dynamic free surface boundary condition (4) had been enforced on the exact free surface. In the present study, however, we shall simplify the solution procedure by applying the nonlinear free surface boundary conditions (3) and (4) on the mean free surface. Since the computational domain and numerical grids remain fixed under this approximation, it enables us to use the same geometric coefficients and PEGSUS interpolation stencils throughout the entire simulation.

In addition to the first-order Euler implicit method used in Chen and Lee (1996) for steady-state wave calculations, the CHAMPS code also incorporated several higher-order integration schemes that are more appropriate for time-domain wave simulations. In particular, we have performed extensive calculations using the fourth-order Adams-Bashforth-Moulton predictor-corrector method, the classical fourth-order Runge-Kutta method, and the fifth-order Runge-Kutta-Fehlberg method to evaluate the general performance of these higher-order methods. The results indicated that the Runge-Kutta-Fehlberg method is the most stable and accurate time-integration scheme for wave-body interaction problems considered here. Therefore, we shall present only the results obtained from the Runge-Kutta-Fehlberg method in the next section.

Results and Discussions

Three numerical simulation results are given in this section to demonstrate the effectiveness of this new method. The first example is a classical diffraction problem of waves through a single gap in a breakwater. This case is similar to wave entrance into a harbor. In this computation, the water depth is set to be constant at 25 meters. The length of the breakwater is 500 meters each side and the harbor opening is 250 meters. The total computation domain is 1250 meters in length and 1250 meters in width.

Figure 1(a) shows the computational grid on the free surface. For each one of the breakwater, there are 30,000 cells (100 x 30 x 10). The entire background computation domain has 144,000 cells (240 x 60 x 10). The advantage of the multi-block chimera scheme is quite obvious here. The background grid and the two breakwater grids intersect each other, and the chimera scheme can automatically blank out the unnecessary grids and perform solution interpolation across the block boundaries.

For actual simulation, a shallow water plane progress wave is sent in from the right-hand-side of the fluid domain. The wave length in this case is 125 meters (half of the harbor opening) and the wave amplitude is 0.5 meters. Figures 1(b) to 1(f) shows the wave diffraction at different time instances of the simulation. Figures 1(c) to 1(f) are plotted with an interval of quarter wave period T. The existence of standing waves in front of the breakwater can be seen clearly by the interesting wave superposition and cancellation phenomena of the reflection and incident waves. The numerical simulation results are quite similar to the classical wave diffraction results of Penny and Price (1952).

The second example is a simple breakwater of rectangular cross-section resting on a bottom of constant water depth at 25 meters. This example is used to demonstrate the efficiency of a finite width breakwater. The breakwater is 25 meters wide and 1000 meters long in this computation. The computation domain is 2500 meters x 2500 meters. Again, the chimera scheme is used with 60,000 cells (200 x 30 x 10) around the breakwater and 144,000 cells (240 x 60 x 10) on the background computation domain.

As shown in Figure 2, a regular plane progressive wave (wave length of 250 meters) is coming from the right (a direction normal to the breakwater). The efficiency of the breakwater, the interaction of the diffraction, reflection, and the incident wave system are very evident. Figures 2(d) to 2(f) are plotted at a interval of quarter wave period T. Again, the superposition and cancellation between the incident and the reflection waves can be seen clearly. In addition, ring waves are formed in front of the breakwater.

The third example is a combined platform and breakwater configuration. As shown in Figure 3(a), the geometry for this combined model is a breakwater in three sections and a large floating platform behind it. The total length of the breakwater is 2500 meters and the floating platform is 1200 meters x 360 meters x 4 meters. The water depth is constant at 20 meters in this computation. In a typical near-shore airfield configuration, there is shoreline behind the platform, although a shoreline is not modeled in this computation.

The complete computation domain is 4800 meters x 4200 meters. Only the grids near the breakwater and the floating platform are shown in Figure 3(a). Five overlapped blocks were used for this complex configuration. The background grid block covering the entire computation domain has 200 x 100 x 10 cells; the second grid block is around the breakwater and has 194 x 34 x 10 cells; the third grid block is around the platform and has 122 x 14 x 10 cells; the fourth grid block is below the platform and has 122 x 16 x 8 cells; and the fifth grid block is a branch cut of the grid block below the platform and has 60 x 3 x 9 cells. These five grids intersect each other and the chimera scheme was used in the CHAMPS method to automatically block out the intersecting grid points in the calculation.

Figures 3(b) - 3(f) present the result of a wave diffraction computation with a plane progressive wave of 600 meters (wave length) x 4 meters (wave height) sent in from one side of the breakwater. Figures 3(b), 3(c), 3(e), and 3(f) show an animation sequence of wave diffraction in front and behind the breakwater as well as the interaction between the breakwater and the large platform. The result gives a clear picture of the wave propagation along the breakwater and the formation of the ring wave due to the interaction of the incident and reflection waves. The most severe wave effect on the platform is on the upper left-hand corner of the platform. The wave height there is about 1/3 of the incident wave height.

A particular interesting result is given in Figure 3(d), where the velocity potential in the flow domain at $t/T = 12.0$ is given. This top view clearly shows that the "ring" waves are actually very close to square in shape. This is the result of interaction between the incident waves and reflection waves by the 45 degree breakwater (the longest piece of breakwater). The dark black lines around the floating platform and around the conbined configuration are the boundaries of the overlapped blocks. The continuity of the solution across the block boundaries can be clearly seen.

Conclusion

A new 3-D potential flow numerical method is presented for studying nonlinear wave diffraction around complex geometries. With the chimera domain decomposition technique, the modeling of complex geometries in advanced numerical simulations is now becoming practical. The examples presented in this study demonstrate the effectiveness of this new numerical method. Some interesting wave interactions are observed, including superposition and cancellation of incident and reflection waves, and the formation of squared-shaped waves. For design assessment of the large floating platform, this method provides complete flow field information, including the interaction between breakwater and platform. Wave loads on the platform can be computed from the pressure field and hydro-elastic effects of the platform can be examined.

Acknowledgments

The research is partially funded by an IR&D project of Science Applications International Corporation. We would like to thank Mr. Kenneth Weems for his help on the preparation of this paper. All computations were performed on the CRAY C-90 and YMP of Cray Research Inc. at Eagen, Minnesota, under the sponsorship of Mr. Chris Hempel.

References

1. Chan, R.K.C. (1977), "Finite-Difference Simulation of the Planar Motion of a Ship," *Proceedings of the 2nd International Conference on Numerical Ship Hydrodynamics*, pp. 39-52, University of California, Berkeley, CA.
2. Chen, H.C., Patel, V.C., and Ju, S. (1990), "Solutions of Reynolds-Averaged Navier-Stokes Equations for Three-Dimensional Incompressible Flows," *Journal of Computational Physics*, Vol. 88, No. 2, pp. 305-336.
3. Chen, H.C., and Lee, S.K. (1996), "Interactive RANS/LAPLACE Method for Nonlinear Free Surface Flows," *Journal of Engineering Mechanics*, Vol. 122, No. 2, pp. 153-162.
4. Chen, H.C., and Lee, S.K. (1998), "RANS/LAPLACE Simulations of Nonlinear Waves Induced by Surface-Piercing Bodies," *Journal of Engineering Mechanics*, accepted.
5. Lee, S.K., and Chen, H.C. (1996), "A Multiblock RANS/LAPLACE Coupling Method for Viscous Fully Nonlinear Body Wave Problems," *Hydrodynamics, Theory and Applications*, edited by A.T. Chwang, J.H.W. Lee, and D.Y.C. Leung, Vol. I, pp. 41-46, A.A. Balkema Pubishers, Rotterdam.
6. Orlanski, I. (1976), "A Simple Boundary Condition for Unbounded Hyperbolic Flows," *Journal of Computational Physics*, Vol. 21, pp. 251-269.
7. Penney, W.G., and Price, A.T. (1952), "The Diffraction Theory of Sea Waves and the Shelter Afforded by Breakwater," Philos, Trans. Roy. Soc. A, Vol. 244 (822), pp.236-253.
8. Suhs, N.E., and Tramel, R.W. (1991), "PEGSUS 4.0 Users Manual," Arnold Engineering Development Center Report, AEDC-TR-91-8, Arnold Air Force Station, TN.

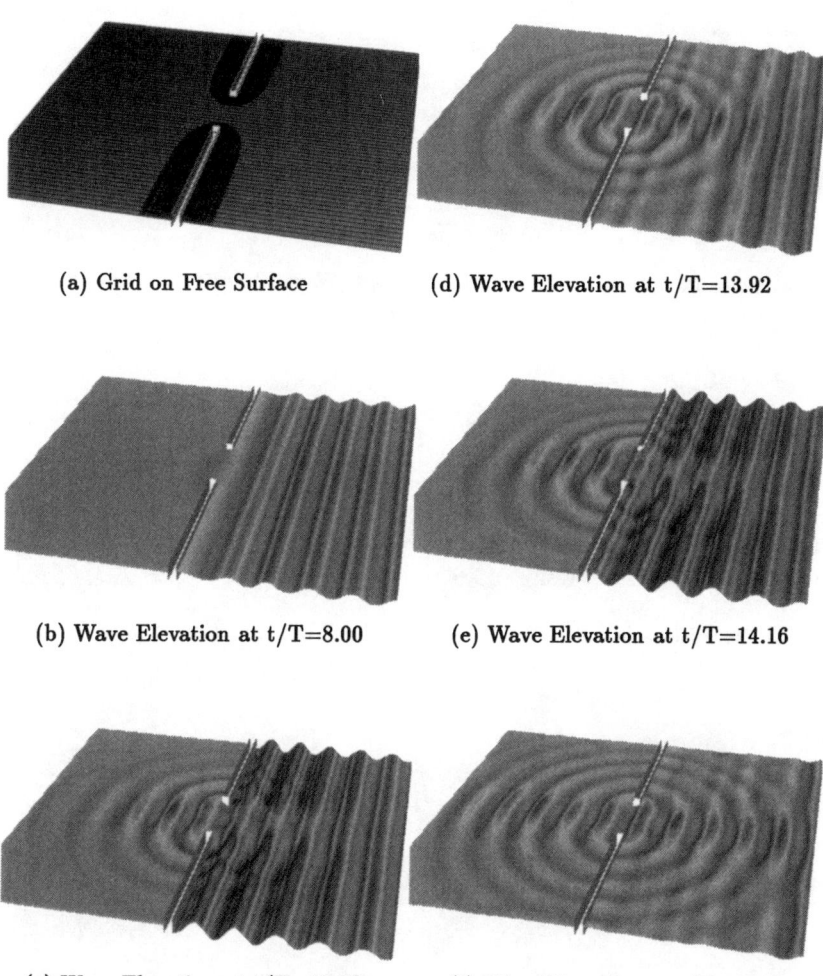

(a) Grid on Free Surface
(b) Wave Elevation at t/T=8.00
(c) Wave Elevation at t/T=13.66
(d) Wave Elevation at t/T=13.92
(e) Wave Elevation at t/T=14.16
(f) Wave Elevation at t/T=16.92

Figure 1: CHAMPS Simulation of Wave Diffraction Through a Single Gap in a Breakwater

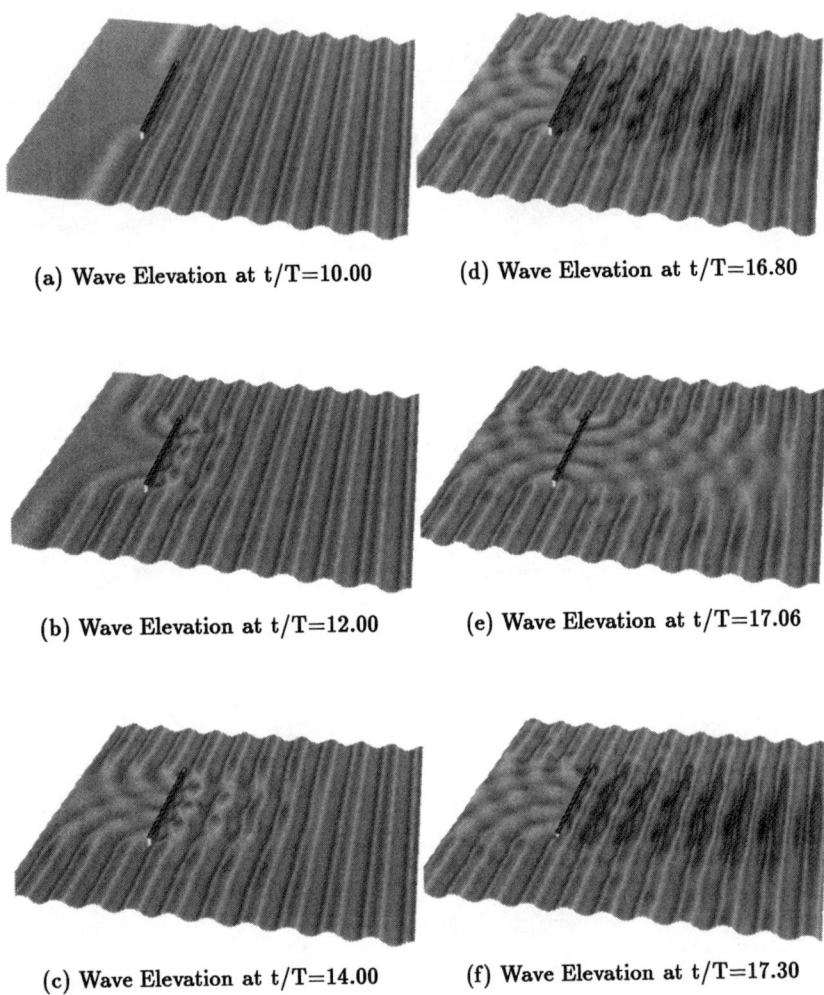

(a) Wave Elevation at t/T=10.00

(b) Wave Elevation at t/T=12.00

(c) Wave Elevation at t/T=14.00

(d) Wave Elevation at t/T=16.80

(e) Wave Elevation at t/T=17.06

(f) Wave Elevation at t/T=17.30

Figure 2: CHAMPS Simulation of Wave Diffraction Behind a Finite Width Breakwater

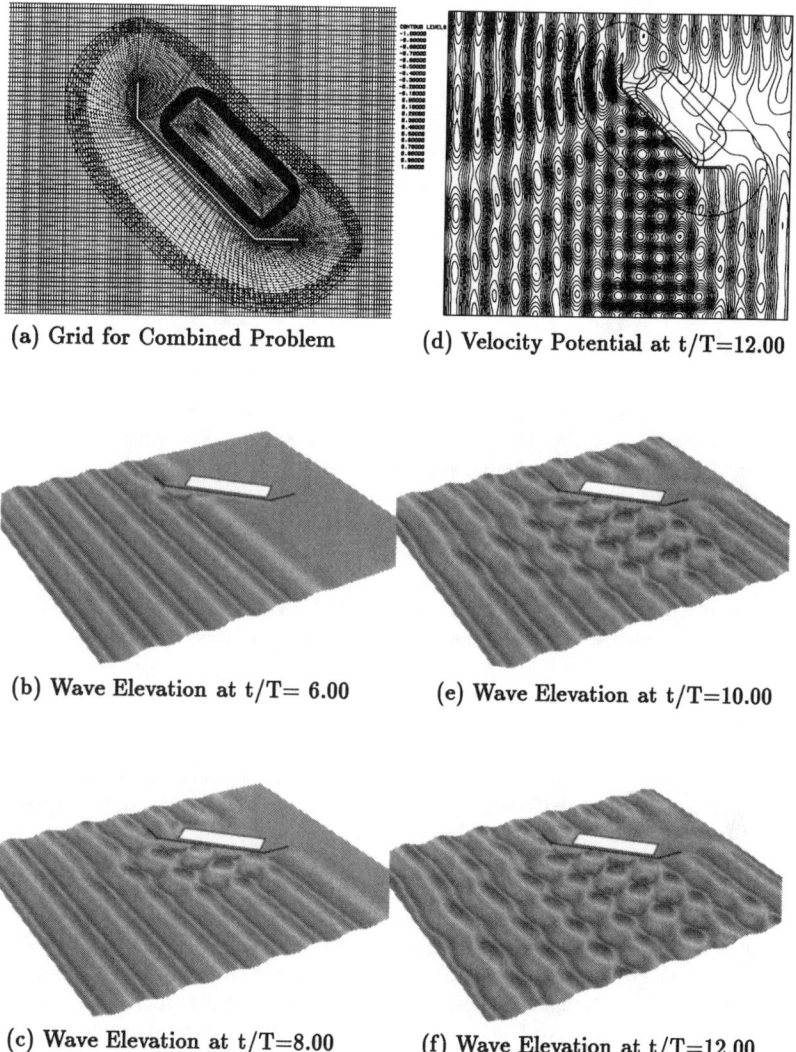

(a) Grid for Combined Problem

(d) Velocity Potential at $t/T=12.00$

(b) Wave Elevation at $t/T= 6.00$

(e) Wave Elevation at $t/T=10.00$

(c) Wave Elevation at $t/T=8.00$

(f) Wave Elevation at $t/T=12.00$

Figure 3: CHAMPS Simulation of Wave Field for a Combined Breakwater and Platform Configuration

Nonlinear Inertial Loading in Steep 2-D Water Waves

C. Swan[*], T. Bashir[*] & O.T. Gudmestad[+]

Abstract

The nonlinear potential loading on a vertical surface piercing cylinder subject to large waves is considered. The water particle accelerations arising beneath an extreme 2-D wave are investigated, and the results show that an effective wave model must incorporate both the nonlinearity and the unsteadiness of a sea state. The implications of these results for the estimation of fluid loads is then addressed. Comparisons between measured and predicted data confirm that the accuracy of any force calculations is largely dependent upon the ability of a wave model to accurately describe the unsteady water particle accelerations. This result is important for the prediction of nonlinear or high-frequency forces which are of particular concern in the design of some deep-water structures.

Introduction

This paper describes a study into the nonlinear accelerations arising beneath steep 2-D water waves and seeks to provide an improved description of the horizontal forces acting on a vertical, surface-piercing, cylinder. In particular, the study addresses the case of a large volume structure, but one which is not sufficiently large to cause significant diffraction of the incident wave field. As a result, the body lies within the drag-inertia regime, with the forcing dominated by potential flow loads. In recent years, structures lying within this range have been the subject of considerable attention. This interest has arisen because a combination of model test data and field measurements have suggested that such structures may be subject to transient structural deflections occurring at frequencies well above those associated with the incident wave field. This phenomena, which is commonly referred to as "ringing", is strongly correlated with the steepest waves in a sea state. The present study contributes to the on-going discussion of "ringing" in that it seeks to define the high-frequency forces believed to be responsible for this response. To date, these forces

---

[*]Dept. of Civil Engng., Imperial College, London & [+] Statoil, Stavanger, Norway.

cannot be accurately predicted. Indeed, although this forcing is undoubtedly nonlinear, the origins of this nonlinearity remain unclear. For example, there is uncertainty as to the relative importance of :
(a) The accuracy of the water particle kinematics models used to define the steepest waves within an irregular or random sea state.
(b) The accuracy of the applied force models and the need to include the nonlinear forces recently identified by Rainey (1995) and Faltinsen et al. (1995).

The present paper will address this uncertainty and will present the results of two separate laboratory investigations. The first concerns the prediction of the velocities and accelerations arising beneath an extreme 2-D wave; while the second considers the horizontal forces acting on a vertical surface piercing cylinder. By combining these results, further insight is provided into the origins of the nonlinearity responsible for the generation of high-frequency 'ringing' forces.

Study I: Kinematic Measurements

The purpose of this investigation was to examine the nonlinear water particle accelerations arising beneath an extreme transient wave. A new set of experimental observations were undertaken within a large deep-water wave channel located in the Civil Engineering Department at Imperial College. Within this facility the water surface elevation was measured using an array of surface piercing wave gauges. Each gauge provides a time-history of the water surface elevation, $\eta(t)$, at a fixed spatial location, and has an accuracy of ±0.5mm. In addition, the water particle velocities were measured, with an accuracy of ±2%, using a laser Doppler anemometer (LDA).

In line with Baldock et al. (1996) the wave group chosen in the present study was based upon a relatively narrow banded spectrum consisting of 50 wave components which were equally spaced in the period range $0.8 \leq T \leq 1.2s$, and were of equal amplitude such that the power spectrum, based on the input signal sent to the wave paddle, decays according to ω^{-4}. A linear analysis suggests that an extreme wave group, measured in terms of the wave slope, arises due to the summation (or focusing) of the zero up-crossings. Generation of the desired wave form, at the centre of the measuring section (13m downstream of the wave paddle), simply required an iterative procedure whereby a linear signal was sent to the wave paddle and the phasing of the individual components adjusted until a focused event was achieved. Further details of this experimental study are provided by Swan et al. (1998).

Discussion of Kinematics Data

To investigate the importance of both the nonlinearity and the unsteadiness of the extreme wave, the measured data is compared to three kinematics models. The first represents a nonlinear steady wave solution (a Stokes' model) based on an equivalent wave height (H) and wave period (T); while the second represents an empirically corrected (Wheeler, 1970) linear random wave theory based on a Fourier transform of the water surface elevation measured at the focal position. These models are representative of typical design solutions in that the first incorporates the nonlinearity but neglects the unsteadiness, while the second includes the unsteadiness but neglects the nonlinearity. In contrast, the third solution is both fully nonlinear and unsteady, and is based upon a time-stepping procedure originally developed by Fenton and Rienecker (1980) and further developed by Johannessen and Swan (1998).

Figure 1a describes the time-history of the water surface elevation, $\eta(t)$, measured at the focal position. Each of the three wave models, noted above, are in reasonable agreement with the measured data. However, both the linear random wave theory and the steady wave theory essentially represent no more than a 'best fit' to the water surface, although in the latter case the periodic constraint leads to very large errors away from the leading face of the largest wave. In contrast, the time-stepping solution is based upon the underlying free waves and thus the agreement between this solution and the measured data is significant. Figure 1b concerns the depth variation in the maximum horizontal velocities arising beneath the largest wave crest (u(z) at t=0.18s). These results confirm that a good fit to the measured water surface elevation is not sufficient to provide a reliable estimate of the underlying velocities. Indeed, figure 1b clearly shows that only the time-stepping solution provides a good description of the measured data throughout the water column. Figure 1c concerns the depth variation in the unsteady component of the horizontal acceleration arising directly beneath the focus location (x=0, t=0). This data was produced by curve fitting and numerically differentiating the time-history of the horizontal velocities recorded at x=0. Although this data shows considerable scatter, the time-stepping solution again provides the best description.

Having demonstrated that both the nonlinearity and the unsteadiness of the wave must be included if reliable estimates of the water particle kinematics are to be achieved, the importance of the present results in relation to the estimation of fluid loading is examined. To determine the significance of the various nonlinear contributions, we sought to define the forces acting on a single vertical cylinder, of diameter D=0.1m, which extends from the bed up through the water surface. In relation to the extreme wave identified in figure 1a, the cylinder has a diameter to wavelength ratio of $D/\lambda=0.1$, and a corresponding Keulegan-Carpenter number of UT/D=3.0. In this case the following loads are applicable:

$$F_{I(t)} = \int_{-d}^{\eta} C_M \rho \frac{\pi D^2}{4} \frac{du}{dt} dz \quad (1) \qquad F_{I(x)} = \int_{-d}^{\eta} C_M \rho \frac{\pi D^2}{4} \left(\frac{udu}{dx} + \frac{wdu}{dz} \right) dz \quad (2)$$

$$F_{AD} = \int_{-d}^{\eta} M_x u \frac{dw}{dz} dz \quad (3) \qquad F_{SI} = \frac{-1}{2} \frac{d\eta}{dx} M_x u^2 \quad (4) \qquad F_{SD} = \frac{7}{2g} \frac{du}{dt} M_x u^2 \quad (5)$$

where d is the water depth, ρ is the density, and M_x the added mass per unit length.

These forces respectively correspond to the standard unsteady Morisons' inertia load; the convective Morisons' inertia load; the second-order axial divergence load; and the third-order surface intersection and surface distortion forces. Whilst the first two forces are commonly applied in design practice, the remaining loads correspond to the additional nonlinear forces discussed by Rainey (1995) and Faltinsen et al. (1995).

Figure 2a concerns the force spectra produced by taking a fast Fourier transform of the time-history of the total base-shear predicted using the Wheeler solution and the time-stepping solution. This figure clearly suggests that if the fluid motion is predicted by an appropriate nonlinear wave model, large fluid loads (accounting for approximately 10% of the maximum) arise at high frequencies within the range 2-3 times the peak spectral frequency. In contrast, the force spectra based upon a Wheeler solution predicts significantly lower forces within this range (approximately 1% of the maximum). Furthermore, figure 2b provides four separate force spectra, each calculated using the time stepping procedure, corresponding to the unsteady inertia force, the convective inertia force, the axial divergence force, and the total free surface force (ie.the sum of equations 4 and 5). In this case the contribution arising from the unsteady inertia force is dominant over the entire frequency range. For example, if one considers the high-frequency range noted above, which is appropriate to 'ringing' calculations, the standard Morisons inertia term is again of the order of 10% of the maximum; while the contribution due to the axial divergence term is of the order of 1%; and the convective and free surface terms contribute less than 0.2%.

Although these results highlight the importance of the nonlinearity associated with the wave motion, and emphasise the dominance of the unsteady Morisons' inertia load, they only represent one highly nonlinear wave case. As such there remains some uncertainty as to whether these results are representative of a large deep-water wave. Indeed, Newman (1996) has considered a similar case, but undertook his calculations on the assumption that in deep-water the nonlinearity of the wave (at least to a third-order of approximation) could be neglected. Although this latter approach ignores the nonlinear wave-wave interactions and local energy transfers identified by Baldock, et al. (1996), the nonlinear force components (identified in equations 3-5 above) were shown to be significant.

Study II: Force Measurements.

To further clarify the relative importance of the force components, previous work (undertaken as part of the Norwegian Joint Industry Project on "ringing") has provided an appropriate data set. Although this data includes a wide range of wave conditions, the present investigation will consider one example of a steep irregular sea state corresponding to a JONSWAP spectrum with a peak period of $T_p=2.4s$, a significant wave height of $H_s=0.279m$, and a peak enhancement factor of $\gamma=1.70$. This spectrum represents a 1:55 scaled model of a 100 year storm arising in the Norwegian Sea. The model study was undertaken in a large deep-water towing tank in

Trondheim, Norway. Three vertical cylinders of diameter D=0.2m (case I), D=0.326 (case II) and D=0.626 (case III) were individually mounted 38.6m downstream of the wave maker, and in each case time-histories of the water surface elevation η(t), the total horizontal force F(t), and the pitch moment M(t) were recorded. Further details of this study are given by Stansberg et al. (1995). Within the above noted sea state seven large wave events were considered. However, in the present paper we will only present data relating to the largest wave event since this was shown to be generally representative of all seven cases.

Discussion of Force Data.

Before presenting the force data it is worth noting that comparisons between the predicted kinematics based on the wave theories discussed above are in general agreement with the trends identified in figures 1b-1c. This was taken as further evidence of the need to apply an appropriate wave model. Figures 3a-3b concern the horizontal forces, F(t), recorded on cylinder I and II, and provide comparisons with the various force components calculated using the time stepping solution. Within this comparison the Morrison's inertia forces (equations 1 and 2) were calculated using an inertia coefficient of C_m=1.8 for cylinder I and C_m=2.0 for cylinder II. These values were determined by comparing the measured and predicted forces arising due to smaller (linear) waves, where the nonlinear force components (equations 2-5) are effectively zero. It is clear from figures 3a-3b that the unsteady Morrison's inertia load, integrated up to the instantaneous water surface, is at least one order of magnitude larger than the remaining nonlinear force components. However, it is also apparent that the nonlinear free surface force components are larger in this case, although their maxima is out of phase with the measured data.

The importance of the nonlinear forces arising at cylinder I are further examined in figures 4a- 4b. In figure 4a the frequency spectra of the measured force is compared to a similar spectra derived from the predicted force based on equations 1-5, with the kinematics calculated using a time stepping solution. The agreement between the measured and predicted force spectra is clearly very good. In figure 4b, the calculations were again undertaken using a time stepping solution, but in this case the frequency distribution and magnitude of the individual force components are compared. Once again, the unsteady Morrison's inertia force is dominant, although the contribution from the other nonlinear loads is significantly larger than that observed in the previous narrow-banded study. Indeed, if one considers the nonlinear forces arising at approximately three times the spectral peak (i.e., f≈1.4Hz), the free surface intersection force accounts for approximately 15% of the total; the free surface distortion force 25%; and the unsteady Morrison's load the remaining 60%.

Conclusions

The present study has shown that in relation to the potential flow forces generated by an extreme 2-D deep water wave, the accuracy of any force calculations is largely dependent upon the ability of a wave model to describe the unsteady water particle accelerations, particularly those arising close to the water surface. Indeed, comparisons between the measured and predicted forces indicate that the nonlinearity associated with an extreme wave field produces the dominant contribution to the nonlinear or high-frequency forces, and that without a fully nonlinear and unsteady wave model these important force components cannot be adequately predicted.

From a practical point of view the present results demonstrate that provided the fluid motion is well modelled, the unsteady Morrison's inertia force integrated up to the instantaneous water surface provides a reasonable description of the measured data. However, it is important to note that the third-order forces attributed to Rainey (1995) and Faltinsen et al. (1995) become more significant in a broad-banded sea state. Nevertheless, these terms are smaller than the nonlinear contribution due to the standard unsteady Morrison's inertia load, and their inclusion in the predicted force does not consistently provide an improved description of the measured data.

Acknowledgements

The authors gratefully acknowledge the financial support provided by Statoil.

References

Baldock, T.E., Swan, C., & Taylor, P.H. (1996). A laboratory study of nonlinear surface waves on water. Phil. Trans. Roy. Soc., Ser. A. **354**, pp 649-676.
Faltinsen, O.M., Newman, J.N. & Vinje, T. (1995). Nonlinear wave loads on a slender vertical cylinder. J. Fluid Mech. **289**, pp 179-98.
Fenton, J.D. & Rienecker, M.M. (1980). Accurate numerical solutions for nonlinear waves. Proc. 17th. Intl. Conf Coastal Engng., ASCE, **1**, pp 50-69.
Johannesson, T.B. & Swan, C. (1998). Numerical calculations of 2-D transient waves. Part I: comparisons with laboratory data. To appear in Appl. Ocean Res.
Newman, J.N. (1996). Nonlinear scattering of long waves by a vertical cylinder. In: Waves and nonlinear processes in hydrodynamics, (Eds: J.Grue et al.), Kluwer Academic Publishers.
Rainey, R.C.T. (1995). Slender-body expressions for the wave load on offshore structures. Proc. Roy. Soc., Lond. Ser. A., **450**, 391-416.
Stansberg, C.T., Huse, E., Krokstad, J.R., & Lehn, E. (1995). Experimental study of nonlinear loads on vertical cylinders in steep random waves. Proc. 5th. ISOPE conf., The Hague, The Netherlands. **1**, pp. 75-82.
Swan, C., Bashir, T. & Gudmestad, O.T. (1998). Accelerations in steep 2-D water waves, with implications for nonlinear wave loading. Submitted to J. Fluid Mech.
Wheeler, J.D. (1970). Method of calculating forces produced by irregular waves. Proc. Offshore Tech. Conf., Houston, USA. 1970. **1**, pp 71-82.

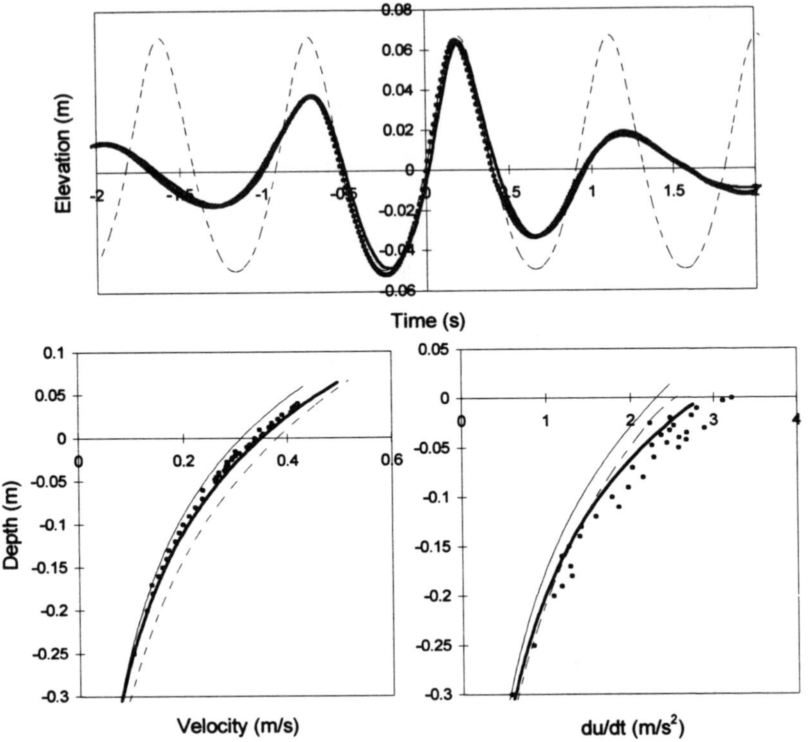

132   OCEAN WAVE KINEMATICS, DYNAMICS AND LOADS ON STRUCTURES

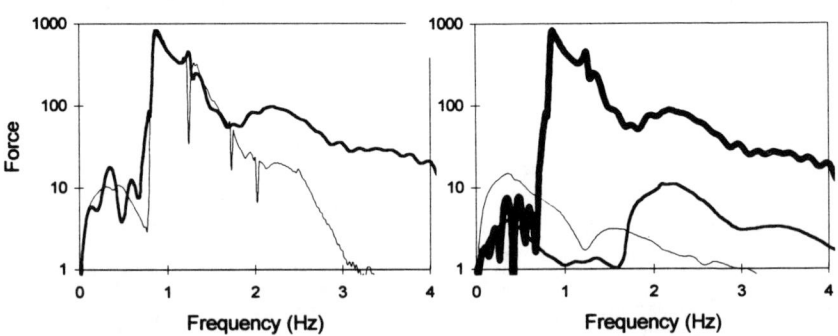

Figs 1a-1c. (a) Surface profile, $\eta(t)$. (b) Velocity, $u(z)$, (c) Local acceleration, $du/dt(z)$
• data, ---- Stokes' solution, —— Wheeler solution, —— Time-stepping solution.

Fig 2a. Total force spectra.
—— Time-stepping solution.
—— Wheeler solution.

Fig 2b Component force spectra.
—— Morisons' inertia load.
—— Axial divergence load
—— Free surface force

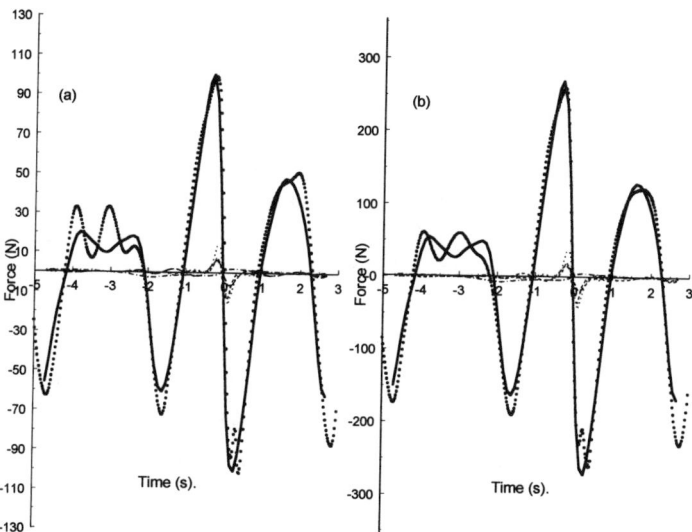

Figures 3a-3b. Force components calculated using the time-stepping solution. (a) Cylinder I (D=0.2m, C_m=1.8), (b) Cylinder II (D=0.326m, C_m=2.0).
• data, ——— eq 1, —·— eq. 2, — — eq 3, - - - - eq 4, •••• eq 5

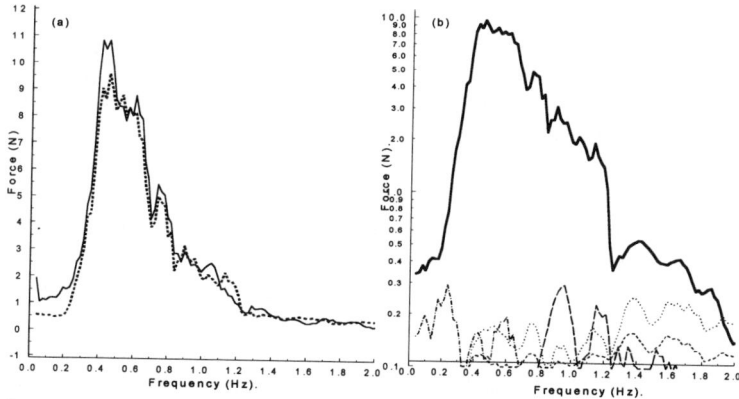

Figure 4a. Total force spectra.
——— measured data,
- - - - Time-stepping solution.

Figure 4b. Component force spectra.
——— eq 1, —·— eq. 2, — — eq 3,
- - - - eq 4, •••• eq 5.

Non-linear Wave Loading and Dynamic Response of Drag-Dominated Offshore Platforms

Daniel Karunakaran[1], Nils Spidsøe[1] and Ove T. Gudmestad[2]

Abstract

For slender marine structures like jack-ups and deep-water jackets, the response to wave loading is frequently of a dynamic character. Due to the drag-type of loading on these structures, pronounced non-Gaussian statistical properties of the response are observed for the highest sea states. In this paper, present design procedures are reviewed and an efficient procedure for non-linear dynamic response analysis of fixed offshore platforms is described. The procedure is exemplified by application to two jack-up platforms designed to operate in the Central part and Southern part of the North Sea, respectively and to a slender deep water jacket platform installed in the Central part of the North Sea. It is also discussed how this procedure may be used in practical design as a supplementary method to traditional design wave analysis

Introduction

As the quest for oil resources continues, the offshore activities have moved from shallow water conditions to moderate and deep-water conditions. Hence, there has been a continuous development of new structural concepts for new and challenging applications. Furthermore, traditional concepts like jackets and jack-ups have been extended to meet new and challenging conditions.

In shallow or moderate water depths these platforms will show a dominantly quasistatic behaviour. In such cases the traditional design wave approach may be successfully applied in design. The small or moderate dynamic effects experienced by the platforms in these waters are accounted for by applying dynamic amplification factors derived from simplified dynamic analysis. In deep waters, however, the platforms will respond highly dynamically to the irregular wave loading process. A full stochastic dynamic response analysis is thus required, either for direct estimation of extreme response or for estimation of reliable dynamic amplification factors to be used in combination with a design wave analysis

---

[1] SINTEF Civil and Environmental Engineering, Trondheim, Norway

[2] STATOIL, Stavanger, Norway.

These platforms consist of slender cylindrical members as basic components. This implies that the wave and current induced loading is of a highly non-linear nature due to non-linear drag force and free surface effects. These nonlinearities in the load process may introduce non-linear structural response even if the structure acts as a linear system. Furthermore, second order effects in the case of large structural motions introduce non-linear behaviour. The highly dynamic behaviour also implies that the hydrodynamic drag damping caused by the relative velocity between the structure and the surrounding fluid is very important for these concepts. For some of these structures, in particular for the jack-ups, the soil-structure interaction is often non-linear, at least for extreme loading conditions, and exhibits hysteretic behaviour.

All these nonlinearities will influence the dynamic response significantly and can not be modelled properly by a linear frequency domain analysis. A proper estimation of dynamic response for these structures therefore requires the application of a time domain analysis, as indicated by several authors, for example, Gudmestad et al. (1990), Chakrabarti (1991), Hoyle (1991), Manuel and Cornell (1993) and Karunakaran et al. (1992 a). Such analysis may be implemented in several ways.

In this paper a step-wise procedure is outlined and demonstrated for two types of platforms: a jack-up and a jacket. This procedure is tailored for these types of structures and has been shown to be computationally very efficient and also capable of modelling the non-linear response effects in a reliable manner. Furthermore, this paper summarises the authors' experience with the non-linear dynamic behaviour of such drag-dominated and dynamically sensitive structures, see also Karunakaran (1993).

The non-linear dynamic response analysis procedure

The procedure for prediction of extreme dynamic response consists of the following steps:

(1) Modelling of the structural system and the hydrodynamic loading including the wave process and wave and current loading
(2) Simulation of dynamic response samples
(3) Statistical modelling of the simulated response and estimation of extremes

In the following, the basic elements of those steps are described. The description is limited to short term analysis, as this is normally applied for practical structural analysis. For a complete overview, reference is made to Karunakaran (1991). The procedure may also be extended to long term response analysis, see Karunakaran et al. (1992 b). The load effects are estimated by solving the dynamic equation of motion for a structural model, which typically is based on the Finite Element Method (FEM). In the incremental form this equation is given as:

$$M\Delta \ddot{x}_t + C\Delta \dot{x}_t + K\Delta_{x_t} = \Delta F_t \tag{1}$$

Here; M is the mass matrix, C is the damping matrix, K is the stiffness matrix, ΔF_t is the vector of incremental hydrodynamic nodal forces, Δ_{x_t} is the incremental structural displacement vector, $\Delta \dot{\mathbf{x}}_t$ is the incremental structural velocity vector and $\Delta \ddot{\mathbf{x}}_t$ is the incremental structural acceleration vector

The mass matrix contains topside mass, structural mass and hydrodynamic added mass and is established according to either the lumped or consistent mass principle.

The stiffness matrix includes structural stiffness as well as the stiffness contributions from soil-structure interaction. The structural stiffness is modelled using beam elements. The soil-structure interaction is either modelled by linear springs at base nodes or by non-linear hysteretic model based on kinematic hardening.

The damping matrix including structural and soil damping is modelled using the Rayleigh model, i.e. as a linear combination of the stiffness and mass matrices. Concentrated dampers may alternatively model the soil damping. Furthermore, for structural members with relatively large displacements, the hydrodynamic damping is accounted for in the load vector by using relative velocities in the drag load model.

The in-line hydrodynamic forces are calculated using an extended form of Morison equation. According to this force model the load per unit length of a member of a vibrating system may be written as:

$$dF_t = \frac{\pi}{4}(C_M - 1) D^2 \rho (\dot{u}_t - \ddot{x}_t) + \frac{\pi}{4} D^2 \rho \dot{u}_t + \frac{1}{2} C_D D \rho (u_t + U - \dot{x}_t) | u_t + U - \dot{x}_t | \qquad (2)$$

Where; C_M is the inertia coefficient, C_D is the drag coefficient, ρ is mass density of water, D is the diameter of member, u_t is wave induced water particle velocity, $\dot{u}_t$ is wave induced water particle acceleration, U is current speed, $\dot{x}_t$ is velocity of the member at the actual location and $(\ddot{x}_t)$ is acceleration of the member at the actual location

Surface elevation time series are simulated from the model wave spectrum assuming independent random phase angles uniformly distributed between 0 and 2π by Monte-Carlo simulation. Since Eqs. (1) and (2) imply that the force vector depends on the response process, an iterative procedure should normally be employed at each time step. However, if small time steps are used, the following step-wise procedure may be applied:

- The phase angles are generated by Monte-Carlo simulation.
- The water particle kinematic time series are calculated according to applying Fast Fourier Techniques.
- The force vector at each time step is established based on Eq.(2), applying structural velocities and accelerations for the previous time step, with start value equal to zero and in the mean position of the structure.
- Eq.(1) is solved stepwise using the Newmark-β method, $\beta = 0.25$ and $\gamma = 0.5$, i.e., the so-called constant average acceleration algorithm.

This procedure has shown to be very computer efficient compared to an integrated iterative procedure. Furthermore, if small time steps are applied and the structural motions are small compared to the dominating wavelengths, it has proven to give accurate results.

Formally the extreme dynamic response could be estimated directly from a set of simulated samples applying standard extreme value statistics directly to the sample extremes. However, this would require sets of samples with lengths corresponding to the storm duration. For large structural systems this is impractical, even with today's computing efficiency. An alternative procedure is therefore established. This is based on modelling of sample distributions by analytical probabilistic models and extrapolation of the samples to the storm return period probability level by these models. For a detailed description of this procedure reference is made to Karunakaran et al. (1993 a), where also the statistical properties and the practical use of this procedure is discussed.

The Weibull models are fitted to the sample maxima by the method of moment. From the established Weibull model parameters, extrapolated extreme response for each simulated sample can be estimated. Based on these extrapolated extreme estimates, the final estimate of the extreme response is taken as the average extremes from many samples. The accuracy of this estimate depends on the number of samples and sample size. Generally, short samples require more samples than long samples in order to obtain the same accuracy. This is further discussed by Karunakaran et al. (1993 a), where also recommendations for choice of sample size and sample length is given.

Description of example structures

Two drag-dominated and dynamically sensitive structures are studied extensively by Karunakaran (1993). These are the TPG500 jack-up platform and the Veslefrikk jacket platform installed at a water depth of 176 m. These platforms are shown in Figure 1 and the main parameters for load calculations are presented in Table 1.

Table 1 Characteristics of example structures

	TPG 500 Jack-up	Veslefrikk Jacket
Water depth	110 m	175 m
Sea sate	H_S=15.5m, T_P=16.5 s	H_S=15.5m, T_P=16.5 s
Current	MWL - 1.15 m/s	MWL - 1.15 m/s
	-30 m - 0.70 m/s	-30 m - 0.70 m/s
	-50m to mudline - 0.60 m/s	-50m to mudline - 0.60 m/s
Drag coefficient, C_D	1.0	0.7 – members with antimarine growth coating 0.9 - elsewhere
Mass coefficient, C_M	2.0	2.0
First natural period	5.3 sec	3.3 sec
Damping	2 % structural damping + hydrodynamic damping	2 % structural damping

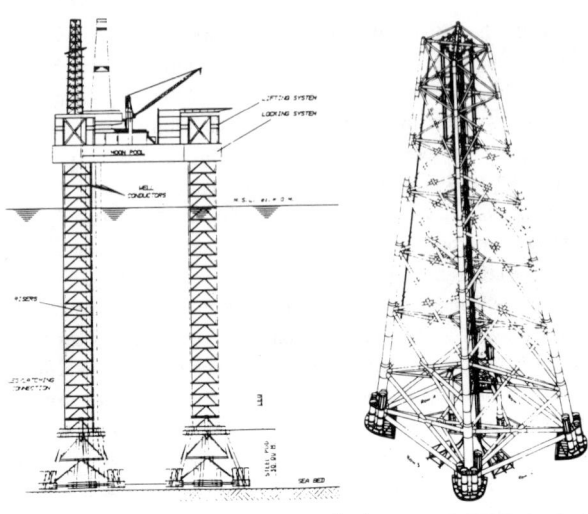

Figure 1 (a) TPG 500 Jack-up   (b) VESLEFRIKK Jacket
Example structures

Non-linear dynamic response

All the response analyses are performed with a simulation length equal to 2048 sec. The wave process and the wave kinematics are simulated with a sampling period of 0.5 sec and the time step for time integration in dynamic response calculations is 0.25 sec. For all cases seven independent simulations are carried out and the average extreme response and response statistics, such as standard deviation of response, skewness and kurtosis coefficients of response are calculated.

The estimated average of extreme dynamic response and statistical parameters of the response process and the dynamic amplification factors (DAF) for TPG500 jack-up platform are presented in Table2. Furthermore, the results from one of the seven samples in term's sample of probability of response maxima for some of the response quantities are shown in Figure 2.

Table 2   Nonlinear dynamic response - TPG 500 Jack-up

Response Quantity	Average response statistics				Dyn. Amp. Factors	
	St. Dev. of response	Skewness	Kurtosis	Extreme response	St. dev.	Ext. resp.
Base Shear (MN)	3.187	0.966	5.35	24.07	1.22	1.12
Overturning moment (MNm)	313.4	0.678	4.73	2180.0	1.50	1.19
Deck displacement (m)	0.082	0.647	4.66	0.562	1.55	1.20

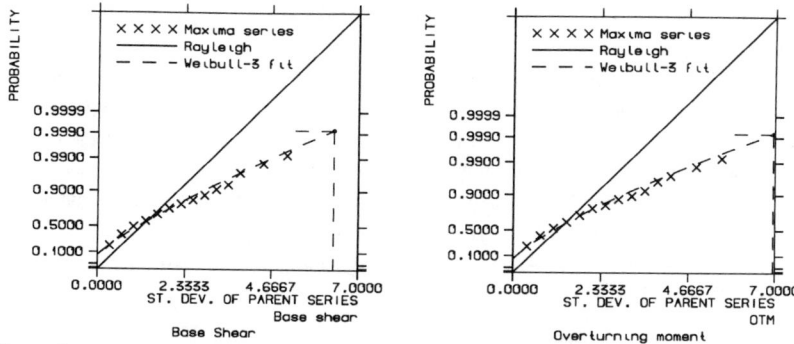

Figure 2 Nonlinear dynamic response – TPG 500 jack-up

The sample probability distributions of the response and the estimates of skewness and kurtosis coefficients show that the dynamic response is clearly non-Gaussian with higher extremes than in a Gaussian process. The sources of this high nonlinearity are drag forces, which are amplified due to the current, and the surface effects (integration to actual surface level). The current and the integration of forces to exact surface primarily cause the skewness, while kurtosis (peakedness) is mainly caused by the wave drag forces. The sample probability distributions also demonstrate that the 3-parameter Weibull distribution fits well with the response maxima.

The DAFs presented in Table 2 indicate that the DAF for the standard deviation of response is in general higher than the DAF for extreme response. The reason for this is that the resonant response tends to *Normalize* (make more Gaussian) the overall response. This effect is due to the nonlinear properties of the applied damping model. The total damping for this structure is due to the structural and soil damping, which is described by a linear viscous model, and the hydrodynamic drag damping arising from use of relative velocities in the Morison equation. A detailed discussion of these damping models and their influence on the structural response is found in Spidsøe et al. (1992) and a summary is reproduced below.

The statistical nature of the resonant response due to linear viscous damping will be Gaussian, independent of the statistical nature of the excitation, provided that the excitation is broad-banded compared to the resonance peak of the system, Brouwers (1982). This effect is referred to as a *Normalization* of the response process i.e. it tends to make the response a Gaussian process with Normal parent distribution. The wave excitation of these platforms is composed of both direct wave excitation in the region around the resonant frequency and superharmonic excitation caused by extreme waves. The direct wave excitation is broad-banded, provided the resonant frequency does not coincide with the hydrodynamic cancellation or amplification frequencies.

This *Normalization* effect is demonstrated in Figure 3, where the sample probability distributions of the quasistatic response, the total dynamic response, the dynamic response in the wave peak frequency region (f < 0.16 Hz) and the resonant response are shown. The figure shows that the quasistatic response and the dynamic response in the wave peak frequency region have similar statistical nature, i.e. they are equally non-Gaussian. On the other hand, the total dynamic response is less non-Gaussian than the quasistatic response, whereas the resonant response is almost Gaussian.

The strong non-Gaussian nature of the dynamic response indicate that a linearized frequency domain analysis based on Gaussian models would fail with respect to prediction of extreme response. This is demonstrated by the results shown in Table 3 where the extremes predicted by linear frequency domain analysis is compared to the extremes predicted by time domain analysis. It is seen that the extreme response from the linear frequency domain analysis is significantly lower than the extreme response predicted by the stochastic time domain analysis. Furthermore, it is very complicated to describe the wave surface effects and the influence of current in the linear analysis.

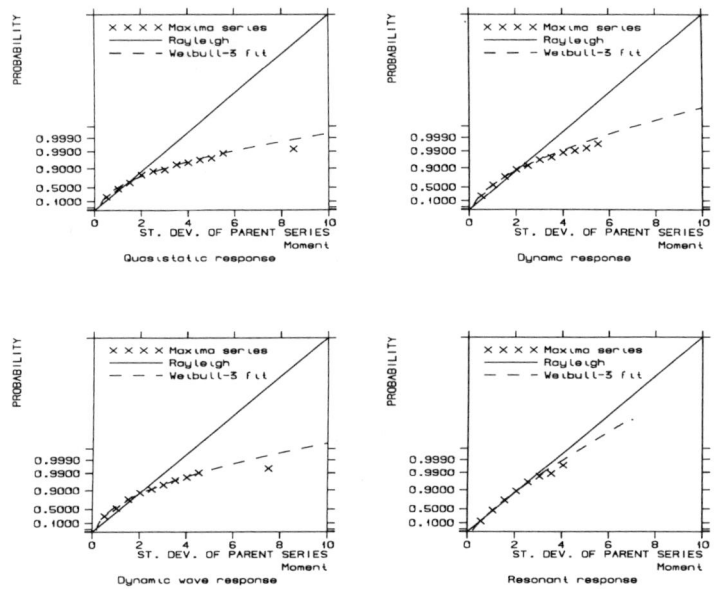

Figure 3   TPG 500 Jack-up – Overturning moment

Table 3   Comparison between nonlinear and linear analyses - TPG 500 Jack-up Extreme sea state without current

Response Quantity	Time domain analysis				Linear frequency domain analysis			
	St. deviation		Extreme		St.deviation		Extreme	
	Resp.	DAF	Resp.	DAF	Resp.	DAF	Resp.	DAF
Base Shear	2.35	1.20	14.72	1.07	2.38	1.24	9.36	1.24
Overturning moment	225.1	1.47	1405	1.16	231.7	1.53	923.2	1.53
Deck displacement	0.058	1.52	0.366	1.18	0.060	1.58	0.241	1.58

A very similar behaviour is seen for the Veslefrikk jacket, see Table 4. However, for this jacket, the linearized analysis estimates significantly lower standard deviation of response than the non-linear analysis, Karunakaran (1993). The explanation for this is that the non-linear drag force generate superharmonic load components with frequency equal to 5 times the wave peak frequency, Spidsøe and Karunakaran (1989), which in this sea state coincide with the natural period of the structure. This effect can not be included in the linear frequency domain analysis.

Table 4   Nonlinear dynamic response - Veslefrikk jacket - Base Case

Response Quantity	Average response statistics				Dyn. Amp. Factors	
	St. Dev. of response	Skewness	Kurtosis	Extreme response	St. dev.	Ext. resp
Base Shear	4.03	0.749	4.78	30.00	1.10	1.15
Overturning moment	484.6	1.041	5.93	4040.7	1.13	1.16
Ax. force in brace at MWL	0.52	1.406	7.37	4.77	1.11	1.12
Ax. force in leg at MWL	0.81	1.474	9.45	8.32	1.50	1.22

Important parameters for dynamic response prediction

There are many parameters that are applied in a non-linear dynamic response analysis. The following are the few important parameters, the effects of which will be discussed in this paper:
- Wave kinematics and wave current interaction
- Damping
- Non-Gaussian waves

Wave kinematics and wave current interaction

The modelling of irregular waves is especially important in the time domain analysis since it forms the basis for prediction of the highly nonlinear drag loading process in the surface zone which is a major source of non-Gaussian response. A thorough review of different wave kinematic models that can be applied in time domain non-linear analysis of drag-dominated offshore platforms is found in Gudmestad and Karunakaran (1994). That review also contains the effects of different wave models on the estimated extreme response for these two example structures. There is a large variability in the estimated extreme response depending on the wave model used.

For drag-dominated platforms, the wave loading around mean water level (MWL) is very important. The wave kinematics around and above MWL can not be properly evaluated by linear Airy wave model, and this requires some type of stretching or extrapolation above MWL. Among the available stretching models, Wheeler stretching, Wheeler (1970), is simple to use even though it does not satisfy the set of basic hydrodynamic equations. Furthermore, based on the wave tank experiments presented by Skjelbreia et al. (1991) it is concluded that this model predicts wave kinematics reasonably well. However, for steep waves, the Wheeler model underpredicts the kinematics below the crest, Gudmestad and Haver (1993). A correction factor for Wheeler model based on wave steepness is therefore proposed in Gudmestad and Haver (1993).

Analysis of full scale measurements from a jacket platform in North sea from a severe storm conditions indicated that for a very steep wave, an increase in Wheeler kinematics of 15% was needed to match the extreme quasistatic load level, Karunakaran et al. (1997 a). On the other hand for a similar wave condition, with no steep waves, the Wheeler wave model gave a reasonably good fit to the loading and response. A similar comparison was found from an analysis involving full-scale measurement data from a large jack-up in central North sea, Karunakaran et al. (1997 b). However, it is emphasised that the above mentioned comparisons with full scale measurements did not include a systematic evaluation of wave kinematics, but an evaluation of over all loads and response of the jacket and jack-up platforms.

The wave-current interaction includes the superposition of wave induced velocities and current velocity and stretching of the current profile in the surface region. The most commonly used model is a direct addition of the wave-induced velocity for the non-current situation and the current velocity. However, due to the wave-current interaction, there is a change in wave kinematic spectra (the Doppler effect). The effect of this wave-current interaction on the response of the jacket platform is investigated and reported in Gudmestad and Karunakaran (1990). The study indicated that the wave-current interaction does not have a significant influence (less than 10%) on the extreme dynamic response.

Damping

As for any bottom fixed offshore platform, the damping of a jack-up platform consists of structural damping, soil damping and hydrodynamic damping. There are thus three different damping mechanisms involved:
- The linear viscous mechanism modelled by the Rayleigh model
- The nonlinear drag damping mechanism, modelled by the extended Morison equation
- The hysteretic soil damping mechanism, modelled by the kinematic hardening model

These mechanisms have different properties and effects on the dynamic response, which may have significant impact to the response extremes.

The most important and least known property of the linear viscous damping mechanism is the so called Normalization effect, i.e. its ability to generate Gaussian resonant response independent of the statistical nature of the excitation process, Brouwers (1982). This effect is demonstrated in Figure 3 and Table 5, which summarize results from a study on damping for the TPG 500 jack-up. This example shows that even if the resonant response is very high in the rms. sense, its contribution to the extreme response is significantly less because the total response is less non-Gaussian than the quasistatic response due to the nearly Gaussian resonance generated by the damping mechanism. The drag damping mechanism has a similar effect, though the mechanism is different. As also shown by Brouwers (1982) this damping mechanism generates non-Gaussian resonant response with a lower kurtosis coefficient than in a Gaussian process, i.e. with a statistical nature opposite to the quasistatic response. When the resonant and quasistatic response is added, an even stronger Normalization effect to the total response occurs than generated by the linear viscous damping model.

Table 5 Dynamic amplification factors (DAF) for various damping combinations TPG 500 Jack-up

Case Identification	Estimated damping	Base Shear		Overturning moment	
		DAF for St. deviation	DAF for extremes	DAF for St. deviation	DAF for extremes
Linear damping only.					
0.5 % Lin. damping	0.52 %	2.47	1.39	3.89	1.83
1.0 % Lin. damping	1.03 %	1.93	1.24	2.90	1.51
2.0 % Lin. damping	2.07 %	1.58	1.18	2.22	1.34
3.0 % Lin. damping	3.11 %	1.43	1.18	1.93	1.29
4.0 % Lin. damping	4.15 %	1.36	1.18	1.76	1.29
5.0 % Lin. damping	5.19 %	1.31	1.17	1.65	1.30
Linear damping and drag damping.					
No Lin. damping	2.76 %	1.51	1.12	2.09	1.24
1.0 % Lin. damping	3.10 %	1.41	1.12	1.90	1.22
2.0 % Lin. damping	4.16 %	1.34	1.12	1.73	1.22
3.0 % Lin. damping	5.21 %	1.29	1.12	1.63	1.23

The nonlinear properties of the hysteretic soil damping mechanism depend highly on the shape of the backbone curve. The more nonlinear this curve is the wider will the hysteretic curve be at the high load cycles. Consequently, the damping in these response cycles will be higher than in the lower. This is similar to the drag damping mechanism. Though it has not been investigated so far, this means that one may assume that a Normalization effect similar to what is found for the drag damping also will occur for this soil damping mechanism

The different damping mechanisms are not independent, but work together. This is demonstrated by the results given in the last part of Table 5. From these it is seen that when the linear damping reduces the dynamic amplification of the rms. response increases. This is as expected since the total damping level reduces. However, it is also noted that the dynamic amplification of the extremes is almost independent of the linear damping level. The explanation of this is that the reduction of linear damping is compensated for by the Normalization effect both from the linear damping itself but also from the drag damping which becomes more important as the linear damping level reduces. From a design point of view this is interesting because it implies that as long as both these damping models are involved, it is not important with an accurate specification of the linear damping level. However, since neither the linear damping mechanism nor the drag damping mechanism may be regarded as verified, this effect should be utilized with care. For further discussion on the nonlinear damping properties, reference is made to Spidsøe et al. (1992).

Non-Gaussian waves

The first step of the time domain analysis is to simulate the irregular surface elevation from a wave spectral model. The state-of-the-art concerning this simulation is to assume that the sea surface can be modelled as a Gaussian random process. However, it is known that in the real seas the surface elevation is not a perfect Gaussian process, Spidsøe and Karunakaran (1987) and Haver and Karunakaran (1998). However, for reasonably deep-water conditions, it has been assumed that the deviations are rather small. In particular, it has been assumed that the Gaussian wave model is sufficiently accurate regarding dynamic amplification, i.e., it is assumed that the possible deviations do not affect the dynamic behaviour. Vinje and Haver (1994) investigate the non-Gaussian structure of ocean surface waves in details.

An investigation was performed to study whether or not a possible non-Gaussian sea surface is important concerning the assessment of dynamic behaviour of drag-dominated structures, Karunakaran et al. (1994).

The study concluded that if nonlinear wave models giving non-Gaussian surface elevation, consistent with field observation are used in stochastic dynamic analysis, this will give significantly higher extreme response than if Gaussian waves are used. Also the non-Gaussian effects increases the dynamic response, but the effects are less significant than in the quasistatic response. Furthermore, there is a slight decrease in the dynamic amplification factors.

Design practice for jack-up platforms is based on the design wave method. The extreme quasistatic response is then calculated by deterministic design wave analysis and the dynamic response found by multiplying this with a dynamic amplification factor (DAF) obtained from a stochastic response analysis or from some kind of simplified method.

The effects of non-Gaussian waves discussed in this paper concern only the stochastic analysis. The presented results show that if a non-Gaussian wave model is applied a lower drag coefficient should be used than for Gaussian waves, in order to obtain consistent quasistatic response compared to design wave analysis. Furthermore, the results show that if non-Gaussian waves are used, this will give lower DAFs than with Gaussian waves, ie a beneficial effect with respect to design.

Conclusions and Practical application of the time-domain analysis procedure

A computer efficient, stepwise procedure for time domain analysis of the stochastic dynamic response of drag-dominated platforms is presented. This procedure is demonstrated through example studies of three different platforms. It is shown that the response of these platforms are highly non-Gaussian due to wave induced drag loading and that the dynamic behaviour may be highly influenced by superharmonic load components and non-linear damping effects. Furthermore, it is demonstrated that linearized frequency domain analysis will severely underestimate the extreme response and also dynamic effects of platforms of this type. On this basis it may be concluded that non-linear time domain analysis must be involved in dynamic analysis of drag-dominated offshore platforms in order to have reliable analysis.

However, a detailed analysis of this type may not be feasible in a practical design because of its complexity and computational limitations. It requires thus personnel with competence, which is not found with traditional design engineers. It is therefore widely discussed today among offshore engineers, how limited time domain analysis may be used in a simplified way combined with traditional design methods.

One way of doing this is as follows:
- Prepare an idealised model and estimate the non-linear quasistatic and dynamic response for a limited set of response quantities, which is sufficient to establish a picture of the dynamic behaviour of the structure. This analysis should include sensitivity studies on all relevant parameters.
- Estimate the design wave response for those response quantities.
- Estimate the calibration factors (CF) defined as the ratio between the extreme quasistatic response from non-linear time domain stochastic analysis and response from design wave analysis for the simplified structural model.
- Establish the dynamic amplification factor (DAF), which is the ratio between the extreme dynamic response and the extreme quasistatic response from time domain stochastic analysis for the simplified model.
- From the results obtained for the simplified model, establish CF and DAF for the design structural model.
- Calculate design response in the design structural model as:
Design response = CF x DAF x response from the design wave analysis

References

Brouwers, J.J.M. 1982: "Response Near Resonance of Nonlinearly Damped Systems Subjected to Random Excitation with Application to Marine Risers", *Ocean engineering*, 1982.

Chakrabari, S.K. 1991: "Strategies for Non-linear Analysis of Marine Structures", Ship Structures Committee, *Report No. SR-1304*, 1991.

Gudmestad, O.T. and Karunakaran, D., 1990: "Wave Current Interaction", *Advances in Underwater Technology, Ocean Science and Offshore Engineering*, Volume 26, 1990, Kluwer Academic Publishers, Nederlands, pp. 81-110.

Gudmestad, O.T., Spidsøe, N. and Karunakaran, D., 1990: "Wave Loading on Dynamic Sensitive Offshore Structures", *Proc. Offshore Mechanics and Arctic Engineering Conference*, Houston, 1991.

Gudmestad, O.T. and Haver, S. (1993): "Uncertainties in Prediction of Wave Kinematics in Irregular Waves", Wave Kinematics and Environmental Forces, Volume 29, Society of Underwater Tech., 1993.

Gudmestad, O.T. and Karunakaran, D. 1994: "Wave Kinematics Models for Calculation of Wave Loads on Truss Structures", *Proc. 26th Offshore Technology Conference*, Houston, 1994.

Haver, S. and Karunakaran, D. 1998: "Probabilistic Description of Crest Heights of Ocean Waves", *Proc. 5th International workshop on wave hindcasting and forecasting*, Melbourn, Florida, 1998.

Hoyle, M.J.R. 1991: "Jack-up Dynamics, A Review of Analysis Techniques", *Proc. Third Int. Conf. on The Jack-up drilling platform*, London, 1991.

Karunakaran, D. 1991: "Procedure for Nonlinear Dynamic Response Analysis of Offshore Structures - Both for Extreme and Fatigue Response", *SINTEF Report STF71 A91016*, Trondheim, 1991.

Karunakaran, D., Gudmestad, O.T. and Spidsøe, N. 1992 a: "Nonlinear Dynamic Response Analysis of Dynamically Sensitive Slender Offshore Structures", *Proc. 11 th Int. Conf. on Offshore Mechanics and Arctic Engineering*, Calgary, 1992.

Karunakaran, D., Leira, B.J., Spidsøe, N. and Moan, T. 1992 b: "Nonlinear Long Term Response of Dynamically Sensitive Drag-dominated Offshore Platforms", *Proc. Int. Conf. on the Behaviour of Offshore Structures - BOSS*, London, 1992.

Karunakaran, D. 1993: "Nonlinear Dynamic Response and Reliability Analysis of Drag-dominated Offshore Structures", *Dr. ing Thesis*, Div. of Marine Structures, NTH, Norway, 1993.

Karunakaran, D., Spidsøe, N. and Leira, B.J. 1993 a: "Prediction of Extreme Dynamic Response of Jack-up Platforms using a Nonlinear Time Domain Simulation Method", *Proc. 12th Int. Offshore Mechanics and Arctic Engineering Conf.*, Glasgow, Scotland, 1993.

Karunakaran, D., Spidsøe, N. and Haver, S. 1994: "Nonlinear Dynamic Response of Jack-up Platforms Due to Non-Gaussian Waves", *Proc. 13 th Int. Conf. on Offshore Mechanics and Arctic Engineering*, Houston, 1994.

Karunakaran, D., Bærheim, M. and Leira, B.J. 1997 a: "Measured and simulated dynamic response of a jacket platform", *Proc. 16 th Int. Conf. on Offshore Mechanics and Arctic Engineering*, 1997.

Karunakaran, D., Bærheim, M. and Spidsøe, N. 1997 b: "Full-scale measurements from a large deep water jack-up platform", *Proc. Sixth Int. Conf. on The Jack-up drilling platform*, London, 1997.

Manuel, L. and Cornell, C.A. 1993: "Sensitivity of the Dynamic Response of a Jack-up Rig to Support Modelling and Morison Force Modelling Assumptions", *Proc. 12 th Int. Conf. on Offshore Mechanics and Arctic Engineering*, Glasgow, 1993.

Skjelbriea et al. 1991: "Wave Kinematics in Irregular Waves", *Proc. Offshore mechanics and Arctic Engineering*, Stavanger, 1991.

Spidsøe, N. and Karunakaran, D. 1987: "Statistical and Directional Properties of Measured Ocean Waves", *Proc. Offshore Technology Conf.*, Houston, 1987.

Spidsøe, N. and Karunakaran, D., 1989: "Effects of Superharmonic Excitation to the Dynamic Response of Offshore Platforms", *E&P Forum Workshop on Wave and Current Kinematics and Loading*, IFP, Paris, 1989.

Spidsøe, N., Karunakaran, D. and Gudmestad, O. 1992: "Nonlinear Effects of Damping to Dynamic Amplification Factors for Drag-dominated Offshore Platforms", *Proc. 11 th Int. Conf. on Offshore mechanics and Arctic Engineering*, Calgary, 1992.

Vinje, T. and Haver, S. 1994: "On the Non-Gaussian Structure of Ocean Waves", To be presented in *Int. Conf. on the Behaviour of Offshore Structures - BOSS 94*, USA, 1994.

Wheeler, J.D. 1970: "Methods for Calculating Forces Produced on Piles in Irregular Waves", *Journal of Petroleum Technology*, March, 1970.

On the Contribution of the Mooring System to the Damping of the Slow Oscillation of Moored Floating Structure

Ju Fan[1]   Xiaohong Chen[2]   Xianglu Huang[3]

Abstract

In this paper a method in frequency domain, which considered the dynamic coupling between the floating structure and mooring system up to the second order, was used to analyze the contribution of mooring system to the damping of the slow oscillation of a moored structure. The damping coefficient due to the mooring system to the slow oscillation of moored floating structure was calculated by the principle of energy reservation law. The comparison between the results of surge motion obtained by time domain calculation and the corresponding frequency domain calculation conformed well. It seems that the effect of mooring system on the slow oscillation of a moored floating structure is something like a dynamic system, the influence may be more complicated than only consider it as a single damping parameter.

Introduction:

The damping of a mooring system contributing to the slow oscillation of the moored floating structure has been given more and more attention in recent years. This is due to the important role of such damping playing in the mooring system design. But up till now it is hardly to say that the problem has been solved successfully. The difficulty of the problem is the dynamic coupling between the mooring system and the floating structure, which is difficult to be solved by only using the experimental or numerical method. In some of the researches, it seems that

---

[1] Doctor Candidate,College of NAOE,Shanghai Jiao Tong University,China
[2] Lecturer,College of NAOE,Shanghai Jiao Tong University,China
[3] Professor,College of NAOE,Shanghai Jiao Tong University,China

the dynamic calculation has been somewhat simplified (de Kat,and Dercksen, 1994), (Huse,and Matsumoto,1989), (Wichers,and Huijsmans,1990). The most detailed treatment of such problem may be De Kat. et al.(1994). In their paper considering the influence of the shape of wave spectrum on the slow oscillation damping, a method of numerical calculation was suggested which was operated in time domain. In that method the mooring line force was considered as an external force and the tanker low frequency motion was determined by numerical calculation where the mooring line force was obtained from the previous time step. The wave frequency oscillation of the structure, which was superposed on the slow oscillation, was determined before solving the motion equation. Although this method has partly considered some of the couple effects, it still has some problems as to the rigorous coupled calculation between floating structure motion and mooring line in each time step. It seems that the time domain method is not very easy to be implemented in this very complicated problem.

For this reason we suggested to use the frequency analysis method. Frequency domain method has the advantage of analyzing the coupling problem, but it can only be used in linear system. For a moored floating system, the non-linearity of which can not be neglected. So we proposed a perturbation method up to second order. For the moored floating structure, perturbing to the second order is enough. So in our method the frequency domain analysis is extended to bi-frequency domain. The merit of this bi- frequency domain method is that it can handle the couple effect between mooring system and floating structure including the slow drift oscillation easily. As the comparison between the calculated result and the experimental result showed in previously publications(Chen,Huang 1993,1997), it can estimate the main part of the dynamic effect. In this paper we explore it to predict the mooring line damp effect.

Theory:

As we have suggested in the previous section, the oscillations of a moored floating structure on the wave can be calculated by a second-order frequency domain method. The wave as an input could be either simple harmonic sinusoidal wave with a pair of frequencies or the multi-frequency waves as irregular waves. The theory and formulation of the method have been described in the previous papers of authors (Chen,Huang 1993,1997). In this method the dynamic coupling between the floating structure and mooring line are considered up to second order. So, theoretically it is suitable for our purpose. But there are some deficiencies that have to be mentioned. First, this method is based on the method of perturbation in which the systems only have small amplitude oscillation as it perturbs about the original static mooring line configuration. The large amplitude slow oscillation is simply ignored. So it will produce large deviation if the motion became large. Second, the points attach to the sea bottom will still have movement in real condition, which may produce some frictional force between mooring line and bottom, but this effect must be ignored in the perturbation procedure, since it is assumed that the attach point is fixed. But considering that the results obtained by the perturbation calculation up to second order show satisfactory conformity with the corresponding model test for some of the experiments(Chen,Huang,1997), we think at least for such kinds of the mooring systems the precision of this method may be enough as long as the slow drift oscillation is concerned.

According to the relationship of the work and energy, we can calculate the work done by the mooring line force on the structure in one period as:

$$W = \int_0^T F(t)ds = \int_0^T F(t)\dot{X}(t)dt \tag{1}$$

In which F(t) is the mooring line resultant force, ds is the structure displacement under which the force is acted, while $\dot{X}(t)$ is the velocity of floating structure oscillation. T is the period of oscillation. For the first order response, it can be calculated:

$$W = -\int_0^T F\cos(\omega t + \delta)\omega X \sin \omega t\, dt \tag{2}$$

where F is the force amplitude, ω is the frequency that is the same for force and oscillation. δ is the time lag phase angle between the mooring line force and floating structure oscillation. The final form is:

$$W = FX \sin \delta\, \pi \tag{3}$$

The result shows that the work depends not only on the F and X but also on the phase δ. This relation also shows that the work done by the mooring line force on the floating structure may change the sign due to the variation of phase angle.

In the second order condition, the oscillation as well as the mooring line force is expressed in terms of a pair of frequency. The difference of the frequencies is the frequency of the slow oscillation component.

$$X^{(2)}(t) = X^{(2)} \cos((\omega_1 - \omega_2)t + \delta) \tag{4}$$

$$F^{(2)}(t) = F^{(2)} \cos((\omega_1 - \omega_2)t + \delta_F) \tag{5}$$

so, the work done by the mooring line force on the floating structure will become

$$W = \int_0^{T1} F^{(2)}(t)(\omega_1 - \omega_2)\dot{X}(t)dt = \pi F^{(2)} X^{(2)} \sin(\delta_F - \delta) \tag{6}$$

$$T_1 = \frac{2\pi}{\omega_1 - \omega_2} \tag{7}$$

where δ and δ_F are phase angles of the slow oscillation motion and mooring line force separately. T_1 is the period of difference frequency. It should be noted that in second order condition, for a wave train with different components, the calculation must take all of the component into consideration.

As for the moored floating structure in random sea, the contribution of mooring system to the damping of slow drift oscillation can be expressed by the corresponding mean damping coefficient. For the slow drift oscillation, the response as well as the damping coefficient should be a function of a pair of frequencies, the damping coefficient of slow varied oscillation $B(\omega + \mu, \omega)$ can be obtained from the work:

$$B(\omega+\mu, \omega) = \frac{W}{\pi X^2 (\omega_1 - \omega_2)} \tag{8}$$

where: $\omega_1 = \omega + \mu$, $\omega_2 = \omega$, $\omega_1 - \omega_2 = \mu$, μ is the frequency of the slow oscillation.

The product $B(\omega + \mu, \omega) \cdot \dot{X}(\mu)$ has the dimension of force, and all of the pairs of frequencies $\omega + \mu, \omega$ in the motion spectrum have contributions to the same μ.

Then, the weighted mean has to be done to all ω for the damping coefficient, which result in $\overline{B}(\mu)$, here $\overline{B}(\mu)$ is the contribution to the slow frequency motion damping of mooring lines in the frequency μ. The damping coefficient of one specified motion spectrum is:

$$\overline{B}(\mu) = \frac{\sum [(8 \cdot s_\zeta(\omega+\mu) s_\zeta(\omega) \cdot G_{2x}^{\,2}(\omega+\mu, \omega) \cdot B(\omega+\mu, \omega)] \Delta \omega}{\sum 8 \cdot s_\zeta(\omega+\mu) s_\zeta(\omega) \cdot G_{2x}^{\,2}(\omega+\mu, \omega) \cdot \Delta \omega} \tag{9}$$

With the $\overline{B}(\mu)$, a time domain calculation can be carried out, where the mooring line is substituted by a linear spring with no mass. The equation of slow oscillation in surge direction in time domain is:

$$(M + a_{11})\ddot{X} + B_{11}\dot{X} + KX = F^{(2)} \tag{10}$$

where M is the mass of tanker, a_{11} is the added mass at frequency μ, B_{11} include B_{tanker}, B_{wave} and also the mean damping coefficient $\overline{B}(\mu)$, K is mooring stiffness, $F^{(2)}$ is second-order wave drift force at frequency μ.

The damping coefficient calculation of slow oscillation and discussion

We have calculated some of the moored floating structures for which the model tests results were available. In those systems, a semisubmersible platform, which was moored by a spreading catenary mooring system, and a tanker model moored by four chains were included. The arrangement of mooring tanker with mooring lines is as below. The surge motion and mooring line tension both in wave and low frequency were calculated by the second order perturbation method and compared with the corresponding test results. The results of the first order surge frequency response function(RAO) and the surge motion spectrum compared with the experiment analysis results are shown separately:

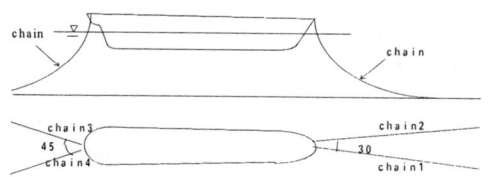

Fig 1. Arrangement of moorings

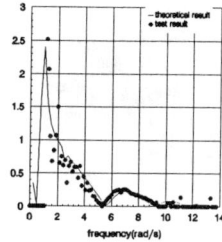 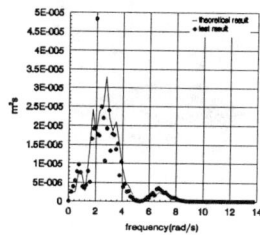

fig 2. RAO of surge(platform)   fig 3. Spectrum of surge (platform)

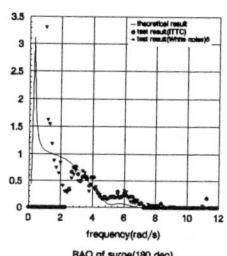 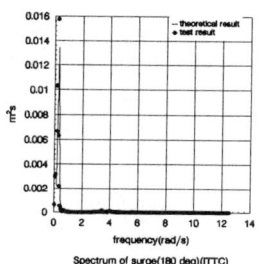

Fig 4. RAO of surge (180 deg)   Fig 5. Spectrum of surge (180 deg)(ITTC)

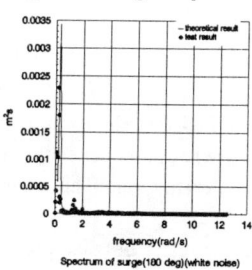

fig 6. Spectrum of surge(180 deg) (white noise)

From the above figures, we can find that the theoretical calculations in the frequency domain conform well with the experiment analysis results. To the tanker, the low frequency peak value is large than the wave frequency value, while to the platform, the results are different.

By using the bi-spectra of the resultant force of all the mooring lines in the mooring system in surge direction obtained in frequency domain results and with the expression obtained above for slow oscillation damping coefficient $B(\omega + \mu, \omega)$, we can easily obtain the slow oscillation damping coefficient:

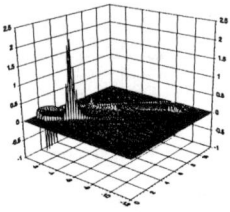
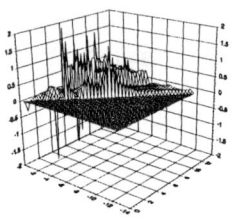

fig7. damping coefficient(tanker)(ITTC)　　fig8.damping coefficient(platform)
(due to the mooring system)　　　　　　　(due to the mooring system)

The corresponding mean damping coefficients of mooring line to the second order slow oscillation are as follows:

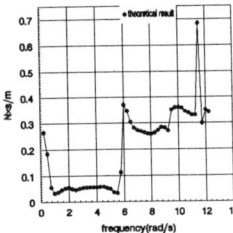
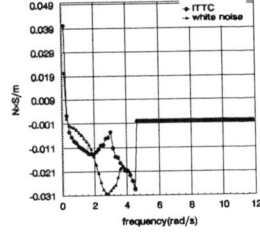

fig 9. mean damping coefficient　　　　fig 10. mean damping coefficient
(platform)　　　　　　　　　　　　　　(tanker) (180 deg)

From fig.10, it should be noted that the mean damping coefficient of mooring line to the second order slow oscillation is depending upon the input wave spectra. There are different values for different input. This clearly show the characteristics of the second order effect.

From the above figures, we can see that the damping coefficients may be negative in some of the frequency regions. This implies that in such area the mooring system will amplify the surge oscillation. We think it may only be explained as the effect of a coupled dynamic system. As it is well known in the mechanical vibration theory, the multi-mass system in some of the frequency region can increase the motion of the main body. It is very difficult to prove analytically in our case, because of the complexity of the mooring line dynamics. But the essentiality may be the same. The mooring system may be a more complex dynamic system than only to be considered as a single damping force, and the motion characteristic of the mooring system is essential in the determination of such damping effect.

In order to show if our consideration is correct or not, we proceeded a time domain calculation as in the current routine, which means that the mooring line is substituted by a linear spring with no mass. The dynamic effect of which is considered only by the damping. The damping coefficients used previously in the frequency domain calculation, such as viscous damping and wave damping of slow oscillation of structure, were used in time domain and also include the mooring system damping coefficient obtained in previous section. The comparison results are as follows:

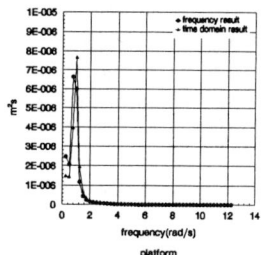

fig 11. Spectrum of surge (platform)

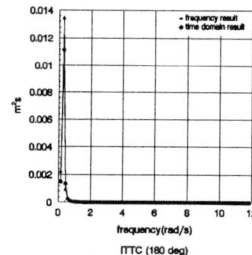

fig 12. Spectrum of surge(180 deg) (tanker)(ITTC)

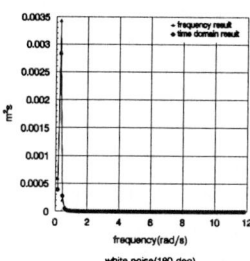

fig 13. Spectrum of surge (180 deg) (tanker)(white noise)

From the figures above, we can find that the comparison between the results of surge motion obtained by frequency domain calculation and time domain calculation conforms well. It implies that, as long as the mooring lines are not too loose, the frequency method is available to predict the damping contribution to the slow oscillation of moored floating structure of the mooring system even if it ignores some details of the problem.

Concluding remarks

From above calculations, several conclusions can be drawn:

1. The effect of mooring system on the slow oscillation of a moored floating structure is dynamically, which means that the mooring system should be considered as a dynamic system. With it's more complicate behavior due to its flexibility, the influence of such effect may have more complexity than only consider it as a single damping parameter.

2. The result of the second order surge determined by perturbation agrees with the time domain result if the damping term includes the mooring line damping in the time domain. So the frequency domain approximate method can be valid in the prediction of such effect, providing that the mooring line is not too loose.

3. The contribution of the mooring system to the damping of the slow oscillation of moored floating structure is not very large, but its effects can not be omitted. So when predicting the floating structure motion, the mooring system damping effect should not be ignored.

References

1. Chen Xiaohong and Huang Xianglu ,1993, Bi-frequency analyses of mooring line dynamics, Proceedings of 1993 OMAE
2. Chen Xiaohong and Huang Xianglu ,1993, The analyses of the dynamics of mooring line in time domain and frequency domain, Proceedings of the third international offshore and polar engineering conference,428-434
3. Chen Xiaohong and Huang Xianglu,1997,The motion and mooring line loads of a moored semisubmersible in waves, Proceedings OMAE'97
4. de Kat,J.O.,and Dercksen,A.,1994, The Influence of Wave Spectrum Formulation on the Dynamics of a Turret-Moored Tanker, Journal of Offshore Mechanics and Arctic Engineering,Vol.116,pp.7-13
5. Huse,E.,and Matsumoto,K.,1989, Mooring line Damping due to First- and Second-Order Vessel Motion,OTC#6137,pp.135-148
6. Wichers, J.E.W., and Huijsmans, R.H.M., 1990, The Contribution of Hydrodynamic Damping Induced by Mooring Chains on Low-Frequency Vessel Motions, OTC#6218, pp.171-182.

NONLINEAR LOADS AND RESPONSES OF SHIPS AND OFFSHORE PLATFORMS IN DETERMINISTIC AND RANDOM WAVES

By

Sclavounos, P. D., Kim, S., Kim, Y. and Kring, D. C.

Department of Ocean Engineering
Massachusetts Institute of Technology
Cambridge MA 02139

Abstract

This article presents recent results on the computation of linear and nonlinear loads and responses of offshore platforms and ships engaged in the exploration of hydrocarbons in monochromatic and random waves. The first set of results have been obtained by the computer program **SWIM**, based on a set of accurate, robust and very efficient algorithms for the study of the slow-drift responses of ships and offshore platforms. The second set is derived from the three-dimensional time-domain Rankine panel method **SWAN** which is shown to be capable to treat a wide range of linear, second-order and nonlinear wave body interaction problems at zero or forward speed, in essence behaving like a "Numerical Wave Tank" relieved of viscous effects.

1. INTRODUCTION

Linear wave body interaction theory and related computational methods have found widespread use by the offshore industry for the analysis and design of offshore platforms [cf 1,2]. The most important nonlinear effects derived from linear theory are the mean drift forces and moments in monochromatic waves. While their theoretical definition is well established, their accurate computation requires the careful discretization of the body surface, particularly near resonance and at high frequencies. This offers a first glimpse into the challenges presented by the simulation of nonlinear loads and responses.

More complex nonlinear loads include the slow-drift excitation Quadratic Transfer Function (QTF) matrices, the sum-frequency excitation QTF's of Tension Leg Platform (TLP) tethers, the surge-sway-yaw wave drift damping coefficients and QTF matrices and the nonlinear extreme loads in steep ambient waves. The corresponding nonlinear responses include the large amplitude slow-drift motions, the flexural vibration of the TLP tethers, the large amplitude motions of ships in steep waves and the wave run-up on the legs of offshore platforms. These effects must be modeled in the time domain due to their nonlinearity, the importance of viscous effects and the random and often nonlinear nature of severe storms.

In response to these challenges, two theoretical/computational methods have been developed. They are of quite different origin and objectives. The **SWIM** method was developed with the objective to predict the large amplitude slow-drift oscillation of ships and offshore platforms in random waves. The **SWIM** hydrodynamics module is based on a set of analytical algorithms for the solution of the linear forces and RAO's, slow-drift QTF matrices, and drift damping coefficients of multi-leg offshore platforms and slender ships [cf. 3,4,5]. The **SWIM** simulation module carries out the large amplitude slow-drift motion simulations in random waves in the time domain, independently of the evaluation of the hydrodynamic effects which are computed *a priori* in the frequency domain. Both modules enjoy good accuracy, ease of use and very high efficiency, permitting the generation several-hour long linear and slow-drift response records in directional random waves on a Pentium PC.

The computational method **SWAN** was developed in response to the emerging need to evaluate nonlinear forces and responses on ships and offshore structures in deterministic and random waves. The use of **SWAN** in practice is complementary to that of **SWIM**. As a three-dimensional Rankine panel method it is capable of treating arbitrary geometries and of evaluating linear, second-order and nonlinear wave loads and responses with an accuracy which is limited by the error in the assumed wave model. **SWAN** therefore may be characterized as a "Numerical Wave Tank" which would be used only when a more detailed and thorough analysis of an offshore structure is necessary. Moreover, in light of the emergence of fast personal PC's its execution time on such systems is quite comfortable.

2. LINEAR WAVE FORCES AND RAO's

SWIM

The solution of the linear wave-body interaction problem in **SWIM** is carried out in the frequency domain. Computations of the surge and pitch motion RAO's for the Troll Olje semi-submersible floater are presented in Figure 1 and compared with computations by **WAMIT**.

The solution of the linear problem for a ship in **SWIM** is carried out by implementing the slender body theory described in [6]. The heave and pitch motions of the Series 60 hull Cb=0.7 are compared to **WAMIT** in Figure 2 in finite depth.

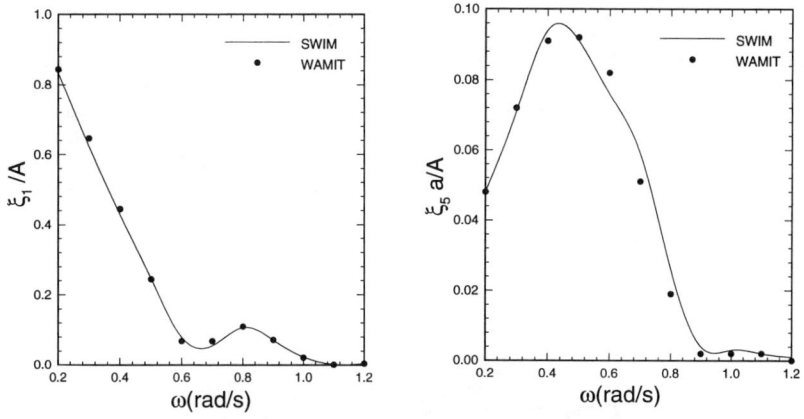

Figure 1: Surge and Pitch RAO's of the Troll Platform

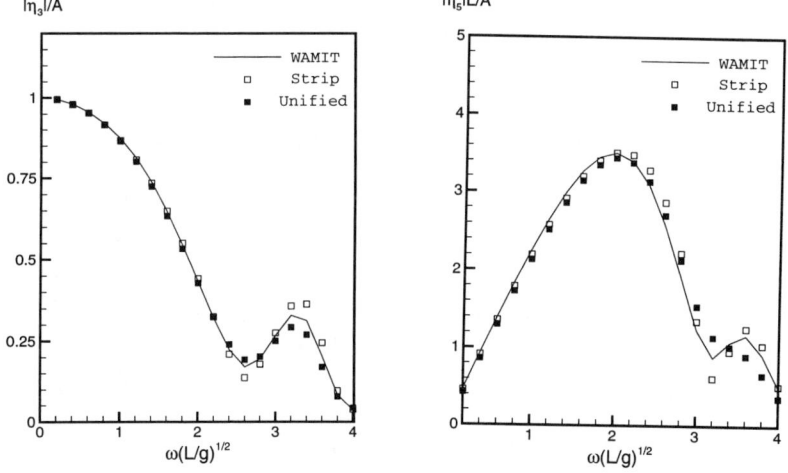

Figure 2: Heave and Pitch RAO's of the Series 60 Cb=0.7 Hull in Finite Depth

SWAN

SWAN is a time-domain three-dimensional Rankine panel method which distributes panels over the body geometry, the mean or actual position of the free surface and an annular dissipative beach designed to absorb the radiation and diffraction wave disturbances.

By its nature **SWAN** can handle geometries of arbitrary shape, enforce linear, second-order and nonlinear free surface conditions at a zero or finite forward speed. All computations are carried out in real arithmetic and the ambient wave disturbance may be monochromatic, polychromatic or random. The output force and response signal may then be processed by Fourier analysis to derive the linear force and response RAO's in the frequency domain or the second-order sum- or difference-frequency QTF matrices by cross-spectral analysis [cf. 7]. Figure 3 presents the body and free-surface mesh around a 4-cylinder structure.

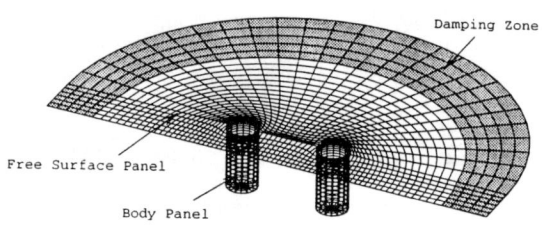

Figure 3: Body and Free-Surface Discretization by **SWAN**

Computations of the surge added-mass and damping coefficients are shown in Figure 4.

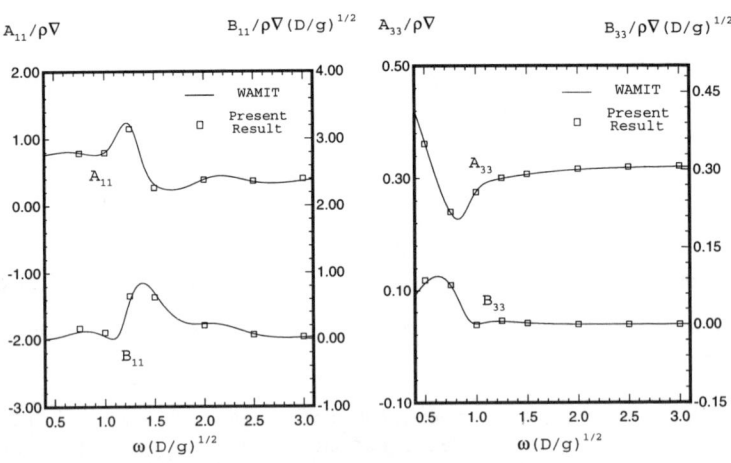

Figure 4: Surge and Heave Added Mass and Damping Coefficients of a 4-Cylinder Array

These forces have been derived by forcing the structure to undergo a surge oscillation with a displacement equal to the summation of the harmonics where the hydrodynamic coefficients are desired. The added-mass and damping coefficients are then obtained from the resulting force signal by Fourier analysis.

SWAN has been extended to the treatment of bodies with zero thickness. In such cases the Rankine panel method allows the representation of the ideal flow around such bodies as a distribution of a sheet of normal dipoles. Figure 5 presents the heave added-mass and exciting force on a submerged circular flat plate.

The case of a vertical cylinder with helical strakes is studied in Figures 6 and 7. The panel mesh in **SWAN** is illustrated in Figure 6. The velocity potential jump across the strake with increasing mesh density is plotted in Figure 7.

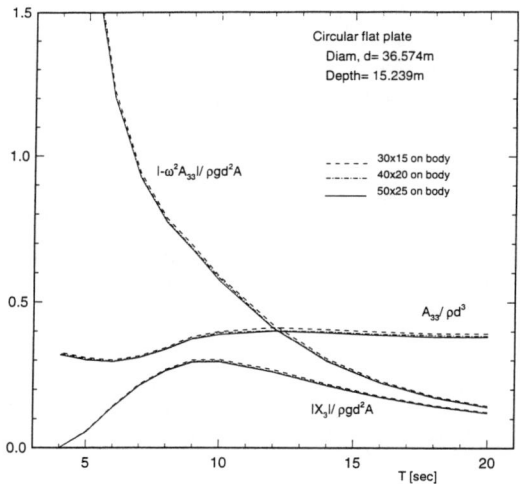

Figure 5: Heave Added-Mass and Exciting Force on a Submerged Circular Flat Plate

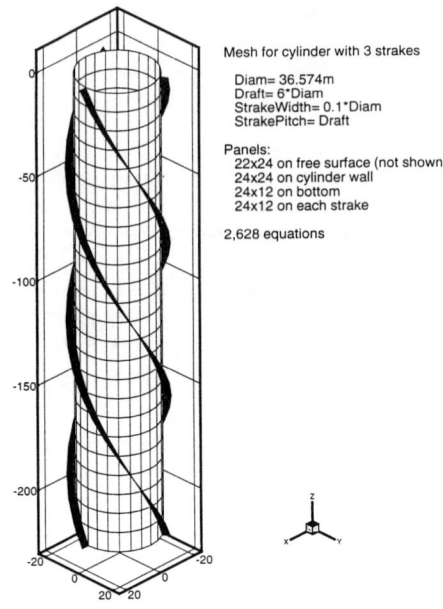

Figure 6: Discretization of a Straked Circular Cylinder

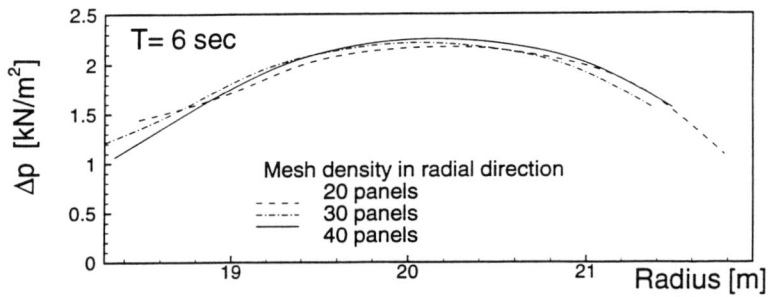

Figure 7: Convergence of the Linear Pressure Jump Across Strakes in a Wave with T=6s

3. SECOND-ORDER FORCES

Sum and Difference-Frequency QTF Matrices

SWIM

The second-order sum- and difference-frequency problem in **SWIM** is solved in the frequency domain, based on the linear multi-cylinder solution presented above. One of the unique features of the **SWIM** solution algorithm is that the infinite integrals over the free surface which enter in the definition of the sum- and difference-frequency forces are evaluated analytically leading to very efficient algorithms. More details on the method efficiency are presented at the end of the article.

Figure 8 presents **SWIM** and **WAMIT** computations of the difference-frequency QTF matrix diagonals for the Troll Olje semi-submersible.

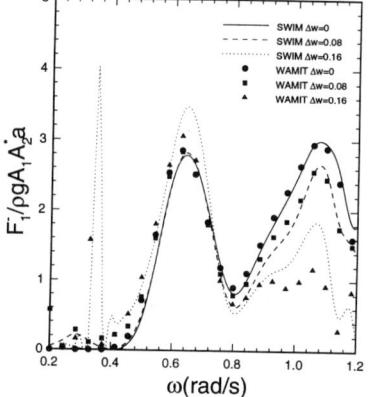 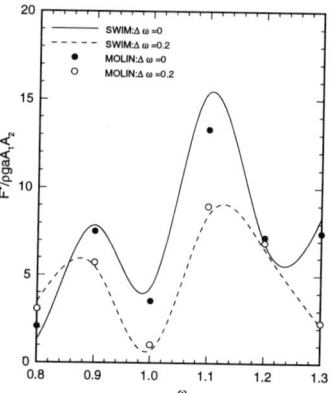

Figure 8: Difference-Frequency QTF Matrix of the Troll Platform - Sum-Frequency QTF Matrix for the Snorre TLP without Pontoons

Computations of the sum-frequency heave QTF matrix of the Snorre TLP (without pontoons) are also presented in Figure 8 and compared to computations by three-dimensional panel method.

SWAN

The solution of the second-order problem in **SWAN** is carried out in the time-domain. Based on the solution of the linear problem and evaluation of the velocity potential and its gradients over the body boundary and free surface, the solution of the second-order problem follows easily. By virtue of the solution algorithm, the evaluation of infinite free surface integrals is circumvented and the second-order forces are evaluated by direct pressure integration. The same solution algorithm is applied to the evaluation of the drift damping coefficient and QTF matrix, discussed below.

Figure 9 plots the sum-frequency heave exciting force on a truncated vertical cylinder. Comparison with benchmark computations show very good agreement.

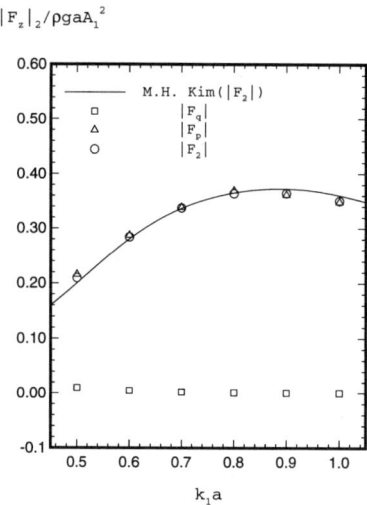

Figure 9: Sum-Frequency Heave Exciting Force on a Truncated Circular Cylinder

One attractive property of the **SWAN** time-domain solution algorithm is that it can accept as input a polychromatic incident wave, and produce as output the second-order force as one signal, in a manner analogous to an experiment in a wave basin. The sum-frequency QTF matrix may then be determined by cross-spectral Fourier analysis. More details are presented in [8].

Drift Damping Coefficients

SWIM

The drift damping coefficients are evaluated in **SWIM** by extending the solution algorithm to allow for a low forward speed as detailed in [9]. The surge drift damping coefficient on the Troll Olje Floater are compared in Figure 10 with computations by the three-dimensional panel method described in [10].

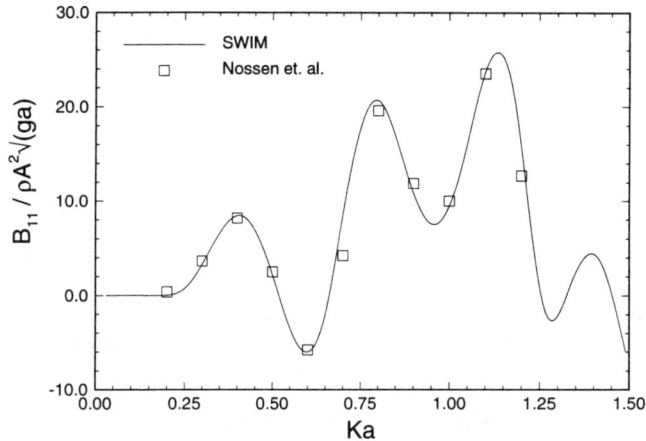

Figure 10: Surge Drift Damping Coefficient on a Four-Leg Platform

SWAN

The drift damping coefficients in **SWAN** are determined along the same lines outlined above, but allowing for a low forward speed. This is an easy extension of the method which was originally developed for the seakeeping problem of ships advancing with significant forward speed. The surge drift damping coefficient of a spar is compared in Figure 11 to predictions by **SWIM** which are regarded as benchmark

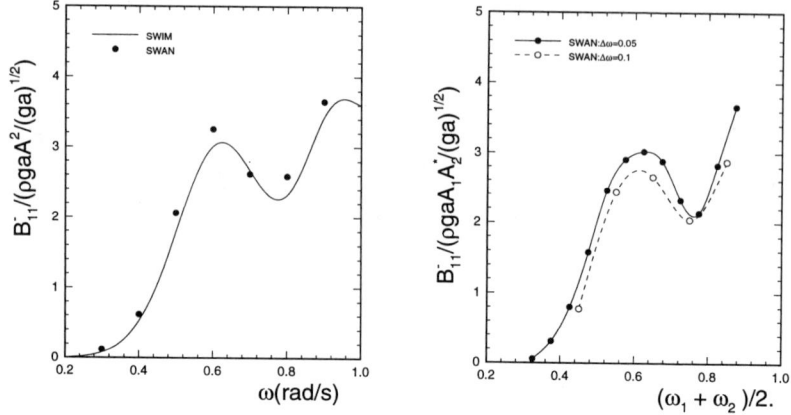

Figure 11: Surge Drift damping Coefficient of a Spar

One attribute of the time-domain solution algorithm in **SWAN** is that it allows the determination of the off-diagonal elements of the drift-damping QTF matrix. They are also plotted in Figure 11 for the same Spar. To our knowledge, no independent three-dimensional panel method computations are available for these quantities. Their use in predicting the slow-drift response of compliant structures is discussed later in this article.

SWAN was recently extended to the evaluation of the surge-sway-yaw drift damping coefficients of ships. Figure 12 presents the yaw drift damping computations by **SWAN** and by the independent three-dimensional panel method computations by Finne and Grue.

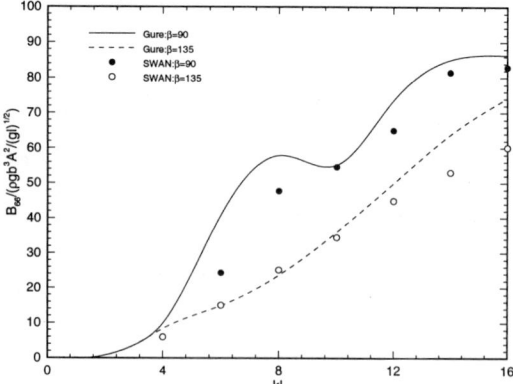

Figure 12: Yaw Drift Damping coefficient for Finne & Grue's Mathematical hull

OCEAN WAVE KINEMATICS, DYNAMICS AND LOADS ON STRUCTURES   165

4. NONLINEAR FORCES AND RESPONSES

The evaluation of the forces and responses of offshore structures in steep periodic or random waves is of considerable interest for the design of a platform in survival storms. Examples include the springing and ringing excitation of TLP tethers, the wave run-up and deck clearance in severe storms and the extreme heave and pitch motion of an FPSO.

The **SWAN** solution algorithm has been extended to the treatment of the nonlinear seakeeping of ships in steep ambient waves. The solution algorithm assumes that the radiation and diffraction wave disturbances are linearized around the *a priori* known ambient wave profile which may be periodic or random. While not a fully nonlinear treatment, this approach nonetheless accounts for the most important nonlinearities for the evaluation of the global forces and responses. Local effects, like the wave run-up for the case of a platform and slamming loads in the case of a ship may be evaluated *a posteriori* by a localized fully nonlinear model. This approach allows the treatment of the nonlinear problems in an accurate and computationally tractable manner for use in design.

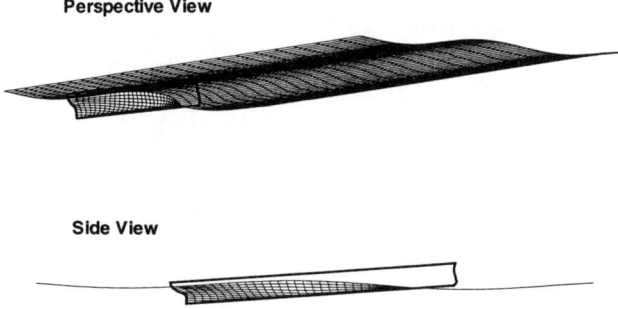

Figure 13: Meshing of the Nonlinear Interaction of a Ship with Ambient Waves

Figure 13 plots a perspective plot of the mesh over the ship hull and the free surface. The solution algorithm allows for the discretization of the instantaneous wetted surface of the ship hull at each time step and for an ambient wave profile which may be random.

Computations of the nonlinear heave and pitch motions on a realistic ship hull advancing in head periodic waves at a Froude number 0.325 are shown in Figure 14. Comparison with linear theory and experiments confirms that the solution algorithm in **SWAN** captures the most significant nonlinearities in the motions which are seen to differ appreciably from three-dimensional linear theory. The wave height for the simulations in Figure 14 is 3% of the ship length.

166 OCEAN WAVE KINEMATICS, DYNAMICS AND LOADS ON STRUCTURES

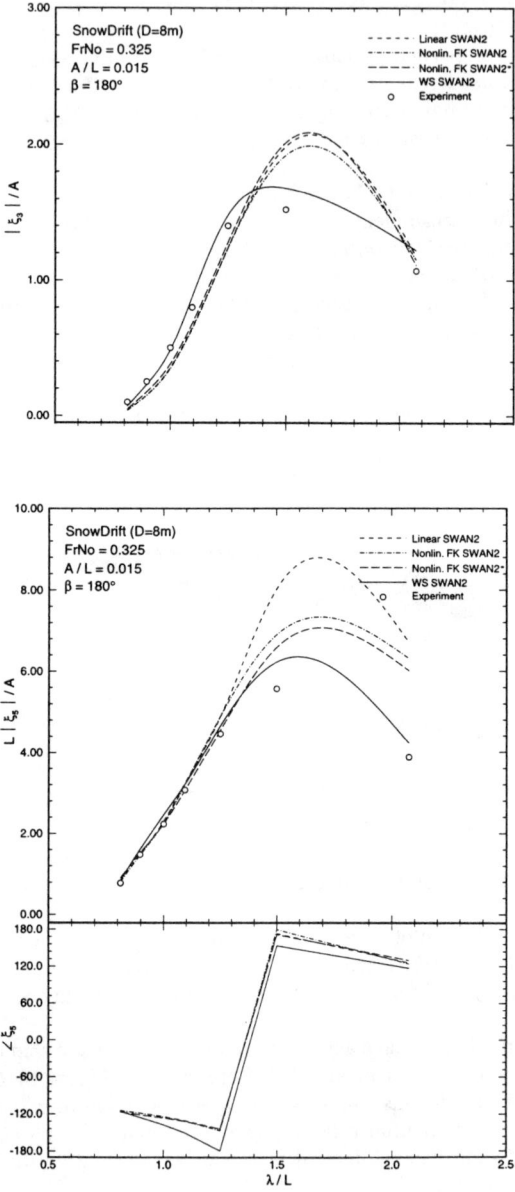

Figure 14: Linear and Nonlinear Heave and Pitch Motion RAO's by **SWAN**

OCEAN WAVE KINEMATICS, DYNAMICS AND LOADS ON STRUCTURES   167

5. SLOW-DRIFT MOTION SIMULATIONS

The preceding sections have outlined two computational algorithms for the evaluation of linear and nonlinear hydrodynamic forces acting on floating structures in controlled wave records. In all cases the structure was assumed to be stationary or to advance at a constant forward speed and the ambient waves were assumed to be either mono- or bi-chromatic.

Realistic conditions are far more complex! A compliant structure moored in a sea state, wind and current undergoes a complex slow-drift motion often of amplitude large relative to its characteristic dimensions. This response is primarily driven by second-order forces and leads to responses with time varying displacement, velocity and orientation. Moreover, the slow-drift responses are affected appreciably by viscous effects arising from the flow (air & water) separation around the structure and the mooring/tether/riser systems. In view of these nonlinearities, the slow-drift motions of compliant structures can only be simulated in the time domain and a method must be devised for the proper input of the frequency domain linear RAO's and second-order QTF wave effects discussed in earlier sections.

Such a method has been developed in [4] by taking advantage of the disparity of the time scales of the linear and slow-drift responses. This slow-drift motion simulation algorithm has been implemented in **SWIM**. It permits the a priori evaluation of all linear and second-order wave effects in the frequency domain by **SWIM, WAMIT, SWAN** or any other three-dimensional panel method. The slow-drift motion simulations may then be carried out in the time domain very efficiently for time intervals which may be over 100 hours long. Details on the implementation of the method may be found in [5].

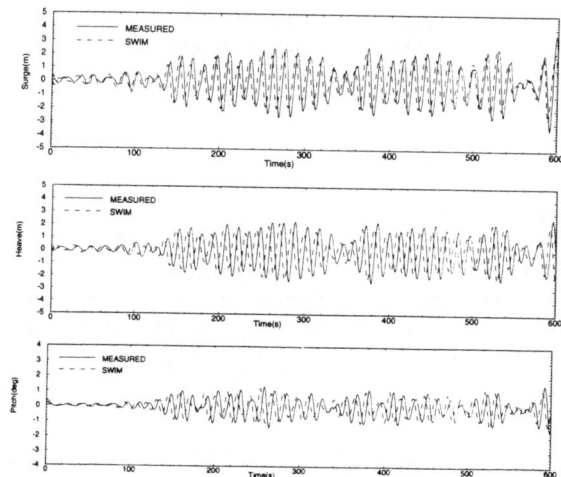

Figure 15. Linear Oscillations of the Troll Platform in Random Waves Computed by **SWIM** and Measured in Experiments

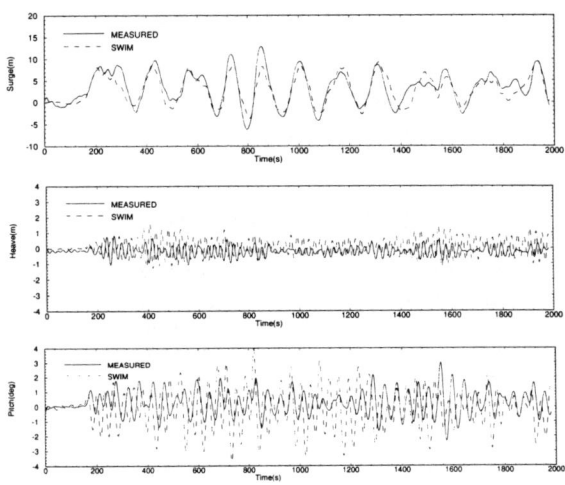

Figure 16. Slow-Drift Oscillations of the Troll Platform in Random Waves Computed by **SWIM** and Measured in Experiments

Figures 15 and 16 compare the linear and slow-drift motion simulations in a sea state by **SWIM** with experiments conducted on the Troll Olje semi-submersible. The agreement for the linear responses is excellent. For the slow-drift responses the agreement for the surge motion is very good. The corresponding correlation for the heave response is less satisfactory, and the discrepancies are attributed to the modeling of viscous effects around the structure and the mooring lines.

5. COMPUTATIONAL EFFORT

This section discusses the computational resources necessary for the execution of **SWIM** and **SWAN** for realistic structures. Both **SWIM** and **SWAN** have been developed on DEC Alpha workstations. However, with the advent of powerful Pentium PC's both programs have been benchmarked on such systems. For uniformity, most of the benchmarks are reported for a 300 MHertz Pentium II PC which a system which is comparably fast to a single processor Unix workstation.

SWIM

The solution of the linear RAO's require about 1 min for 30 frequencies for a ship or an offshore structure. The geometry input is in both cases very simple and does not require the discretization of the body wetted surface. The RAM memory requirements are negligible.

The evaluation of the sum- or difference-frequency full QTF matrices for an offshore platform in one wave heading and about 30 frequencies requires about 30 CPU minutes. The RAM necessary for the solution of the second-order problem in one wave heading is about 23 Mbytes.

The simulation of the slow-drift responses by **SWIM** proceeds very efficiently since it accepts as input all linear and second-order hydrodynamic quantities of interest. The integration of the slow-drift equations of motion in the time domain is carried out by a standard Runge-Kutta time marching scheme. The generation of a slow-drift response record 10 hours long requires about 10 CPU minutes and about 25 Mbytes of RAM. The computational effort and RAM requirements increase linearly with the length of the simulation record.

SWAN

As a three-dimensional Rankine panel method, **SWAN** requires the evaluation of the Rankine influence coefficients over the mesh of panels laid out over the body surface and the free surface. In the linear and second-order problems, the majority of the CPU time and RAM requirement is devoted to the set-up of the respective matrices and the LU decomposition which is carried out once for the linear and second-order problems. The respective problems are then solved in succession over the same panel mesh in real arithmetic.

For the linear problem, up to 2,000 panels are necessary over the body (1,000) and free surface (1,000) for a structure of modest complexity. The resulting RAM requirement is about 75 Mbytes and the simulation time for the linear problem proceeds with an efficiency ratio of about 1/1 relative to real time. So a linear response record 1 hour long in real time is necessary, the resulting CPU time is about 1 CPU hours. This applies equally to the zero and forward speed problems.

The number of panels necessary for the solution of the second order problems is higher, and about 4,000 for a realistic structure. The corresponding RAM requirement is about 250 Mbytes and the simulation time for the generation of a second-order force record is about 20/1 relative to real time. Therefore if a second-order force record 1 hours long is necessary, the corresponding CPU time will be about 20 hours.

Nonlinear SWAN

The computational effort required for the simulation of the nonlinear forces and responses by the nonlinear extension of the method is considerably higher than the linear and second-order versions of the method. The reason lies with the need to discretize the instantaneous position of the wetted hull surface and the ambient wave profile.

Its efficiency ratio relative to real time is about 50/1, which translates into a 50 hour CPU effort, or about 2 days, for the generation of a 1 hour real time simulation record. Work is underway to improve the method efficiency aiming at a ratio of 10/1. Yet, such nonlinear simulation tools would normally be used in conjunction with their linear counterparts to study local nonlinear events. Therefore, the nonlinear method would be introduced for time windows which are a lot smaller that 1 CPU hours, reducing considerably the computational effort necessary for a routine design study.

6. ACKNOWLEDGMENTS

The development of **SWIM** has been supported by a Joint Industry Project consisting of the following sponsors: Amoco, Aker, Conoco, DnV, Exxon, Norsk Hydro, Saga and Statoil. The development of **SWAN** has been supported by the Office of Naval Research and Det Norske Veritas (DnV). The financial support from all sponsors is greatly appreciated.

7. REFERENCES

[1] Breit, S. R., Newman, J. N. and Sclavounos. P. D. (1985). A New generation of Panel Programs for Radiation-Diffraction Problems. BOSS'85 Conference.

[2] Lee C-H and Newman, J. N. (1994). Second-Order Wave Effects on Offshore Structures. BOSS'94 Conference.

[3] Emmerhoff, O. J. and Sclavounos, P. D. (1992). The Slow-Drift Simulation of Arrays of Vertical Cylinders. *Journal of Fluid Mechanics*, Vol. 242, pp. 31-50.

[4] Sclavounos, P. D. (1994). Slow-Drift Oscillation of Compliant Floating Platforms. BOSS'94 Conference.

[5] Emmerhoff, O. J. and Sclavounos, P. D. (1996). The Simulation of Slow-Drift Motions of Offshore Platforms. *Applied Ocean Research*, Vol. 18, 55-64.

[6] Kim, Y. and Sclavounos, P. D. (1998). A Finite-depth Unified Theory for the Linear and Second-Order problem of Slender Ships. Submitted for publication.

[7] O'Dea, J., Powers, E. and Zselecsky, J. (1994). Experimental Determination of Nonlineraities in Vertical Plane Ship Motions. Proc. ONR Conference, Korea.

[8] Kim, Y., Kring, D. C. and Sclavounos, P. D. (1998). Linear and Nonlinear Interactions of surface Waves with Bodies by a Three-Dimensional Rankine Panel Method. To appear in *Applied Ocean Research*

[9] Kim, S., Sclavounos, P. D. and Nielsen F-G. (1997). Slow-Drift Responses of Moored Platforms. BOSS'97 Conference.

[10] Noosen, Grue, J. and Palm, E. (1993). The Mean Drift Force and Yaw Moment on Marine Structures in Waves and Current. *Journal of Fluid Mechanics*, Vol 250, pp. 121-142.

Spectral response surfaces, designer waves and the ringing of offshore structures

Peter S. Tromans[1]
I. Ketut Suastika[2]

Abstract

The purpose of this paper is to apply the spectral response surface method to the ringing problem. We have re-formulated Newman's results for diffraction force on a column in terms of the frequency components of the ocean surface and their Hilbert transforms. The resulting expression, together with a single degree of freedom model for the structure, allows us to generate hyper-surfaces of constant dynamic response as functionals in the Hilbert space of the spectral components of the ocean surface. As the components are uncorrelated, linear processes, it is straightforward to treat the response surfaces as limit states in a FORM (first order reliability method) type of analysis. This provides a very efficient means of calculating the ocean surface history most likely to generate an extreme ringing response (the "designer wave" for ringing) and the ringing response history.

Introduction

Ringing may be described as a response resembling that generated by an impulse excitation of a linear oscillator: it features a rapid build up and slow decay of energy at the resonant frequency of the structure. It may be possible that ocean waves excite ringing in an offshore structure. Indeed, it has been observed in small scale experiments on gravity base structures during the passage of steep waves (Davies et al, 1994). The natural frequency of the structure and the observed response were relatively high compared to the dominant wave frequencies. However, some frequencies in the exciting force must have coincided with the natural frequency of the structure. This implies a non-linear loading model is required to predict the forces at the sum frequencies of the ocean wave field.

Newman (1996) studied the interaction between a body and random waves in the long wavelength regime. He concluded that non-linearity in loading is due more

---
[1] Shell International Deepwater Services, PO Box 3006, 2280 MH Rijswijk, The Netherlands.
[2] TU Delft, Dept. of Civil Engineering, PO Box 5048, 2600 GA Delft, The Netherlands.

to fluid structure interaction than to non-linear features in the incident waves. In particular, if one considers only the loads which occur at the sum of the wave frequencies, the incident wave field is described adequately by linear theory for loads up to third order. We note that the Newman's expression for horizontal force on a column is consistent with the slender body results of Rainey (1995). We shall use Newman's formulation in this paper.

Ringing becomes important if it controls the extreme response of a structure to a random sea. Therefore, a stochastic analysis is required and we shall use the spectral response surface method. In this, a linear random sea is represented by the sum of many frequency components, each obeying a normal distribution. In many cases, a structural response can be expressed as a function of these components and their Hilbert transforms. A FORM type of analysis is then applied, treating a surface of constant response level as a limit state. More technical details will be given in a future paper. The method has use in several ocean engineering problems (Tromans and Taylor, 1998, and Tromans and van Dam, 1996)

Development of the response model - applied load

Consistent with a linear description of the incident waves, we assume that the ocean surface elevation is a sum of many, random, narrow banded components. Each component is normally distributed. All the components are independent and uncorrelated. Thus, the ocean surface at a fixed point is

$$\eta(t) = \sum_j \eta_j = \sum_j \underline{A}_j \sin(\omega_j t - \underline{\vartheta}_j) \qquad (1)$$

where $\underline{A}_j$ is random amplitude, ω_j is frequency, $\underline{\vartheta}_j$ is random phase angle, and t is time. The frequency components can be transformed into standarised (unit-variance, zero mean) variables by dividing each by its standard deviation. Thus,

$$x_j = \frac{\eta_j}{\sigma_j}, \quad \tilde{x}_j = \frac{\tilde{\eta}_j}{\sigma_j} \qquad (2)$$

where η_j is the jth spectral component and $\tilde{\eta}_j$ is its Hilbert transform. The Hilbert transform is the signal phase shifted by 90 degrees. The joint density function of the of the standardized variables, x_j, $\tilde{x}_j$, is then unit-variance normal.

Newman (1996) calculated the diffraction forces generated by such an incident sea on a vertical, cylindrical column in the limit where the wavelength is long compared with the column diameter. In a coordinate system with the x and y axes in the mean water surface, x oriented in the wave direction and z vertical, the first order force on the column is

$$F_1 = \frac{1}{2}\pi D^2 \rho \int_{-\infty}^{0} u_t\, dz \tag{3.1}$$

This is the 'Morison inertia load' below mean water line. The second-order force is

$$F_2 = \frac{1}{4}\pi D^2 \rho \int_{-\infty}^{0}(2ww_x + uu_x)\,dz + \frac{1}{2}\pi D^2 \rho u_t \eta_1 \tag{3.2}$$

The third-order force is

$$F_3 = F_3^{(1)} + F_3^{(2)} \tag{3.3}$$

where $F_3^{(1)} = \frac{1}{4}\pi D^2 \rho \left[\eta_1 \left(u_{tz}\eta_1 + 2ww_x + uu_x - \frac{2}{g}u_t w_t \right) - (u_t/g)(u^2 + w^2) \right]$

and $F_3^{(2)} = (\pi D^2 \rho/g) u^2 u_t + O(\varepsilon^6)$

Both the third order terms are point forces acting at mean water line.

Given linear wave kinematics, we can use (2) to express Newman's forces in terms of the standardized variables, x_j, $\tilde{x}_j$. The following expressions result:

$$F_1 = \frac{1}{2}\pi D^2 \rho g \sum_j \sigma_j \tilde{x}_j \tag{4.1}$$

$$F_2 = \frac{1}{4}\pi D^2 \rho \sum_j \sum_k \left(\frac{2\omega_j \omega_k^3 - \omega_j^3 \omega_k}{\omega_j^2 + \omega_k^2} + 2\omega_j^2 \right) \sigma_j \sigma_k \tilde{x}_j x_k \tag{4.2}$$

$$F_3 = \frac{1}{4}\pi D^2 \rho \frac{1}{g}\left[\sum_j \sum_k \sum_l \left(\omega_k^4 + 2\omega_k \omega_l^3 - \omega_l \omega_k^3 + 2\omega_k^2 \omega_l^2 + 3\omega_j \omega_k^2 \omega_l \right) \sigma_j \sigma_k \sigma_l x_l \tilde{x}_k x_l \right.$$
$$\left. - \sum_j \sum_k \sum_l \omega_j^2 \omega_k \omega_l \sigma_j \sigma_k \sigma_l \tilde{x}_j \tilde{x}_k \tilde{x}_l \right] \tag{4.3}$$

Development of the response model - dynamic response

The dynamic behaviour of the structure is modelled as a single degree of freedom system. The load is applied directly on the mass and the response is the force transmitted by the spring. The dynamic amplification and the phase shift are given by

$$D(\omega) = \cfrac{1}{\left\{\left[1-\left(\cfrac{\omega}{\omega_n}\right)^2\right]^2 + \left[2\zeta\cfrac{\omega}{\omega_n}\right]^2\right\}^{1/2}} \qquad \tan(\theta) = -\cfrac{2\zeta\cfrac{\omega}{\omega_n}}{1-\left(\cfrac{\omega}{\omega_n}\right)^2}$$

The application of this transfer function to equations (4.1) to (4.3) leads to the corresponding responses R_1, R_2, R_3. The first order response is

$$R_1 = \frac{1}{2}\pi D^2 \rho g \sum_j \sigma_j \left[\tilde{x}_j D_j \cos(\theta_j) - x_j D_j \sin(\theta_j)\right] \tag{5.1}$$

where $D_j = D(\omega_j)$ and $\theta_j = \theta(\omega_j)$, are the dynamic amplification factor and the phase shift at frequency ω_j. The second order response is

$$R_2 = \frac{1}{8}\pi D^2 \rho \sum_j \sum_k \left(\frac{2\omega_j \omega_k^3 - \omega_j^3 \omega_k}{\omega_j^2 + \omega_k^2} + 2\omega_j^2\right) \sigma_j \sigma_k \left\{\tilde{x}_j \tilde{x}_k \left[D_{j+k}\sin(\theta_{j+k}) - D_{j-k}\sin(\theta_{j-k})\right]\right.$$

$$+ \tilde{x}_j x_k \left[D_{j+k}\cos(\theta_{j+k}) + D_{j-k}\cos(\theta_{j-k})\right] + x_j \tilde{x}_k \left[D_{j+k}\cos(\theta_{j+k}) - D_{j-k}\cos(\theta_{j-k})\right]$$

$$\left. + x_j x_k \left[-D_{j+k}\sin(\theta_{j+k}) - D_{j-k}\sin(\theta_{j-k})\right]\right\} \tag{5.2}$$

where $D_{j\pm k} = D(\omega_j \pm \omega_k)$ and $\theta_{j\pm k} = \theta(\omega_j \pm \omega_k)$ are the dynamic amplification factor and the phase shift at frequency $(\omega_j \pm \omega_k)$. The expression for the third order response, R_3, is similar, but rather long. It can be found in Suastika (1997). The total response is $R = R_1 + R_2 + R_3$. R = constant defines a surface in the space of the standardized variables, x_j, $\tilde{x}_j$.

The spectral response surface

Using the expressions above, it is possible to generate surfaces of constant response in the space of the spectral components of the ocean surface. We recall that the spectral components and their Hilbert transforms are un-correlated, linear processes, they obey a joint normal distribution with zero cross correlation. Thus, surfaces of constant probability density are concentric spheres in the space of the standardised variables. The probability density is highest at the origin and falls montonically as a function of distance from the origin. Under these circumstances, it is straightforward to treat the response surfaces as limit states in a FORM type of analysis. The point on a surface of constant response where the distance to the origin is shortest and the probability density greatest is called the "design point." This is, to a good approximation, the point where a response maximum is most likely. We find the design point using Lagrange's method of undetermined multipliers (Melchers, 1987).

Since the design point $(x_j^*, \tilde{x}_j^*)$ defines the amplitude and phase of the frequency components at the instant when the extreme occurs, it allows us to deduce the time histories of the response and related variables around the time of the extreme. These histories are those most likely to be associated with a response maximum. Thus, in the case of the ringing problem, we can identify if ringing determines the extreme and identify the type of applied load history that excites it. In addition, we find the ocean surface history that generated it all. Finally, we can estimate the exceedance probabilities of extreme ringing responses very efficiently.

Extreme ringing responses

We studied the case of a uniform 10 m diameter column in a sea with a significant wave height of 15 m and a zero crossing period of 13.5 s that obeys a JONSWAP spectrum. The peak of the surface energy spectrum is approximately 0.36 rad/s. The water depth is 300 m. The dynamics are modelled with a damping coefficient that is 2.5 % of critical and a natural frequency of 1.57 rad/s. We shall discuss the case of the maximum negative or "swing back" response, that is in the direction opposed to the waves. We select this since it is more severe than the swing forward response. We describe here the solutions we found for an extreme response level such that the probability that an individual maximum exceeds it is 10^{-4}.

We investigated the effects of using the Newman load model to first, second and third order. The response time histories are plotted in figures 1 to 3. The contribution to the dynamic response from the higher order forces is very significant. There is some evidence of ringing from second order forcing. With third order forcing the ringing is unmistakable. The corresponding surface elevation is shown in figure 4.

To gain insight into the processes, we look more closely at the third order contribution to the response. From the design point we have calculated, it is possible to reconstruct the time series of all the components of the response and of the excitation. The time history of the third order component of the response is plotted in figure 5 and the corresponding force excitation in figure 6. The third-order component of response resembles the response of a linear oscillator to an impulse excitation: a rapid increase of energy at the natural frequency followed by a slow decay due to damping. The corresponding exciting force in figure 6 supports this view. In fact, the excitation takes the form of a double impulse. The double impulse is associated with a wave crest; the positive impulse immediately precedes the crest and the negative impulse follows it. Apart from the isolated double impulse the third order excitation is remarkably small. Though not plotted here, similar, but less pronounced, effects are seen at second order.

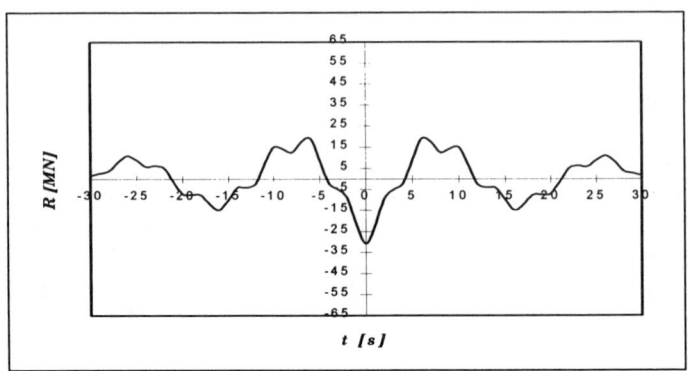

Figure 1. Dynamic response with only first-order excitation

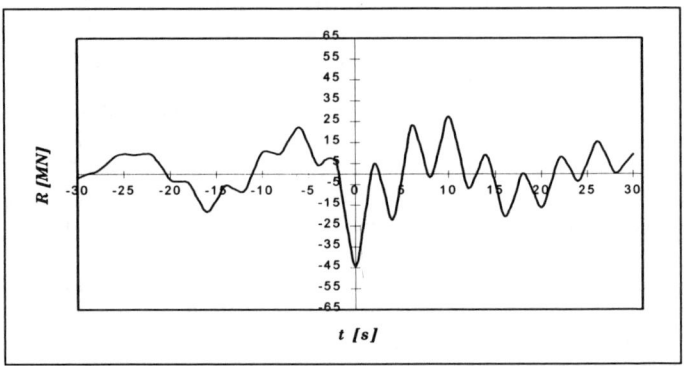

Figure 2. Response with first and second-order excitation

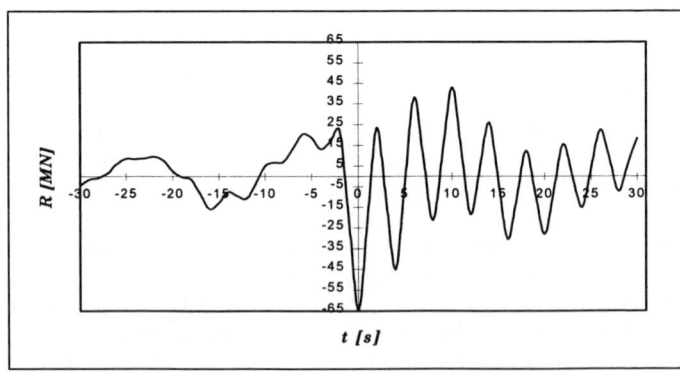

Figure 3. Response with first, second and third-order excitation

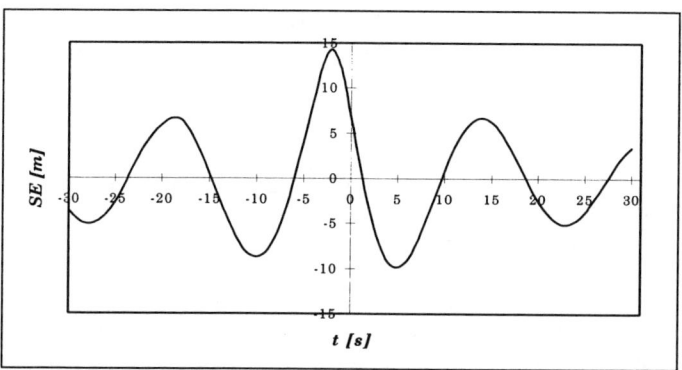

Figure 4. Surface elevation generating the response time series plotted in figure 3

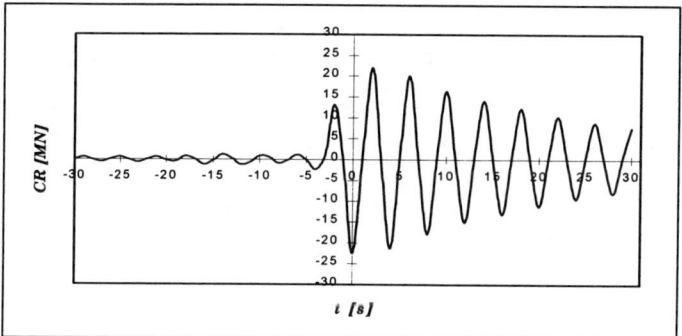

Figure 5. Third-order component of the dynamic response

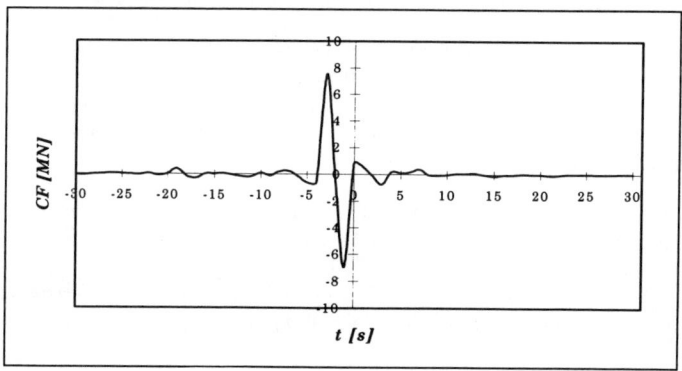

Figure 6. Third-order component of the force excitation

Finally, we note that the results provided here involved a few minutes of computer simulation. To achieve similar results by time domain calculation of random waves would involve many hours of computing.

Conclusions

1. The ringing generated by non-linear terms in Newman's model can have a significant effect on dynamic response.
2. Some ringing effects can be found in the response with only second order excitation.
3. Ringing arises from an impulse generated by non-linear loading.
4. The spectral response surface method is tractable and much faster than random time domain simulation.

Acknowledgements

I.K. Suastika's work formed his graduation project from T.U.Delft, performed at Shell International Deepwater Services. He expresses his gratitude to Delft supervisors: Prof. J.H. Vugts, Prof. J.A. Battjes, Prof. A.C.W.M. Vrouwenvelder and W.W. Massie. He thanks Shell, particularly P.H. Taylor, F. Baar and P.S. Tromans.

References

Melchers, R.E. (1987). Structural reliability analysis and prediction, Ellis Horwood Limited, Chichester, U.K.

Newman, J.N. (1996). Non-linear scattering of long waves by a vertical cylinder, paper presented in the book: J. Grue et al (eds.), Waves and non-linear processes in hydrodynamics, pp. 91 - 102, Kluwer academic publisher, The Netherlands.

Rainey, R. C. T. (1995). Slender body expressions for the wave load on offshore structures. Proc. R. Soc. Lond., A 450, pp. 391-416.

Suastika, I.K. (1997). Ringing of gravity structures. Project Report, Offshore Technology, Dept. Of Civil Engineering, Technical University of Delft.

Tromans, P.S. and Taylor, P.H. (1998). The shapes, histories and statistics of non-linear wave crests in random seas. Proc. of OMAE Conf., Lisbon.

Tromans, P.S. and van Dam, J. (1996). The probability distribution of extreme load on space frame structures. Proc. of OMAE Conf., Florence.

Forced Motions of a Pneumatic Floating Platform

Kwok Fai Cheung[1], Ludwig H. Seidl[2], Amal Phadke[3], and Sing Kwan Lee[4]

Abstract

This paper describes a numerical model to simulate the forced motions of a pneumatic floating platform. The platform is composed of a number of open-bottom cylinders trapping pressurized air that displaces the water. The cylinder diameter is much smaller compared to the wavelength and thus the water inside each cylinder oscillates in the vertical direction like a piston. The numerical solution to the boundary value problem is obtained by the source distribution method. Through a numerical example, the general hydrodynamic characteristics of this pneumatic platform are discussed.

Introduction

Most conventional floating platforms acquire flotation force by directly displacing the water from their hulls. A pneumatic floating platform utilizes indirect displacement, in which the platform has an open bottom trapping pressurized air that displaces the water. The buoyancy force to carry the weight of the structure is provided by air pressure acting on the underside of the deck. This introduces a compressible element between the platform and the water, which might affect the stability of the dynamic system.

---

[1] Assistant Professor and Corresponding Author, Department of Ocean Engineering, University of Hawaii at Manoa, Honolulu, HI 96822
[2] Professor, Department of Ocean Engineering, University of Hawaii at Manoa, Honolulu, HI 96822
[3] Graduate Research Assistant, Department of Ocean Engineering, University of Hawaii at Manoa, Honolulu HI 96822
[4] Post-Doctoral Fellow, Department of Ocean Engineering, University of Hawaii at Manoa, Honolulu, HI 96822

Seidl (1980) experimentally investigated a few design concepts of the pneumatic floating platform, one of which was later applied to study the launching of a large floating structure (Chakrabarti 1995). Both studies provide detailed formulations of the hydrostatic problem and discuss scale effects in modeling the compressibility of air in laboratory studies. Theoretical studies on these structures have been focused on their use as wave-power devices and breakwaters (e.g., Evans 1982, Falnes and McIver 1985, and Kim and Iwata 1991).

The pneumatic floating platform considered in this study is composed of a number of open-bottom cylinders, each trapping a pressurized air cavity and a water column (see figure 1). In a particular sea state, the platform can be tuned to minimize the overall hydrodynamic loads and response. Because of its adaptive characteristics, this pneumatic platform is being considered as a building block for breakwaters, marine terminals, floating airports and offshore military bases. To provide a design tool, the present study develops a numerical model to simulate the dynamic characteristics of this platform.

Theoretical Formulation

The oscillation of the water columns inside the pneumatic cylinders in response to incident wave excitations and forced motions of the platform is considered here. The diameter of the cylinders is small compared to the wavelength and the oscillation of the water columns can be treated as that of pistons. The reaction of the air inside each cylinder is modeled by a linear spring and damper system. This simplifies the formulation of the hydrodynamic problem in that the free surfaces inside the cylinders need not be modeled and the base of the platform can be modeled as a continuous surface on which the hydrodynamic pressure is calculated and applied.

Let t denote time and ω the angular frequency. The relative velocity of the water column oscillation in response to forced motions of the platform and incident wave excitation is given by

$$u_k^m = \begin{cases} \text{Re}\left(i\omega \xi^m \eta_k^m e^{i\omega t}\right) & \text{for } m = 1, \ldots, 6; k = 1, 2, \ldots, N_c \\ \text{Re}\left(i\omega A \eta_k^m e^{i\omega t}\right) & \text{for } m = 7; k = 1, 2, \ldots, N_c \end{cases} \quad (1)$$

where ξ^m denotes the complex amplitudes of the platform oscillation in the six degrees of freedom, A is the incident wave amplitude, η_k^m is the relative displacement of water column k due to a unit amplitude excitation in the mth mode and N_c is the total number of cylinders.

With the assumption of potential flow theory, the fluid motion can be described by a velocity potential satisfying the Laplace equation and the boundary conditions. The linear hydrodynamic problem can be separated into components associated with the diffraction and radiation of waves from the platform with a rigid bottom and the radiation of waves due to oscillation of the water columns relative to the rigid body response:

$$\Phi = \text{Re}\left[A\left(\phi^0 + \phi^7 + \varphi^7\right)e^{i\omega t} + \sum_{m=1}^{6}\xi^m\left(\phi^m + \varphi^m\right)e^{i\omega t}\right] \quad (2)$$

where ϕ^0 is the known incident potential, ϕ^7 is the scattered potential, and ϕ^m are the radiation potentials associated with the forced motions of the platform with a rigid bottom; and φ^m are the radiation potentials due to the oscillation of the water columns.

The solution for the rigid-body problem involving ϕ^m can be obtained by a standard method (e.g., Faltinsen and Michelsen 1974) and is not discussed here. Our discussion focuses on the radiation potentials φ^m associated with the oscillation of the water columns relative to the platform motions. Using the source distribution method, the radiation potential can be expressed as

$$\varphi^m = \frac{1}{4\pi}\int_S f(\mathbf{x}')G(\mathbf{x},\mathbf{x}')dS \quad (3)$$

where S includes the exterior cylinder walls (wetted) and the base of the cylinders, $\mathbf{x}$ and $\mathbf{x}'$ are respectively the position vectors of the field and source points, f is the source strength, and G is the free-surface Green function.

With the surface S discretized into N constant panels, the potential φ^m can be expressed in terms of its normal derivative as:

$$\{\varphi^m\} = [K]\left\{\frac{\partial \varphi^m}{\partial n}\right\} \quad (4)$$

in which $[K] = [J][I]^{-1}$ is an $N \times N$ matrix, where $[I]$ and $[J]$ are the $N \times N$ influence and potential coefficient matrices calculated respectively from the Green function and its normal derivative. The normal velocity in equation (4) can be expressed in terms of the unknown η_k^m through equation (1).

With the flow potentials in equation (2) defined, the total hydrodynamic force acting on the base of cylinder k is obtained by an integration of the pressure and can be written as:

$$f_k^m = (f_k^m)_{\text{rigid}} + \omega^2 \sum_{l=1}^{N_c}\gamma_{kl}\eta_l^m - i\omega\sum_{l=1}^{N_c}\lambda_{kl}\eta_l^m \quad m = 1,\ldots,7 \quad (5)$$

where $(f_k^m)_{\text{rigid}}$ is the hydrodynamic force associated with the rigid-body components, and γ_{kl} and λ_{kl} are the added mass and damping coefficients accounting for the oscillation of the water columns. Both γ_{kl} and λ_{kl} can be evaluated from the matrix $[K]$ in equation (4). The process represents a condensation of the $N \times N$ matrix associated with the motions at all the facets to an $N_c \times N_c$ matrix associated with the oscillation of the water columns.

For a unit amplitude excitation in the mth mode ($m = 1, \ldots, 7$), the equations of motion of the water columns can be obtained by equating the inertia and restoring forces with the hydrodynamic forces

$$-\omega^2 \left([m] + [\gamma]\right) \{\eta^m\} + i\omega([\lambda] + [c])\{\eta^m\} + ([k] + [k'])\{\eta^m\} = \{f^m\}_{\text{rigid}} + \omega^2[m]\{\zeta^m\} - [k']\{\zeta^m\} \quad (6)$$

where $[m]$ is the mass matrix of the water columns, $[k]$ is the air cavity stiffness matrix, $[c]$ is the air cavity damping matrix, $[k']$ is the hydrostatic stiffness matrix of the water columns, and $\{\zeta^m\}$ is the vertical displacement of the cylinders due to the mth mode of platform oscillation. Both the air cavity stiffness and damping matrices are determined from the ideal gas law. All four matrices $[m]$, $[c]$, $[k]$ and $[k']$ are $N_c \times N_c$ and diagonal; and the equations are coupled through the added mass and damping matrices $[\gamma]$ and $[\lambda]$, which are full. Since the quantities on the right-hand side of equation (6) are known, the oscillation of the water columns can be evaluated.

Results and Discussion

The numerical model is applied to a prototype design of the pneumatic platform that consists of 75 open-bottom cylinders packed together in a rectangular pattern of 5 × 15. Each cylinder is 12.2 m high and has an external diameter of 6.1 m. The cylinder wall is 10.2 cm thick. The drafts of the air cavity and cylinder are 3.05 m and 6.1 m respectively. Figure 2 shows the numerical discretization of the platform with 370 quadrilateral facets. Due to symmetry, only one half of the total surface is modeled. The sealed interstices between the cylinders are modeled as infinitely stiff air cavities. The cylinder walls which are very thin compared to the wavelength are not modeled. For the wave periods of interest (8 to 16 seconds), this mesh provides an adequate resolution of the structure configuration and the hydrodynamic pressure.

To illustrate the hydrodynamic characteristics of the pneumatic platform, the response amplitude operators (RAO) of the water column oscillation are shown in figures 3a to 3d for an excitation period of 8 second under deep water conditions. In the figures, only the results on one side of the symmetric plane are shown and S_1 indicates the row of cylinders along the centerline of the platform. For surge and heave, the RAOs are defined as the ratios between the amplitudes of the water column oscillation and the platform forced motion, while the RAO for pitch is defined with respect to a unit rotation. For the incident and scattered wave excitation, the RAOs of the water column oscillation are defined with respect to the incident wave amplitude.

Although the cylinders do not move vertically in the surge motion of the platform, as shown in figure 3a the water columns are excited by the surface waves generated by the motion. The resulting oscillation of the water columns is anti-symmetric in the fore-aft direction. Figures 3b and 3c show the amplitudes of the water column oscillation in response to forced heave and pitch motions of the platform, in which the water columns are not only excited by the surface waves generated, but also by the vertical motions of the cylinders. In figure 3d, the platform is fixed and subject to the incident head waves propagating from left to right. The results indicate a reduction of the oscillation amplitude in the direction of wave propagation, which is consistent with the attenuation of the hydrodynamic pressure along the platform.

The oscillation of the water columns has been observed to closely correspond to the excitation. Because of dynamic effects, a phase lag exists between the oscillation of the platform and water columns, and the phase angle of water column oscillation also varies across the platform. For longer periods, the phase lag between the excitation and response diminishes, and the dynamic responses of the water columns closely resemble to those of static cases.

Conclusions

The theoretical formulation and numerical procedures to simulate the forced motions of a pneumatic floating platform are presented in this paper. The platform is composed of a number of pneumatic cylinders, each trapping a water column and a pressurized air cavity. The linear hydrodynamic problem is separated into components associated with the radiation and diffraction of waves from the platform with a rigid bottom and the radiation of waves due to oscillation of the water columns. In the equations of motion, the cylinders are modeled as spring-mass-damper systems, which are coupled together through the hydrodynamic interaction. The forcing to the dynamic system is derived from the hydrodynamic loading on the platform with a rigid bottom. The capability of the numerical model and the hydrodynamic characteristics of the pneumatic floating platform are demonstrated through a numerical example. Laboratory model studies are currently being conducted at the University of Hawaii to verify the numerical model described in the paper.

References

Chakrabarti, S. (1995). Scale effects on a unique launch sequence of a gravity-based structure. *Applied Ocean Research,* 15, 33–41.

Evans, D. (1982). Water-power absorption by systems of oscillating pressure distributions. *Journal of Fluid Mechanics,* 114, 481–499.

Falnes, J. and P. McIver (1985). Surface wave interactions with a systems of oscillating bodies and pressure distributions. *Applied Ocean Research,* 7(4), 225–234.

Faltinsen, O. and F. Michelsen (1974). Motions of large structures in waves at zero Froude number. In *Proceedings of the International Symposium* on the *Dynamics of Marine Vehicles and Structures in Waves, London,* pp. 3–18.

Kim, D. and I. Iwata (1991). Dynamic behavior of tautly moored semi-submerged structure with pressurized air-chamber and resulting wave transformation. *Coastal Engineering in Japan,* 34(2), 223–243.

Seidl, L. (1980). Development of an air stabilized platform. Technical report, University of Hawaii at Manoa. Department of Ocean Engineering, submitted to US Department of Commerce, Maritime Administration.

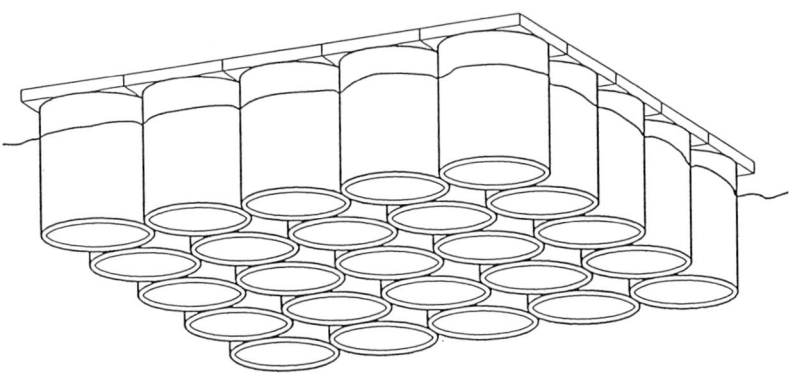

Figure 1. Pneumatic platform module.

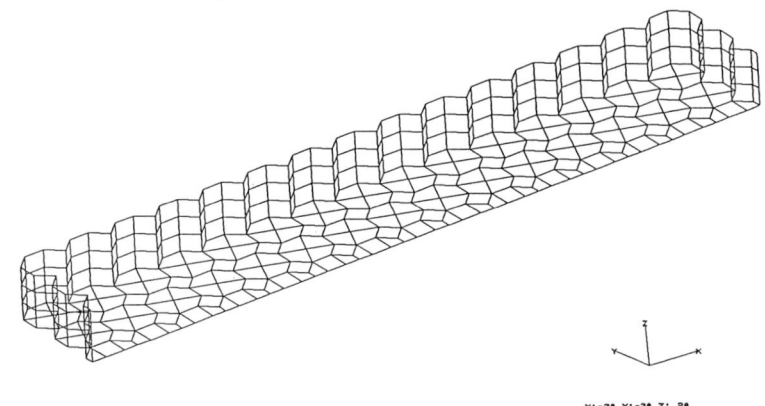

Figure 2. Numerical model with 370 facets.

(a)

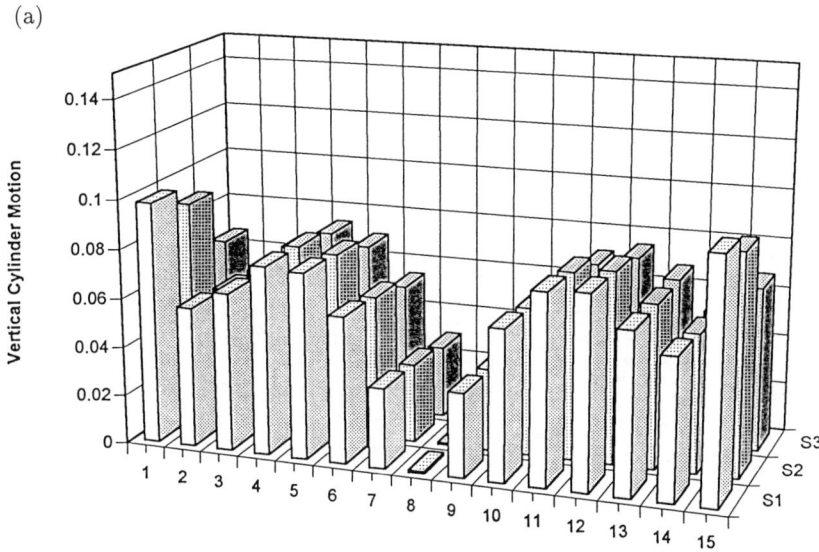

(b)

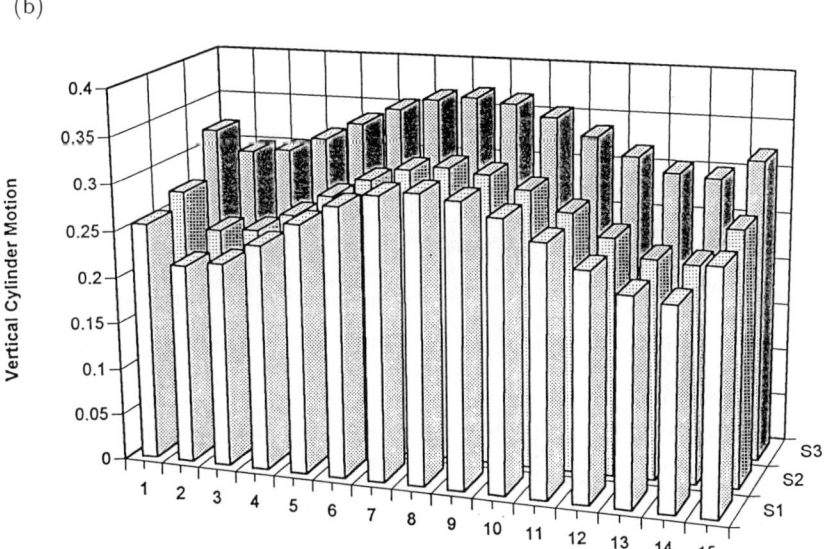

Figure 3. Amplitude of water column oscillation for excitation period of 8 s. (a) Forced surge motion. (b) Forced heave motion.

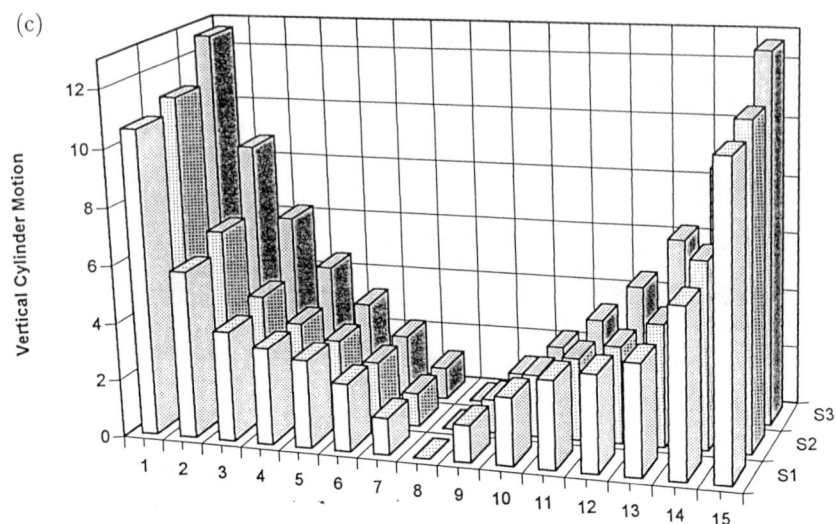

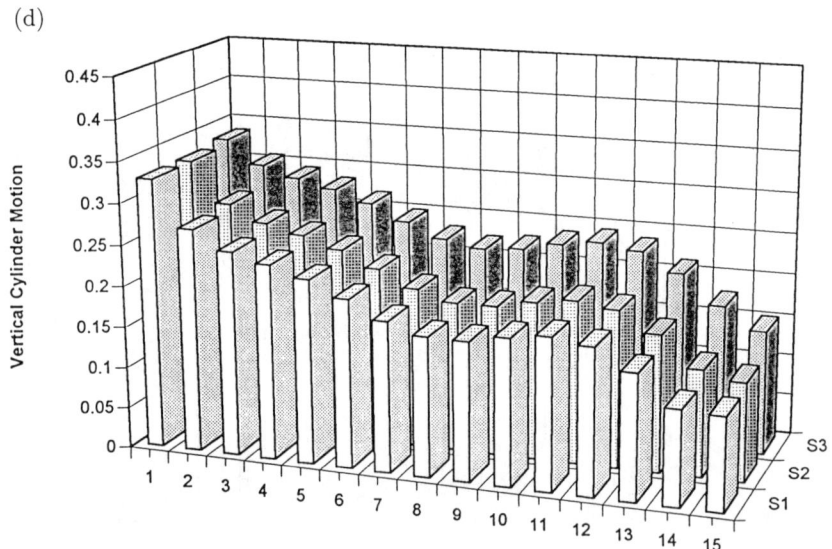

Figure 3 (cont.) Amplitude of water column oscillation for excitation period of 8 s. (c) Forced pitch motion. (d) Incident and scattered waves.

A Study of Plunging Wave Impact Pressures on Placed Block Revetments

Gopinath A.L[1], Eng-Soon Chan[1], Hin-Fatt Cheong[1]

Abstract

An experimental study of plunging wave impacts on placed block revetment was conducted. In the study, a cubic block instrumented to monitor pressures was fixed at different locations along a plane slope. The study was conducted systematically with controlled simulation of the breaking wave using a frequency and amplitude modulated wave packet. The wave packet was programmed to generate a single breaking wave in water of constant depth. However, due to the interaction with the sloping structure, such a wave packet yielded five successive plunging waves. In this paper, the pressures around the cubic block associated with the most impulsive plunger are presented. The characteristics of impact pressures and the mechanics of the impact are elucidated.

Introduction

In recent decades, placed block revetments are commonly used for coastal protection. Many studies including laboratory, field, and numerical studies have been conducted to study the characteristics of wave loading on revetments and the stability of the revetment blocks. Different failure mechanisms including toe failure, block sliding and block lift-outs have been identified. Among these possibilities, the lifting of the blocks is one of the critical failure modes.

In the splash zone, the dislodgements of blocks have been attributed to both the downrush and the impact during wave breaking (Boer et al 1983, Pilarczyk 1990). Due to the impulsive nature of impact, it may be anticipated that uplift pressures due to the impacts are higher. However, pressure measurements of blocks located slightly below the still water have suggested that the net uplift pressures due to downrush are higher (Sparboom and Debus 1992). On the other hand, recent measurements (Gopinath 1994;

---

[1] Department of Civil Engineering, National University of Singapore, 10 Kent Ridge Crescent, Singapore 119260

Gopinath et.al. 1994) have suggested that uplift pressures due to plunging wave pressures can be much higher than that due to down-rush. The difference in these results may perhaps be attributed to the difference in the location of the blocks.

This paper is an attempt to elucidate the mechanics of uplift due to wave breaking. In the study, a single block prefixed at selected locations in the breaking wave zone along the sloping boundary was considered. For each of the cases examined, the sloping boundary was impermeable except in the vicinity of the block. This would correspond to a scenario in which the revetment block-layer is relatively impervious except in the vicinity of the block. Pressures at the boundaries of the block were obtained and analyzed. The net uplift pressures were also estimated. It is well known that wave impact loads could vary significantly even for almost identical conditions. Consequently, the experiments in this study were also repeated to yield an idea of the variability.

Experiments

The model used in the study consisted of a sloping structure (1V:2H) with a 10cm cubic block fixed at a prescribed location (Figures 1 and 2). The spacing between the surfaces of the block and the sloping structure was 3mm. The top face of the block was flushed with the surface of the sloping structure and the block was held rigidly with bolts at the sides. The toe of the structure was fixed at a distance of 13.48 m (6.314 L_c where L_c =2.1352 m is the characteristic wavelength) from the mean position of the wave paddle.

The cubic block was instrumented to measure pressures on four faces of the block as shown in Figure 1 (at locations PT1, PT2, PT3, and PT4). Pressures were measured using a system consisting of four pressure transducers (PDCR-200), four amplifiers (Flyde 351 Uni-Amp), and a high-speed data acquisition system. A H.P. Multiprogrammer (model No. 6942A) was used to capture the pressure signals from the four transducers. A sampling rate of 20 kHz per channel was used. In order to synchronise the data acquisition, the Multiprogrammer was triggered by a signal from the main computer controlling the wave generation. Further details of the data acquisition can be found in Gopinath (1994).

An important feature of the study was the controlled simulation of wave breaking. Following the procedure of Chan and Melville (1988), a single breaking wave in the absence of the sloping structure was simulated using a frequency and amplitude modulated wave packet. The latter was generated by superimposing 28 sinusoidal wave components with frequencies ranging from 0.56 Hz to 1.1 Hz. The phase of each component was chosen such that all the different components would interfere constructively at a prescribed location. In the experiments, however, non-linearity was strong and the waves broke ahead of the prescribed location.

Figure 2 shows the prescribed block locations used in the experiments. The co-ordinates of these locations relative to the still water level (SWL) mark on the slope are presented in Table 1. In the study, the block was moved along the slope in steps of 5 cm (0.0234 Lc). The zone covered in 11 steps was 55 cm (0.2576 L_c).

Table 1. Co-ordinates of block locations in normalized units

Block Location	X/L_c	Z/L_c
1	0.042	0.021
2	0.021	0.0105
3	0.000	0.000 (SWL)
4	-0.021	-0.0105
5	-0.042	-0.021
6	-0.063	-0.0315
7	-0.084	-0.042
8	-0.105	-0.0525
9	-0.126	-0.063
10	-0.147	-0.0735
11	-0.168	-0.084

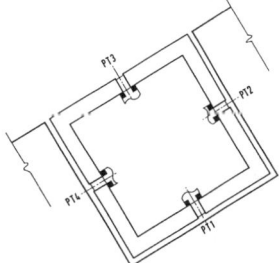
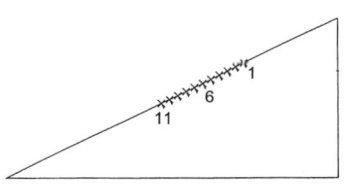

Figure 1. Locations of pressure transducers on the cubic block

Figure 2. Prescribed block locations along the slope

Results

Throughout the entire study, the wave packet was tuned to yield a single plunger at a specified location in the absence of the structure. Due to wave-structure interactions, five successive plungers were generated. It was observed, however, that impact pressures from the second of the five plungers were higher. The results presented in this paper were those resulting from the second plunger.

Typical impact pressure time-histories obtained with different block locations are presented in Figure 3. The results were obtained from repeated experiments but with the block fixed at different locations. Pressures are normalised by ρC^2, where ρ is the density of water (1000 kg/m³) and C is the wave speed corresponding to the central frequency component (1.772 m/sec). Time scales are normalised by the characteristic wave period, T_c (1.205 s).

The kinematics of the impinging waves are depicted by the series of video prints presented in Figure 4. These were obtained with the block fixed at location 5 (below SWL). From the results in Figure 4, it can be seen that the zone of wave impact was approximately around locations 5, 6 and 7. At these locations, the pressure time histories were characterized by impulsive pressures followed by damped oscillations. This signature faded away when the block was located further away at deeper locations (i.e. towards location 11). The simultaneous pressure records on the other sides of the block were similar. Pressure levels, though not identical, were comparable and the occurrences of peak pressures were in phase.

The impact zone was the most dynamic zone. From the results obtained in repeated experiments (Table 2), it was noted that peak pressures varied from 0.66 ρC^2 to 2.09 ρC^2 at the top face and from 0.93 ρC^2 to 1.86 ρC^2 at the bottom face. From the video recordings, it can be seen that the wave impinged directly onto the blocks in this region, hence the high peak pressures and short pressure rise times (as small as 0.006 T_c at location 5).

The video recordings also showed that a sizeable amount of air was entrapped between the plunger and the structure at the impact zone. This region was also the region with significant pressure oscillations. The number of oscillations varied from 3 to 7 and the oscillation frequency varied from 25 Hz to 75 Hz. The oscillations may be attributed to the compression and expansion of the entrapped air. The oscillation amplitudes were high at the impact locations but decay with distance from the impact region.

In the region below impact zone, the pressures were generally lower and the pressure rise times were longer. Closer to the impact zone, there were essentially two main components of the pressure rise times: a slowly varying component associated with the wave evolution and a component that was comparable to that of the impulsive pressures at the impact locations.

In the region above the impact zone, the pressures were also lower than those in the impact zone. The pressures varied significantly between the repeated experiments possibly due to the complex turbulent spray ahead of the breaking wave. Well-defined peaks and pressure oscillations such as those in the impact region were absent. The peak pressures varied from 0.35 ρC^2 at the bottom of the block at location 1 to 1.78 ρC^2 at the bottom side face of the block at location 3. The pressures were mainly associated with the up-rush of the air-water mixture along the slope. Pronounced pressure

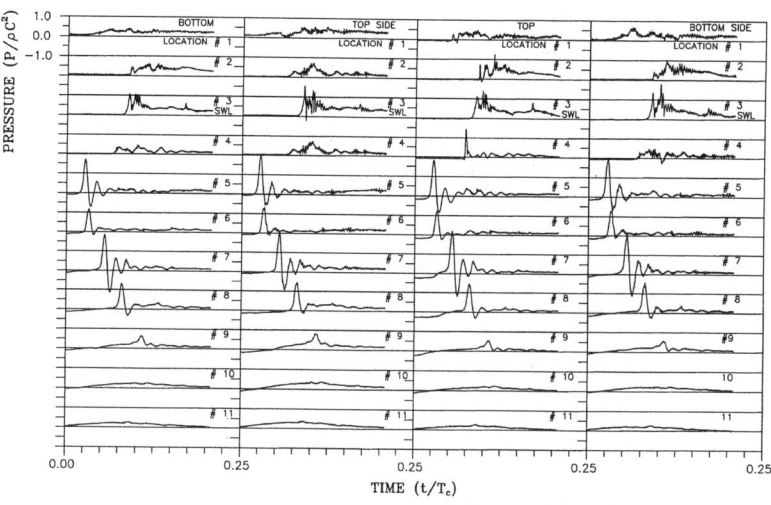

Figure 3. Pressure-time histories at prescribed block locations

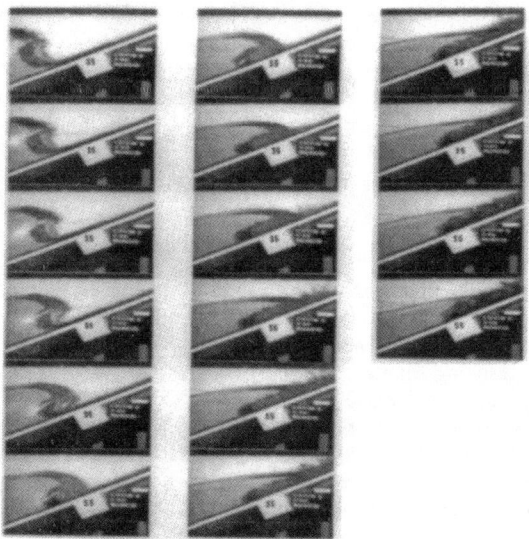

Figure 4. Incident wave profiles during impact

Table 2. Range of impact peak pressure (in ρC^2).

Block location	PT1 Bottom	PT2 Top Side	PT3 Top	PT4 Bottom Side
1	0.3542-0.416	0.571-0.66	0.473-0.736	0.555-0.703
2	0.431-0.639	0.768-0.85	0.953-1.506	1.132-1.783
3 (SWL)	0.797-1.059	1.06-1.555	1.216-1.648	1.132-1.783
4	0.400-0.616	0.704-0.882	1.324-1.432	0.474-0.762
5	1.756-1.860	1.138-1.973	1.918-2.087	2.213-2.316
6	1.182-1.694	1.307-1.907	1.236-1.824	1.332-1.931
7	0.928-1.856	1.028-2.119	0.655-1.824	1.058-2.153
8	0.824-1.309	0.971-1.440	0.946-1.456	0.984-1.569
9	0.639-0.693	0.838-0.907	0.554-0.615	0.599-0.659
10	0.262-0.285	0.381-0.400	0.236-0.257	0.259-0.326
11	0.246-0.273	0.362-0.374	0.257-0.279	0.259-0.289

Oscillations were absent except for location 4 which was closer to the entrapped air region. At higher block locations, the air pocket would have collapsed and there was only air-water mixture moving up the slope.

The above results suggest that high peak pressures were present on the underside (i.e. the bottom) of the block at several locations (Figure 3). A careful comparison of the simultaneous pressure records showed that there were moments when the bottom pressures were in fact higher than those at the top. This would mean a net uplift force. To further examine the uplift, the difference between the bottom and top pressures for the block were computed and presented in Figure 5. A positive value would mean an uplift and a negative value would mean a net pressure in the downward direction. It should be noted, however, that the results were only indicative of the net uplift since they were based on pressures at a single point both on the top and the bottom faces of the block.

Evidently, net uplift pressures were obtained at and above the impact zone. The uplift pressures were highest at location 7. Low level uplift pressures were also present before the occurrence of the peak pressures. These net uplift pressures may be attributed to the down-rush resulting from the previous wave. In the video record, it was also evident that there was a thin wedge of water on the slope before impact. At several locations, the net pressure time histories were oscillatory. On careful examination, it was noted that the oscillations were not in phase with pressure oscillations on either side of the block. In other words, the peak uplift pressure was not the difference of the simultaneous peak pressures on the top and bottom of the block. The peak uplift was mainly associated with the pressure variation in between the peaks.

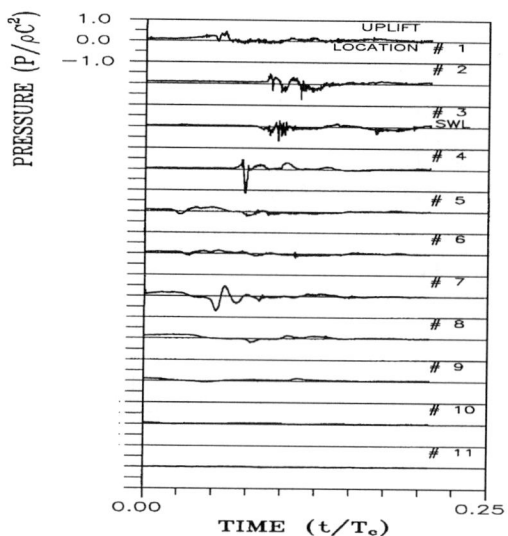

Figure 5. Uplift pressure-time histories at prescribed block locations

Summary and Implications of the Results

In this study, a systematic variation of the impact pressure characteristics with block locations has been obtained. In the impact region the peak pressure are the highest and in this region the peak pressures varied from 0.7 ρC^2 to 2.3 ρC^2. The maximum impact pressures occured at block location 7 which was approximately 0.4 H below SWL, where H was the breaker height. This is in agreement with the results of Fuhrboter and Sparboom (1988), Fuhboter (1986), and Stive (1984).

Characteristic pressure oscillations were found to be dominant in the zone of impact. The oscillation frequency ranged from 52 Hz to 75 Hz and the amplitude of oscillation varied from 0.1 ρC^2 to 0.4 ρC^2. These oscillations may be attributed to the compression and expansion of the entrapped air in the impact region.

Peak pressures occured almost simultaneously on all the faces of the block even though the corresponding pressure rise times were different. However, the amplitude of the peak pressures varies. Oscillatory uplift pressures were possible, suggesting that a single block can be unstable under wave impact. In retrospect, this study has demonstrated the importance of plunging wave impacts on placed block revetments, especially in showing that the uplift from impact could be higher than that due to downrush.

References

1. Boer, Kentet C. J. and Pilarczyk K. W.: 1983. Large Scale Model Tests on Placed Block Revetments. Coastal Structures'83. pp. 307-319.
2. Chan E. S. and Melville W. K.: 1988. Deep Water Plunging Wave Pressures on a Vertical Plane Wall. Proceedings Royal Society of London, A417. pp. 95-131.
3. Fuhrboter A.: 1986. Model and Prototype Tests for Wave Impact and Run-up on a Uniform 1:4 Slope. Coastal Engineering, 10, pp. 49-84.
4. Fuhrboter A. and Sparboom U.: 29 Aug - 2 Sept 1988. Shock Pressure Interactions on Prototype Sea Dykes Caused by Breaking Waves. Proceedings of the International Symposium on Modelling Soil-Water Interactions. Delft. pp. 209-217.
5. Gopinath A. L.: 1994. Plunging Wave Impact on a Placed Block Revetments. M. Eng. Thesis. National University of Singapore, Singapore.
6. Gopinath A. L., Cheong Hin Fatt, and Eng-Soon Chan: 1994. Breaking Wave Loads on Placed Block Revetments. Ninth Congress of the Asian and Pacific Division of the International Association for Hydraulic Research. pp. 115-122.
7. Pilarczyk K. W.:1990. Design of Seawalls Including Overview of Revetment. Proceedings of the Short Course on Coastal Protection, Delft University of Technology. pp. 197-288.
8. Sparboom, U. and Debus, W.: 1992, Wave-Induced Uplift Characteristics on Concrete Block Slope Revetments. Coastal Engineering, pp. 1573-1586.
9. Stive R.J.H.: 1984. Wave Impact on Uniform Steep Slopes at Approximately Prototype Scale. Proceedings Symposium on Scale Efffects in Modelling Hydraulic Structures.

Hurricane Wave Forces on the Decks of Offshore Platforms

By Robert Bea[1], James Stear[2], Tao Xu[3], and Rafael Ramos[4]

Abstract

Many platforms designed according to early generations of oceanographic criteria for the Gulf of Mexico have lower deck elevations that bring these decks into the crests of waves specified in the recently issued Supplement 1 to API RP 2A (1997). Given the API guidelines to determine wave forces acting on the decks of these platforms, it is clear that the platforms can not survive such loadings. The decks must either be removed or the decks raised to elevations that clear the specified wave crests.

A variety of laboratory tests have been performed to address this problem. Several approaches have been developed to compute these loadings and the responses of the platforms to the loadings. There have been many instances in which platforms have experienced significant wave loadings in their lower decks during hurricanes. Some of these platforms have survived and some have failed.

This paper summarizes results from a detailed analytical study of the performance characteristics of platforms in the Gulf of Mexico that have survived and failed during hurricane wave loadings in their decks. A modification to the API Supplement 1 deck wave force guidelines is discussed and validated with platform field performance during past hurricanes.

Introduction

In development of criteria to requalify platforms in the Bay of Campeche, it became obvious that additional work was needed to refine criteria to determine wave crest forces on the lower decks of platforms. With that objective, modifications were developed to the API Supplement 1 guidelines to determine wave loadings on decks of offshore platforms (API, 1997). These modifications were founded on recent experiences with platforms whose lower decks had been subjected to loadings from hurricane wave crests (e. g. Botelho, et al, 1994), results from laboratory testing (e. g. Finnigan, Petrauskas, 1997) and the associated analytical developments. These modifications were intended to remove the conservative bias incorporated into the API Supplement 1 guidelines (Finnigan, Petrauskas, 1997) and develop unbiased estimates

---

[1] Prof., Dept. of Civil & Environmental Engrg., Univ. of California, Berkeley, CA 94720-1712
[2] Doctoral Graduate Student Researcher, Dept. of Civil & Environmental Engrg., Univ. of California
[3] Post-Doctoral Researcher, Dept. of Civil & Environmental Engrg., Univ. of California
[4] Senior Engineer, Office of Special Studies, Instituto Mexicano de Petroleo, 07730 Mexico, D.F.

of the maximum total lateral wave-in-deck forces developed on platforms in severe hurricanes.

The Modified Procedure for calculation of maximum lateral forces developed on platform decks is as follows:

• Determine wave crest elevation for requalification Ultimate Limit State (ULS) condition wave whose steepness (H / L) is 1/10 using Stokes Fifth Order wave theory and the expected maximum wave height together with the storm surge that is expected at the time of occurrence of the maximum wave height .

• Determine the wave crest maximum horizontal velocities using Stokes Fifth Order Theory. Modify these velocities to recognize wave directional spreading by multiplying the velocities by 0.85. Do not include hurricane current velocities in the wave crest velocities.

• Use the frontal projected area of the lower deck and the equipment securely (will not be swept from the deck by the waves) attached to this deck that is inundated by the wave crest and one velocity head above the wave crest (total area = A). Use the corrected wave crest velocity (Uc) to compute the wave run-up (RU) from $RU = Uc^2 / 2g$. Use the crest velocity Uc in the run-up zone to compute lateral forces.

• Use a drag coefficient, Cd = 1.0 for tubular elements and Cd = 2.0 for rectangular elements at a depth of two velocity heads (Uc^2 / g) below the wave crest and run-up free surface. Use a drag coefficient at the free surface of Cd = 0.. Use a drag coefficient between the free surface and a depth of two velocity heads that is a linear function of depth below the free surface.

Verification of the Modified Procedure

Verification of the Modified Procedure was based on a detailed analysis of the performance characteristics of platforms in the Gulf of Mexico that have survived, been seriously damaged, and failed during hurricane wave loading in their decks. The computer program ULSLEA (Ultimate Limit State Limit Equilibrium Analysis) was used to perform all analyses (Bea, Mortazavi, Loch, 1997). ULSLEA is based on a set of methods to characterize the lateral load carrying capacities of the elements that comprise the platform structure and foundation, to determine the storm lateral loadings, to determine the lateral load capacities of the critical lateral loading carrying elements, and to determine the reliability of the platform elements. The procedures incorporated in ULSLEA to determine forces and capacities have been verified with detailed analyses of more than 40 platforms. ULSLEA is able to produce essentially unbiased estimates of the maximum lateral loadings and capacities.

In ULSLEA, aerodynamic and hydrodynamic loading are calculated basically according to API RP 2A guidelines (API 1997). Wave horizontal velocities are determined based on Stokes fifth-order theory. The ULSLEA user can specify directional spreading correction for the wave velocities and blockage corrections for the currents. The specified variation of current velocities with depth is stretched to the wave crest. The total horizontal water velocities are taken as the sum of the wave horizontal velocities and the current velocities. The maximum hydrodynamic force acting on the portions of structure below the wave crest are based on the fluid velocity pressure or drag component of the Morison equation.

All the structural elements are modeled as equivalent vertical cylinders that are located at the wave crest. Appurtenances (conductors, boat landings, risers) are modeled in a similar manner. For inclined members, the effective vertical projected area is determined by multiplying the product of member length and diameter by the cube of the cosine of its angle with the horizontal to resolve horizontal velocities to normal to the member axis.

For wave crest elevations that reach the lower decks, the wave-in-deck forces acting on the lower decks are computed based on the projected area of the portions of the structures that would be able to withstand the high pressures. Analyses of the performance characteristics of six platforms are summarized in this paper based on application of the API Supplement 1 guidelines (API, 1997) and the Modified Procedure.

Platform A

Platform A is one of three similar platforms located in the South Pass region of the Gulf of Mexico, offshore Louisiana. It was installed in 1967. The platform is an 8-leg jacket structure sited in 104 m of water. It is a drilling and production platform, and it supports eighteen well conductors.

In 1969, hurricane Camille passed over the South Pass region, subjecting Platform A and its two sister platforms (B and C) to very severe wind and wave loads. Measurements and inferences from displaced equipment indicate that platform A was struck by 22 to 24 m high waves which struck and inundated the lower of the platform. Nearby, platform B was struck by waves between 22 m and 23 m high and suffered some minor damage to the lower deck and some equipment. Platform C was loaded by a wave between 21m and 22 m high based on platform damage and wave height measurements. All three platforms experienced wave-in-deck forces, and they survived these extreme loads without any noticeable structural damage.

Application of the API procedure for the critical loading direction (Fig. 1) indicated that the platform would fail or be seriously damaged. Comparison of the storm shear profile and the platform capacity profile indicates that failure would be initiated in the second bay of jacket diagonal braces. The Modified Procedure did not predict that there would be any damage to the platforms (Fig. 2). The Modified Procedure produced results that were in agreement with the observed performance of these three platforms. The API procedure predicted damage that did not occur.

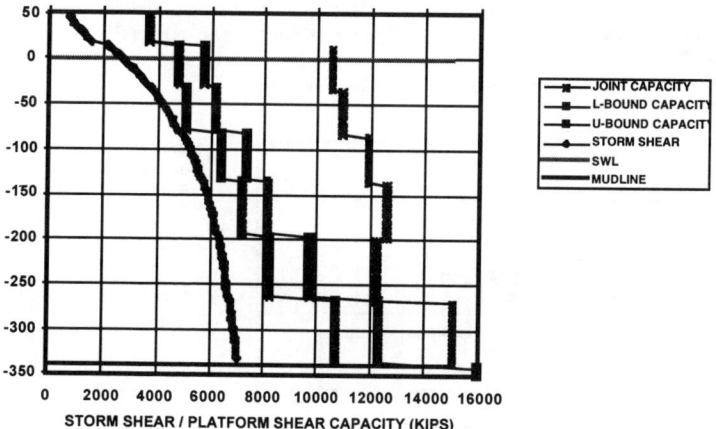

Fig. 1 - API procedure applied to platform A in hurricane Camille

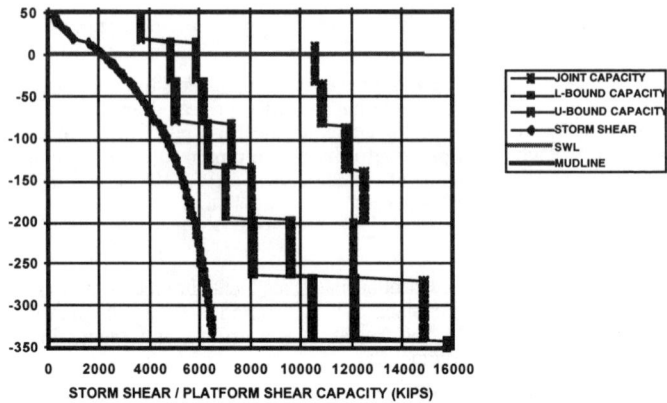

Fig. 2 - Modified Procedure applied to Platform A in hurricane Camille

Platform B

Platform B was installed in the Ship Shoal region of the Gulf of Mexico off Louisiana in 1964. The platform is an 8-leg jacket structure sited in 65 m of water. It is a self-contained drilling and production platform, and it supports twelve conductors. In 1964, hurricane Hilda passed over the Ship Shoal region, subjecting platform B and nearby structures to very severe wind and wave loads. Inferences from miscellaneous damage indicate that platform B was struck by a wave or waves in the range of 16 m to 17 m high. Damage to the lower deck was slight. The platform jacket sustained slight damage as a result of this loading.

Results from the ULSLEA analyses are summarized in Fig. 3 and Fig. 4. The API procedure predicts that the platform would fail in the first bay of diagonal bracing with damage likely in the second bay. The Modified Procedure predicts that the platform would survive without significant damage. The performance of Platform B is in substantial agreement with the Modified Procedure. The API procedure predicts failures that did not occur.

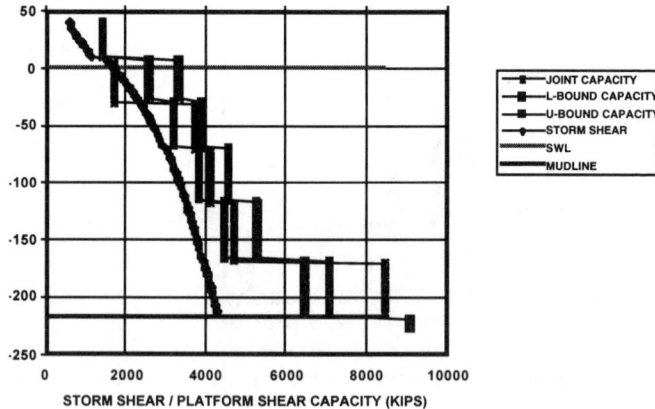

Fig. 3 - API Procedure applied to platform B in hurricane Hilda

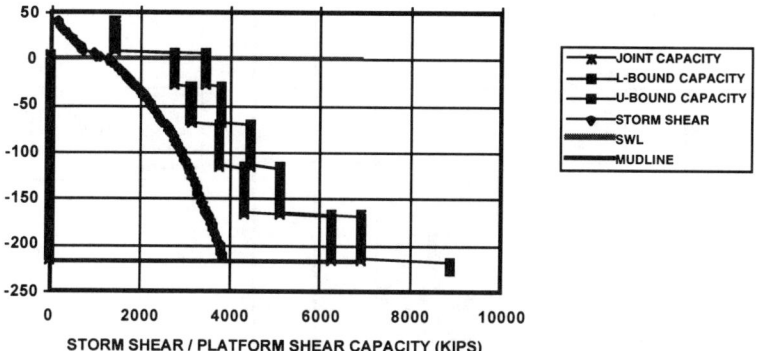

Fig. 4 - Modified Procedure applied to platform B in hurricane Hilda

Platform C

Platform C is an 8-leg self-contained drilling and production platform that was located in the same area as Platform B and whose deck was clearly inundated by waves in hurricane Hilda (Bea, DesRoches, 1993). Platform B was a standard (for its time, 1962) 25-year return period wave design. The platform was located in a water depth of 52 m. The platform supported three wells.

The platform was about 15 miles east of the path of hurricane Hilda (1964). The platform had been in place for less than one year. Based on damage observations and wave staff recordings from nearby platform, the maximum wave height was estimated to be 17 m.

Platform C failed due to the lower deck wave loadings during Hurricane Hilda. Inspections of the platform following its failure indicated that the deck legs parted at the deck leg to jacket connections and the top bay of diagonal bracing and legs failed.

Fig. 5 and Fig. 6 summarize the ULSLEA results for the API Procedure and the Modified Procedure. Both procedures predict failure of the platform in the manner that was observed.

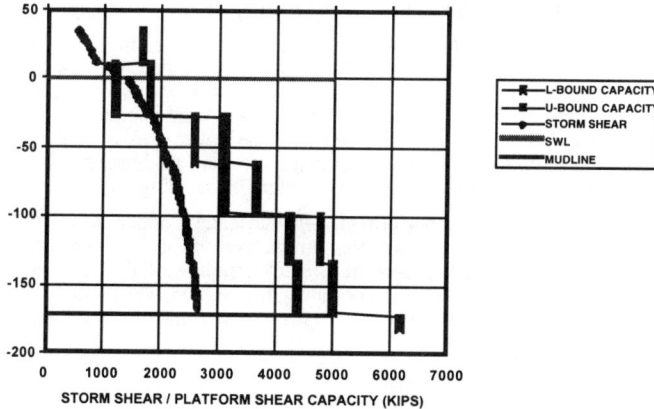

Fig. 5 - API Procedure applied to platform C in hurricane Hilda

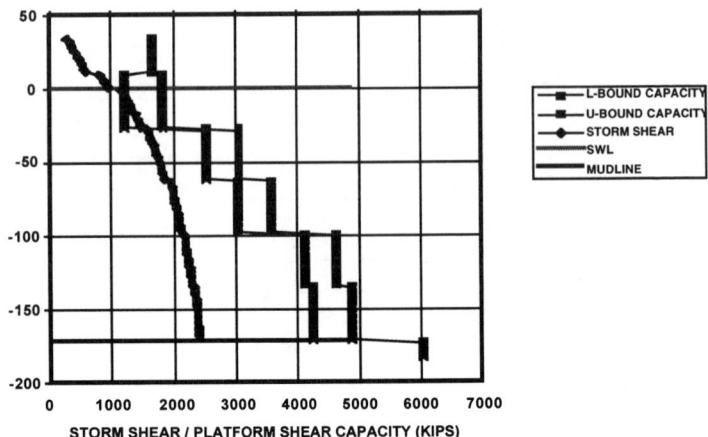

Fig. 6 - Modified Procedure applied to platform C in hurricane Hilda

Platform D

Platform D is an eight-leg structure sited in 36 m of water in the South Timbalier region of the Gulf of Mexico (Imm, et al, 1994). Platform D was subjected to two major hurricanes during its service life. Hurricane Carmen passed close by 1974; damage to the lower deck suggested that the structure has been subjected to wave up to 18 m high. A post hurricane platform condition assessment revealed significant green water damage to equipment located on the lower decks and some damage to the vertical diagonal joints at the top of the uppermost jacket bay. In 1988, the platform was subjected of a comprehensive risk analysis. Consequent risk mitigation measures include removal of the conductors and all equipment from the lower decks. In 1992, hurricane Andrew passed within a few miles of this platform; damage to cellar deck and hindcast studies performed following the passage of Andrew suggested waves with heights of 18m to 19 m had struck the platform from approximately 15 degree off the broadside loading direction. The platform survived with some yielding of the K-joints at the top of the uppermost jacket bay.

Platform D was evaluated for two loading conditions: 1) the hurricane Carmen case, and 2) the hurricane Andrew case. For hurricane Carmen, the Platform D was analyzed with lower deck equipment in place. Based on damage reports and a hindcast of hurricane Carmen, the maximum wave height was characterized as 18 m. For hurricane Andrew, Platform D was analyzed with the lower deck equipment removed and the conductors removed. Based on damage reports and hindcasts of hurricane Andrew. The maximum wave height was characterized as being 18 m.

Results for the platform performance in hurricane Carmen based on the API Procedure and the Modified Procedure are summarized in Fig. 7 and 8, respectively. For hurricane Carmen, the API Procedure predicted failure of the platform, and the Modified Procedure predicted no failure or damage. The Modified Procedure predicted what was observed and the API procedure did not. The same observations applied to the results for hurricane Andrew.

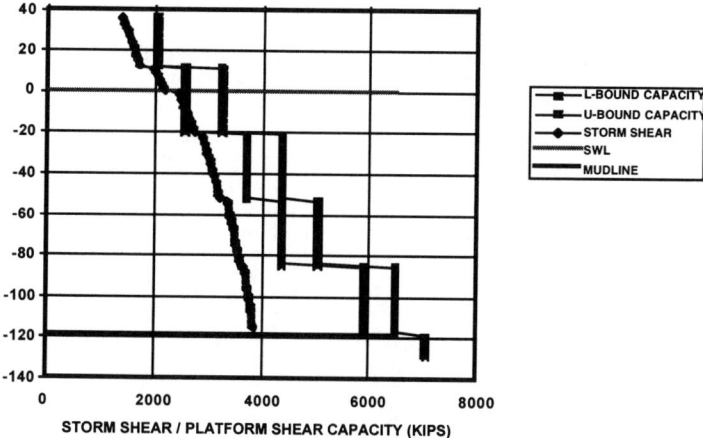

Fig. 7 - API Procedure applied to platform D in hurricane Carmen

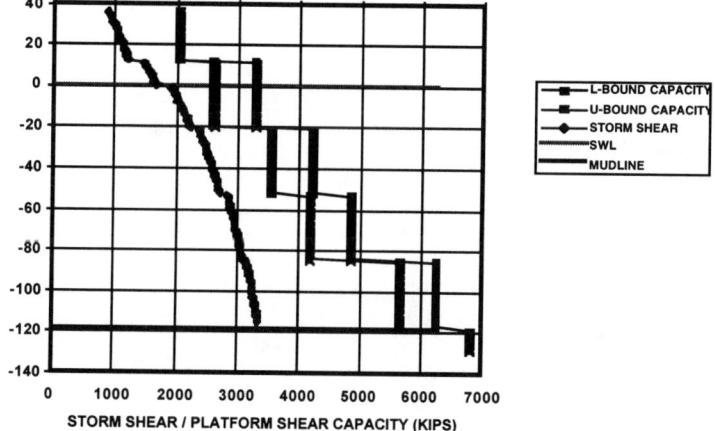

Fig. 8 - Modified Procedure applied to platform D in hurricane Carmen

Platforms E and F

Platform E is located in the Gulf of Mexico's South Timbalier region. Platform E is an eight pile drilling and production platform with eleven well conductors (Petrauskas, et al, 1994). The platform was designed in 1964 and installed in 42 m of water. Platform E collapsed during hurricane Andrew (Petrauskas et al, 1994). Based on hindcast data and damage to adjacent platforms, it was estimated that the platform was loaded between end-on and diagonally with a maximum wave height of 16 m to 17 m (Botelho, et al, 1994; Petrauskas, et al, 1994). The platform appeared to have failed in the end-on direction (Petrauskas et al, 1994).

Platform F was bridge-connected to Platform E. Platform F is also an eight pile drilling and production platform with sixteen conductors. The platform was designed and installed in the same year as Platform D (1964). Platform F is similar in geometry to platform D except that platform F is battered at 1:10 in both broadside and end-on directions. Platform F experienced significant loading during Hurricane Andrew. However, unlike platform E, platform F did not collapse (Petrauskas et al. 1994).

Both procedures predicted the observed failure of Platform E. However, the API Procedure predicted the failure of Platform F and the Modified Procedure did not predict the failure of Platform F. Platform F did not fail. The API Procedure did not produce results that conformed with the observed performance of this platform during hurricane Andrew. The Modified :Procedure did produce results that were consistent with the observed performance of Platform E and Platform F in hurricane Andrew.

Conclusions

Based on experimental data, the API Procedure generally produces conservative results (Finnigan, Petrauskas, 1997) for the total maximum lateral wave-in-deck forces. Based on observations and analyses of platform performance in hurricanes, the API procedure also produces conservative results and predicts failures when only damage or no damage was observed. The Modified Procedure produces results for all of the platforms studied that are in conformance with the observed performance of these platforms. The Guideline Procedure predicts failures, damage, and lack of damage when this performance was observed in the field.

References

American Petroleum Institute (1996). Recommended Practice for Planning, Designing, and Constructing Fixed Offshore Platforms - Working Stress Design, Supplement 1, Washington, DC.

Bea, R. G., Mortazavi, M. M., and Loch, K. J. (1997). "Evaluation of Storm Loadings on and Capacities of Offshore Platforms," *Journal of Waterway, Port, Coastal, and Ocean Engineering*, American Society of Civil Engineers, Vol. 123, No. 2, New York, NY.

Botelho, D. L. R., Ullmann, R. R., Chancellor, D. P., and Versowsky, P. E. (1994). "A Survey of the Structural Damage Caused by Hurricane Andrew on Some of the Platforms Located in the South Timbalier Area," *Proceedings Offshore Technology Conference*, OTC 7470, Society of Petroleum Engineers, Richardson, TX.

Finnigan, T. D., and Petrauskas, C., (1997) "Wave - in Deck Forces", *Proceedings of the 6th International Offshore and Polar Engineering Conference*, Honolulu, International Society of Offshore and Polar Engineers, Golden, CO.

Imm, G. R., O'Connor, J. M., Light, J. M., and Stahl, B., (1994). "South Timbalier 161A: A Successful Application Platform Requalification Technology. " *Proceedings Offshore Technology Conference*, OTC Paper No. 7471, Society of Petroleum Engineers, Richardson, TX.

Petrauskas, C., Botelho, D. L. R., Krieger, W. F., and Griffin, J. J. (1994). "A Reliability Model for Offshore Platforms and its Application to ST151 "H" and "K" Platforms During Hurricane Andrew (1992). " *Proceedings of the Behavior of Offshore Structure Systems, Boss'94*, Massachusetts Institute of Technology, Cambridge, MA.

Nonlinear Loads on Vertical Columns in Laboratory High Seas

C.H. Kim[1] and J. Zou[2]

Introduction

This paper presents a semi-empirical model for nonlinear forces and moment of vertical truncated column affected by laboratory irregular waves. Given the measured wave elevation time series, we predict the wave loads in the time domain. Vertical columns are very important members of column-supported offshore units, since the nonlinear wave loads are mainly due to the surface-piercing of columns. Because the theoretical approach currently appears to take too long a time to succeed, pragmatic tools are needed for industry, especially for the ringing load prediction. To create such models, we had to conduct experimental investigations of the kinematics of nonlinear waves and forces including impact and non-impact. Through numerous experiments and analyses (Kim, et al., 1992; Zou and Kim, 1995, 1996; Kim et al., 1996, 1997), we found a universal linear system model ULSM (Kim and Zou, 1995) for the nonlinear horizontal velocity field necessary to determine the horizontal force and moment. We also have investigated ULSM combined with linear diffraction to predict the horizontal force and moment. Another application of ULSM/diffraction is to use a modified linear diffraction boundary condition for the vertical nonlinear force.

Third-order diffraction models for ringing loads have been developed and still are in the development stage. Krokstad et al. (1996) proposed FNV/QTF model which appears to be promising.

[1,2]Department of Civil Engineering, Ocean Engineering Program, Texas A&M University, College Station, Texas 77843-3136, USA.

ULSM for Horizontal Velocity Field

Our intention is to use Morison equation for which we need a model to predict the horizontal velocity field. Given the measured wave elevation time series, we want to predict horizontal particle velocity field just below the measured wave surface. We regard the time series of the wave elevation and horizontal velocity field as the nonlinear input and output. Taking FFT of the pair of samples, we have complex amplitude spectra of the wave and particle velocity. IFFT of these reproduces exactly the same measured velocity time series in the same sampling interval. The amplitude spectra correspond to each other at each frequency. As the amplitude spectra are all expressed by sinusoids (but the phase angles are not random), the input and output are regarded as linear systems in the frequency domain. The ratio of the velocity amplitude to the wave amplitude at the corresponding frequency is a measured linear transfer function LTF*. Then the product of wave amplitude spectrum and LTF* at the corresponding frequency is the output complex spectrum, IFFT of which yields the same measured output in the time domain. Therefore, if we can develop or find an appropriate formula for LTF*, i.e. the particle velocity amplitude per unit wave amplitude, theoretically or empirically, we will be able to predict the particle velocity. The LTF* is linear in the form but includes nonlinearity.

IFFT of the wave amplitude spectrum is given by

$$\eta(x,t) = R_e \left\{ \sum_{j=1}^{n} a_j \exp\left[i(k_j x - \omega_j t + \phi_j) \right] \right\} \qquad (1)$$

where η is wave elevation at x and a_j, ω_j, k_j and ϕ_j are amplitude, frequency, wave number and phase angle (not random) of the jth component, respectively. The IFFT of velocity amplitude spectrum is expressed as follows,

$$u(x,z,t) = R_e \left\{ \sum_{j=1}^{n} a_j \exp\left[i(k_j x - \omega_j t + \phi_j) \right] G_u(z,\omega_j) \right\}, \quad z \leq 0 \qquad (2)$$

with $\quad G_u(z,\omega_j) = \dfrac{gk_j}{\omega_j} \dfrac{\cosh k_j(z+h)}{\cosh k_j h} \qquad (3)$

$$\dfrac{\omega_j^2}{g} = k_j \tanh(k_j h) \qquad (4)$$

where G_u is the ordinary LTF for the horizontal velocity at z, and h is the water depth, but it will become an appropriate LTF after the following process.

Use of Eq. 3 makes the velocity at the MWL diverge. Wheeler (1970) introduced a stretching method to avoid the divergence and determine the horizontal particle velocity in the wave crest. Our stretching model previously developed (Kim et al., 1996) creates a slight abrupt change on the MWL. The improved model is given by,

$$z_e = az_a^3 + bz_a^2 + cz_a + d \quad \text{for} - h \le z_a \le H_c \quad (5)$$

with
$$a = \left[(-h + H_c) + k(h + H_c)\right]/(h + H_c)^3 \quad (6)$$
$$b = \left[-2(h^2 - hH_c + H_c^2) - k(h + H_c)(H_c - 2h)\right]/(h + H_c)^3$$
$$c = \left[H_c(H_c^2 - H_c h + 4h^2) + kh(h + H_c)(h - 2H_c)\right]/(h + H_c)^3$$
$$d = \left\{h^2 H_c\left[k(h + H_c) + 2H_c\right]\right\}/(h + H_c)^3$$
$$k = (2.00 - \lambda)H_t / H$$
$$\lambda = T_f / T_r$$
if $H_c/H_t \le 1.0$, then $\lambda = 1.0$, and if $\lambda > 1.95$, then $\lambda = 1.95$.

where H_c, H_t and H stand for the crest height, trough height and height of the wave, respectively. A wave of one period is taken between two zero down crossing points. z_e is the effective vertical coordinate ($-h \le z_e \le 0$), whereas z_a is the actual vertical coordinate ($-h \le z_a \le H_c$). Eqs. 5 and 6 are the mapping equations between z_e and z_a. Substitution of the effective coordinate into Eq. 3 gives an appropriate LTF G_{11} as a function of the actual coordinate. The appropriate LTF is linear in the form but contains nonlinearity. This procedure is not applicable to breaking waves.

Modified Morison Equation

The above horizontal velocity distribution is used to predict the horizontal force using the following equation.

$$F(t) = \rho C_M \frac{\pi}{4} D^2 \int_{-d}^{\eta} \frac{\partial u}{\partial t} dz + \frac{1}{2} \rho C_D D \int_{-d}^{\eta} u|u| dz \quad (7)$$

where C_M and C_D are assumed to be uniformly distributed along the cylinder axis. We call the above model ULSM/Morison.

ULSM/Diffraction for Horizontal Force

Given the nonlinear incident wave elevation at the center of the column, we want to predict the nonlinear horizontal force and moment of the column, and regard the wave and force as the nonlinear input and output. Taking FFT of both of these, we have complex amplitude spectra. Because the spectra consist of many sinusoids, both systems are linear in the frequency domain; but the phase angles are not random. The ratio of the force amplitude to wave amplitude at the corresponding frequency is a linear transfer function LTF*, and the product of the wave spectrum and LTF* becomes the force amplitude spectrum, IFFT of which gives the measured wave force. If we can find an appropriate LTF, theoretically or empirically, then the nonlinear wave force (output) will be predicted using IFFT, which is expressed by

$$F(x,t) = R_e \sum_{j=1}^{n} a_j \exp[i(k_j x - \omega_j t + \phi_j)] G_{fh}(x, \omega_j), \quad z \le 0 \qquad (8)$$

where G_{fh} is the LTF for the linear horizontal wave-exciting force of the column which can be computed by a 3-dimensional linear diffraction code. Similarly we can predict the nonlinear moment. We call the above technique ULSM/Diffraction.

ULSM/Diffraction for Vertical Force

Given the nonlinear measured wave elevation at the center of the column, we want to predict the vertical nonlinear force. The wave and force time series are the nonlinear input and output respectively. FFT of these are the complex amplitude spectra of the wave and force. The ratio of the force amplitude to the wave amplitude is the measured LTF*. The product of the wave amplitude spectrum and LTF* at the corresponding frequency is the force amplitude spectrum, the IFFT of which gives the measured wave force time series. Thus, the vertical force is given by

$$F(x,t) = R_e \sum_{j=1}^{n} a_j \exp[i(k_j x - \omega_j t + \phi_j)] G_{fv}(x, \omega_j), \quad z \le 0 \qquad (9)$$

where, G_{fv} represents mathematical LTF available from linear potential theory. We consider two LTFs. One is the conventional and the other is derived satisfying a modified diffraction boundary condition. In the present work, for the convenience sake, the diffraction theory by Fenton (1978) is employed.

The modified linear diffraction boundary condition is

$$\frac{\partial \phi}{\partial n} = \mathbf{v}^* \bullet \mathbf{n} \qquad \text{on the mean wetted body surface} \qquad (10)$$

where $\frac{\partial \phi}{\partial n}$ is the normal velocity of the diffraction flow, and $\mathbf{v}^*$ is the measured nonlinear particle velocity of the free incident nonlinear wave and $\mathbf{n}$ is unit outward normal from the fluid to the body surface. The diffraction boundary condition Eq. 4.2 on page 246 (Fenton, 1978) is modified as follows; the ratio of the measured nonlinear velocity per unit measured wave amplitude to the linear velocity per unit wave amplitude is multiplied to the linear velocity per unit wave amplitude in Eq. 4.2 on page 246 (Fenton, 1978), at the center point of the bottom of the column. The modification of the velocity at the bottom center point automatically satisfies the boundary condition on the entire mean wetted body surface.

Computations and Discussions

We compute the horizontal forces and moments of single columns with different diameters 0.11 m, 0.18 m and 0.34 m measured in MARINTEK (Stansberg et al., 1995). The length scale ratio used in the present paper is 1/100. The high sea was generated employing JONSWAP spectrum (H_s = 0.154 m, T_p = 1.78 s and gamma = 1.7) for 0.3 hour. Fig. 1 illustrates a sample of the wave containing a large wave group, ringing wave event, which is the input in our computations. Figs. 2a and 2b represent the horizontal force and moment of the column of 0.18 m diameter, computed by ULSM/Morison, which show slight overpredictions. This model was used in the previous work (Kim et al., 1997) including prediction of weak impact.

The same horizontal force and moment were computed by ULSM/ Diffraction for each column. Figs. 3a and 3b show those of the column of diameter 0.11 m, Figs. 4a and 4b those of 0.18 m diameter and Figs. 5a and 5b those of 0.34 m diameter. The model appears to predict very well but there is a tendency of slight underprediction as the diameter increases. This may be attributed to the use of linear diffraction forces which cannot take into account the effect of nonlinear deformation of the waves over the large body surface. The wave (Fig. 1) produced ringing of Heidrun TLP. Therefore it is important to investigate the high-pass filtered force. Figs. 6a and 6b illustrate the high-pass filtered force (above 8.4 rad/s) for 0.18 m diameter column. The former is due to ULSM/Morison while the latter due to ULSM/Diffraction. The former model overpredicts, while the latter slightly underpredicts. More research is needed to improve the above models. ULSM/Diffraction will be useful for loadings due to weakly nonlinear waves such as those in Fig. 1 (MARINTEK). Another advantage of this model is that it can be used to determine the pressure distribution over the body surface.

The vertical wave force on the column of 0.164 m diameter and 0.30 m draft (measured in TAMU wave tank) was computed by ULSM/Diffraction with the linear diffraction force and the modified linear diffraction force. A typical high sea measured in TAMU wave tank is shown in Fig. 7. The particle velocity was measured at 0.30 m depth with an acoustic Doppler velocimeter. The measured particle velocity per unit measured wave amplitude and the linear particle velocity per unit wave amplitude are shown in Figs. 8a and 8b, from which the ratio of the above are computed. Multiplying the above ratio by the linear velocity per unit wave amplitude in the computer program we satisfy the modified diffraction boundary condition Eq. 10. The conventional LTF and modified LTF of the vertical forces are shown in Fig 9. The former is slightly less than the latter, particularly in the high frequency region. This explains that there are more nonlinearities in the shorter waves than in the longer ones. Using the ULSM/Diffraction with the conventional and modified LTF, we predict the vertical forces as shown in Figs. 10a and 10b. It is clearly seen that the use of conventional LTF gives a fair agreement while that of modified LTF yields a very good result. The above is as expected, because the former does not include the effect of the nonlinear particle velocity. The modified is the appropriate LTF which is linear in the form but contains nonlinearity. The above study suggests that one could derive a method to generate an appropriate LTF without using the measured particle velocity.

Concluding Remarks

We have investigated semi-empirical models of nonlinear wave loads of single columns, with a variety of diameters, fixed in the laboratory storm seas. The comparison with the experimental data indicates that ULSM/Morison slightly overestimates, while ULSM/Diffraction slightly underpredicts. Morison model is useful for impacting load due to strongly nonlinear waves, whereas diffraction model works well for the loads due to weakly nonlinear waves. In fact, the wave measured in MARINTEK belongs to weakly nonlinear waves. An advantage of diffraction model is that it can predict the pressure distribution on the body surface. A lot more work has to be done to make the models practically useful.

Acknowledgment

We are thankful to ABS and Koje Maritime Research Institute of SAMSUNG for their strong support for the research. We are also thankful to Statoil for kindly permitting us to access the wave and force data measured in MARINTEK.

References

Fenton, JD (1978). "Wave Forces on Vertical Bodies of Revolution," *J Fluid Mech*, Vol 85, P 2, pp 241-255.

Kim, CH, Randall, RE, Krafft, MJ, and Boo, SY (1992). "Kinematics of 2-D Transient Water Waves Using Laser Doppler Anemometry," *J Waterway, Port, Coastal and Ocean Eng*, ASCE, Vol 118, No 2, pp 142-165.

Kim, CH and Zou, J. (1995). "A Universal Linear System Model for Kinematics and Forces Affected by Nonlinear Irregular Waves," *Int J Offshore and Polar Eng*, Vol 5, No 3, pp 166-170.

Kim, CH, Xu, Y, Zou, J and Won, YS (1996). "A Model for Weak Impacting Force on Vertical Truncated Cylinder Due to Steep Asymmetric Wave," *Proc 6th Int Offshore and Polar Eng Conf*, Los Angeles, ISOPE, Vol 3, pp 215-220.

Kim, CH, Xu, Y and Zou, J. (1997). "Impact and Non-Impact on Vertical Truncated Cylinder Due to Strong and Weak Asymmetric Waves," *Int J Offshore and Polar Eng*, Vol 7, No 3, pp 161-167.

Krokstad, JR, Stansberg, CT, Nestegard, A and Marthinsen, T (1996). "A New Non-Slender Ringing Load Approach Verified Against Experiments," OMAE, Vol 1, Part A, pp 371-387.

Stansberg, CT, Huse, E, Krokstad, JR and Lehn, E. (1995). "Experimental Study of Non-Linear Loads on Vertical Cylinders in Steep Random Waves," *Proc Int Offshore and Polar Eng Conf*, Vol 1, pp 75-82.

Wheeler, JD (1970). "Method for Calculating Forces Produced by Irregular Waves," *J Pet Tech*, pp 359-367.

Zou, J and Kim, CH (1995). "Extreme Wave Kinematics and Impact Loads On a Fixed Truncated Circular Cylinder," *Proc 5th Int Offshore and Polar Eng Conf*, ISOPE, Vol III, pp 216-225.

Zou, J and Kim, CH (1996). "Experimental Study of Impacting Wave Force on Vertical Truncated Cylinder," *Int J Offshore and Polar Eng*, Vol 6, No 4, pp 291-293.

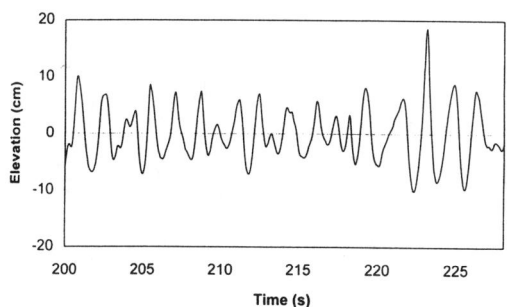

Fig. 1 Ringing wave measured in MARINTEK

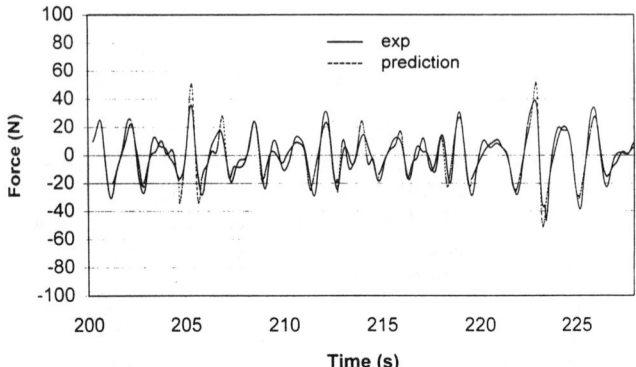

Fig. 2a Prediction of horizontal force by ULSM/Morison model (0.18 m)

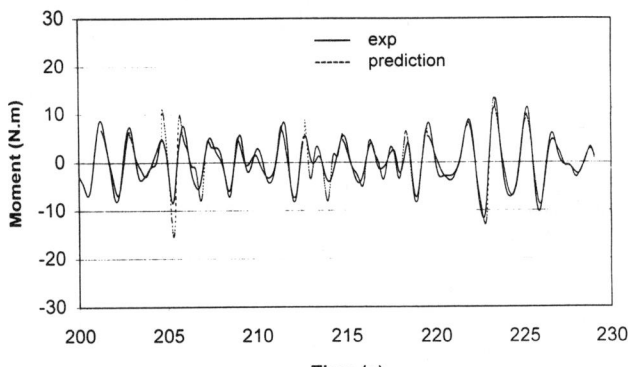

Fig. 2b Prediction of moment by ULSM/Morison model (0.18 m)

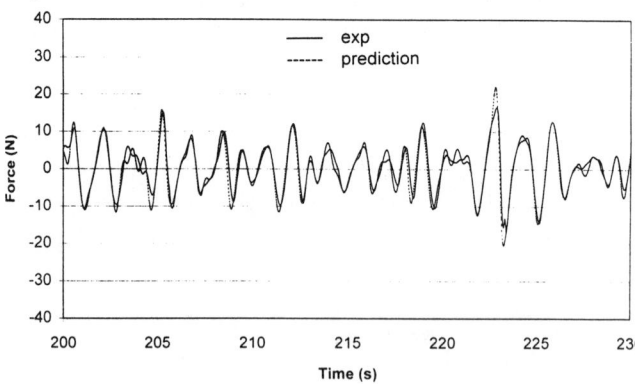

Fig. 3a Prediction of horizontal force by ULSM/Diffraction model (0.11 m)

OCEAN WAVE KINEMATICS, DYNAMICS AND LOADS ON STRUCTURES 211

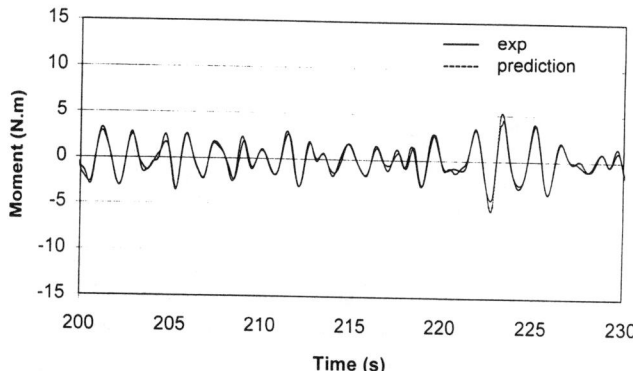

Fig. 3b Prediction of moment by ULSM/Diffraction model (0.11 m)

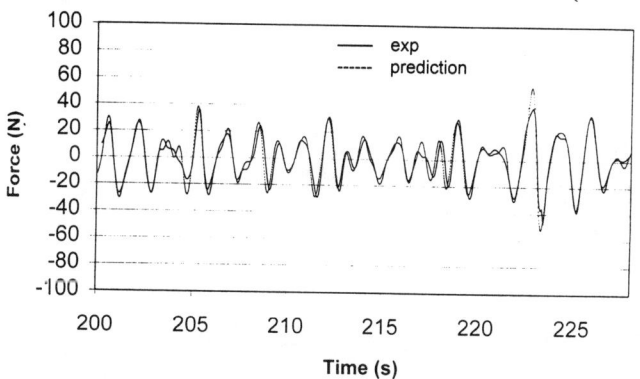

Fig. 4a Prediction of horizontal force by ULSM/Diffraction model (0.18 m)

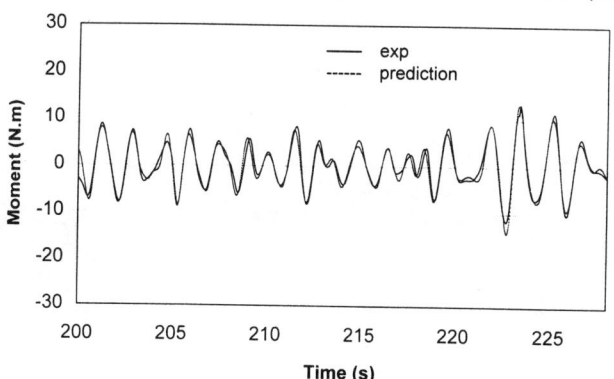

Fig. 4b Prediction of moment by ULSM/Diffraction model (0.18 m)

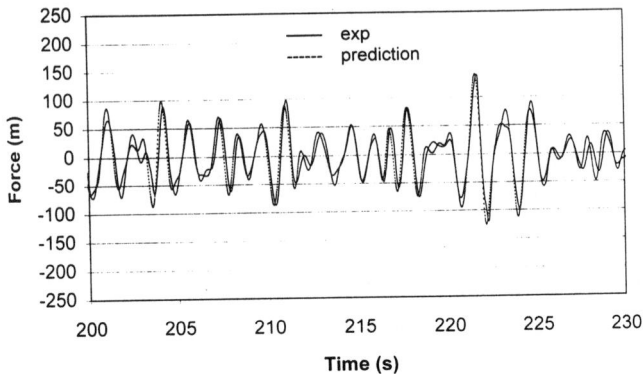

Fig. 5a Prediction of horizontal force by ULSM/Diffraction model (0.34 m)

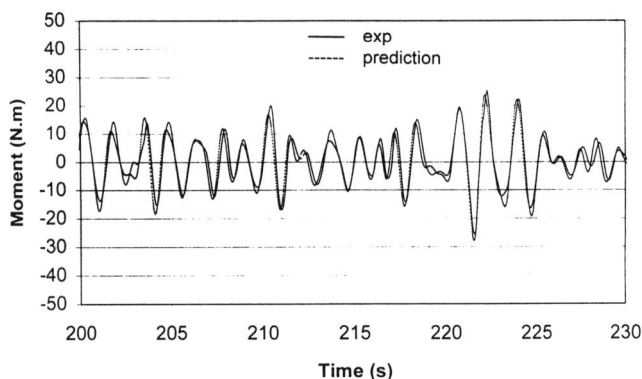

Fig. 5b Prediction of moment by ULSM/Diffraction model (.34 m)

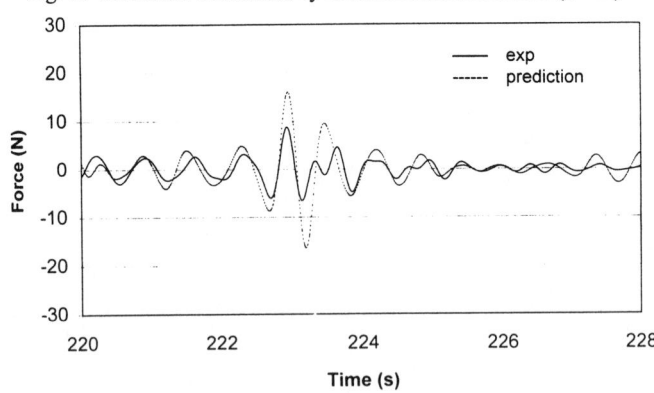

Fig. 6a Prediction of filtered high frequency force by ULSM/Morison model (0.18 m)

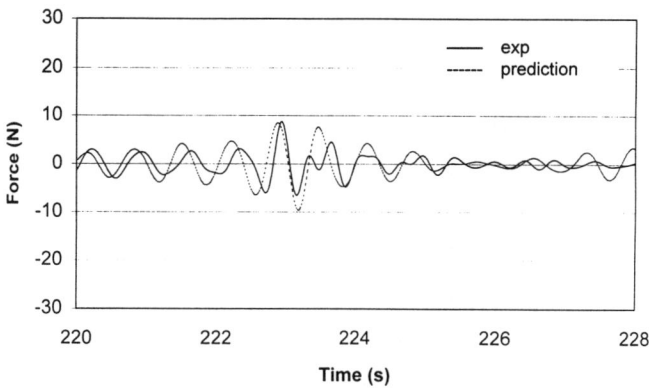

Fig. 6b Prediction of filtered high frequency force by ULSM/Diffraction model (0.18 m)

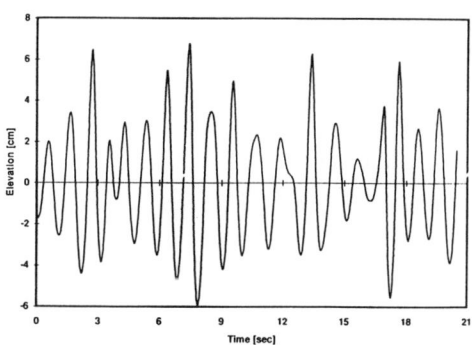

Fig. 7 A high sea measured in TAMU wave tank

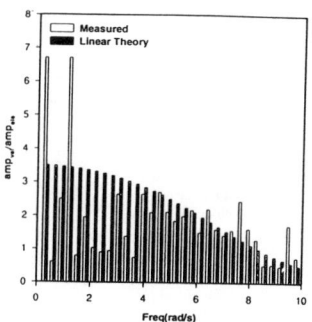

Fig. 8a Horizontal particle velocity per unit wave amplitude

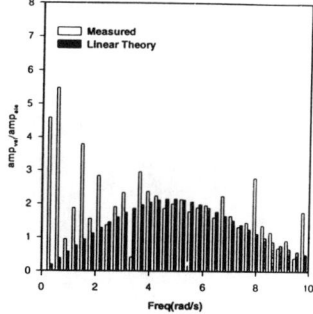

Fig. 8b Vertical particle velocity per unit wave amplitude

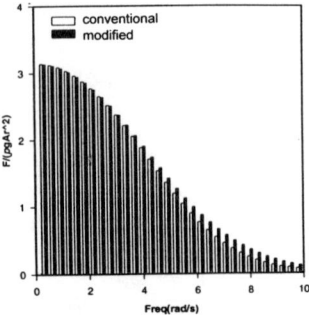

Fig. 9 Comparison of conventional and modified LTF for vertical force

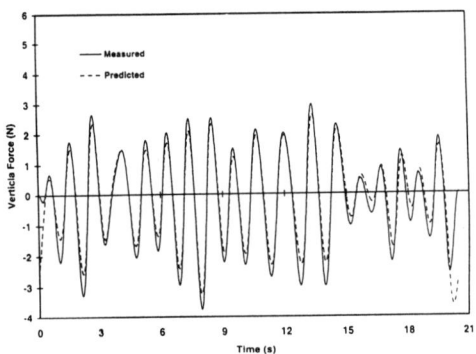

Fig. 10a Vertical force by ULSM /Diffraction model with conventional LTF

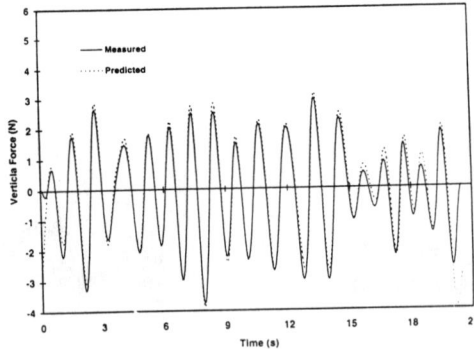

Fig. 10b Vertical force by ULSM/Diffraction model with modified LTF

An Investigation of Wave Forces for Design of a Cruise Ship Pier Bridgetown, Barbados

Mark Mattila, M.A.Sc., P.Eng.[1]
Michael Tarbotton, M.Sc., P.Eng.[2]
Andrew Cornett, Ph.D., P.Eng.[3]

ABSTRACT

One of the main engineering challenges in the design of coastal structures is the determination of design waves, wave forces and the wave structure interaction. A pile supported pier designed for the Port of Bridgetown, Barbados, in the Eastern Caribbean Sea, is on a section of open coastline exposed to waves generated by tropical storms and hurricanes. Through literature review, numerical analysis and physical modelling the effects and forces developed by these waves were assessed and quantified for the design of the structure. Wave uplift forces were of particular significance as operational constraints limited the pier deck elevation, thereby placing it in the wavefield during storm events. Perforation of the pier deck and other methods of mitigating the effects of wave uplift were also examined.

INTRODUCTION

In recent years, Barbados has seen a steady rise in the number and size of cruise ships calling at the Port of Bridgetown. This growth has greatly benefited the Island's tourism industry and is expected to continue well into the next century. The existing Port of Bridgetown does not have room to accommodate more, and larger, cruise

---

[1] Project Engineer, Delcan Corporation, 300 - 604 Columbia Street, New Westminster, British Columbia, Canada V3M 1A6 (email: markm@delcan.com)

[2] Principal, Triton Consultants Ltd., 3530 West 43rd Avenue, Vancouver, British Columbia, Canada V6N 3J9 (email: triton@rogerswave.ca)

[3] Project Manager, Canadian Hydraulics Centre, National Research Council, Ottawa, Ontario, Canada K1A 0R6 (email: andrew.cornett@nrc.ca)

ships. New cruise ship designs currently under construction and scheduled for commissioning in the year 1999 and beyond are in excess of 130,000 GRT, almost double the gross registered tonnage of existing vessels. The pier will accommodate these new vessel designs and alleviate congestion in the Port, *Figure 1*.

The design criteria for the pier includes survival of severe wave conditions which may be generated by tropical storms and hurricanes. The pier will have a pile supported platform and access trestle. The platform will be a total of 130m long by 40m wide and stand in water ranging from 14 to 20m deep, *Figure 2*. The access trestle will be 184m long by 10m wide and stand in water ranging from 0 to 17m deep. Pile bents are spaced at 8m and have eight piles per bent in the platform and three piles per bent in the trestle. To accommodate existing cruise ships, the pier deck is constrained to an elevation which places the design water level at the deck underside. As a consequence, during extreme wave events the platform and access trestle decks will be inundated by waves.

A further complicating situation facing the designers was the geotechnical properties of the seabed. The seabed is largely coral and sand, a weak formation for foundations. The coralline seabed formation limited the uplift and compression capacity of piling used to support the pier. Thus a thorough understanding of hydrodynamic forces on the platform and access trestle during extreme wave events was required to complete the design of the structures and their foundations.

INVESTIGATIVE PROGRAMME

Design Wave Climate

Extreme design waves for the Cruise Ship Pier have been estimated from deep water GSOWM (Global Spectral Ocean Wave Model) data for the period from June 1985 to May 1991 at a location close to Barbados, combined with shallow water wave transformations to bring the offshore waves into the nearshore area at the Pier. Information on wave period during extreme storms was not adequate to develop a wave period statistic, however, calculated offshore deepwater wave heights for a 100 year return period storm are consistent with a fully arisen sea with a peak wave period (T_p) of 12.0s and a significant wave height (H_s) of 7.5m.

Table 1: Design Wave Cases for Cruise Ship Pier, Bridgetown, Barbados.

Design Case	Direction at Cruise Ship Pier	H_s (metres)	T_p (seconds)
1	South	4.3	12.0
2	South West	4.6	10.8
3	West	3.2	10.5

OCEAN WAVE KINEMATICS, DYNAMICS AND LOADS ON STRUCTURES 217

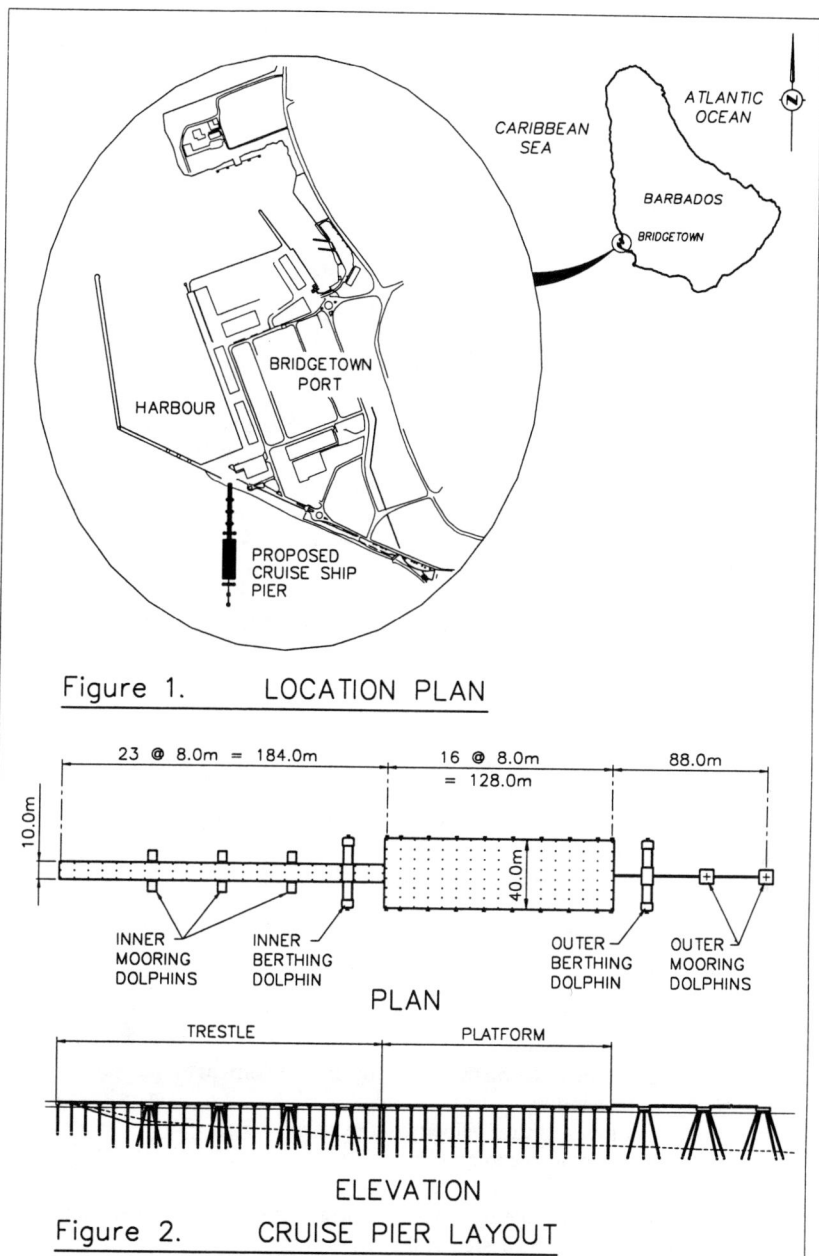

Figure 1. LOCATION PLAN

Figure 2. CRUISE PIER LAYOUT

Design wave cases are shown in **Table 1** in terms of statistical wave parameters H_s and T_p. A maximum wave height (H_{max}) of 8.5m and wave period (T) of 12.0s was used for wave force calculations, such as pile wave forces, that were evaluated using regular waves rather than random waves. A maximum wave height of H_{max} = 8.5m yields a ratio H_{max}/H_s equal to 1.93. The storm duration (t_d) is 12 hours.

Literature Review

The literature review focused on wave forces on platforms in a wavefield. Some key findings were:
- most of the existing research focused on 2 dimensional numerical and physical model studies;
- vertical wave forces were the largest hydrodynamic force on platforms, often an order of magnitude greater than horizontal wave forces;
- vertical wave forces on platforms have three characteristic components: (i) an initial large upward impact pressure followed by (ii) a slowly varying uplift pressure and (iii) a slowly varying downward or suction pressure; and
- upward impact pressures act over a small area and the slowly varying uplift and suction forces act over a large area with the frequency of these pressures matching that of the incident wave.

Research (Lee and Lai, 1989, Patarapanich, 1984 and Shih and Anastasiou, 1992) expressed the maximum slowly varying uplift and suction pressure in the form:

$$P_{up} = C_p * \gamma * (h_{cr} - h_{sf}), \quad h_{cr} > h_{sf} \tag{1}$$

where:
- P_{up} = maximum slowly varying uplift or suction pressure;
- C_p = pressure coefficient for uplift or suction;
- γ = unit weight of fluid (10kN/m³ assumed for ocean water);
- h_{cr} = amplitude of wave crest above still water level (SWL); and
- h_{sf} = height of platform soffit above SWL.

Work by Wang (1970) presented a formulation for impact pressure as:

$$P_{imp} = \pi * \gamma * c * v_{max} / g \tag{2}$$

where:
- P_{imp} = impact pressure on the underside of a platform at the surface;
- c = wave celerity;
- v_{max} = maximum vertical velocity of the wave;
- g = acceleration due to gravity.

These items from the literature review subsequently directed numerical analysis and physical model studies for the design of the cruise pier.

Numerical Analysis

From the design wave climate, the wave conditions used for numerical analysis were $H_{max} = 8.5$m; $H_s = 4.3$m; $T_p = 12.0$s; water depth at structure, $d = 15.0$m; wave direction: south. Using stream function wave theory the H_{max} wave profile was calculated yielding crest amplitude, $h_{cr} = 6.4$m; trough amplitude, $h_{tr} = 2.1$m; wave length, $L = 151$m. The use of stream function theory to develop the wave profile was important as the waves were very close to breaking.

Work by Patarapanich (1984) was selected from the literature reviewed as the most appropriate for application to preliminary design of the cruise ship pier. Assuming the stream function theory wave surface profile with still water level at deck underside, calculations yielded maximum uplift pressures in the range of 48 to 82kPa. The corresponding pressure coefficients C_p range from 0.75 to 1.27 and mean forces on the design deck section of 8 x 40m range from 15,400 to 25,600kN. The work by Patarapanich provided no analysis of suction pressures.

Suction forces investigated by Lai and Lee (1989) were in the order of 40% of the uplift pressure forces. This resulted in a provisional estimate of maximum suction pressures in the range of 18 to 32 kPa. Corresponding forces on the design deck section were 6,100 to 10,200kN.

The response of the pile foundation to wave loading was then assessed by calculating the net vertical force acting on an 8 x 40m platform deck area:

$$F_{net} = F_{by} + F_{dl} + F_{up} + F_{wd} \tag{3}$$

where:

F_{net} = net vertical force on pile foundation, $\wedge$ or $\vee$ (+ or -)
F_{by} = structure buoyancy, $\wedge$ (+)
F_{dl} = structure deadload, $\vee$ (-)
F_{up} = hydrodynamic force on structure, $F_{up} = P_{up} * A_{deck}$, $\wedge$ or $\vee$ (+ or -)
F_{wd} = water on the platform deck, $\vee$ (-)
A_{deck} = design deck section area (8 x 40m) encompassing a pile bent

The calculation of the net uplift force per pile bent was checked against pile tension and compression capacity to ensure appropriate factors of safety against failure. This procedure was primarily used to develop the preliminary design of the cruise ship pier.

Impact pressures estimated using the method of Wang (1970) gave a value of 80kPa. Shih and Anastasiou (1992) formulate impact pressure in the same manner as the slowly varying pressure (equation 1) except that their impact pressure coefficients range from 1.8 to 7.6 resulting in calculated pressures ranging from 115 to 486kPa. The variability of these pressure coefficients was attributed to the effects of roughness and entrained air in the physical models used and the characteristics of the waves tested. As a result of these uncertainties a more detailed investigation of impact and slowly varying pressure was performed in the physical model study.

Pile forces were estimated using Morison's equation. The maximum total wave induced drag and inertial force on an eight pile bent was estimated at 880 kN; a value an order of magnitude less than the calculated uplift forces on the same bent.

Physical Model

The physical model study was conducted at the Canadian Hydraulics Centre, Ottawa, Ontario, Canada, during May and June, 1997. The tests were conducted in the CHC's multidirectional wave basin. The model was constructed at a horizontal and vertical scale of 1 : 50. Since wave motion is governed by a balance between gravitational and inertial forces acting on water particles, Froude scaling was used to relate conditions in the model to those at full scale. The seaward, central and landward areas of the platform deck had moveable 18 x 40m instrumented sections for pressure and force measurement.

The majority of the testing was undertaken with regular and irregular unidirectional waves. Parameters varied during the tests included wave direction, water level, wave height and wave period. The range of test conditions produced data which encompassed both the design wave criteria for the pier and broad range of other severe wave loading scenarios. An investigative sideline examined the effect of deck perforation on the magnitude of uplift forces. These tests were run with five percent areal perforation of the 18 x 40m deck section.

JONSWAP spectral shapes were used in the study. The peak enhancement factor of the JONSWAP spectrum was 3.3. Relatively long time series containing at least 500 waves for the longest wave period (approximately 120 minutes prototype) were generated to ensure that statistical variability was minimized.

A sampling rate of 200Hz was used for force and pressure transducers in order to capture loading fluctuations that could have a significant effect on the stability and dynamic response of the pier. Anti-aliasing filters were used to isolate the force measurements from frequencies higher than the chosen sampling rate.

The model study produced a vast amount of data which returned results that were generally in agreement with the findings of the numerical analysis and the literature review. However, upon completion of the model study, it became apparent that some extrapolation of the model results for application to design was required to take into account two important factors discussed below.

storm duration

Statistics of measured model parameters were based on a 2 hour time series. These statistics were extrapolated to a time series representative of a 12 hour storm duration. This was accomplished by establishing ratios for both uplift and suction forces using the statistical properties F_{max}, F_{95} and standard deviation from a series of 10 confidence band trials run for H_s = 5.0m, T_p = 12.0s and t_d = 2 hours. The calculated ratio $R = F_{max}$ *(design, t_d = 12 hours)/ F_{95} (measured, t_d = 2 hours)* varied from 1.4 to 2.5 depending on the location on the platform and wave direction. For the critical south section of the platform, R_{uplift} = 2.17 and $R_{suction}$ = 1.59. The corresponding ratios $R' = F_{max}$ (design, 12 hours)/ F_{max} (measured, 2 hours) are about 1.09 and 1.07 for uplift and suction respectively. The resulting design wave pressures for a 12 hour storm duration averaged over an 18 x 40m deck area for the range of wave heights tested for the south wave direction are presented in *Figure 3*.

Figure 3: Maximum Wave Pressures, South Waves, 12 hour Storm Duration, Averaged over an 18 x 40m Section of Platform Deck.

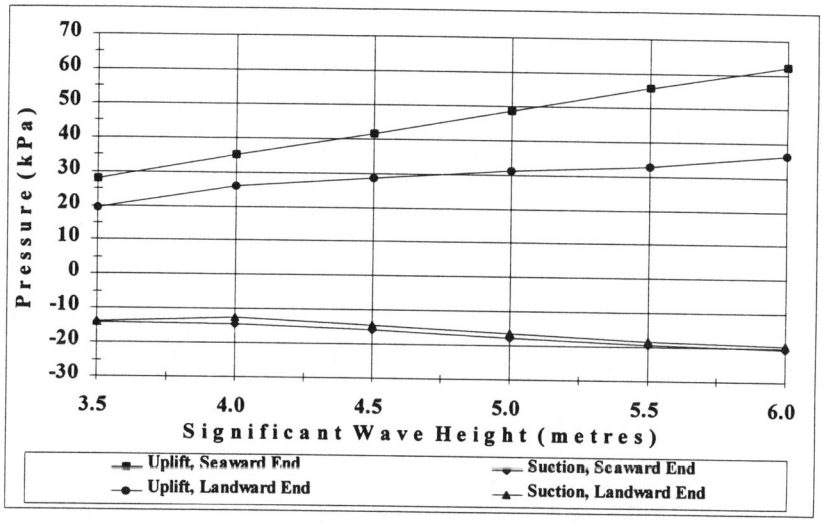

spacial distribution of pressure

The uplift and suction pressures acting on the platform deck are known to be spacially and temporally distributed. A typical time history of the uplift, water on deck and net pressure difference acting on the platform, *Figure 4*, shows considerable fluctuation in pressure. It is clearly apparent that local wave pressures greatly exceed those averaged over a larger area of deck. Model force measurements made on the 18 x 40m deck section were extrapolated to the design deck section of 8 x 40m by using a series of schematized pressure distributions. These pressures distributions were developed from a reanalysis of the work of Shih and Anastasiou (1992) and Wang (1970). The re-analysis took into account the effect of wave steepness on the maximum wave impact pressure and its duration. The result was an areal scaling factor of 1.27 for extrapolation to the 8 x 40m design deck section. These extrapolated results, *Table 2*, were used to complete the design of the cruise ship pier.

Details of the extrapolation of the model results to account for storm duration and spacial distribution of pressure will be presented in a future paper by the authors.

Figure 4: Wave Induced Pressure on the Platform Deck as a function of Time, H_{max} = 8.5m, T_p = 12.0s, South Wave Direction.

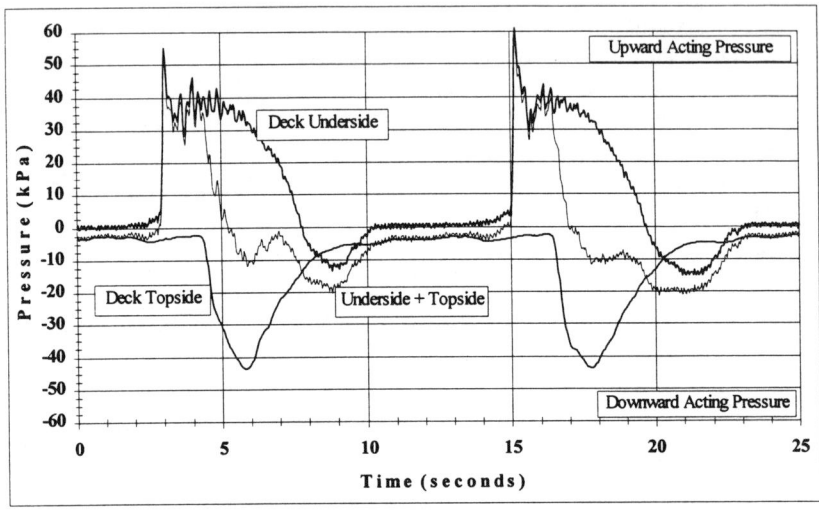

Table 2: Maximum Average Design Pressure and Force Over an 8 x 40m Deck Section, H_s = 4.3m, T_p = 12.0s, t_d = 12 hours, South Wave Direction (see Design Case 1, Table 1).

Location on Platform	Pressure (kPa)	Force (kN)
Seaward	49	15,700
Central	42	13,400
Landward	35	11,200

MITIGATION

Mitigation of net wave uplift is of interest for structure design and economy. In general, this can be achieved by configuring the structure to decrease the magnitude of the uplift forces and by counteracting the uplift forces:

structure configuration

- *deck elevation.* Keeping the deck soffit elevation as high as possible is the most practical means to minimize wave uplift forces;
- *deck underside roughness.* Pile caps descending below the deck soffit and aligned transverse to the wave direction add roughness to the deck underside. Physical modelling (Suchitra, N. and Koola, P.M., 1995) has shown that pile caps can reduce wave uplift force by as much as forty percent;
- *deck underside profile.* Deck underside profiles which create air voids (ie. precast double tee beams, perimeter edge beams) can, in the absence of appropriate venting, trap air and increase buoyancy forces;
- *structure width.* Physical modeling (Shih and Anastasiou, 1992) suggests that structure width is a factor in the magnitude of uplift forces. Platforms that present a narrow width to incident waves appear to develop lower average uplift pressures as they allow lateral escape of wave energy and air whereas wide structures are less able to do so.
- *deck perforation.* Depending on the wave conditions tested in the physical model, perforation of the deck decreases the magnitude of wave uplift force by 10 to 50 percent. This reduction in uplift force is thought to result from venting of air from beneath the platform; relieving hydrodynamic pressure through the deck; and increasing the volume of water on the deck.

counteracting uplift forces

- *structure mass.* A simple approach to counteracting wave uplift is to design a massive structure. Two obvious problems with using mass to resist uplift are economy and seismic risk.
- *foundation anchoring.* Foundation piling can be anchored to the seabed by various means. The weak coralline soils of the cruise ship pier site made foundation anchoring impractical.

The Cruise Ship Pier design incorporates many of these ideas for the mitigation of net wave uplift forces.

CONCLUSIONS

Waves and wave forces for design of a proposed Cruise Ship Pier for the Port of Bridgetown, Barbados have been assessed by means of literature review, numerical analysis and a physical model study. Of particular significance for foundation and structural design is storm wave induced vertical pressure on the platform and access trestle decks. The vertical wave pressure has a characteristic signal which consists of three components: (i) an initial large upward impact pressure of short duration acting over a small area followed by (ii) a slowly varying uplift pressure and (iii) a subsequent slowly varying suction pressure, the latter two which act over a broad area. The vertical wave forces on the deck are an order of magnitude larger than horizontal wave forces on the deck or supporting piles. Physical model results were extrapolated for application to design by accounting for two important factors: (i) storm duration and (ii) spacial distribution of pressure on the underside of the deck. Net wave uplift and suction pressures on platform decks can be mitigated by (i) configuring the structure so as to minimize the induced forces and (ii) by counteracting the forces.

REFERENCES

Lai, C.P., J-J. Lee. 1989. Interaction of Finite Amplitude Waves with Platforms or Docks. Journal of Waterway, Port, Coastal, and Ocean Engineering, Vol. 115, No.1: pp. 19-39.

Patarapanich, M. 1984. Forces and moment on a horizontal plate due to wave scattering. Coastal Eng., 8: pp. 279-301.

Shih, R.W.K., Anastasiou, K., 1992. A laboratory study of the wave-induced vertical loading on platform decks. Proc. Instn Civ. Engrs Wat., Marit. & Energy, 96, Mar., pp. 19 - 33.

Suchitra, N. and Koola, P. M. 1995. A study of wave impact an horizontal slabs. Ocean Engng, Vol. 22, No. 7, pp. 687- 697.

Wang, H., 1970. Water Wave Pressure on Horizontal Plate. Journal of the Hydraulics Division, ASCE, Vol. 96, No. HY10, pp. 1997- 2016.

Forecasting Rig Heave for Drilling Operations in Harsh Environments

Colin Grant [1], Stéphane Cornut [1], Roger Dyer [2], Martin Holt [3], and John Mitchell [4]

Abstract

Vessel motions are a key parameter for operational decision making offshore. With the move to deeper waters and harsher environments, the direct assessment of drilling rig heave has become more important. Traditional offshore forecasts give partial guidance only and are often difficult to interpret by offshore personnel. A new service, jointly developed by BP and the UK Meteorological Office, directly predicts drilling rig heave for up to 5 days ahead. The background to the development is described, together with some initial results and plans for future work.

Introduction

The offshore oil industry has only recently commenced year round drilling operations in areas where persistent swells are a predominant feature. One such area is West of Shetland where BP, together with partner Shell U.K., has actively engaged during the last few years in bringing the first oilfield, Foinaven, into production. Significant research was performed in order to establish the design basis for these fields, as described in Grant et al. (1995). During the drilling operations associated with these developments, several weather-related incidents occurred that were not adequately predicted by the traditional offshore weather forecast service. The main limitation is the accurate forecasting of the arrival of Atlantic swell at the drilling rig, in order that decisions such as riser disconnection and hang off can be made in a timely and safe fashion. As noted by one of BP's drilling engineers, "experience gained anticipating likely rig motion from a conventional weather forecast in the North Sea is simply not relevant West of Shetland."

An example of such an event occurred in February 1996. The "Sonat Arcade Frontier" 4th generation rig (now the "Paul B Loyd Jnr") was drilling at Foinaven

---

1. BP Exploration, Chertsey Road, Sunbury-on-Thames, Middlesex, TW16 7LN, UK
2. BCD Offshore, Worting House, Basingstoke, Hampshire, RG23 8PY, UK
3. UK Meteorological Office, London Road, Bracknell, Berkshire, RG12 2SZ, UK
4. Aberdeen Weather Centre, Seaforth Centre, Lime Street, Aberdeen, AB2 1BJ, UK

when it experienced serious heave motions as swell from a storm generated to the SW of Iceland travelled across the location. The significant wave height reached 13m during this event, but more importantly the peak wave period (Tp) changed from 12s to 20s in a couple of hours.

Although the worsening seastate had been predicted in the routine forecast, the fact that the seastate contained significant energy levels with periods of 18 to 20 seconds was not captured. To some extent, this is an artefact of the traditional forecast output that tends to use the mean swell period as an indicator of the energy frequency. However, in the harsh Atlantic swell environment, a greater degree of accuracy is required in defining the energy present at all frequencies in the seastate as the drilling rig motions are particularly susceptible to periods above 18 seconds. With events such as this in mind, meteorological and oceanographic (metocean) specialists in BP and the UK Meteorological Office (UKMO) met to discuss the potential for developing a forecasting service to address these limitations.

UKMO Wave Modelling
Examination of the UKMO numerical wave model output for the few days leading up to the storm at Foinaven in February 1996 immediately revealed that this model contained useful information about the long period swell present in the seastate. This offered encouragement for the development of a forecast service based on the wave model output. The UKMO operational wave model is a 2nd generation model based on that first developed and described by Golding (1983), but there has been significant development since that time. At each grid point in the model, the wave energy spectrum is divided into 13 frequency components and 16 direction components, giving a resolution of 22.5° for the direction of wave propagation. The 13 frequency components are spaced logarithmically between 0.04Hz (25s, 976m wavelength) and 0.324Hz (3.1s and 15m wavelength).

Input of wave energy from the winds is calculated from two terms: a linear term allows growth from rest in a calm sea, and a term allowing pre-existing waves to grow exponentially uses the formulation due to Snyder (1982). For wind speeds below 7 m/s the wave growth is calculated using a parametric expression as the explicit calculation is limited by model frequency resolution (Holt 1994). Advection of wave energy is calculated using the split explicit scheme due to Gadd as described in Golding (op cit.). Because the global wave model runs on a regular latitude-longitude grid, a term to account for 'Great Circle' propagation of swell is also included. Dissipation of wave energy, due to white-capping (wave breaking) and other processes, follows the formulation of Komen et al. (1984). The coefficient for swell dissipation is reduced by a factor of one half, in line with the latest understanding of the wave energy balance (Holt op cit.). This allows wind sea growth at the correct rate, corresponding to the Snyder input expression, but dissipates swell energy at a slower rate. Non linear transfer of energy due to wave-wave interactions is parameterised by fitting the growing wind sea to a spectrum from the JONSWAP family. Also parameterised is the effect of the wave–wave interactions in delaying the response of waves to a turning wind.

The UKMO wave model is run operationally over two areas, the global model and a European model. For the purposes of the work reported here only the global version is relevant. The global model covers 80.42°N to 77.017°S on a regular latitude-longitude grid, at a resolution of 1.25° longitude and 0.833° latitude. This is the same resolution as the operational atmospheric forecast model that provides the input wind data. The model is run twice daily, at around 0430 and 1630UTC, and provides a forecast out to 5 days ahead. It is driven by hourly values of surface winds from the global numerical weather prediction (NWP) model. Each run begins with a hindcast starting from the wave conditions of 12 hours earlier and running forward with wind data from the NWP assimilation. Observations of wave height and surface wind speed from the radar altimeter on ERS-2 are assimilated in the model, using a scheme developed by Thomas (1988). Before use the observations are grouped into a 20-second average, giving a value every 140 km – approximately the model grid spacing. Each observation influences a region of radius 250 km, but careful QC checks are applied to ensure unrealistic observations (e.g. near coasts) are filtered out.

Heave Calculations

The UKMO global wave model outputs the wave energy spectrum at 6 hourly intervals for the 13 frequency and 16 directional bins previously noted. In order for the offshore user to gain maximum value from this potential source of wave information, the challenge for BP's metocean specialists was to take this information and generate a product of immediate use to an offshore decision-maker. The key parameter for most operational decisions is the rig heave or vertical motion. Ideally a product is required which gives a direct forecast of rig heave without the offshore personnel having to interpret complex wave spectral information from the global model.

A post-processor has been developed by BP which takes the wave energy predicted by the Met Office model to give a direct calculation of vessel heave response. The initial version of the program did not account for the directionality of the waves. A more recent version has taken directionality into account. For clarity, the equations will be given without directional information. The three main operations carried out by the program are described in Figure 1:

- A wave spectrum S is built from the energy E in each frequency bin:

$$S(\omega) = \sum_i E(\omega, \theta_i)$$

The energy coming from the different directions has been summed to obtain the total wave energy at frequency ω. The energy at the different frequencies can then be interpolated to get a continuous function for the wave spectrum S.

- This spectrum is combined with the vessel heave RAO to build the response spectrum R.

  $R(\omega) = S(\omega) \cdot RAO^2(\omega)$

- Assuming this response spectrum is Rayleigh distributed, the significant and maximum heave responses (Z_{sig} and Z_{max}) of the vessel are determined for the 6-hour sea state:

$$m_0 = \int_0^\infty R(\omega) \cdot d\omega \qquad m_2 = \int_0^\infty \omega^2 \cdot R(\omega) \cdot d\omega$$

$$Z_{sig} = 4 \cdot \sqrt{m_0} \quad (peak\ to\ peak\ in\ m)$$

$$T_Z = 2 \cdot \pi \cdot \sqrt{\frac{m_0}{m_2}}$$

$$Z_{max} = \left[\sqrt{2 \cdot \ln\left(\frac{T}{T_Z}\right)} + \frac{\gamma}{\sqrt{2 \cdot \ln\left(\frac{T}{T_Z}\right)}} \right] \cdot \sqrt{m_0}$$

m_0 and m_2 are the moments of order zero and two for the response spectrum. T_z is the upcrossing period. γ is the Euler constant. T is the time over which the maximum is to be calculated (taken as 6 hours in this case). This procedure is carried out for each of the predicted sea states, out to 5 days ahead.

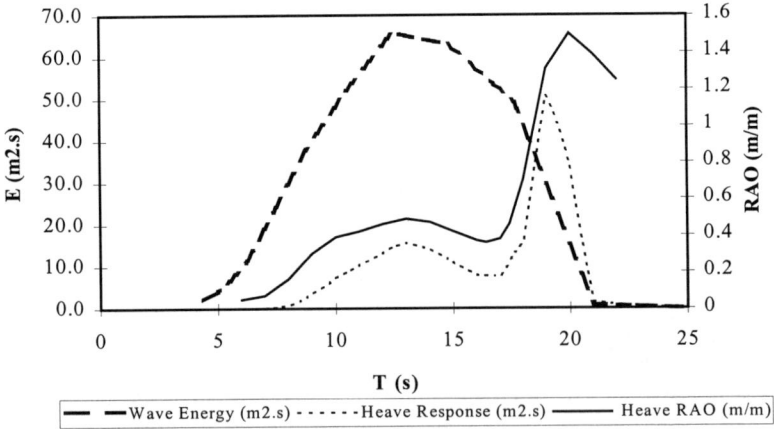

Figure1: Building heave response from the wave model

Figure 1 clearly shows that the swell energy has a very large impact on the heave response spectrum. The software developed by BP, which interfaced the rig RAO and wave model output, was passed to the UKMO in order for it be included in the operational suite of forecast software. The heave forecast products could then be routinely generated and passed to the rig as a direct, computer-generated output. It should be stated that the heave service is an additional service to the traditional weather forecast and not a replacement. A typical format for the rig heave forecast is seen in Table 1.

Time	00:00	06:00	12:00	$\Rightarrow$	12:00	18:00	00:00
Date	14-Feb	14-Feb	14-Feb	$\Rightarrow$	19-Feb	19-Feb	20-Feb
Sig. Response Peak to Peak (m)	1.47	1.46	1.41	$\Rightarrow$	2.73	2.79	2.66
Max Response Peak to Peak (m)	2.83	2.81	2.70	$\Rightarrow$	5.18	5.30	5.06

Table 1: Example of Offshore Heave Prediction Format

Applications

The first trial of the new heave forecast service commenced in mid September, 1996. It was for the "Sedco Explorer", a 2nd generation rig performing appraisal drilling and well testing operations, also West of Shetland, at the Clair Field. This trial lasted until late October of that year when the rig went off location. The service was reinstated for this rig at the same location during the 1997 drilling programme from April to October. In addition, the service was extended to the "Paul B Loyd Jnr" 4th generation rig, working on developing the new Schiehallion Field, from June 1997. This service is continuing. In October 1997 a third rig was added to the list of clients - the "Stena Forth", working in the northern North Sea at BP's Bruce Field.

Results

Feedback from the offshore personnel on the Sedco Explorer during the first trial in 1996 was positive and encouraging. Unfortunately, a quantitative comparison of the forecast and measured rig heave was not possible as the heave monitoring on the rig is essentially a qualitative assessment made by visual means. The following quotation (D. Saul, pers. comm.) is relevant in that the success (or otherwise) of the technique is whether the service adds value to the operational drilling staff. *"Finally we started to receive the rig heave forecast, and right from the start it appeared to be remarkably accurate. Offshore the heave predictions were accepted very quickly. They allowed us take advantage of short operational periods which without the prediction we would have been hung off."*

The limitations in validating the heave predictions during the first trial were addressed in 1997 by installing a wave and rig motion sensor package on the Sedco Explorer. The intention was to obtain quantitative measurements of the wave heights and rig heave in order to compare these with the forecast values. In the event, despite several modifications, the installed package did not perform satisfactorily and only limited comparative data were obtained. However, it can be noted that the rig personnel again reported favorably on the heave forecast service.

A quantitative comparison was made possible by the efforts of the rig personnel on the Stena Forth rig in the North Sea. Over the period 17 December 1997 to 6 January 1998, the heave was measured routinely and recorded to compare against the forecast values (Figure 2).

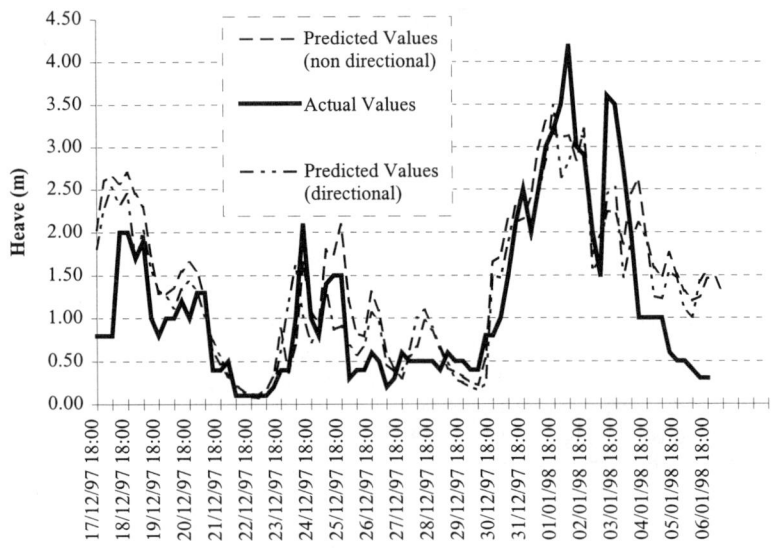

Figure 2: Measured and Forecast Heave - Stena Forth at Bruce Field

It was noted that, although generally following the measured events, there was a tendency to overpredict the estimates. This was suspected to be due to the original heave algorithm not incorporating directionality. In order to test whether this would have a significant effect on the results, the predictions were re-run using the directional version of the algorithm. The directional results are also shown in Figure 2. There is some improvement to the predictions, especially early in the period. We are currently investigating and quantifying the probable causes of the smaller differences which are related to:
- Accurate definition of rig RAO's.
- The operational accuracy of the method used to measure rig heave.
- Global wave model resolution, in near coastal areas.

The Stena Forth rig RAO's were provided by the rig owners but have not been confirmed to take account of the latest minor rig modifications. The essentially visual method of measuring rig heave means quantitative comparisons are also limited. However, rig personnel base their "belief" in the accuracy of the system on such measurements, so they cannot be disregarded. Finally, the global wave model resolution is relatively coarse. For example it does not resolve the Shetland Islands. Therefore, model locations in the North Sea, such as the Bruce Field, may not be "sheltered" from westerly swells in the Atlantic. This lack of sheltering may be one of the causes for the overprediction. This is being addressed using information from the European wave model, which has higher spatial resolution in the North Sea and does include the Shetland Islands in the domain.

Further Work
One of the main limitations of the work to date is the lack of good quality spectral wave and rig heave measurements with which to validate the procedure. Further effort will be put into this during 1998.

As further program developments, it is planned to predict motions of other types of vessels including monohull vessels. The model will then be enhanced to cater for moored vessels that have weathervaning capability. Some discussions are taking place on these issues with potential BP clients such as the Schiehallion and Foinaven production vessel operators.

At the moment, the model only predicts responses at the centre of gravity of the vessel. To calculate the motions at other important locations (such as the helideck), the roll and pitch motions will need to be predicted as well.

The UKMO are now commercialising the service, and making it available to other offshore operators. W S Atkins will provide the hydrodynamic support to generate appropriate RAO's and algorithms, a role performed by BP Sunbury staff in the developments described here.

Conclusions
An operational heave forecast service has now been established by BP and the UKMO which provides forecasts of rig heave out to 5 days ahead in a format that is readily interpretable by offshore personnel.

The service is tailored specifically to each rig at each location through the use of the rig RAO's. A recent development has seen the inclusion of the directional information from the UKMO global wave model in the heave prediction algorithm.

Feedback from offshore has been positive. However, accurate and quantitative validation has proved difficult.

The value of the heave forecasting service for offshore operations can be summarised under two main headings:

- The forecast can be used as an alarm to improve safety during offshore operations. The rig can be warned in advance when operations should be stopped for the safety of the personnel and the environment.

- Losses of production time due to waiting for a weather window for offshore operations are improved with the reliability of the planning; the heave forecast giving more direct information than the standard weather forecast. Offshore operations are an important part of the cost of field developments. The new service can lead to substantial savings by achieving better rig productivity.

The impact of the heave forecast will become greater as offshore exploration goes to harsher environments and deeper waters. In these conditions, the offshore operations become more complex and schedule sensitive. Reliable information on forecast rig motions then becomes critical for safety and good planning. The service was developed in UK waters, for exploration and appraisal drilling and well testing. We are confident it can now be implemented for any part of the world - for exploration rigs, drill ships and floating production vessels.

References

Golding, B. (1983). "A wave prediction system for real time seastate forecasting." *QJR Meteorol Soc.*, 109, 393-416.

Grant, C.K., Dyer, R.C., and Leggett, I.M. (1995). "Development of a new metocean design basis for the NW Shelf of Europe." *OTC 7685, Offshore Tech. Conf.* 415-424.

Holt, M.W. (1994). "Improvements to the UKMO wave model swell dissipation and performance in light winds." *FR Tech Report 119*, UK Met Office,.

Komen, G.J., Hasselmann, S, and Hasselmann, K. (1984). "On the existence of a fully developed wind sea spectrum." *J Phys Oc.*, 14, 1271-1285.

Snyder, R.L., Dobson, F.W., Elliott, J.A. and Long, R.B. (1982). "Array measurements of atmospheric pressure fluctuations above surface gravity waves." *J Fluid Mech.*, 102, 1-60.

Thomas, J.P. (1988). "Retrieval of energy spectra from measured data for assimilation into a wave model." *QJR Meterol Soc.*, 114, 781-800.

ON WAVE AND FORCE UNCERTAINTY FOR STEEL STRUCTURES

Léon A. Harland[1] and Paul H. Taylor[2]

Abstract

Waves in the open sea are random in both elevation and direction of motion. However standard design guidelines such as API-LRFD recommend the use of regular uni-directional wave theories. This paper reviews the influence of wave randomness on the reliable prediction of extreme loads on drag-dominated space-frame structures in severe storms.

Observation and modelling of large waves

Lindgren (1970) was the first to examine the statistics of the shapes of large excursions in a linear Gaussian process. He showed that the average shape of a sufficiently large crest is a scaled-up version of the usual auto-correlation function for the process, but that for smaller waves the average shape is more complex. Various authors have re-invented parts of this result over the years (Boccotti (1981), Phillips et al. (1993), Tromans et al. (1991)), presenting some comparisons to measured data. The most detailed comparisons have been described by Jonathan and Taylor (1995), based on field data gathered in severe storms in the northern North Sea. For sufficiently large crests (and troughs), the scaled auto-correlation function, often referred to as NewWave, fits offshore data remarkably well after due allowance is made for the bound harmonics. These make the crests higher and sharper and the troughs smaller and more rounded. Apart from the work of Phillips, all these

---

[1] Delft University of Technology, Faculty of Civil Engineering and Geosciences,
   Stevinweg 1, P.O. Box 5048, 2600 GA Delft, The Netherlands
[2] Dept. of Eng. Science, University of Oxford, Parks Road, Oxford OX1 3PJ, U.K.
   Formerly with Shell International Exploration and Production B.V., Rijswijk,
   The Netherlands

analyses were based on measurements of surface elevation in time at a fixed location.

Of course, time records of surface elevation at a single point yield no information on the directionality or spreading of the wave field. However, so long as the linear Gaussian model of the wave statistics is appropriate, Lindgren's analysis carries straight over into the spatial structure of large crests. The average shape of a large crest in space would be a scaled-up version of the spatial auto-correlation function. This is an isolated large hump surrounded by troughs and wiggles decaying rapidly with distance from the highest point. Of course, being the correlation function for an oscillatory process, the volume of water in the humps above m-s-l is balanced by the air in the troughs up to m-s-l.

Usually in offshore engineering, discussions of spreading are restricted to estimation of the kinematics reduction factor on fluid velocity. However, the shape of the surface and the internal velocity field are strongly coupled. Thus, for values of the kinematics factor commonly used by industry ($\phi \sim 0.9$), even large waves within a random field are remarkably short crested, on average. This result is confirmed by simultaneous estimates of wave spreading by various methods by Jonathan & Taylor (1995, 1996). The Tern oil production platform in the northern North Sea was instrumented with a fluid velocity meter at −41m below the surface as well as two surface elevation sensors separated by 40m horizontally. For one of the large storms, the line connecting the surface elevation sensors was almost exactly normal to the mean wave direction. This allowed a direct estimate of the cross-correlation between the two sensors in the form 'given a large crest at sensor 1, what is the averaged elevation time history at sensor 2?' For a wave field consistent with a spreading value of $\phi = 0.9$ as deduced from the velocity measurements, the average height of a large crest at the 2nd sensor is only ~75% of the crest height at the 1st (conditioning) sensor. This drop in average crest height occurs in a distance of only 40m along the average along-crest direction, perpendicular to the mean wave direction. It is, however, consistent with the spatial correlation function of a linear Gaussian model of the wave field with a spreading value of $\phi = 0.9$.

Thus, wave fields with a realistic amount of spreading are short-crested. This also implies that the variation in the height of an extreme wave crest across the width of a large structure is important, an effect which is not recognised in the standard design process (API LRFD 1993). This provides an explanation for the effect of wave attack direction, broadside versus end-on, on the hydrodynamic loads measured on Tern, as discussed but not explained by Heideman and Weaver (1992) in their wave-by-wave force analysis. Tern is a large, elongated space-frame with an aspect ratio at m-s-l of roughly 2 ½ to 1. Thus, for waves broadside on to the structure, the average crest elevation over the width of the structure will be lower than if the same waves approached the structure end-on. A reduced average crest elevation will result in lower hydrodynamic forces, although this will depend on the distribution of steel within the structure.

The results of analysis of field data from Tern are entirely consistent with the statistical ideas originated by Lindgren (1970) for a linear process. This implies that virtually all the non-linearity of ocean waves is local in the sense that the 2nd and higher order bound-waves (familiar from a Stokes perturbation expansion) do not affect the underlying statistics and dynamics of the random field apart from the obvious corrections. Of course, this conclusion is only supportable given sufficient data – and the most extreme waves in a storm are so infrequent that it is impossible to collect enough data!

Constrained random simulation

The realisation that NewWave is a good description for the shape of a large wave leads to the idea of constrained random time domain simulation. Simply put, it is possible to embed a large wave of selected size in a random sequence in such a way that it is in practice not possible to tell that this wave didn't arise by chance. In this way, little simulation effort is required to estimate the distribution of a structural response variable given a particular wave crest height, F(response | crest). The distribution of the extreme response in a storm, F(response | storm), can be obtained by convoluting the F(response | crest) distribution with the well-known distribution of crests in a sea-state - the simple Rayleigh distribution is a suitable approximation for this. This is a straightforward calculation to perform numerically. The use of constrained simulations speeds up the calculation of the probability density function for the extreme response to a reasonable level of accuracy by a factor of ~100x over simple Monte Carlo simulation. The technique is powerful and ideally suited to the treatment of dynamically sensitive structures (Harland et al. 1996). It is also useful for looking at global force variability on fixed structures.

Extreme force given a wave crest

Simultaneous offshore measurements of both the wave field and the resultant total forces have been made on several platforms (e.g. OTS, Magnus, Tern - see Heideman and Weaver (1992)). The most obvious method of treating the data is to sort the data by crest elevation, then to examine the associated peak forces: (force | crest). This method is generally referred to as a wave-by-wave analysis. In this way coefficients of variation (CoV) of the order of 30% are produced even for forces associated with crests heights up at a 1 in 3-hour level (Jonathan et al. 1996). This is such a large level of uncertainty that one might conclude that attempting to predict extreme forces accurately is useless. However, the results from analysis of this type depend on a hidden variable - the position of the wave elevation sensor. As the wave elevation sensor is moved away from the centre of the structure in any direction the mean value of peak force given a measured wave crest decreases whereas its variability increases. The limit of this process would be force measurements on one structure and wave crest measurements on a second, the two structures being sufficiently far apart that individual waves on each are completely uncorrelated. However, the structures are sufficiently close that the overall statistics of the sea-state at each are identical (H_s, T_p, directional spreading etc.).

The large force CoVs observed in measured data on a wave-by-wave basis are only slightly larger than those predicted using constrained simulations of a computer model of Tern (Harland et al. 1997). The only source of randomness in these simulations is the Gaussian model for the ocean surface. The force coefficients in Morison's equation are treated as deterministic. Thus, in the field data, the additional variability of global force due to other sources must be relatively small. One obvious source would be local variation in the Morison coefficients along individual structural members. However, the correlation length-scale for local spatial variation in these coefficients is likely to be a few member diameters, very small compared to the dimensions of the structure and the length-scales of the wave field.

Jonathan and Taylor (1996) tackled the problem of force variability analytically. They started by looking at the statistics of the velocity at one point given a wave crest at a 2nd point. The theory is a straightforward extension of Lindgren's analysis, yielding a simple cross-correlation for large events instead of an auto-correlation (NewWave). Good comparisons with the Tern data were obtained, again after making the necessary corrections for the bound-wave structure. Having the statistics of the velocity field conditional on a crest height, a simple quadratic form was used to represent the drag-part of the Morison equation. Although this analytical work did require some simplifications, the estimates of force CoVs are in excellent agreement with Harland's random constrained time-domain simulations (1997).

Extreme forces given a storm

Engineers do not design structures to withstand the largest force associated with a large wave crest, (extreme force | crest) say. Instead, structures are designed to withstand the peak force in a given period. Thus, the most important force quantity for design is not (force | crest). Instead, we are interested in (extreme force | storm). For this type of analysis, it is irrelevant where the wave field is measured relative to the structure, so long as both waves and forces are measured at two points such that the overall statistics of the sea-state at both are identical.

The use of the constrained simulation technique, checked against full random simulations, demonstrates that the CoV of extreme force on a full model of the Tern structure in a particular period is small. For a severe sea-state producing drag-dominant loading, the CoV is 2x that for the CoV of the largest crest in the same period (typically 8% at a 1 in 3 hour level). For an inertia-loaded structure, the equivalent CoVs are virtually identical. The doubling of the CoV for the drag-case simply reflects the effect of the velocity-squared term in the Morison equation. Thus, all the uncertainty associated with (force | crest) is integrated out of the final result. The variability based on the size of the largest crest in a random sea is inevitable.

These results were compared to global loads measured on the Tern platform in one of the largest storms. Of all the storms recorded at Tern, this storm is unique in that the main characteristics of the wave field were very close to constant for a period of 9 hours. Thus, each individual hour can be regarded as a separate realisation of the same random process. The connection between the CoV of the extreme force within an hour and the CoV of the size of the largest wave within an hour is borne out by the results from constrained simulations. Unfortunately, although measurements were made at Tern for several years, during which some major winter storms occurred, the waves were never large enough to get in the drag-dominant regime for the Tern structure. Despite this limitation, this work indicates that the variability in the extreme load on a structure is controlled to a large extent by the possible variability of the size of the extreme wave in that storm.

Extreme force in an interval given a spread sea

We have already referred to the discrepancy in loads end-on versus broadside for the Tern platform as recognised by Heideman and Weaver (1992). This was on a wave-by-wave basis. A similar discrepancy exists in the extreme loads in a given period. This is not properly accounted for in standard design recipes since deterministic uni-directional wave models are used with a kinematics reduction factor to account for spreading. Unfortunately, full spread sea random simulations are too complex for the routine design and re-assessment of space-frame structures. However, it would be good to account in an approximate manner for the short crested nature of real waves.

In recent work by van Weert and Harland (1998), a simple procedure is proposed to account for wave spreading on elongated structures. This has been validated against multiple simulations using fully 3-D random wave kinematics for structures of various sizes and shapes. During the design process, a simple uni-directional and deterministic model of the wave field (such as NewWave or Stokes) is used. However, the value of the wave spreading parameter used in this design calculation is modified from the free-field value (applicable for the treatment of wave spreading for a single vertical stick). It is now dependent not only on the physical spreading of the waves but also on the size and orientation of the structure. This treats spreading in a comparable way to the current - for design the free-stream current in the open ocean is reduced by a blockage factor dependent on the geometry of the platform (Taylor 1991).

Force variability and structural reliability

The variability in the largest waves and most extreme loads within a sea-state has been reviewed in this paper. The combination of results for several sea-states to obtain the statistics of extremes in a storm is straightforward given the storm history. However, the significance of these results to the assessment of the overall structural reliability of an offshore structure depends on whether the largest force in a given period is likely to occur within the most severe storm in the same period.

For areas where the expectation (or most probable value) of maximum storm severity observed or predicted in a given period increases significantly as the period is increased, the short-term variability of extreme loads within each storm is relatively less important. This is the case for tropical cyclone environments, the North Atlantic west of Shetland, and the northern North Sea. In contrast, if the maximum expected storm severity is only increasing very slowly, then more storms of comparable intensity are observed as the exposure period is lengthened. The largest load could occur in any one of these many, very similar storms. Now, the short-term variability of extreme load within a storm is much more important - how many waves from comparable sea-states does the structure see? The southern North Sea is an example of this situation.

Efthymiou, Van de Graaf and Tromans (1996) provide a detailed discussion of the overall structural reliability of offshore structures under extreme storm loading,

As we have discussed, the variability in the extreme force on a space-frame structure in a sea-state can be traced back to the (close to Rayleigh) distribution of wave crests. However, all this analysis is predicated on the tallest wave not reaching the deck of a structure. For all the wave crests measured at Tern, there is little if any evidence that the highest crests diverge significantly from the tail of a Rayleigh distribution if the usual crest-trough asymmetries are accounted for. However, the problem of estimating the height of the highest crest is key to assessing the air-gap requirements for platforms – an area of active study at present.

Conclusion

This paper has discussed how the randomness of waves on the open sea is reflected in the extreme forces exerted on offshore structures. The statistical ideas are traceable back to the analysis of Lindgren (and through him to Kac and Slepian, and Leadbetter). The overall conclusion is that, for structures where the extreme response is associated with the passage of a single large wave, the degree of variability of that extreme response is effectively set by the variability in size of the largest wave in a random sea-state.

Acknowledgements
The authors are grateful to their colleagues Peter Tromans, Phil Jonathan, George Forristall, Mike Efthymiou and Jan Vugts for extensive discussions and to Shell International Exploration and Production B.V. for financial support.

References

API RP2A-LRFD (1993) *Recommended practice for planning, designing and constructing fixed offshore platforms – load and resistance factor design*, 1st edition, July1, 1993.

Boccotti, P. (1981) *On the highest waves in a stationary Gaussian process.* Atti. Acc. Lig. 38, 45-73.

Efthymiou, M. van de Graaf, J.W. and Tromans, P.S. (1996). *Reliability based criteria for fixed steel offshore platforms.* 15th Offshore Mechanics and Arctic Engineering Conf. (OMAE), 1-A, 129-141, Florence, Italy.

Harland, L.A., Vugts, J.H., Jonathan, P. and Taylor, P.H. (1996) *Extreme responses of non-linear dynamic systems using constrained simulations.* 15th Offshore Mechanics and Arctic Engineering Conf. (OMAE), 1-A, 193-200, Florence, Italy.

Harland, L.A., Taylor, P.H. and Vugts, J.H., (1997) *The variability of extreme forces On offshore structures.* 8th International Conference on the Behaviour of Offshore Structures (BOSS97) 2, 221-235. Delft, The Netherlands.

Heideman, J.C. and Weaver, T.O. (1992) *Static wave force procedure for platform design.* ASCE Civil Engineering in the Oceans V. 496-517.

Jonathan, P. and Taylor, P.H. (1995) *Irregular, non-linear waves in a spread sea.* 14th Offshore Mechanics and Arctic Engineering Conf. (OMAE), 1-A, 9-16, Copenhagen,Denmark.

Jonathan, P. and Taylor, P.H. (1996) *Wave-induced loads on fixed offshore structure – an assessment of wave-by-wave load variability and bias.* 15th Offshore Mechanics and Arctic Engineering Conf. (OMAE), 1-B, 179-190, Florence.

Lindgren, G. (1970) *Some properties of a Normal process near a local maximum.* The Annals of Mathematical Statistics, 41(6), 1870-1883.

Phillips, O.M., Gu, D. and Walsh, E.J. (1993) *Expected structure of extreme waves in a Gaussian sea. Part 2: SWADE scanning radar altimeter measurements.* J. Phys. Oceanog. 23, 2297-2309.

Taylor, P.H. (1991) *Current blockage : Reduced forces on offshore space-frame structures.* Offshore Technology Conference Paper OTC6519,Houston, USA.

Tromans, P.S., Anaturk, A. and Hagemeijer, P. (1991) *A new model for the kinematics of large ocean waves – application as a design wave.* 1st Offshore and Polar Engineering Conf. (ISOPE), 3, 64-71, Edinburgh.

van Weert P.J. and Harland L.A. (1998), *Wave spreading and extreme wave loading.* Paper submitted for presentation at 17th Offshore Mechanics and Arctic Engineering Conf. (OMAE), Lisbon, Portugal.

LONG TERM RESPONSE ANALYSIS OF FIXED AND FLOATING STRUCTURES

Sverre Haver[1], Gro Sagli[2] and Tone M. Gran[1]

Introduction

The aim of the design load calculation is in principle to establish loads or responses corresponding to a required return period, e.g. 100 years. This calls for simultaneous probabilistic models for the load generating processes wind, current and waves. In order to predict consistent estimates of 100-year loads and responses, some sort of a long term analysis is to be carried out. Various approaches are briefly reviewed in the next chapter. Thereafter the results from the various methods will be compared for idealized response problems, which are assumed to be qualitatively representative for a broad range of practical offshore problems. In this paper, emphasis is given to the wave induced response. Further details of the study presented herein can be found in Haver et al. (1998).

Methods for long term response predictions

A long term response analysis can be carried out in a number of ways, and the main principles of the various approaches are briefly reviewed below. Uncertainties related to the long term analysis will be discussed by considering several idealized structural concepts. Both linear and non-linear cases will be considered. All analysis will be carried out for environmental data representing the Northern North Sea or the Norwegian Sea.

All sea states approach

The observed sea surface at a fixed location, $\xi(t)$, can be considered as a realization of a non-stationary stochastic process with slowly varying parameters: significant wave height, H_{m0}, and the spectral peak period, T_p. The long term variability of surface process can then be modeled by a joint probabilistic model of the adopted sea state characteristics, i.e. $f_{H_{m0}T_p}(h,t)$. For a more detailed review of the long term response analysis, see e.g. Haver and Nyhus (1986).

Under the Gaussian assumption, the global response maxima, y_Γ, (i.e. largest crest between adjacent zero-up-crossings) are well modeled by the Rayleigh distribution. Assuming all global maxima to be statistically independent, the distribution function of the largest value in τ hours is found by raising the Rayleigh distribution to a power equal to the number of zero up-crossing cycles in this period. The value corresponding to a probability of $1 - a$ of being exceeded is then given by:

$$y_{\Gamma,a}^{(\tau)} = \sigma_\Gamma \left[-2\ln\left(1 - a^{\left(1/3600\,\tau\,v_{\Gamma,0}^+\right)}\right) \right]^{0.5} \tag{1}$$

---

[1] Statoil, Exploration and Production Technology
[2] NTNU, Department of Marine Structures

where σ_Γ is the standard deviation and $v^+_{\Gamma,0}$ is the zero-up-crossing frequency.

This result is valid for a stationary sea state characterized by the significant wave height, the spectral peak period, and, if a forward speed problem is considered, a given forward speed, u, and the ship heading relative to the waves, β. The Rayleigh distribution can be considered as a conditional distribution given these characteristics. The marginal distribution, i.e. the long term distribution, can then be calculated as follows:

$$F_{Y_\Gamma}(y) = \frac{1}{\overline{v^+_{\Gamma,0}}} \int_h \int_t \int_\beta \int_u v^+_{\Gamma,0}(h,t,\beta,u) F_{Y_\Gamma|H_{m0}T_pBU}(y|h,t,\beta,u) \qquad (2)$$

$$\times f_{H_{m0}T_p}(h,t) f_{B|H_{m0}T_p}(\beta|h) f_{U|H_{m0}}(u|h) \, dh \, dt \, d\beta \, du$$

In this expression it is assumed that forward speed and heading angle are independent variables, and that the effect of peak period is negligible concerning the maneuvering in heavy weather. The long term mean zero up-crossing frequency is given by:

$$\overline{v^+_{\Gamma,0}} = \int_h \int_t \iint_{u\beta} v^+_{\Gamma,0}(h,t,u,\beta) f_{H_{m0}T_pUB}(h,t,u,\beta) \, dh \, dt \, dh \, d\beta \qquad (3)$$

This method is very convenient for linear systems where the response problem is characterized by a transfer function. All that is needed is then a procedure for selecting a proper wave spectrum for the various combinations of the sea state characteristics, see e.g. Haver and Nyhus (1986) or Torsethaugen (1996), and a joint probabilistic model for these characteristics, see e.g. Haver and Nyhus (1986). Uncertainties related to the choice of wave spectrum will be demonstrated later on.

Regarding the forward speed case, the conditional distribution of speed and direction are simplified by assuming $f_{B|H_{m0}}(\beta|h) = \delta(0)$ and $f_{U|H_{m0}}(u|h) = \delta(c(h))$, where $\delta()$ is the Dirac delta function. These simplifications mean that we merely consider head sea, and we assume a deterministic relation between the significant wave height and the forward speed. In the numerical examples we will also include a case where the speed is considered as a random variable with a mean and standard deviation being functions of h. Some information about effects of maneuvering in severe sea states is given in Soares (1984). In the present study, the forward speed is taken to be the largest value that can be used in order to fulfill some operational requirements regarding motions, Nordforsk (1987)

If non-linearities are involved, the Rayleigh distribution may no longer be a proper model. Replacing the Rayleigh model by a more adequate model, the procedure shown above can still be used. A simple and empirically based approach is to assume that the short term distribution of global maxima can be modeled by a Weibull distribution, i.e.:

$$F_{Y_\Gamma}(y) = 1 - \exp\left\{-\left(\frac{y}{\rho_\Gamma}\right)^{\lambda_\Gamma}\right\} \qquad (4)$$

For this model the $1 - \alpha$ - fractile of the largest value read:

$$y^{(\tau)}_{\Gamma,a} = \rho_\Gamma [-\ln(1 - \alpha^{(1/q_\tau)})]^{1/\lambda_\Gamma} \qquad (5)$$

where q_τ is the number of maxima in τ hours.

Regarding a long term analysis, the main problem is now to establish λ_Γ and β_Γ as functions of the slowly varying parameters. In view of the illustrative purpose of the present paper, the following expressions are adopted for the shape and scale parameter, respectively:

$$\lambda_\Gamma(h) = \gamma_0 + a_1 h^{a_2} \qquad (6)$$

$$\rho_\Gamma(h,t) = c_o\left[1 + \cos^{c_1}\left(\frac{2\pi\,(t-t_0)}{c_2}\right)\right]h^{c_3} \tag{7}$$

$a_1 = 0.0$ is used for all cases, $\gamma_0 = 2$ is used for linear cases, while $\gamma_0 = 1$ is used for quadratic response cases. The scale parameter parameters are selected such that we obtain cases with different period sensitivity. The cases included are shown in Fig. 1 and corresponding results are presented in the case studies.

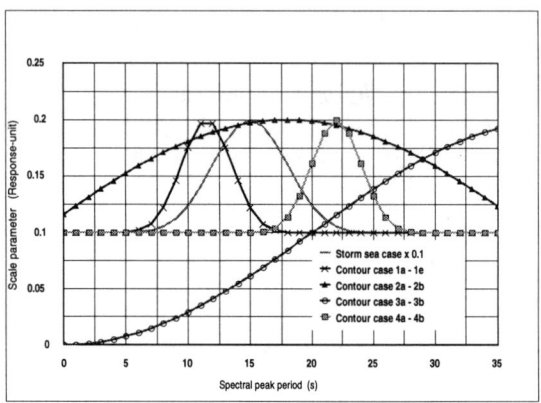

Fig. 1 Parameterized scale parameter of the Weibull distribution.

Storm approach
If joint occurrence is to be utilized, the structures has to be exposed to a realistic joint environment. This can of course be done by the approach reviewed above, but this will require quite some effort in establishing reliable joint environmental models. The storm approach is very convenient when lack of full correlation between the various environmental processes are to be utilized. This method has been used by a number of authors and regarding a review of various interpretations and applications, reference is made to Jahns and Wheeler (1972), Haring and Heideman (1979), and Tromans and Vanderschuren (1995). The storm approach requires that simultaneous data from a large number of storms are available.

In this paper we will mainly focus on the importance of having a sufficient amount of data. For this purpose we have assumed that the response maxima within each stationary time period follow a Weibull distribution, see Eq. (4). The shape parameter is assumed to equal 1, e.g. an exponent, that could represent a slow drift problem, and the scale parameter is taken to be the first curve shown in Fig. 1.

The distribution of the largest maximum within the short term event can be approximated by a Gumbel distribution with parameters which can be calculated from the Weibull parameters, see e.g. Bury (1975). A realization of the largest maximum is now obtained by Monte Carlo simulation. Extreme values are predicted by fitting an analytical model to the data, See Haver et al. (1998).

Contour line approach
The contour line concept is mainly suggested as a method for predicting load- and response maxima corresponding to a prescribed return period without having to carry out a full long term analysis. In Statoil, we have on a routine basis established contour lines for the significant wave height and the

spectral peak period since the mid-eighties. The contour lines, which followed constant probability functions, were earlier determined such that the most probable largest wave crest during 6 hours, for the most unfavorable sea state along the contour line, did equal the 100-year crest height established from a long term analysis. Thus, the maximum significant wave height along the contour line exceeded the marginal 100-year value. At present, we do not inflate the contour lines, so instead of predicting the most probable maximum in, say, 3 hours, the impact of the short term variability is accounted for by estimating a higher fractile of the extreme value distribution of the largest value in a 3-hour sea state. For a further explanation of the contour line approach, see Haver et. al. (1998).

Over the years, we have always determined contour lines such that they follow lines of constant probability density. A more consistent approach is probably to determine contour lines corresponding to constant exceedance probability. For a more thorough discussion of this contour interpretation, reference is made to Winterstein et al. (1993). However, a comparison for the Northern North Sea and the Norwegian sea shows that the difference is rather minor, see Fig. 1.

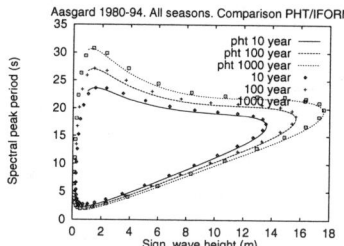 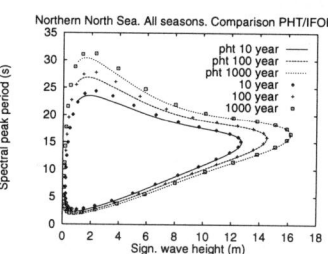

Fig. 2 Contour lines for the Åsgard field and for the Northern North Sea (Statfjord field).

The contour line approach can probably also be used for predicting the 100-year response for a ship with forward speed. This will however in principle introduce maneuvering characteristics to the problem and the adequacy remains to be demonstrated. These parameters should in principle be included as new dimensions of the contour problem, i.e. the contour line should be generalized to a contour surface. Herein we will study the application of contour lines in a somewhat simplified manner. We will investigate the adequacy of $H_{m0} - T_p$ contour lines for various choices of forward speed, and we will study the use of $H_{m0} - U$ contour lines for given spectral periods.

Case studies
Wave Climate
Environmental contour lines for the Haltenbanken and Northern North Sea will be used in the following, see Fig. 2. The wave climate of these areas are described by a joint distribution for the significant wave height, H_{m0}, and the spectral peak period, T_p. The probabilistic models together with the fitted parameters for the Northern North Sea are given by Haver and Nyhus (1986), while the parameters for the Haltenbanken area are given in Haver et al. (1998).

Regarding the storm climate, hindcast data for the Haltenbanken area are used. The hindcast data cover the years 1955 - 1996, i.e. a period of 42 years.

Effects of spectral models
The effects of different spectral models on long term analysis are exemplified by calculating the long term statistics using a Pierson-Moskowitz, a JONSWAP and a Torsethaugen spectrum. For this purpose, three response problems characterized by their transfer functions are considered. Sensitivity to various period bands in the scatter diagram is obtained by varying the natural period and the

damping level. The selected response amplitude operators are shown in Fig. 3. The results for the three cases are shown in Table 1.

Impacts of the data amount used for the long term analysis
Using the all sea state approach, long term results are obtained by using the smoothed joint density functions for the sea state parameters and also by simply using the observed scatter diagram. The results are shown in Table 2 for the three response cases shown in Fig. 3. Case 1 (t_0=10s and λ=1%) is sensitive to steep extreme sea states, Case 2 (t_0=10s and λ=5%) is driven by the extreme sea states in a central band of the scatter diagram and Case 3 (t_0=23s and λ=5%) is governed by severe long period sea states. The observed scatter diagram represents about 8 years of data. A JONSWAP wave spectrum is chosen for all these cases. The two cumulative distributions for Case 3 are shown in Fig. 4.

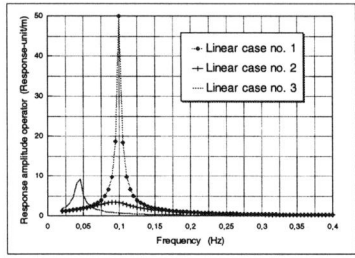
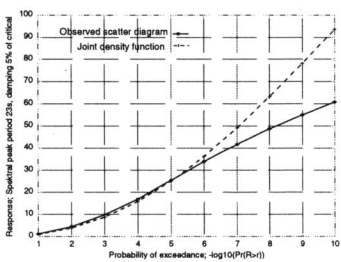

Fig. 3 RAO's with different peak periods t_0 and damping coefficients λ used in the case studies.

Fig. 4 Effect of smoothed vs. observed scatter diagram on long term statistics, Case no. 3.

Storm sea approach
Concerning the storm approach we will demonstrate both the importance of including the short term variability and the effect of the amount of data used. By choosing different parts of the 42 year data set for the simulation, the effect of increasing the amount of data and the variation between data sets of the same size can be studied. The results are given in Table 3, which shows the importance of including the short term variability when predicting the dynamic extreme value.

Table 1 Long term results for various spectra. (P-M=Pierson-Moskowitz, J=JONSWAP and T=Torsethaugen)

Return period	Case 1			Case 2			Case 3		
	PM	J	T	PM	J	T	PM	J	T
1	148.7	170.5	140.5	31.9	31.8	31.6	43.2	43.2	37.1
10	168.2	195.2	157.9	36.7	36.4	36.5	56.7	56.9	49.0
100	187.0	218.5	174.9	41.5	41.0	41.4	71.0	71.5	61.8
10 000	222.7	261.0	208.0	51.0	49.9	51.5	101.1	102.2	87.5

Table 2 Results for smoothed and non-smoothed scatter diagram. (Sm=Smoothed, Obs.=Observed and Diff. =Difference in %)

Return period	Case 1			Case 2			Case 3		
	Sm.	Obs.	Diff.	Sm.	Obs.	Diff.	Sm.	Obs.	Diff.
1	170.5	159.7	6.8	31.8	30.9	2.9	43.2	38.4	12.5
10	195.2	184.2	6.0	36.4	35.0	4.0	56.9	45.7	24.5
100	218.5	207.3	5.4	41.0	38.7	5.9	71.5	52.3	36.7
10 000	261.0	249.1	4.8	49.9	45.6	9.4	102.2	63.8	60.2

OCEAN WAVE KINEMATICS, DYNAMICS AND LOADS ON STRUCTURES   245

Table 3 Statistical values for five-, ten-, twenty- and forty-year periods of hindcast data.

Years	100-year response		Years	100-year response	
	short term var.	no short term var.		short term var.	no short term var.
55-60	2853.6	1702.8	55-65	2836.8	1693.1
60-65	3180.2	2072.0	65-75	2674.7	1639.2
65-70	2528.8	1689.5	75-85	3275.8	1926.4
70-75	2739.6	1658.2	85-95	3462.0	1886.7
75-80	3118.1	1970.0	55-60	2853.6	1702.8
80-85	3549.0	1949.1	55-65	2836.8	1693.1
85-90	3425.4	2040.7	55-75	2706.6	1576.5
90-95	3368.7	1986.6	55-95	3079.3	1685.0

Adequacy of the Contour Line Approach
At first we will indicate the adequacy of the contour line concept for Gaussian sea surfaces and linear response problems. The response cases are shown in Fig. 3. The long term result is obtained by means of the all sea states approach. The wave spectrum is modeled by a Pierson-Moskowitz type of spectrum for all sea states. Short term extremes are calculated for various sea states along the contour lines and the duration of the contour line sea states are taken to be 3 hours. The best fractile to adopt in order to obtain a good estimate of the 100 year value vary for the three cases. For Case 1, the 80% fractile is the best choice, for Case 2 the best estimate is obtained using the 85% fractile, while for the long period case, Case 3, the 90% fractile seems to be the most adequate choice. Out of the three selected cases, the second one is the one being most representative for most practical applications. This suggests that the 85% fractile is an adequate recommendation if linear response problems are considered.

For non-linear cases, the Weibull model is assumed to represent an adequate short term model and the scale parameter are taken according to Eq. (7) and Fig. 1. For case 1c the shape parameter is taken to be 2 (i.e. the Weibull model reduces to the Rayleigh model) and for the cases 1d and 1dd it is set to 3, i.e. these cases represent very non-linear cases. For all other cases the shape parameter equals 1, which is in consistence with a quadratic system driven by a Gaussian input. Two levels of the zero up-crossing frequency are included, slow drift problems (s.f.) and wave frequency problems (w.f.). The results are shown in Table 4. Again it is seen that for the most realistic non-linear cases (from a practical point of view), reasonable estimates for the 100-year response are obtained for fractiles in the range 85 - 90%.

Table 4 Long and short term results for the included cases - non-inflated contour line concept.
All values are normalized with respect to the 100-year value obtained from the long term analysis.

Case	Long Term	Contour Line Results					
		M.P.M:	Median	80%	85%	90%	95%
1a (s.f.)	1	0.68	0.74	0.9	0.95	1.01	1.12
1b (w.f.)	1	0.78	0.83	0.95	0.99	1.04	1.12
1c (w.f.)	1	0.93	0.95	1.02	1.04	1.07	1.11
1d (s.f.)	1	0.57	0.63	0.84	0.91	0.99	1.14
1dd (w.f.)	1	0.61	0.66	0.82	0.86	0.93	1.04
2a (s.f.)	1	0.71	0.77	0.94	0.99	1.06	1.17
2b (w.f.)	1	0.73	0.78	0.9	0.93	0.98	1.06
3a (s.f.)	1	0.7	0.76	0.93	0.98	1.04	1.15
3b (w.f.)	1	0.71	0.76	0.89	0.92	0.97	1.05
4a (s.f.)	1	0.68	0.73	0.89	0.94	1	1.11
4b (w.f.)	1	0.66	0.7	0.82	0.86	0.9	0.98

Forward speed case.
To study the adequacy of the contour line concept in the forward speed case, the contour line concept is applied to the ITTC container vessel, S-175. The data and the hull form is given in Haver et al. (1998). The long term wave climate for the Northern North Sea given by Haver and Nyhus (1986) is adopted. The environmental contour line for the Northern North Sea corresponding to constant probability of exceedance will be adopted, see Fig. 2.

The velocity profile used herein is shown in Fig. 5, i.e. at first we assume that there is a deterministic relation between forward speed and the significant wave height. Assuming that the response is linear, the transfer functions can be calculated by using ordinary strip theory, Salvesen *et al.* (1971). In this example, the vertical bending moment midship was chosen. The calculation was performed for three different Froude numbers, namely Fn=0.1,0.2 and 0.3, i.e. U=4.14, 8.29 and 11.39m/s, respectively. This response quantity is not very sensitive to speed, therefore rather large speeds are chosen for illustrative purposes.

Figure 6 shows 85% fractiles for the bending moment for the three considered speeds. The long term result corresponds to the speed profile given in Fig. 5. The extreme value obtained by adopting a fractile of 85%, estimate the long term extreme value (100-year value) with good accuracy for Fn=0.2 (U=8.29m/s).

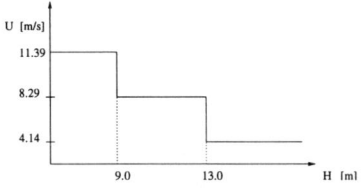

Figure 5 Velocity profile for the S-175 container ship. Established by use of the motion criteria 0.15g in the center of gravity.

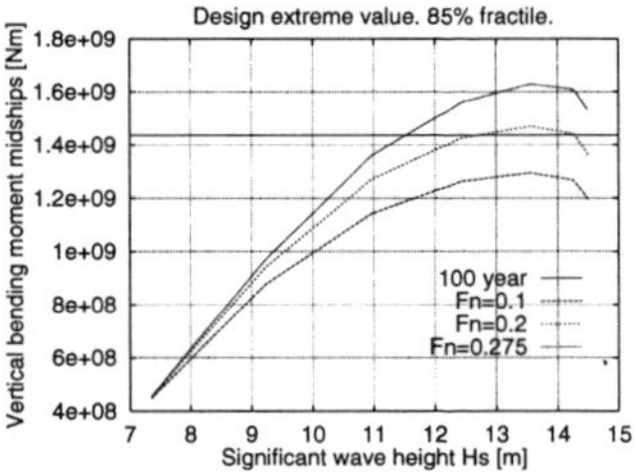

Figure 6 The design extreme value (85%) for the vertical bending moment mid ship.

Table 7 shows the adequacy of various fractiles for various speed cases. All numbers are normalized with respect to the long term extreme value. One may notice that in the case of a constant velocity profile, the 90% fractiles seems to be adequate for all speeds. For the "step" profile, the fractile is dependent on the velocity, and the best fractile seems to lie between 75-80% for the Froude number 0.2. Of course this is case dependent; other results may be obtained for other heading angles, speeds and vessels.

Table 7 Normalized design extreme values. The values are normalized by using the life time extreme value obtained by using the "step" profile and constant profiles. (by constant profiles it is meant that one velocity is used for all significant wave heights.)

Fractile	75%		80%		85%		90%		95%	
Profile	Step	Const.	Step	Const.	Step	Const.	Step	Const.	Step	Const.
Fn=0.1	0.876	0.957	0.887	0.967	0.901	0.986	0.921	1.007	0.954	1.043
Fn=0.2	0.995	0.961	1.007	0.973	1.023	0.989	1.045	1.010	1.082	1.046
Fn=0.275	1.102	0.964	1.116	0.967	1.134	0.992	1.158	1.013	1.199	1.048

In order to study the use of $H_{m0} - U$ contour lines, the distribution $f_{H_{m0}U}(h,u) = f_{H_{m0}}(h)f_{U|H_{m0}}(u|h)$ must be established. The randomness of U given H_{m0} is assumed to be well described by a log-normal distribution. A smoothed mean velocity versus significant wave height is, in view of the illustrative purposes of this study, assumed to be given by (in most cases a much lower mean speed should be used):

$$\mu_U = -3.0 Arc\tan[(h-9)/4] + 7 \tag{10}$$

The coefficient of variation is set constant to 20% for all wave heights. The expected value and the standard deviation of $\ln U$ (i.e the parameters in the log normal distribution) in terms of μ_U and the coefficient of variation can be found in Bury (1975). The contour lines are generated such that they correspond to constant probability of exceedance. The resulting lines for return periods of 10, 100 and 1000 years are shown in Fig. 7.

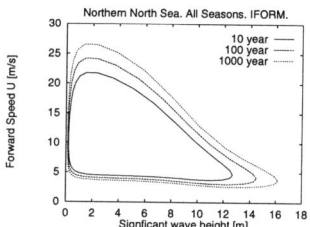
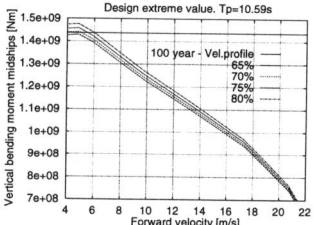

Figure 7 Contour lines following constant exceedance probability. Northern North Sea.

Figure 8 The design extreme value for the vertical bending moment midship, T_p chosen conservatively.

Two cases are presented. First the peak period is set equal to the first harmonic eigen period of the vessel, i.e. a conservative choice. In the second case, the peak period is equal to the conditional mean value of the period given H_{m0}, see Haver and Nyhus (1986). For the first case the results are shown in Fig. 8. The 70% fractile seems to give proper results. When the mean period is adopted, proper estimates can not be obtained without using extremely high fractiles.

The forward speed case represent a case where all three parameters are of importance. Accordingly, contour surfaces in three dimensions should have been established. Further studies should therefore be done to study this problem in more detail.

Acknowledgment
Statoil is acknowledged for the permission to present this paper. The opinions expressed herein are those of the authors and should not be construed as reflecting the views of Statoil.

References

Bury, K. V. (1975): "Statistical Models in Applied Science", John Wiley & Sons, New York.
Haver and Nyhus (1986): "A wave climate description for long term response calculations", OMAE-86, Tokyo.
Haver, S., Sagli, G., and Gran, T.M. (1998): "Uncertainties related to long term response calculations", Statoil, Stavanger
Haring, R.E. and Heideman, J.C. (1980): "Gulf of Mexico Rare Wave Return Periods", OTC 3230, Houston.
Jahns, H.O. and Wheeler, J.D. (1972): "Long Term Wave Probabilities Based on Hindcasting of Severe Storms", OTC 1590, Houston.

Madsen, H., Krenk, S. and Lind, N. (1986): "Structural safety", Prentice Hall Inc., Eglewood Cliffs, New Jersey, 1986.
NORDFORSK (1987). "Assessment of ship performance in a seaway".
Salvesen N., Tuck, E.O., Faltinsen, O. M. (1971). "Ship motions and sea loads", SNAME annual meeting, New York, November 12-13, 1970, Dnv report no. 75, March 1971.
Soares, C. G. (1984). "Probabilistic models for load effects in ship structures", Phd. Thesis, NTH, Norway
Thorsethaugen, K. (1996): "Model for a Double Peaked Wave Spectrum", Sintef Civil and Environmental Engineering, Rep. No.: STF22 A96204, Trondheim.
Tromans, P.S. and Vanderschuren, L. (1995): "Response Based Design Conditions in the North Sea: Application of a New Method", OTC 7683, Houston.
Winterstein, S.R., Ude, T., Cornell, C.A., Bjerager, P. and Haver, S. (1993): "Environmental parameters for extreme response: Inverse FORM with omission factors", ICOSSAR'93, Insbruck, 1993.

Advanced Volterra Models for Fluid-Structure Interactions

Sungbin Im[1], Edward J. Powers[2], and In-Seung Park[2]

Abstract

In this paper we present advanced, orthogonal, sparse, Volterra-like models useful for modeling fluid-structure interactions. The ability of a sparse model to capture the nonlinear aspects of such interactions is demonstrated by applying it to model test data for the surge response of a spar in random seas.

Introduction

For the purpose of modeling various nonlinear fluid-structure interaction phenomena in offshore engineering, the well-known Volterra series model has long been utilized. Since the Volterra series approach is a nonparametric method, it does not require *a priori* information about the unknown system to be modeled (Schetzen 1980). Furthermore, with aid of well-developed higher-order statistical signal processing techniques (Nam and Powers 1994), the Volterra kernels can be determined from the various higher-order moments of the experimental excitation-response measurements.

In most cases, discrete-time or discrete-frequency Volterra models are characterized by a large number of Volterra kernel coefficients. The large number of coefficients required to model nonlinear phenomena results in an increased computational load and requires a large amount of raw experimental data. Such data is necessary to estimate the various higher-order moments which are required, in turn, to estimate the Volterra kernels. For this reason, a procedure, based on orthogonal search techniques, was recently put forward in Im and

---
[1] Soong Sil University, Seoul, Korea
[2] Offshore Technology Research Center, The University of Texas at Austin, Austin, Texas 78712

Powers (1996) for identifying the most significant discrete frequency-domain Volterra kernels. Since the procedure identifies the most significant Volterra kernel coefficients and assumes the remaining coefficients to be zero, the model constructed by this procedure is sparse. The procedure is based on relatively well-known mathematical concepts given in Im and Powers (1996). The objective of this paper is to demonstrate that such sparse models are capable of capturing the physics of nonlinear fluid-structure interactions. Model basin data of a spar in random seas is utilized.

Frequency-Domain Volterra Model

In the discrete frequency domain, the second-order Volterra model can be represented as follows:

$$Y(m) = H_1(m)X(m)$$
$$+ \sum_{p=-M}^{M} \sum_{q=-M}^{M} H_2(p,q)X(p)X(q)\delta_0(m-p-q) + \mathcal{E}(m) \quad (1)$$

where $X(m)$ and $Y(m)$ are the discrete Fourier transforms of the observed wave excitation and surge response and $\delta_0(m)$ is 1 for $m = 0$, and zero, otherwise. $\mathcal{E}(\cdot)$ represents the frequency-domain modeling error. Let M denote the maximum frequency of the system input $X(m)$. In Eq. 1, $H_1(p)$ and $H_2(p,q)$ denote the linear and quadratic transfer functions (i.e., the frequency domain Volterra kernels), respectively. The idea of sparse models is to include only those coefficients at discrete frequency (p,q) which contribute significantly to the output response.

Sparse Volterra-like Model

The frequency-domain Volterra model Eq. 1 can be rewritten using matrix notation as follows;

$$Y(m) = \mathcal{H}(m)\mathcal{P}(m) + \mathcal{E}(m) \quad (2)$$

where

$$\mathcal{H}(m) = [H_1(m), \ldots, I(m_{q1}, m_{q2})H_2(m_{q1}, m_{q2}), \ldots] \quad (3)$$

and

$$\mathcal{P}(m) = [X(m), \ldots, X(m_{q1})X(m_{q2}), \ldots,]^T \quad (4)$$

where the superscript T denotes the transpose of the vector, and $I(i,j)$ is a symmetry factor.

The next step is to develop an orthogonal model. The orthogonalization transformation is carried out on the input vector $\mathcal{P}(m)$ utilizing the Gram-Schmidt orthogonalization procedure (Golub and Van Loan 1989). That is,

$$\mathcal{P}(m) = \mathbf{L}(m)\mathcal{Z}(m) \quad (5)$$

where $\boldsymbol{\mathcal{Z}}(m)$ is the orthogonal input vector and $\mathbf{L}(m)$ is the Gram-Schmidt transformation matrix. By this transformation, the estimated model can be rewritten as follows;

$$\hat{Y}(m) = \hat{\boldsymbol{\mathcal{H}}}(m)\boldsymbol{\mathcal{P}}(m) = \hat{\boldsymbol{\mathcal{H}}}(m)\mathbf{L}(m)\boldsymbol{\mathcal{Z}}(m) \tag{6}$$

We define the orthogonalized Volterra transfer function $\boldsymbol{\mathcal{G}}(m)$ associated with the orthogonal input $\boldsymbol{\mathcal{Z}}(m)$,

$$\boldsymbol{\mathcal{G}}(m) = \hat{\boldsymbol{\mathcal{H}}}(m)\mathbf{L}(m) = [\; g_1 \;\; g_2 \;\; \cdots \;\; g_n \;] \tag{7}$$

Since each term of the vector $\boldsymbol{\mathcal{Z}}(m)$ is orthogonalized with respect to the others, the power spectrum predicted by the model $P_M(m)$ will be equal to the sum of the squares of the orthogonal terms $g_k z_k$. That is, $P_M(m)$ is given by,

$$P_M(m) = \sum_{k=1}^{n} |g_k|^2 E\{|z_k|^2\} \tag{8}$$

Note the absence of any cross-product or interference terms due to the orthogonal nature of the model. Furthermore, by defining the "orthogonal higher-order coherence" $\gamma_k^2(m)$ as follows;

$$\gamma_k^2(m) = |g_k|^2 E\{|z_k|^2\}/E\{|Y(m)|^2\} \tag{9}$$

the total coherence $\gamma^2(m)$ of the model can be represented by

$$\gamma^2(m) = \sum_{k=1}^{n} \gamma_k^2(m) = \gamma_L^2(m) + \gamma_Q^2(m) \tag{10}$$

where $\gamma_L^2(m)$ is the fraction of the observed response power spectrum at discrete frequency m which can be accounted for by the linear model, and $\gamma_Q^2(m)$ represents the fraction which can be accounted for by the quadratic model. The sparse Volterra-like model method utilizes the orthogonal higher-order coherence $\gamma_k^2(m)$, defined in Eq. 9, as a criterion to select, from the candidate terms which satisfy the frequency selection rule, significant terms to be included in a sparse model. The reason for choosing the orthogonal higher-order coherence $\gamma_k^2(m)$ as a criterion is that $\gamma_k^2(m)$ represents the fraction of the model output power spectrum at frequency m, which is accounted for by the term $g_k z_k$. That is, the sparse Volterra-like model method selects the terms with the most significant contributions to the model output. "Stop" criteria are discussed in Im and Powers (1996). Generally speaking we stop adding coefficients when the addition of such coefficients has negligible effect in modeling the observed response power spectrum.

Experimental Results

The experiments were conducted at the Model Basin of the Offshore Technology Research Center (OTRC), Texas A & M University. In a series of experiments, a 1:55 scale model spar was utilized. The data were collected at a sampling frequency very close to 2.7 Hz. The run time was one hour for each experiment. The surge motion of a spar was observed using random waves generated from a JONSWAP spectrum representative of a 100-year storm in the Gulf of Mexico. Since the major interest was in the low-frequency band, the data were resampled using a reduced sampling rate very close to 1.35 Hz. Thus, the actual data points per each one-hour run used for the modeling is 4855. We divided those data points into 37 segments (128 points per segment). Therefore, we have total 74 segments (37 segments per run × 2 runs) of 128 data points for the higher-order spectral estimation. Since each data segment is 94.8 sec (128/1.35), the corresponding frequency resolution is 0.0105 Hz.

Figure 1 shows the power spectra of the observed wave excitation and surge response. Note that the significant power of the wave excitation is located about 0.07 Hz while the surge response power spectrum has two peaks at 0.02 Hz and 0.07 Hz, respectively. Because the wave and surge power spectra are down by more than two orders of magnitude for frequencies greater than 0.25 Hz, our plots extend to 0.25 Hz rather than half the sampling frequency of 0.675 Hz. Thus there are 24 frequency bins (0.25/0.0105) in the interval 0 to 0.25 Hz.

Full Volterra Model: First, we applied the orthogonalized Volterra-like *full* (24 linear, 1392 quadratic coefficients) model approach to the surge response. The 24 linear coefficients correspond to the 24 frequency bins in the 0.0 to 0.25 Hz interval. The coherence spectra of this model are plotted in Figure 2. According to the results in Figure 2, in the band centered at 0.07 Hz where the response has its maximum, the linear component of the model accounts for $\sim$ 90 % of the observed response. In the same band, the quadratic component accounts for $\sim$10% of the observed response. Above 0.1 Hz, the contribution of the quadratic component clearly dominates over that of the linear component. Finally, we note that, at very low frequencies (< 0.05 Hz), quadratic effects contribute 80% or greater to the observed response power spectrum. In addition, the overall power spectrum of the surge response predicted by the orthogonalized model is displayed in Figure 3, in which the experimentally observed power spectrum is also plotted for the purposes of comparison. We see excellent agreement in Figure 3.

Figure 4 displays the distribution of the orthogonal higher-order coherence $\gamma_Q^2(m)$ of the quadratic terms. According to Figure 4, we can observe that the main wave power spectra located about 0.07 Hz contribute significantly to the very low frequency response through second-order difference frequency interactions. This fact is represented by the large peak at approximately (0.07,

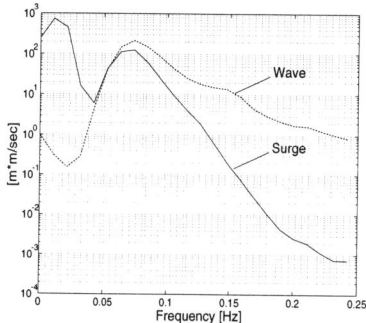

Figure 1: Power Spectra of Wave Excitation and Surge Response.

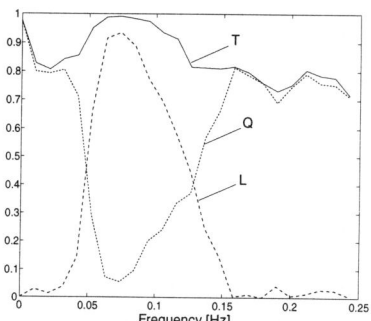

Figure 2: Coherence of the Orthogonalized Volterra Model.

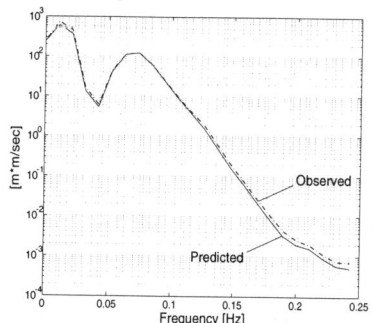

Figure 3: Observed Surge Power Spectrum and Surge Power Spectrum Predicted by the Orthogonalized Volterra Model. This is a "full" Volterra model consisting of 24 linear and 1392 quadratic coefficients.

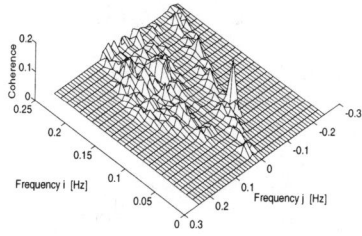

Figure 4: Orthogonal Higher-Order Coherence Distribution $\gamma_Q^2(m)$ ($m = i + j$).

-0.07). The ridge lying along the line $i+j \sim 0.02$ Hz in the difference frequency region ($i > 0$, $j < 0$) indicates that quadratically nonlinear difference interactions generate the low frequency components of the surge response. For the response frequencies at or near 0.07 Hz, however, the quadratic components (satisfying $i+j \sim 0.07$) are characterized by less significant contributions since the linear component accounts for ~ 90 % of the observed response. Finally we note that Figure 4 indicates significant coherence in the sum-frequency region ($i > 0$, $j > 0$). These terms indicate that the relatively weak surge response at frequencies greater than 0.07 Hz are due to quadratic sum-frequency effects, which is consistent with Figure 2.

Sparse Volterra-Like Model: Since the orthogonal higher-order coherence represents the contribution of each discrete frequency pair $\hat{g}_k^Q z_k^Q$ to the model output, we can identify which quadratic terms are most significant and thus construct a sparse Volterra-like model. Specifically, we consider a *sparse* model consisting of 24 linear and 625 quadratic transfer function coefficients. These represent approximately 45 % of the total coefficients associated with the *full* model (24 linear and 1392 quadratic coefficients). Figure 5 shows the numbers of the selected and total coefficients for each output frequency, for both the full and sparse models. In the case of the sparse model, for low frequencies (< 0.05 Hz) and high frequencies (>0.10 Hz), relatively many quadratic components are selected, while a small number of quadratic components are selected in the frequency band (about 0.07 Hz) where the main wave power is located. This is the case because in the band centered around 0.07 Hz, linear effects dominate. The power spectrum of the surge response predicted by the sparse Volterra-like model (24 linear and 625 quadratic coefficients) is displayed in Figure 6, in which the observed power spectrum is also plotted for the purpose of comparison. We note the agreement is quite good.

In the previous development of the full and sparse models, the frequency resolution is 0.0105 Hz (1.35 Hz / 128 points). When we use 256 points per segment to more precisely model the low frequency response of the spar surge motion, the frequency resolution becomes 0.0053 Hz (1.35 Hz / 256 points). For this resolution, however, we need more data points to determine the full Volterra model, because the full second-order Volterra model requires 47 linear and 5568 quadratic coefficients (see Figure 7). In this case, it is impossible to determine the full Volterra model, given the amount of raw experimental data available to us. However, we can construct a sparse model consisting of 47 linear and 375 quadratic coefficients, which are less than 10 % of those of the full model. The number of the total coefficients for each output frequency is shown in Figure 7. The power spectrum of the surge response predicted by the sparse Volterra-like model (47 linear and 375 quadratic coefficients) is displayed in Figure 8, in which the observed power spectrum is also plotted.

OCEAN WAVE KINEMATICS, DYNAMICS AND LOADS ON STRUCTURES 255

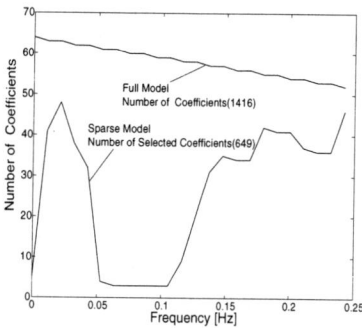

Figure 5: Number of Volterra Model Coefficients vs. Output Frequency

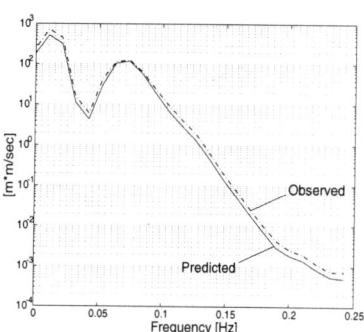

Figure 6: Observed Surge Power Spectrum and Surge Power Spectrum Predicted by the Sparse Volterra Model with the 24 Linear and 625 Most Significant Quadratic Coefficients for Surge Motion Subject to Random Sea Waves.

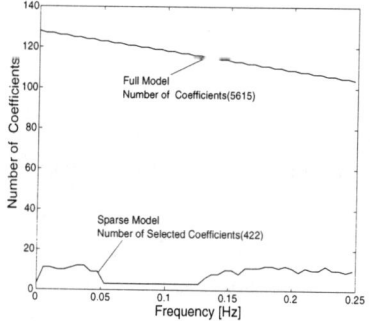

Figure 7: Number of Volterra Model Coefficients vs. Output Frequency for 0.0053 Hz Resolution Case.

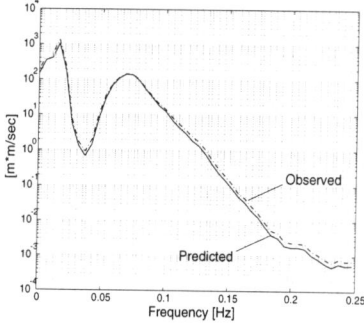

Figure 8: Observed Surge Power Spectrum and Surge Power Spectrum Predicted by the Sparse Volterra Model with the 47 Linear and 375 Most Significant Quadratic Coefficients for Surge Motion Subject to Random Sea Waves.

We note good agreement, which is particularly encouraging in view of the fact that not enough data was available to evaluate the full (47 linear, 5568 quadratic coefficients) model.

Summary

The concept of orthogonal, sparse, Volterra models (Im and Powers 1996) is based on well known mathematical methodologies such as Gram-Schmidt orthogonalization and orthogonal search techniques. Since the sparse second-order models used in this paper yield response power spectra (see Figures 6 and 8) which are in good agreement with the response power spectrum predicted by a full Volterra model (see Figure 3), and are in reasonably good agreement with the observed response power spectrum, we conclude that such sparse models are capable of capturing the quadratically nonlinear physical mechanisms whereby energy is extracted from the random sea excitation and down-converted through second-order "mixing" processes to re-appear in the low-frequency surge resonant motion of the spar.

Acknowledgements

This work is supported by NSF Grant CDR-8721512 through the Offshore Technology Research Center. The sparse Volterra-like model methodology was also developed under the auspices of the Joint Services Electronics Program AFOSR Contract F-49620-95-C-0045, Department of Navy Grant N00014-92-J-1046, Texas ATP Project 003658-392.

References

G. H. Golub and C. F. Van Loan (1989), *Matrix Computations*, Baltimore: The Johns Hopkins University Press.

S. Im and E.J. Powers (1996), "A Sparse Third-order Orthogonal Frequency-domain Volterra-like Model," *J. of the Franklin Institute*, vol. 333(B), pp. 385-412.

S.W. Nam and E.J. Powers (1994),"Application of Higher-Order Spectral Analysis to Cubically-Nonlinear System Identification," *IEEE Trans. on Signal Processing*, vol. 42, no. 7, pp. 1746-1765.

M. Schetzen (1980), *The Volterra and Wiener Theories of Nonlinear Systems*. New York:John Wiley and Sons.

Hurricane Wave Conditions for Design and Requalification of Platforms in the Bay of Campeche, Mexico

By Robert Bea[1], Joseph Suhayda[2], Zhaohui Jin[3], and Rafael Ramos[4]

Abstract

The Bay of Campeche, Mexico, is affected by intense hurricanes that originate in the Caribbean Sea and propagate into the Gulf of Mexico. The tracks of these storms generally cross the Yucatan Peninsula before they enter the Bay of Campeche. This geography has important effects on the storm intensities and wave conditions in the Bay of Campeche. In addition, the sea floor of the Bay of Campeche is overlain by a 10 m to 20 m thick blanket of soft clays. The soft sea floor soils have important effects on the wave characteristics in shallow water.

This paper discusses these conditions and their effects on hurricane waves of importance to design and requalification of platforms in the Bay of Campeche. The effects of the soft sea floor soils in modifying the hurricane wave characteristics are detailed. The analyses of these effects are correlated with data from hurricane Roxanne (1995). Hurricane Roxanne was the most intense hurricane to affect this region during this century. In addition, the effects of the unique geography and soil conditions on a 'maximum credible' hurricane that could be experienced in the Bay of Campeche are evaluated. The implications of these hurricane conditions on forces used to design and requalify platforms in the Bay of Campeche are assessed.

Introduction

At the present time, there are more than 200 platforms located in the Bay of Campeche (Fig. 1). These platforms produce in excess of 2 million barrels of oil and 1.5 billion cubic feet of gas per day. The majority of the platforms are located in water depths between 30 m and 50 m. Most of these platforms were installed in the 1980's and 1990's, with some platforms installed in the late 1970's.

In October 1995, hurricane Roxanne formed in the western Caribbean Sea, crossed the Yucatan Peninsula, and entered the Bay of Campeche. Due to a southward moving front, the hurricane did not follow the normal northerly path of most hurricanes. It was forced back into the Bay of Campeche and the eastern coast of Mexico where it did considerable damage. Roxanne was the most severe hurricane to affect the Bay of Campeche during this century. It generated environmental conditions

---

[1] Prof., Dept. of Civil & Environmental Engrg., Univ. of California, Berkeley, CA 94720-1712
[2] Prof. and Director, Natural Systems Engineering Laboratory, Louisiana Water Resources Research Institute, Louisiana State University, Baton Rouge, LA 70803
[3] Doctoral Graduate Student Researcher, Dept. of Civil & Environmental Engrg., Univ. of California
[4] Senior Engineer, Office of Special Studies, Instituto Mexicano de Petroleo, 07730 Mexico D.F.

which approximated those of 100-year return period hurricanes (Oceanweather 1996a; 1996b; 1996c; Glenn, 1996).

Offshore, the majority of damage was confined to pipelines. There was some damage to platforms, but in general, these structures performed very well. Following Roxanne, Petroleos Mexicano (PEMEX) initiated an extensive inspection of the platforms in the Bay of Campeche. Some damage to the platform structures were found. Most of the damage was confined to the underwater portions of the structures (cracked joints). In general, the platforms were found to be in good condition.

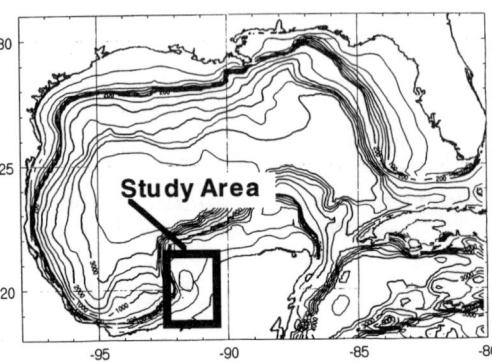

Fig. 1 - Bay of Campeche and Gulf of Mexico bathymetry

Given the results from the platform and pipeline inspections and fitness for purpose studies, PEMEX and Instituto Mexicano del Petroleo (IMP) initiated development of Risk Assessment and Management (RAM) based criteria for design and requalification of the platforms and pipelines in the Bay of Campeche (Bea, 1997a - 1997d). This paper summarizes the hurricane wave characterizations that were developed as a result of this study.

Hurricane Conditions

There have been extensive studies of oceanographic conditions in the Bay of Campeche performed by A. H. Glenn and Associates and Oceanweather Inc. (Glenn 1977; 1996; Oceanweather 1996a; 1996b). Fig. 2 shows a comparison of the expected maximum hurricane wave heights as a function of the average return period associated with these wave heights as developed by Glenn and Oceanweather for water depths in the range of 40 m to 60 m. At an average return period of 100 years the Glenn maximum wave height is about 16.8 m while those for Oceanweather are in the range of 13.7 m to 15.2 m.

While the Glenn hurricane wave heights are larger than those projected by Oceanweather for return periods less than about 1,000 years, they become less at longer return periods. This reflects a lower natural variability in the expected maximum wave heights projected by Glenn and a higher natural variability projected by Oceanweather. This variability has very important ramifications for platform design and requalification criteria. The Glenn results indicate a natural variability in the expected annual maximum wave heights expressed as a Coefficient of Variation (V) of $V = 33\%$. This is about the same value as that for the northeast Gulf of Mexico. The Oceanweather results indicate a natural variability in the expected annual maximum wave heights of $V = 40\%$ to 45%. Review of the background for the Oceanweather results indicates that this greater variability is due to the hurricane track - land crossing effects unique to the Bay of Campeche. Most of the hurricane tracks cross the Yucatan Peninsula before they can enter the Bay of Campeche. As they cross the Yucatan Peninsula, they loose strength. As they enter the Bay of Campeche, they intensify and continue on a generally northward course. However, very rarely, the hurricanes can encounter higher pressure areas or fronts that keep them from propagating northward.

They can become effectively trapped in or near the Bay of Campeche until the hurricanes move into the eastern coast of Mexico or are deflected eastward. It is a very unusual condition that involves the coincidence of the occurrence of a hurricane, its track into the Gulf of Mexico and a southerly moving front or high pressure cell. Because of the extensive database of calibrations of predicted and measured maximum wave heights in hurricanes associated with the Oceanweather hurricane hindcast models, it was elected to base characterization of the oceanographic conditions in the Bay of Campeche on the Oceanweather results.

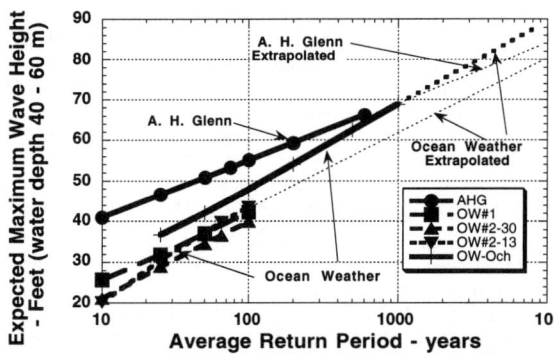

Fig. 2 - Comparison of Glenn and Oceanweather expected annual maximum hurricane wave heights

Fig. 3 summarizes expected maximum wave heights in various water depths for average annual return periods of 100 years, 1,000 years, and 10,000 years. These results incorporate only the wave attenuation processes associated with rigid or non-deformable sea floor characteristics.

Note that there are two 10,000 year characteristics shown in Fig. 3. The upper 10,000 year results (deep water wave height of 32 m) is based on an extrapolation of the Oceanweather results. This extrapolation was based on extremal or lognormal distribution functions (they gave the same results). The lower 10,000 year results (deep water wave height of 27 m) is based on an assessment of the characteristics of a 'maximum credible hurricane' in the Bay of Campeche.

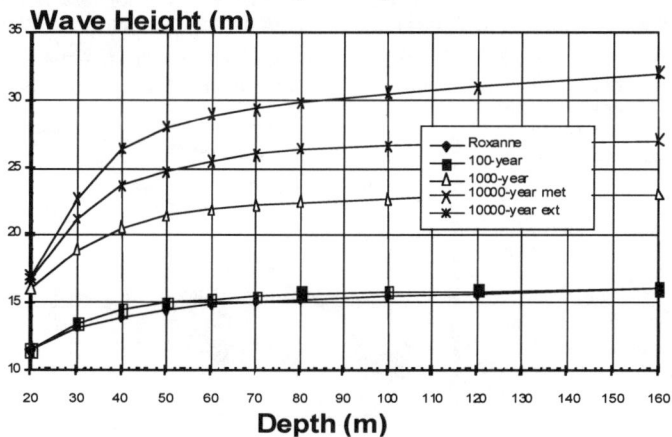

Fig. 3 - Rigid sea floor expected maximum wave heights based on Oceanweather analyses and maximum credible hurricane analyses

On the basis of meteorological considerations, the 'maximum credible' hurricane was assessed to have a maximum central differential pressure of 110 millibars, a radius to the maximum winds of 30 nautical miles, and an average forward speed of 15 knots. Fully developed wave conditions were assessed based on wave generation by the stable wind field for more than 6 hours. A wind field - wave field analytical model similar to that used by Oceanweather was applied. A maximum significant wave height of 16 m and an expected maximum wave height of 27 m resulted from this model. This maximum credible storm was evaluated to have an average return period of the order of 10,000 years.

Comparison of the two 10,000 year characteristics, one from extrapolation of the Oceanweather hindcasts (1996b) and the other from the meteorological considerations, indicates that it is likely that there is a physical limit to the maximum wave heights that can be generated by hurricanes in the Bay of Campeche. Similar results have been developed for hurricanes in the Northeastern Gulf of Mexico. For the Northeastern Gulf of Mexico, the 'maximum credible' hurricane expected maximum wave height in deep water is in the range of 28 m to 29 m.

Effects of Deformable Sea Floor on Hurricane Maximum Wave Heights

Fig. 4 summarizes the results of application of the Sea Wave Bottom Interactions (SWBI) analytical model to the Oceanweather hurricane Roxanne conditions (Suhayda, 1997; Zhaohui Jin, Bea, 1997). This analytical model is based on more than three decades of experimental and analytical work associated with evaluations of sea floor movements and wave conditions in the delta of the Mississippi River (Clukey, et al, 1990; Forristall, et al, 1985; Gu, Thompson 1995; Kraft, et al, 1990; Suhayda, 1977; 1996; Shapery, Dunlap 1978). SWBI is able to account for the wave energy dissipated by hysteretic energy losses in the deformable sea floor. SWBI has been subjected to extensive verifications involving measured and predicted hurricane wave heights in the Mississippi River Delta (Forristall, et al, 1985).

Best estimate soil and bathymetry characteristics were used to produce these results. There is reasonably good agreement between the observed and hindcast maximum wave heights. Note that the primary effects of the soft sea floor soils in modifying the maximum wave heights are in water depths less than about 80 m.

The differences between the observed and hindcast maximum wave heights could be attributed to several sources. First, there is an uncertainty in the observed wave heights (this uncertainty is reflected in the observed ranges shown in Fig. 4). Second, there are uncertainties in the Oceanweather wave heights in both deep and shallow water. And third, there are uncertainties in the SWBI analyses (e. g. wave height, period, travel paths, soil characteristics).

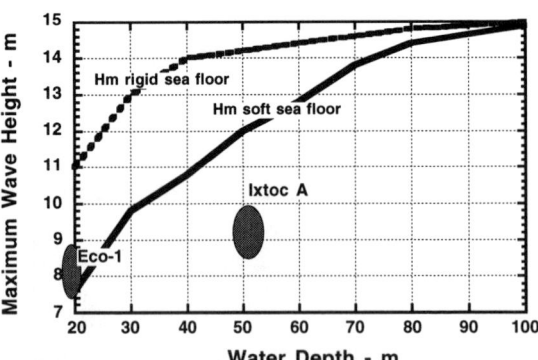

Fig. 4 - Hurricane Roxanne maximum wave heights for rigid and deformable sea floor conditions

Fig. 5 summarizes the best estimate expected

maximum wave heights for the soft sea floor conditions in the Bay of Campeche as a function of return periods in the range of 100 years to 10,000 years (Suhayda, 1997; Zhaohui Jin, Bea, 1997). In Fig. 5, note that for water depths less than about 80 m, the 1,000 year return period maximum wave heights and 10,0000 year maximum wave heights (based on meteorological limits) are about the same. The effect of the soft sea floor soils is to effectively 'truncate' the wave heights for hurricane conditions having return periods in excess of 1,000 years. For the soft sea floor shallow water areas of the Bay of Campeche, the 1,000 year return period wave heights could be regarded as the Ultimate Limit State wave height conditions for all platform Serviceability and Safety Classifications.

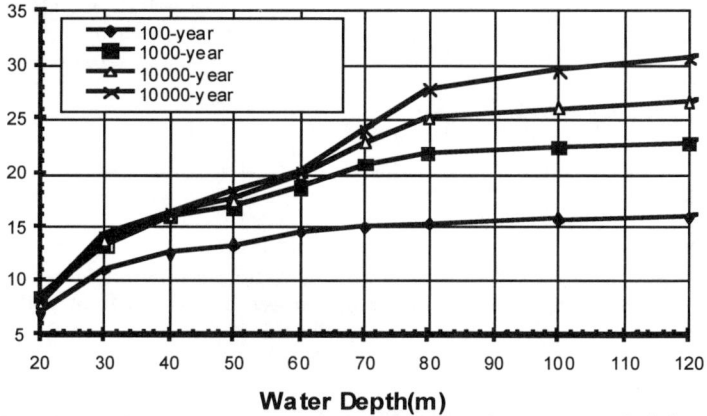

Fig. 5 - Deformable sea floor expected maximum wave heights

These results indicate that the soft sea floor soils in the Bay of Campeche have three important effects on the expected maximum wave heights. The first is to reduce the maximum wave heights in shallow water. The second is to reduce the natural variability in the expected annual maximum wave heights. And, the third is to truncate or limit the maximum wave heights that can be developed in shallow water.

There is a negative effect of the wave - sea floor interaction process: large movements of the near-sea floor soils. It is believed that such movement was responsible for some of the pipelines that failed in the Bay of Campeche during hurricane Roxanne (Valdez, et al, 1997). These movements will generate forces and displacements on all elements that contact or penetrate the sea floor including piles, conductors, mudline braces, and pipelines.

Based on the results of the SWBI analyses, Fig. 6 summarizes general guidelines for modification of the Oceanweather 100-year return period expected maximum wave heights in the deformable sea floor shallow water locations in the Bay of Campeche. The ordinate is the factor that should be multiplied times the Oceanweather expected maximum wave heights (based on rigid bottom conditions) to determine the expected maximum wave height for the Bay of Campeche soil and bathymetric characteristics. These guidelines are based on the best estimate characteristics for the wave travel paths, wave heights and periods, bathymetry, and soil characteristics in the Bay of Campeche.

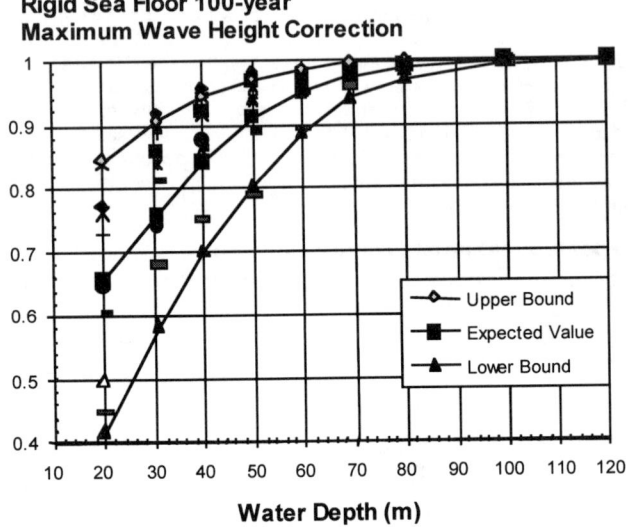

Fig. 6 - 100-year hurricane wave height reduction factors based on best estimate and upper and lower bound sea floor soil characteristics

Probability Characterizations

The foregoing results were used to determine the Type I (natural, inherent, aleatory) uncertainties associated with the expected annual maximum wave heights in the Bay of Campeche. Results for deep water (water depths greater than 70 m) and two shallow water locations (30 m and 50 m) are summarized in Fig. 7. Lognormal distributions were used for the probability function characterizations.

The results indicate that for water depths less than about 70 m, the standard deviation of the logarithms of the expected annual maximum wave heights is in the range of $\sigma_{lnHm} = 0.25$ to 0.30. This can be compared with values of $\sigma_{lnHm} = 0.38$ to 0.47 determined for rigid sea floor conditions. For deep water conditions (water depths greater than about 100 m) values in the range of $\sigma_{lnHm} = 0.40$ to 0.45 were determined during this study. Note that these values and characterizations have not taken advantage of the effective 'truncation' of the wave heights at very long return periods (reflected by change in slopes at return periods greater than 1,000 years in Fig. 7).

For the maximum wave heights influenced by the soft sea floor soils, there is an additional Type I uncertainty introduced by the natural variabilities associated with the sea floor soils. These are variabilities that are due to bathymetry (bottom slopes and geometry) and soil characteristics (strengths, shear modulus, hysteretic behavior). The potential effects of these variabilities were studied by systematically analyzing the effects of changes in the sea floor characteristics (Fig 6). The variability decreases with increasing water depth. In the subsequent RAM based criteria developments, the variability in 50 m water depth was used for the shallow water conditions. This was a value of $\sigma_{lnHm} = 0.10$.

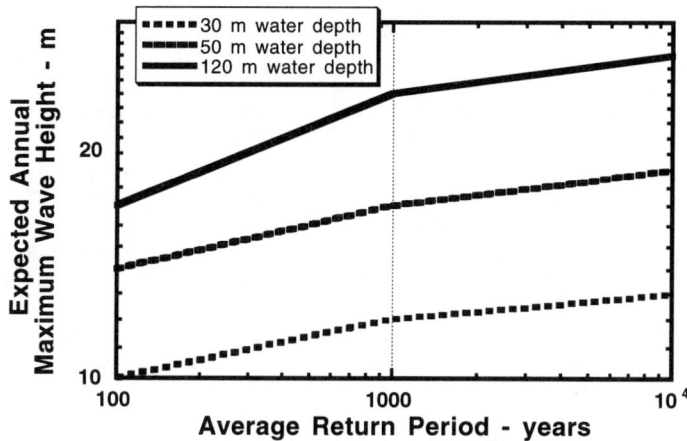

Fig. 7 - Expected annual maximum wave heights for different average return periods and water depths

Conclusions

In the subsequent RAM based design and requalification criteria analyses, the foregoing results were used to determine the total Type I (natural, inherent, aleatory) uncertainties associated with the expected annual maximum wave heights in the Bay of Campeche. Lognornal distributions were used for these characterizations. The results indicated that for water depths less than about 80 m, the standard deviation of the logarithms of the expected annual maximum wave heights was in the range of σ_{lnHm} = 0.10 to 0.30 (Suhayda, 1997; Zhaouhi Jin, Bea, 1997). A value of σ_{lnHm} = 0.30 was used in the criteria development for shallow water platforms and pipelines (Bea, 1997b, 1997c). For deep water conditions (water depths greater than about 100 m) a value of σ_{lnHm} = 0.45 was used in the criteria development for deep water platforms (Bea, 1997d).

References

Bea, R. G. (1997a). *Risk Based Oceanographic Criteria for Design and Requalification of Platforms in the Bay of Campeche*, Report to Petroleos Mexicanos and Instituto Mexicano del Petroleo, Marine Technology & Management Group, University of California at Berkeley.

Bea, R. G. (1997b). *Risk Based Hurricane and Earthquake Criteria for Design and Requalification of Platforms in the Bay of Campeche*, Report to Petroleos Mexicanos, Instituto Mexicano del Petroleo, and Brown & Root International, Marine Technology & Management Group, University of California at Berkeley.

Bea, R. G. (1997c). *Risk Based Criteria for Design and Requalification of Pipelines and Risers in the Bay of Campeche*, Report to Petroleos Mexicanos, Instituto Mexicano del Petroleo, and Brown & Root International, Marine Technology & Management Group, University of California at Berkeley.

Bea, R. G. (1997d). *Risk Based Hurricane Criteria for Design of Floating and Subsea Systems in the Bay of Campeche*, Report to Petroleos Mexicanos,

Instituto Mexicano del Petroleo, and Brown and Root International, Marine Technology & Management Group, University of California at Berkeley.

Clukey, E. C., Maller, A. V., Murff, J. D., Goodwin, R. H., Miller, M. C., and Ebelhar, R. J. (1990). "Wave Attenuation, Mudslide, and Structural Analyses for Mississippi Delta / Main Pass Caisson, *Transactions of the American Society of Mechanical Engineers*, Vol. 112, New York, NY.

Forristall, G. Z., and Reece, A. M. (1985). "Measurements of Wave Attenuation Due to a Soft Bottom: the SWAMP Experiment," Journal of Geophysical Research, Vol. 90, American Geophysical Union, New York, NY.

Gu, G. Z., and Thompson, G. R., (1995). "Wave Induced Mudslide Analyses for Offshore Structure, Design," Mobil Research and Development Corporation Paper, Dallas, TX.

Kraft, L. M., Suhayda, J. N., Helfrich, S. C., and Marin, J. E. (1990). "Ocean Wave Attenuation Due to Soft Seafloor Sediments," *Marine Geotechnology*, Vol. 9, Crane, Russak & Col, Inc., UK.

Oceanweather Inc. (1996a). *Final Report, Update of Meteorological and Oceanographic Hindcast Data and Normals and Extremes Bay of Campeche*, Report to Brown and Root International Inc., Cos Cob, CT.

Oceanweather Inc. (1996b). *Final Report Hurricane Roxanne / 1996 Hindcast Study*, Report to Brown and Root International Inc., Cos Cob, CT.

Schapery, R. A., and Dunlap, W. A. (1978). "Prediction of Storm-Induced Sea Bottom Movement and Platform Forces," *Proceedings of the Offshore Technology Conference*, OTC 3259, Society of Petroleum Engineers, Richardson, TX.

Suhayda, J. N., (1977). "Surface Waves and Bottom Sediment Response," Marine Slope Stability, *Marine Geotechnology*, Vol. 2, Crane, Russak & Co, Inc., UK.

Suhayda, J. N. (1996). "Oceanographic Design Data for South Pass Area Block 47," *Proceedings of the Offshore Technology Conference*, OTC 7952, Society of Petroleum Engineers, Richardson, TX.

Suhayda, J. N. (1997). *Development of Oceanographic Design Criteria for Platforms, Wave / Sea Bottom Interaction Effects*, Report to Petroleos Mexicanos Exploration and Production, Instituto Mexicano del Petroleo, and Brown and Root International Inc., J. N. Suhayda Consulting Coastal Engineer, Baton Rouge, LA.

Zhaohui Jin, E., and Bea, R. G. (1997). *Risk Based Criteria for Design & Requalification of Offshore Platforms in the Bay of Campeche, Analysis of Wave Attenuation in the Bay of Campeche*, Report to Petroleos Mexicanos, Instituto Mexicano del Petroleo, and Brown & Root International Inc., Marine Technology & Management Group, Department of Civil & Environmental Engineering, University of California at Berkeley.

Reliability and Risk Considerations in Offshore Design and Operations.

Cyril E Arney, Marathon Oil Company and David J Wisch, Texaco Inc.

Abstract

Traditional engineering design practice developed over many years incorporates experienced based industry standards along with design and operating practices. This general process yields facilities based on implicit levels of reliability which the standards and practices have incorporated. Hence, structures with reliability levels are designed and constructed without the engineer explicitly choosing any reliability level. Even when target reliabilities have been specified for oil industry operations, the calculation of risk has usually been based only on a subjective consideration of the consequences of failure. Since acceptable risk is largely socially determined, a possible disconnect arises between the development and content of standards, and public understanding of the risks which are necessarily involved in all kinds of industrial operations. This paper reviews present links between reliability and risk in the offshore arena, and indicates areas for further research in order better to predict reliability, assess risk and to ensure that the likelihood of a low probability high consequence event remains within acceptable societal norms.

Introduction

Risk combines the two separate notions of the probability of an undesired event occurring, and the severity of the consequence if it does (after Baker, 19**). In order to do the mathematics, the following equation is necessary:

$$\text{risk} = \text{probability} * \text{consequence}$$ (see footnote 1)

The difficulty of working with risk is as follows. Whereas reliability is determined by scientists and economists, other drivers in society determine the acceptable level of risk. Therefore, if the present move towards risk based codes is to continue and flourish, then it is essential that engineers design systems which meet the expectations of their societies with regard to risk.

Footnote 1: Risk and consequence are measured in the same units, which is often a source of confusion.

The problem of the N year design event

With any system there is always a probability that the design condition may be exceeded, and that the consequence of this will be failure. The probability of a design event occurring during its return period is important but frequently overlooked. As a result, reliability is often not well understood, and therefore not used as often in design as it might otherwise be.

The situation might be significantly improved if the procedure proposed by Court (1952) and recommended by Gumbel (1958) were directly written into industry standards (see footnote 2). This would be easy to achieve, simply by redefining the N year environmental design event so that it is always associated with a probability of exceedance during its N years. For instance, the 100 year design wave would no longer simply be H_{100}, but instead the bracketed term (H_{100}, p=0.63) (see footnote 3).

This approach was adopted when the recommended practices for wind loading on buildings were revised in the UK, as the result of widespread damage to buildings from severe storms in the winters of 1961/62 and 1968 (Wilson, 1968). BRS Digest 101 (1969) gives the clear option of designing to the normally recommended 50 year wind speed (U_{50}, p=0.63), or if greater reliability is required then tables are provided for the (U_{50}, p=0.1) and the (U_{50}, p=0.01) wind speeds, which are 1.1 and 1.2 times greater.

A further example of the benefit of making the proposed change is when considering the reliability of equivalent systems in areas with different wave regimes. Consider the example of a fixed platform installed in the North Sea, which is dominated by winter storms, and in the Gulf of Mexico, which is dominated by hurricanes. Winterstein and Kumar (1995) estimate that the 1000 year design wave in the North Sea is 1.11 times greater than the 100 year wave height traditionally used for design, and from 1.22 to 1.26 times higher in the Gulf of Mexico. Therefore, the platform in the Gulf is likely to experience a greater overload than is the facility in the North Sea. If both have been designed to the same event without consideration of probability of exceedance and by how much, then statistically the platform in the Gulf is also less reliable.

Footnote 2: "If T is large, the probability W(w) that the event with the return period T(w) happens before N(w) years (is) 1-W(w)=exp[-N(w)/T(w)]. For practical purposes we choose N(w), want W(w) to be small, and ask for the design value X(w) corresponding to the return period T(w)."

Footnote 3: This changes the definition of the 100 year wave from "The wave height which occurs on average once in every 100 years" to "The wave height which has a 0.63 chance of being reached or exceeded one or more times in any 100 year period". A probability of exceedance directly introduces the concept of reliability into the definition. As long as p=0.63, the specified wave height is, of course, unchanged. The treatment is still simplistic for ultimate calculations of reliability, since it ignores the variability surrounding determination of the extreme event.

The probability of structural failure of offshore platforms

Overload leads directly to an increased probability of structural failure, and much has been written concerning development of the API Fixed Platform Structures Code API/RP 2A, and design practices in the North Sea. Available literature indicates an historic annualized probability of structural failure for fixed platforms in the US Gulf of Mexico in the order of $1*10E-3$ to $1*10E-4$, and in the North Sea in the order of $1*10E-5$.

Bea (1991) determined the annualized structural failure rate to be $1*10E-3$ for major drilling and production platforms in the Gulf of Mexico, based on actual failures during the period 1948 through 1982. A recent study carried out by the E&P Forum in London (1997) puts the annualized failure rate for modern US Gulf of Mexico platforms generally at $1*10E-4$. With the possible exception of one platform in a mud-slide area, all the fixed steel platforms which have failed to date in the US Gulf of Mexico were designed to early editions of API/RP 2A and installed before 1973, and all collapsed during major hurricanes.

Dalane and Haver (1995) have calculated the probability of failure of a typical steel oil platform in the North Sea due to environmental overload to be in the range of $1*10E-5$. This is supported by the E&P Forum study.

The consequence of structural failure of offshore platforms

The obvious possible consequences of failure of an offshore platform are loss of life, financial loss of capital and income, and damage to the environment. Traditional practice in the Gulf of Mexico on the approach of a hurricane has been to shut in production to guard against pollution, and then to deman the platforms to guard against loss of life. This is contrasted with the North Sea, where a considerable number of platforms are large and fully manned, and where (Lord Cullen, quoted in the Guidance Notes to the UK Safety Case [HMSO, 1992]) "the workforce both live and work offshore (and) rapid evacuation in the case of emergency is impractical". The lower calculated probability of structural failure for platforms in the North Sea compared with the Gulf of Mexico therefore has a rational basis, although perhaps unintended by the original designers, in equalizing the risk of loss of life.

This is shown by the following table, which calculates a hypothetical fatality accident rate (FAR) (see footnote 4) for platforms with one, ten and one hundred personnel on board, and assuming that the consequence in the event of the loss of a platform is

Footnote 4: The Fatality Accident Rate (FAR) was originally developed by Kletz (1992) so that it had typical values in the range of 1 to 100, and corresponds to the risk of death per 100 million hours of exposure to a particular activity or hazard. Hambly and Hambly (1994) equate it to the probable number of fatalities from 1000 people involved in an activity or exposed to a hazard for the whole of their working lives, and in their paper provide a table showing the risks of death in the United Kingdom they have calculated for a number of different activities.

that these people all lose their lives. This is an extremely conservative assumption for risk as the consequence of structural failure, and especially so in the Gulf of Mexico where platforms are shut-in and evacuated when hurricanes approach.

Calculated Hypothetical Fatal Accident Rate

Annualized Risk of Structural Failure	People on Board (see footnote 5)		
	1	10	100
10E-3	10	100	1000
10E-4	1	10	100
10E-5	0.1	1	10

This table seems to confirm two things. The first is that the risk of loss of life due to structural failure of a platform in the Gulf of Mexico and the Northern North Sea from environmental overload may indeed be identical. The second is that this risk is no greater, and possibly is significantly less, than that calculated by Hambly and Hambly of the average man in the UK in his 30s dying in an accident or from disease.

Also, if as suggested by Aggarwal, Dolan and Cornell (1996) and others there is still an appreciable, unrecognized conservatism in the Gulf of Mexico designs, such that the risk of structural failure is closer to 10E-4 or 10E-5, and if Langen's (1995) numbers for the North Sea are more correct than Dalane and Havers', then the risk to individual life from structural failure due to environmental overload of a fixed offshore platform is very small indeed, especially when compared to the unavoidable risks from drilling and production operations and from maintenance activities and travel to and from the facility.

The introduction of Quantitative Risk Assessment

In the same way that industry standards are developed and modified as a result of experience, many times industry regulations are drafted only as a result of industrial accident. Examples abound. In the UK, the CIMAH Regulations were drafted in response to major accidents that took place in chemical plants during the 1970s, notably the Flixborough accident in the UK in 1974 and the disaster at Seveso, Italy in 1976. Loss of the jack-up rig Sea Gem, while drilling an exploration well for gas in the Southern North Sea in 1965, gave direct rise to the 1971 Mineral Workings Act which enabled regulations to be introduced covering offshore oil and gas exploration and production activities in the UK North Sea (McLean and Birkinshaw, 1996).

Footnote 5: A positive denominator is needed, and evacuation is not without risk. Therefore, one person is assumed to lose their life in the Gulf of Mexico. 100 people on board corresponds to a large platform in the North Sea.

On 6 July 1988 an explosion on the Piper Alpha offshore installation led to the loss of 167 lives. The Piper Alpha disaster in 1988 raised a new set of issues, concerning personnel safety, total facility reliability (not just of the structure but also of the overall platform system), and what is an acceptable fatality accident rate for offshore operations. The Exxon Valdez incident in 1989 broadened the debate to include risk of pollution. Both were low probability, high consequence events which have significantly shaped the way in which industry operates, and the regulations and standards covering these operations.

The direct result of Piper Alpha was the UK Safety Case and its requirement for use of QRA when appropriate to measure risk. The Valdez incident gave rise to OPA 90 and double hull tankers. Both incidents have been much studied, and industry practices are still changing as a result of this work. Although it has been argued that both incidents were caused predominantly by human error, rather than by equipment failure which has to date been the primary focus of industry standards, the consequence of both incidents has nonetheless still been far reaching.

More than 70000 offshore platforms have been installed around the world since 1947, when the first platforms were installed out of sight of land in the Gulf of Mexico. Most are located in less than 600 ft waterdepth, and are typical of the Gulf of Mexico type facilities with open and comparatively simple topsides. Less than 100 are large North Sea type platforms, with enclosed modules for protection from the weather, heavy topsides for combined drilling and production, with processes more closely akin to chemical plants than traditional oilfield production practice, and with accommodation for as many as 200 people at one time. It is therefore not surprising that it is in the North Sea, with the limited Norwegian use of Quantitative Risk Assessment, and since 1992 with the UK Safety Case, that consideration of risk based codes and the desire for explicit reliability numbers has gained most momentum.

The move into deep water

The uncertainty with regard to the ability accurately to calculate forces on, and the likely ultimate failure modes of, offshore structures increases as we move into deeper water. Developments based on large floating structures moored to the seabed, with wells drilled and production flowing from seabed to the surface through long slender tubes, increase the importance of an accurate understanding of metocean design conditions, hydrodynamic forces and the interactions of structural, station keeping, safety and process systems and the likely result of human error. Until present uncertainties in these areas have been studied and are better understood, it will not be possible to develop sound risk based codes and standards for deepwater development.

The problem of perception

Whereas the probabilities of technical failure may not have increased over the past 30 years, the levels and intensity of social conflict over the probabilities certainly have.

This, and the problem of perception versus reality of the risk, are subjects about which much is beginning to be written, but the science is at a young age and as yet the field has no integrating theory (Clarke and Short, 1993). Engineers respond by seeking more information. However, it is possible that information alone will be unable to resolve what may be essentially a value conflict. Daly for instance (personal communication) indicates that only 20% of human decisions are based on considered knowledge of the facts, but that 80% are based on perception, and (Slovic, 1987 discussed in Clarke and Short) that the public may judge on non-rational (ie non-traditionally scientific) grounds.

On the positive side, it does seem that, while there is yet no agreed universal threshold of acceptable risk, there is a history of precedents and a set of numbers that are gradually gaining status as legitimate objectives (Pate-Cornell, 1994). It does seem that individual risk is becoming one accepted measurement, and that societal risk aversion is often included by further reducing what is considered an acceptable individual risk when a large number of people is exposed.

The inclusion of reliability and risk in standards

The traditional guidance provided by prescriptive codes does not specifically address the likelihood of failure. Even so, prescriptive codes do seem, based on observation, both to set requirements at levels which available technology can deliver at reasonable cost, and to set risk at levels which are acceptable to contemporary society. This is because codes, standards and safe industry practices historically have evolved in a step-wise function, with each major accident or failure giving rise to enquiry which has first determined cause, and then revised the governing code or standard.

The concept of learning from failure is not new to civil engineers. Many European cathedrals built during the Middle Ages fell down during construction and were completed to different and stronger designs. Today these same structures draw worshipers and tourists from all over the world. The Tay Bridge disaster (mid 1800s) and the Tacoma Narrows bridge failure in Washington in 1940 are only two of many more recent examples of failures which increased our engineering knowledge. Such learnings are now included into Codes of Practice, so that the information shall be available to all engineers, the mistakes shall not be repeated, and so that each new design shall not have to start from first principles.

The most glaring weakness to date in industry codes and standards is lack of a firm link between risk, defined as the probability of failure multiplied by the consequence of failure, and societal acceptance of that risk, followed by reiterating the number back into the design process to ensure that the reliability specified in codes and standards meets engineering and economic realities as well as societies' expectations. Moving design practice into an era which explicitly accounts for risk would seem to be a step to providing industry and society a more rational means of allocating resources in an appropriate manner.

Conclusions

Engineers would today be more likely to use risk-based design techniques if the probabilities of exceedance of design events were more explicitly stated in Codes and Standards.

The risk of loss of life due to the structural collapse of an offshore platform from environmental overload may be similar in the Gulf of Mexico and the North Sea, even though this was not any specific intent on the part of the original designers.

There is a dichotomy at work in the modern world which seems to account for much of the present conflict between industry and society. Engineers work to improve reliability but society measures according to risk. The yardsticks are different.

The first tentative steps are being taken to resolve the dichotomy. On the one hand, engineers are slowly moving towards reliability and risk based design methods. On the other hand, a new science is developing centered on the public perception of acceptable risk.

Because present prescriptive codes and standards include learnings from past failures, they do seem to set risk at levels which are generally acceptable to contemporary society. However, this may not hold true for risks associated with the low probability high consequence events.

In order to help ensure that the reliability for which engineers aim is consistent with the risks which society is willing to accept, the ultimate goal is to develop sound risk-based Codes and Standards which specifically set target reliabilities. These target reliabilities must be set at levels which are economic to industry, which result in prices for good and services which society can afford and is willing to pay, and such that the risks from the consequences of failure meet societies' expectations.

References

Aggarwal, R. K., Dolan, D. K., and Cornell, A. C., 1996. Development of Bias in Analytical Predictions Based on Behavior of Platforms During Hurricanes. OTC 8077.

Baker, M. J., 19**. Course Notes for Hazards Forum Undergraduate Lecture Program. Professor Michael J Baker, University of Aberdeen.

Bea, R. G., 1991. Offshore Platform Reliability Acceptance Criteria. SPE Drilling Engineering, June 1991, pps 131-137.

BRS (UK Building Research Station) Digest 101, 1969. Wind Loading on Buildings- 2. HMSO, 1969.

Clarke, L., and Short, J. F. Jr., 1993. Social Organization and Risk: Some Current Controversies. Annu. Rev. Sociol. 1993.19:375-99.

Court, A., 1952. Some new statistical techniques in geophysics. In: Advances in Geophysics, vol.1 (edited by H. E. Landsberg). New York, Academic Press, Inc.

Dalane, J. I., and Haver, S., 1995. Requalification of an Unmanned Jacket Structure Using Reliability Methods. OTC 7756.

Daly, J. A., Personal communication from John A Daly, Amon G Carter Centennial Professor in Communication, The University of Texas at Austin.

E&P Forum, 1997. QRA Datasheet Directory, Extreme Weather Risk for Fixed Units. The Oil Industry International Exploration and Production Forum. London.

Gumbel, E.J., 1958. Statistics of Extremes. New York, Columbia University Press.

Hambly, E. C., and Hambly, E. A., 1994. Risk Evaluation and Realism. Proc. Instn Civ. Engrs, Civ. Engng, 102, May 1994, pps 64-71.

HMSO, 1992. A Guide to the Offshore Installations (Safety Case) Regulations, 1992. UK Health and Safety Executive.

Kletz, T. A., 1990. An Engineer's View of Human Error. Institution of Chemical Engineers, Rugby.

Langen, H. van, Swee. J. L. K., Efthymiou, M., Overy, R., 1995. Integrated Foundation and Structural Reliability Analysis of a North Sea Structure. OTC 7784.

Mclean, I, H., and Birkinshaw, M., 1996. The Offshore Installations and Wells Regulations, The Regulatory Scene is Set. Proc. ERA Technology Conference on Offshore Structures- Hazards and Integrity Management, London 1996.

Pate-Cornell, M. E., 1994. Quantitative Safety Goals for Risk Management of Industrial Facilities. Structural Safety 13 (1994) pps 145-157.

Slovic, P., 1987. Perception of Risk. Science 236:260-285.

Wilson, P. H., 1968. Glasgow Wind Damage. BRS Current Paper 76/68 UDC 699.83.

Winterstein and Kumar, 1995. Reliability of Floating Structures: Extreme Response and Load Factor Design. OTC 7758.

A Second Order Numerical Algorithm for Simulation of Breaking Wave Impacts on Structures

Eng-Soon Chan [1], Zhi Zong [2] and Chih-Young Liaw [1]

Abstract

Among all the environmental forces acting on ocean structures and marine vessels, those resulting from wave impacts are likely to yield the highest loads. Being highly nonlinear, transient and complex, a theoretical analysis of the impact forces would be impossible without any numerical simulation. In this paper, a second order numerical algorithm for the simulation of wave impact is developed using the Volume of Fluid (VOF) concept. The general algorithm and the applications to wave impacts on horizontal structures are presented.

Introduction

Among all the environmental forces acting on ocean structures and marine vessels, those resulting from wave impacts are likely to yield the highest loads [Chan (1988),Chan (1993)]. Being highly nonlinear, transient and complex, a theoretical analysis of the impact forces would be impossible without any numerical simulation. Numerical simulations of nonlinear transient free surface problems date as far back as the 1960s. A comprehensive literature survey can be found in Hirt (1981).

Attempts to simulate nonlinear water waves using Boundary Element Method (BEM) and Boundary Integral Method (BIM) based on the concept of fully nonlinear potential flow have also been successful [Grilli (1995)]. Relatively accurate results have been reported for simulations up to the stage just

---

[1]Department of Civil Engineering, National University of Singapore, Kent Ridge Crescent 10, Singapore 119260
[2]CCM-MINDEF, Department of Mecanical and Production Engineering, Kent Ridge Crescent 10, National University of Singapore, Singapore 119260

before wave breaking. The main limitations of these approaches are that the simulations fail after wave breaking.

The last three decades have seen the advent of the methodology of Volume Of Fluid (VOF) and its application to the description of free surface probelm. VOF is defined as the fraction of fluid in a cell in an eulerian difference scheme and hence is able to account for free surface fluctuations. The combination of VOF concept with finite-difference methods has given rise to two methods: the SOLA-VOF developed at Los Alamos Scientific Laboratory by C. W. Hirt and B. D. Nichols [Hirt et al(1981)] and the General Hydrodynamics (GH) pioneered by J. C. W. Rogers [Rogers et al (1990)] and W. G. Szymmzack [Szymmzack et al(1993)] at the Naval Surface Warfare Center, USA. Because of its simplicity, the former is gaining wider acceptance. However, SOLA-VOF is limited by its simplified treatment of scenarios when the VOF function goes beyond the range of 0 to 1. In particular, the VOF function is truncated to 1 when the function exceeds 1 and rounded off to 0 when it is negative. The approximations are in contradiction to the conservation of mass. The above limitations have been resolved in the general hydrodynamics by Rodgers and Szymmzack [Rogers et al(1990), Szymmzack et al(1993)].

All these methods, however, are first-order schemes and are not quite able to fully describe wave breaking problems. This is also due to the sensitivity and complexity of the wave breaking phenomenon. In this paper, a second order numerical algorithm in both time and space is developed using VOF concept based on Rodgers' work to solve free surface problems [Rogers et al (1990)]. VOF function and velocities are first estimated by neglecting the pressure gradients. These are then corrected with the pressure included so that the incompressibility condition is guaranteed. Because only one intermediate step (correction of VOF and velocity) is included in the algorithm developed here, the algorithm is the simplest since any algorithm solving incompressible fluid problems requires at least two steps. Numerical experiments using the methodology suggest that the scheme is stable. The general methodology and the applications to wave impact problem are presented in this paper. The presented algorithm extends the standard VOF method to the second order accuracy in time and space while still maintaining its simplicity as much as possible.

The VOF Model

Consider a 2-dimensional spatial domain $\Omega(x,y)$, fixed for all time, which contains an incompressible fluid with density ρ_0. The domain Ω is discretised by a finite difference mesh. In this fixed domain Ω, a Volume Of Fluid (VOF) function f may be defined as

OCEAN WAVE KINEMATICS, DYNAMICS AND LOADS ON STRUCTURES

$$f = \frac{\text{fluid volume in a cell}}{\text{cell volume}} \tag{1}$$

where

$f = 1$ when the cell is full of fluid,
$0 < f < 1$ when the cell is partially filled with fluid, and
$f = 0$ when the cell is empty.

From the definition, it can immediately be concluded that $0 \leq f \leq 1$. The VOF function f gives the approximte free surface variations of the fluid field. Thus, the free surface can be approximately described by the field f. In terms of the VOF function, the governing equations are:

$$\frac{\partial f}{\partial t} + \nabla \cdot (f\mathbf{u}) = 0 \tag{2}$$

$$\frac{\partial (f\mathbf{u})}{\partial t} + \nabla \cdot (f\mathbf{u}\mathbf{u}) = -\frac{1}{\rho_0}\nabla P + f\mathbf{G} \tag{3}$$

$$f \leq 1 \tag{4}$$

$$\nabla \cdot \mathbf{u} = 0 \tag{5}$$

where ρ_0 is the water density, $\mathbf{G} = -\mathbf{k}g$ and $-\mathbf{k}$ is the the unit vector in the direction of the gravitational force; $\mathbf{u}$ and P are velocity vector and pressure.

The Numerical Algorithm

A procedure has been developed to proceed from a known description of the flow $(f^n, \mathbf{u}^n)$ at a given time $t = t^n$ to a description of the flow $(\overline{f}, \overline{\mathbf{u}})$ at a later time $t^n + \Delta t$ where Δt is the time step. The treatment may be regarded formally as generating an approximate flow which converges to an exact flow as $\Delta t \to 0$. The governing equations and the constraints are treated using a *time split* approach which approximates their solution at the new time step in distinct stages. In this section, a second order algorithm in time and space is presented.

The half time step predictions for f and φ may be approximated by:

$$\frac{f^{n+1/2} - f^n}{\Delta t/2} + \nabla \cdot \varphi^n = 0 \tag{6}$$

$$\frac{\varphi^{n+1/2} - \varphi^n}{\Delta t/2} + \nabla \cdot (\varphi^n \mathbf{u}^n) = f^n \mathbf{G} \tag{7}$$

where $\varphi = f\mathbf{u}$. The above equations are solved using a second-order finite difference scheme [Rogers et al(1990)]. The full time step predictions are given by:

$$\frac{f^{n+1} - f^n}{\Delta t} + \nabla \cdot \varphi^{n+1/2} = 0 \qquad (8)$$

$$\frac{\varphi^{n+1} - \varphi^n}{\Delta t} + \nabla \cdot (\varphi^{n+1/2}\mathbf{u}^{n+1/2}) = f^{n+1/2}\mathbf{G} \qquad (9)$$

The predicted values f^{n+1} and $\mathbf{u}^{n+1} = \varphi^{n+1}/f^{n+1}$ are obtained by dropping the pressure terms and hence do not satisfy the incompressibility condition. With pressure term taken into accout, we have

$$\begin{aligned} -\nabla^2 P^{n+1/2} &= -\frac{2\rho_0}{\Delta t^2}(1 - f^{n+1}) \quad \text{if} \quad P^{n+1/2} > 0 \\ -\nabla^2 P^{n+1/2} &\geq -\frac{2\rho_0}{\Delta t^2}(1 - f^{n+1}) \quad \text{if} \quad P^{n+1/2} = 0 \\ P^{n+1/2} &\geq 0 \end{aligned} \qquad (10)$$

The above equations constitute a linear complementarity problem in mathematics, the detailed theoretical study and basic algorithm of which are available in Cottle (1990). By solving equation (10), $P^{n+1/2}$ can be obtained and the corrected VOF function is given by

$$\overline{f} = f^{n+1} + \frac{\Delta t^2}{2\rho_0}\nabla^2 P^{n+1/2} \qquad (11)$$

The corrected VOF function $\overline{f}$ which is smaller than or equal to one yields the approximate form of the free surface.

The momentum is corrected by using

$$\frac{\overline{\varphi} - \varphi^{n+1}}{\Delta t} = -\frac{1}{\rho_0}\nabla P^{n+1} \qquad (12)$$

where P^{n+1} is found through the following equations

$$P^{n+1} = 0 \quad \text{if} \quad \overline{f} < 1 \qquad (13)$$

$$\nabla \cdot \varphi^{n+1} = \frac{\Delta t}{\rho_0}\nabla^2 P^{n+1} \quad \text{if} \quad \overline{f} = 1 \qquad (14)$$

The above equations can be solved using iterative methods. Once P^{n+1} is found from the above equations, $\overline{u}$ can be extracted from $\overline{u} = \overline{\varphi}/\overline{f}$.

Because central difference scheme is used in the time marching, the above scheme is second order accurate in time. At the same time, only one intermediate step is needed, so the algorithm is simple in the sense that a second order algorithm, if any, would at least require two steps to solve an incompressible fluid problem.

Simulation of Wave Impacts on Structures

Example 1 *Breaking Wave Impacts on a Horizontal Cylinder*

Consider a horizontal cylinder subjected to wave action as shown in Fig. 1. The breaking wave is simulated with an initially sinusoidal wave of length $\lambda = 2\ m$ and a ratio of wave height to water depth of 0.8. The circular cylinder is approximated by a polygon. The profiles of incident waves plotted in Fig. 1 (a). Water motion in two directions can be observed in Fig. 1: one rising up along the cylinder surface and the other flowing down below the cylinder. Similar results have been obtained by Chan (1993) in an experimental study of breaking wave impacts on horizontal cylinders. These experiments have clearly elucidated the process of wave breaking and the wave-structure interaction.

Fig. 1(b) shows the experimental results for a scenario comparable to that of the numerical problem. Comparison of Fig. 1 (a) and (b) shows that the simulated results are in good agreement with the experiments, testifying the capability of the present method.

Example 2 *Breaking Wave Impacts on A Square Structure*

To further explore the capability of the VOF methodolody, the circular cylinder in the former example is replaced by a square structure. The profiles of the impinging wave are plotted in Fig. 2.

When the wave front hits the structure (at $t = 0.661$ second), the impinging water mass splits in two directions (upwards and downwards). Meanwhile, the free surface below and ahead of the wave front is still rising. A closed volume (bubble) is also formed ($t = 0.676$ second). The bubble eventually disappears at t = 0.691. On the whole, the kinematics are consistent with experimental observations by Chan et al (1995). It should be noted however that, in the above simulations, the compressibility of the entrapped air during impact has been neglected.

Conclusions

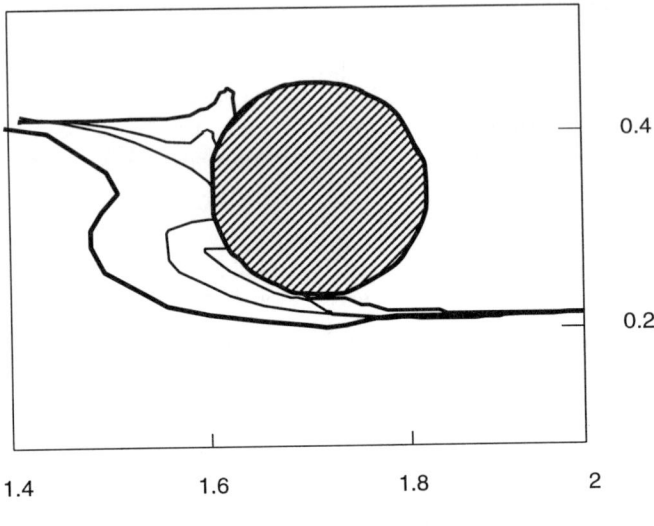

(a): Numerical simulations

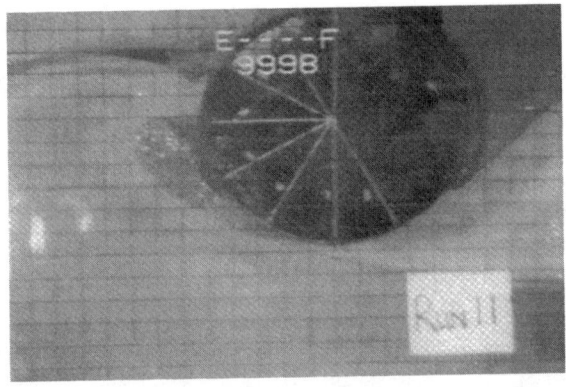

(b): Experiment

Fig. 1: Breaking wave impacts on a circular cylinder

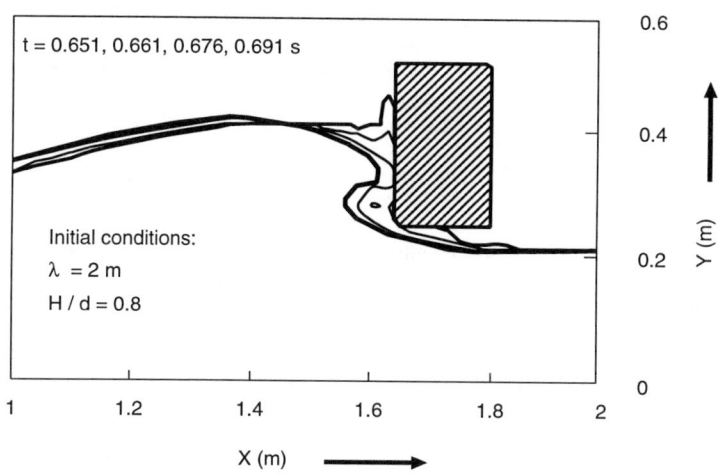

Fig. 2: Breaking wave impacts on a square cylinder

A VOF methodology which is second order accurate in time and space has been successfully developed for the simulation of wave impact. Because only one intermediate step is included, the algorithm is simple in the sense that any algorithm solving incompressible fluid problems would require at least two steps.

The methodology produced reasonable results when applied to the simulations of breaking wave and wave impacts on square and circular cylinders. The numerical simulations have been qualitatively verified with experimental results. Wave breaking and breaking wave impacts on structures are complicated processes. Further efforts are needed, especially the inclusion of models to treat air entrainments and turbulence.

References

[1] E. S. Chan and W. K. Melville (1988), *Deep Water Plunging Wave Pressures on a Vertical Plane Wall*, Proc. R. Soc. Lond., A417, pp. 95-131.

[2] E. S. Chan (1993), *Extreme Wave Action on Large Horizontal Cylinders Located Above Still Water Level*, Proc. of 3rd International Offshore and

Polar Engineering Conference, Singapore, pp. 121 - 128.

[3] E. S. Chan, H. F. Cheong and K. Y. H. Gin (1995), *Breaking Wave Loads on Vertical Walls Suspended above Mean Sea Level*, Journal of Waterway Port Coastal and Ocean Engineering, vol. 121, No. 4, pp. 195-202.

[4] D. R. Cottle (1990), *The Linear Complementarity Problem*, Academic Press.

[5] C. W. Hirt and B. D. Nichols (1981), *Volume Of Fluid (VOF) for Hydrodynamic Free Boundaries*, Journal of Computational Physics, vol. 39, pp. 201-225.

[6] S. T. Grilli and R. Subramanya (1995), *Recent Advances in the BEM Modelling of Nonlinear Water Waves*, in H. Power ed., *BE Applications in Fluid Mechanics*, Computational Mechanics Publications, pp. 92-122.

[7] J. C. W. Rogers, W. G. Szymczak, A. E. Berger and J. M. Solomon (1990), *Numerical Solution of Hydrodynamic Free Boundary Problems*, International Series of Numerical Mathematics, vol.95.

[8] W. G. Szymczak, J. C. W. Rogers, J. M. Solomon, A. E. Berger (1993), *A Numerical Algorithm for Hydrodynamic Free Boundary Problems*, Journal of Computational Physics, vol. 106, pp. 319-336.

Breaking Wave Geometry with Emphasis on Steepness and Curvature

Pål F. Lader[1], Guttorm Grytøyr[2], Dag Myrhaug[3] and Bjørnar Pettersen[3]

Abstract

The spatial geometry of three different types of breaking waves are examined. The geometry is measured by using image analysis. The time development of thirteen individual wave parameters describing the steepness, curvature and asymmetry of the free surface are presented.

Introduction

Knowledge of the mechanisms leading to the generation of breaking waves in deep water is still not fully understood, and breaking of waves can not be defined uniquely by any parameter at present. A breaking criterion is of fundamental importance, and also enters in practical applications, e.g. in estimating the probability of occurrence of breaking waves at sea. Several criteria have been suggested, but none have proven to give a unique definition of the inception of breaking (Tulin and Li, 1992). Many of the proposed breaking criteria are based on the geometry of the wave. This is natural because the wave geometry is easiest to measure at sea, and thus statistical models can be made e.g. from sea measurements and used with the criterion to predict breaking. The wave geometry is commonly described by parameters based on zero-downcross analysis, and Kjeldsen and Myrhaug (1978) suggested a set of such parameters. Bonmarin (1989) used these parameters studying the geometry of breaking waves in detail. None of the investigators have been able to derive a unique breaking criterion based on these parameters.

The purpose of this paper is to present measurements of the spatial geometry of three cases of breaking waves. The measurements are analysed using zero-downcross parameters as well as additional parameters proposed here.

---

1. PhD Student, Department of Marine Hydrodynamics, Norwegian University of Science and Technology (NTNU), N-7034 Trondheim, Norway, Phone: +47 73 59 56 07, fax: +47 73 59 55 28, e-mail: lader@marin.ntnu.no
2. PhD Student, Department of Marine Hydrodynamics, NTNU
3. Professor, Department of Marine Hydrodynamics, NTNU

The work presented here is the first part of a more comprehensive study. In addition temporal measurements with wave gauges of the wave geometry is made. The geometry will be correlated with PIV (particle image velocimetry) measurements of the wave kinematics. A complete report of the work will be given in Lader (1998).

Experimental Arrangement and Analysis Procedure

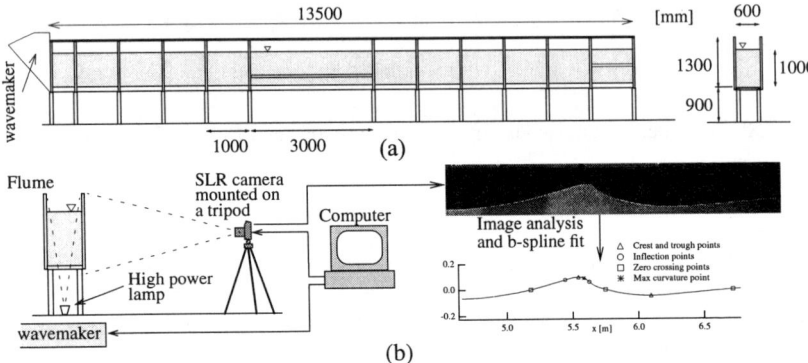

Figure 1. (a) The wave flume with dimensions in mm. (b) Experimental arrangement and analysis procedure.

The experiments were conducted in a narrow wave flume, 13.5m long, 1m deep and 0.6m wide, with walls and bottom of glass, yielding good conditions for using optical measurements techniques (Fig. 1a). The waves were generated by a single hinged paddle, and damped with a passive wave absorber. Figure 1b shows the experimental arrangement for measuring the spatial geometry of the waves. It consists of high power lamps under the flume, and a 35mm SLR (Single Lens Reflex) camera mounted on a tripod on the side of the flume, taking pictures perpendicular to the flume walls. The camera was triggered by the computer controlling the wavemaker, and could therefore be adjusted accurately relative to the wave motion. Fluorescent dye were used in the water to give good contrast between water and air, making the free surface easily detectable. Pictures were taken as the wave approached breaking, through the breaking event and into the post-breaking phase. The camera could not take more than one frame per second, and to get the wanted sampling interval of 1/32s, subsequent runs were made, and the trigger time was shifted for each run. The images were digitized with a real world resolution of 2mm/pix, giving approximately 1000 measurement points on the free surface. The free surface coordinates were extracted from the image by using computerized image analysis routines. A b-spline was fitted to the measurements of the free surface, and the free surface parameters were calculated from the b-spline. A b-spline is a semianalytic function, and it is strait forward to find the derivatives of the b-spline function. For a description of b-spline theory, see Rogers and Adams (1990). There are two parameters that are affecting the b-spline behaviour; the order of the basis functions K, and the number of points in the knot vector N. Optimal values of K and N were obtained from a test case constructed from polynomial expressions. To get optimal results two b-spline fits with different parameters were used. The free surface coordinates and the first derivatives were found from a b-spline fit with parameters $K=4$ and $N=40$, while the second derivative was found from a b-spline fit with parameters $K=4$ and $N=20$. Further details of the experiments and analysis are given in Lader (1998). The b-spline fits were utilized to identify the points on the free surface (Fig. 2) which were used in the subsequent analysis.

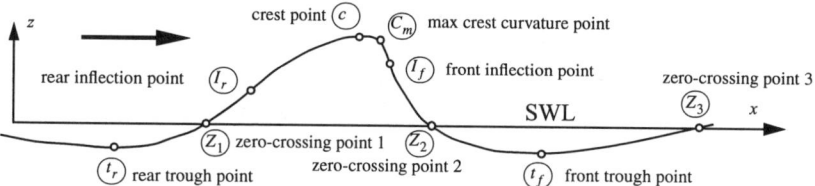

Figure 2. Points on the free surface used in the analysis.

Results and Discussion

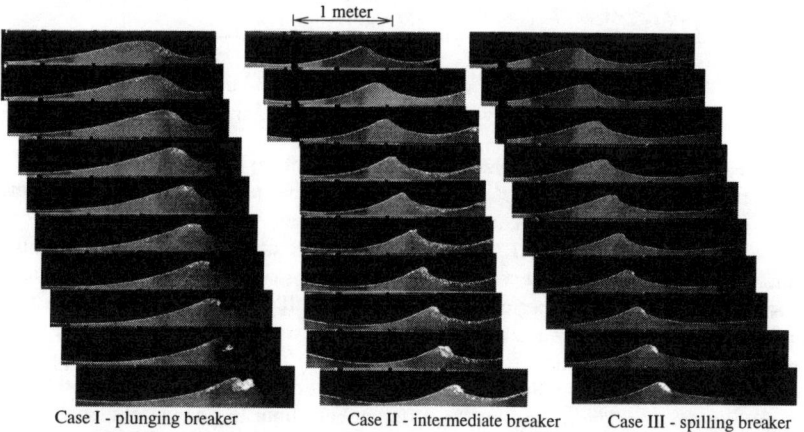

Figure 3. The three cases of breaking waves examined. The time between each exposure is 1/16 s.

Figure 3 shows the three different cases of two-dimensional deep water breaking waves which are examined: A plunging breaker (Case I), a spilling breaker (Case III), and an intermediate breaker (Case II). The intermediate breaker is a breaker that is initiated by a microscale plunging event, and evolves further as a spilling breaker. The classification of breakers follows Kjeldsen and Myrhaug (1978). In Cases I and II the breaking wave is generated by using the method of phase focusing (see e.g. Longuet-Higgins (1974) and Kjeldsen and Myrhaug (1979)). In Case III the breaking is generated due to the increase in wave height at the front of a wave group (Longuet-Higgins, 1974).

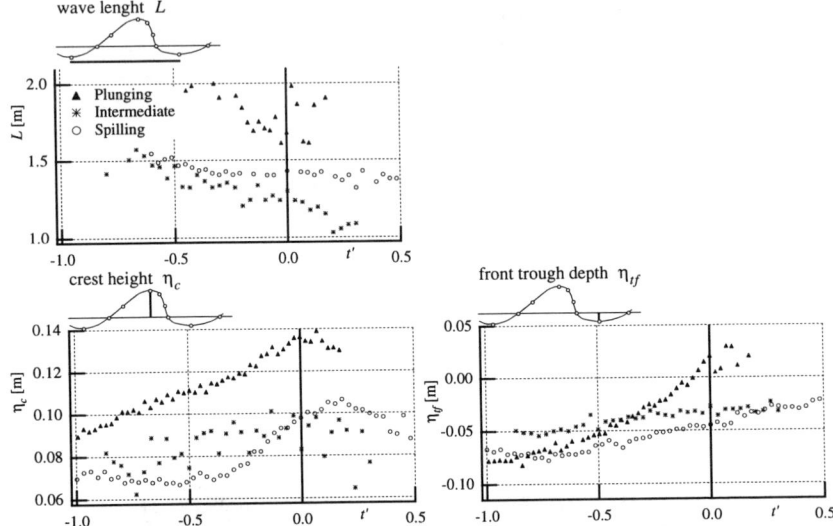

Figure 4. Wave length, crest height and front trough depth versus time. The abscissa value is time made dimensionless with the wave period T, thus $t' = t/T$.

Figure 4 shows the wave length (L), crest height (η_c) and trough depth (η_{tf}) versus the dimensionless time $t' = t/T$. T is the wave period calculated from the speed of the crest point in each of the three cases, using linear wave theory ($T_{pl} = 1.26\,\text{s}$, $T_{int} = 0.96\,\text{s}$, $T_{sp} = 1.03\,\text{s}$). t' is zero at the breaking point, i.e. the point where foam is first observed, or where the front of the wave crest is vertical. Thus $t' = -1$ is one wave period ahead of breaking. It appears that the L has a tendency to decrease as the wave approaches breaking, which is consistent with Bonmarin's (1989) measurements for a plunging breaker. The decrease is most apparent for the plunging breaker, while there is only a small decrease around $t' = -0.5$ for the spilling breaker. The reason for the different behaviour in L may be related to the different breaking wave generation methods, i.e., the plunging and intermediate breakers were generated by using phase focusing, while the spilling breaker were generated at the front of a wave group.

The linear wave length (L_0) calculated from the speed of the crest, are for the three cases: $L_{0\,pl} = 2.48\,\text{m}$, $L_{0\,int} = 1.37\,\text{m}$ and $L_{0\,sp} = 1.67\,\text{m}$. Thus L is significantly smaller at the breaking point than L_0.

The measured values of η_c and η_{tf} show that the crest becomes higher, and the trough shallower as the waves approach breaking in all three cases. This is interpreted as a local increase in the mean water level for the wave. Bonmarin (1989) reported the same results for a plunging breaker, and concluded that the potential energy increases in the crest and decreases in the trough.

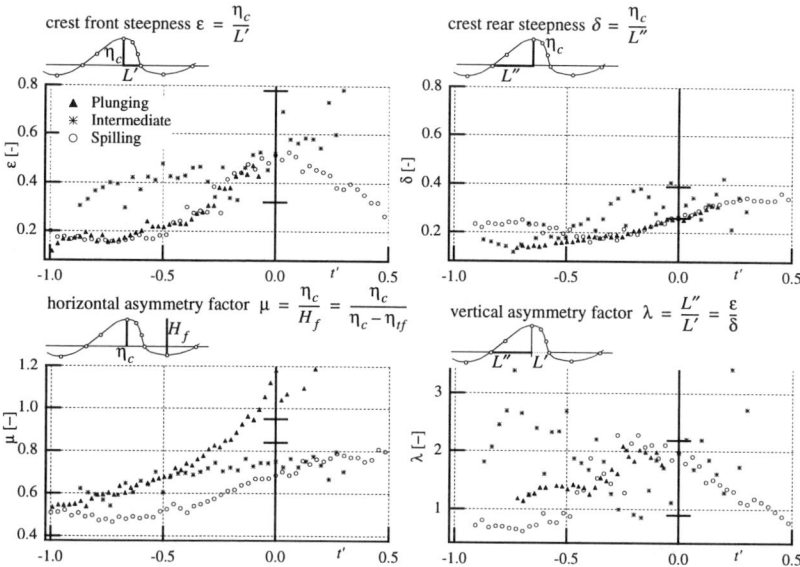

Figure 5. Crest steepnesses and asymmetry factors versus time. The marks on the $t' = 0$ line indicate the interval found to be typical for breaking waves according to Kjeldsen and Myrhaug (1978).

Kjeldsen and Myrhaug (1978) derived parameters from zero-downcross analysis to better describe important features of steepness and asymmetry of a wave that approaches breaking. Figure 5 shows the development of these parameters, together with the typical values found in their study. The crest front steepness experiences a large increase up to breaking, while the crest rear steepness increases moderately. Thus the wave becomes vertically asymmetric, which is observed from the vertical asymmetry factor as well. It is noticed in particular that the spilling breaker has a high degree of vertical asymmetry. Kjeldsen and Myrhaug (1978) characterized a spilling breaker to be nearly symmetric, which thus is not necessarily the case. Otherwise these three parameters have values in the same range found by Kjeldsen and Myrhaug (1978), and do also agree with Bonmarin's (1989) measurements for a plunging breaker.

The horizontal asymmetry factor (μ) illustrates the horizontal position of the local wave geometry relative to the still water level (SWL). For linear waves $\mu = 0.5$. Higher values can be interpreted as a local setup, lower values as setdowns. For the plunging breaker it appears from Fig. 5 that μ increases significantly giving values exceeding 1, as the front trough point elevates above SWL. In Kjeldsen and Myrhaug (1978) $\mu < 1$. However, regardless of the actual values of μ, it appears that for a plunging breaker the local setup increases significantly up to the instant of breaking.

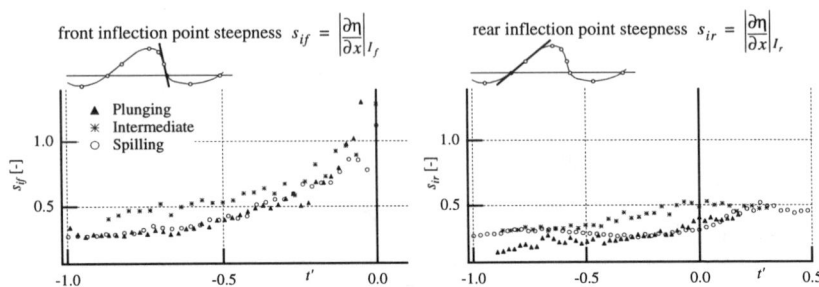

Figure 6. Inflection point steepnesses versus time.

A more detailed description of the surface steepness is obtained by using the derivative at the inflection points as defined and shown in Fig. 6. By comparing Fig. 6 with Fig. 5, it appears that the qualitative behaviour of the front and rear inflection point steepnesses, and the crest front and crest rear steepnesses, respectively, are the same, i.e., the same increase in steepness at the front, with accompanying smaller increase at the rear. By comparing the front inflection point steepness (s_{if}) and the crest front steepness (ε), it is observed that the evolution is similar for the spilling and plunging breakers. However, for the intermediate breaker s_{if} increases more distinctly than ε. Further, for the plunging and spilling breakers the crest rear steepness and the rear inflection point steepness both reach their maximum at $t' = 0.25$ after the breaking point. For the intermediate breaker the same parameter reach its maximum at the breaking point.

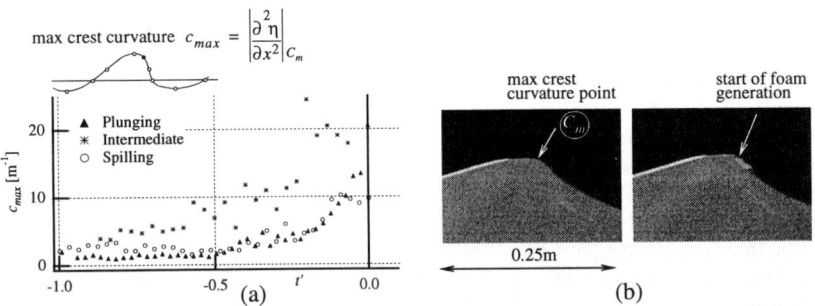

Figure 7. (a) Maximum crest curvature versus time. (b) Details of the crest at the start of breaking for the spilling breaker (case III). Time between each exposure is 1/32 s.

The point on the crest with the highest second derivative is referred to as the max crest curvature point, and the value of the second derivative at this point is denoted max crest curvature (c_{max}). Figure 7a shows c_{max} versus time up towards breaking. Detailed analysis of the crest of the spilling breaker shows that the foam generation starts at the max curvature point (Fig. 7b). For a spilling breaker this suggests that the point of max curvature plays an important role in the breaking process. From Fig. 7a it appears that c_{max} increases up to breaking for all three cases. The spilling and plunging breakers have almost the same c_{max} values except close to breaking. From about $t' \approx -0.1$ the plunging breaker experiences a larger increase up to breaking where $c_{max} \approx 20 \text{m}^{-1}$. For the spilling breaker c_{max} is nearly constant with the value 10m^{-1} from $t' \approx -0.1$, and up to breaking. The intermediate breaker shows generally higher values. Thus the value of c_{max} is not an indication on whether the wave breaks or not, since for the value for which the spilling breaker produces foam, the plunging and the intermediate breakers remain unbroken.

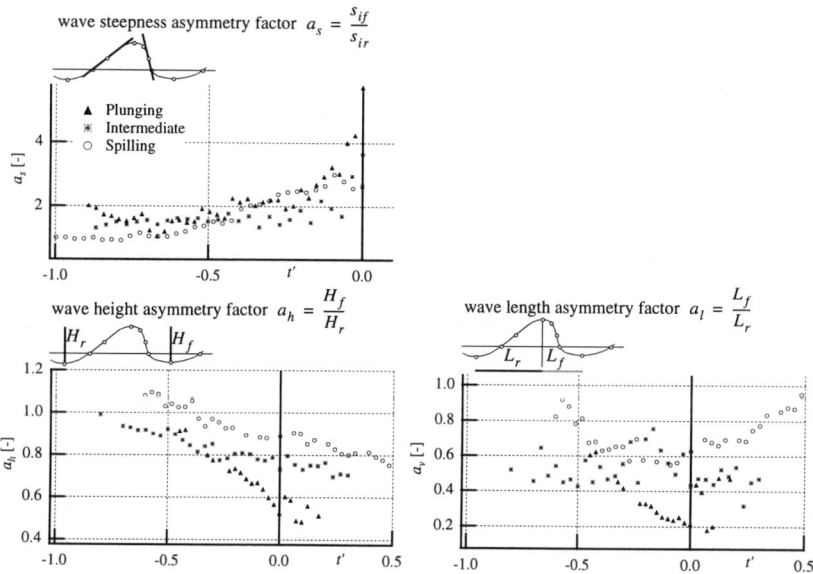

Figure 8. Asymmetry factors based on wave steepness, wave height and wave length versus time. Notice that the still water reference point does not influence these parameters.

Figure 8 shows asymmetry factors that are independent of the zero-crossing points. The wave steepness asymmetry factor (a_s) is analogous to the vertical asymmetry factor (λ) in Fig. 5. However, for a_s it appears that the spilling and plunging breakers have a similar degree of steepness asymmetry up to about $t' = -0.1$, but beyond this a_s increases for the plunger, and slightly decreases for the spiller. Comparison of Fig. 8 and Fig. 7a shows that a_s and c_{max} have the same qualitative behaviour for the spilling and plunging breakers, i.e., an increase of wave steepness asymmetry gives an increase in curvature.

Furthermore, from the wave height and wave length asymmetry factors it is clear that the plunging breaker is more asymmetric at the point of breaking than the intermediate and spilling breakers. However, both the spilling and intermediate breakers also have a high degree of asymmetry in steepness and length.

Summary

The measurements of the zero-downcross parameters presented here are in good agreement with previous results. They show that the steepnesses at the rear and front of the wave crest increase as the wave approaches breaking. However, the front steepness increases more than the rear steepness, resulting in an asymmetric wave. In particular, it is found that the spilling breaker is highly asymmetric, and should not be anticipated to be symmetric.

The rear and front inflection point steepnesses (s_{if} and s_{ir}) and the crest rear and front steepnesses (ε and δ) show similar behaviour.

For the spilling breaker it is observed that the foam generation starts at the point on the crest with the maximum curvature. The second derivative at this point (c_{max}) does not give a criterion for when the foam generation starts. A correlation between c_{max} and the wave steepness asymmetry factor (a_s) is also observed, suggesting a connection between the steepness and the curvature at the crest.

References

Bonmarin, P. (1989). "Geometric properties of deep-water breaking waves." *Journal of Fluid Mechanics*, 209, 405-433.

Kjeldsen, S. P., and Myrhaug, D. (1978). "Kinematics and Dynamics of Breaking Waves." *Report nr. STF 60 A 78100*, River and Harbour Laboratory (NHL) at The Norwegian Institute of Technology.

Kjeldsen, S. P., and Myrhaug, D. (1979). "Breaking Waves in Deep Water and Resulting Wave Forces." *Proc. 11th Offshore Technology Conference*, Paper no. 3646, Houston, Texas.

Lader, P. (1998). "Geometry and Kinematics of Breaking waves," *PhD Thesis, Norwegian University of Science and Technology*, Trondheim, in preparation.

Longuet-Higgins, M. S. (1974). "Breaking Waves - in Deep or Shallow Water." *10th Symposium on Naval Hydrodynamics*, Cambridge, Mass., 597-605.

Rogers, D. F., and Adams, J. A. (1990). Mathematical Elements for Computer Graphics, McGraw-Hill.

Tulin, M. P., and Li, J. J. (1992). "On the Breaking of Energetic Waves." *International Journal of Offshore and Polar Engineering*, 2(1), 46-53.

Numerical Modeling of Breaking Waves over Porous Structures

Pengzhi Lin, [1] Kuang-An Chang [1], Philip L.-F. Liu [2], and Tsutomu Sakakiyama [3]

Abstract

Porous armour units are frequently employed to protect coastal structures, such as seawall or caisson breakwater, from wave attack. To determine the stability of the armour layer requires the knowledge of flow motions in porous media and the corresponding pressure and force fields. A numerical model has been developed recently by Lin & Liu (1998) to study breaking waves in surf zones. The model is further extended in this paper to simulate flow fields in porous media. A simple experiment of flow passing through a porous dam is first used to calibrate the model. With properly calibrated coefficients, the numerical model is then employed to simulate breaking waves overtopping a caisson breakwater protected by porous armour units. The calculated free surface displacement near the caisson breakwater is compared with measured data and reasonably good agreement is obtained.

1. Introduction

Porous armour units such as rubble mound breakwaters are often used in coastal regions to prevent beach erosion or scouring in front of seawall. When waves pass through porous media, significant amount of energy is dissipated. These porous structures could become unstable under continuous wave loadings. The stability analyses for coastal structures have been done primarily by studying small scale physical models (e.g., Aminti & Lamberti, 1996; Donnars & Benoit, 1996), which are in general costly and cannot avoid scaling problems. Recently, as the rapid advance of computational fluid dynamics, numerical studies for waves passing through porous armour units have received increasing attentions. These numerical models are in general more efficient and less expensive than hydraulic

---
[1]Graduate research assistant, [2]Professor, School of Civil and Environmental Engineering, Cornell University, Ithaca, NY 14853 U.S.A. Phone: (607) 255-5090; Fax: (607) 255-9004; [3]Senior Researcher, Central Research Institute of Electric Power Industry, 1646, Abiko, Abikoshi, Chiba, 270-11, Japan. Phone: +81 471 82 1181; Fax: +81 471 84 7142.

models. Furthermore, numerical approaches have no difficulty in modeling large scale prototypes, contrast to physical model approaches.

The early attempts for modeling numerically wave and porous structure interaction were made by Kobayashi & Wurjanto (1990) and Wurjanto & Kobayashi (1993). Their models are based on shallow water equations. van Gent (1994) presented a similar numerical model based on long wave equations. The treatment of infiltration and seepage was discussed in van Gent's paper.

Another approach for this problem is to formulate the flow in porous media based on spatially averaged Navier–Stokes equations (NSE). Liu et al. (1996) proposed a set of governing equations for flows in permeable bed and then conducted perturbation analysis to obtain the semi-analytical solution to wave-induced boundary layer flows above permeable bed. van Gent (1995) proposed another set of equations based on NSE to describe the unsteady turbulent flows in porous media. The parameters used in his model were calibrated by experimental data. The mean motion of turbulent flows outside the porous media, on the other hand, can be described by Reynolds equations. Recently, Lin & Liu (1998) proposed a numerical model to solve the mean motion of breaking waves in surf zones. The turbulence field was obtained by solving the improved k-ϵ model.

In this study, a new approach is used to model breaking waves interaction with a permeable structure. The flow outside the permeable materials are solved using the same approach as the one in Lin & Liu (1998). The flow in porous media is modeled by solving the spatially averaged NSE. The boundary conditions for turbulence energy at the interface of porous bed are derived. The new model is first calibrated by a simple problem of flow passing a porous dam. Then a practical problem of breaking waves overtopping a caisson breakwater with a protective porous armour layer is investigated.

2. Mathematical Model

2.1. Reynolds equations and k-ϵ model

For turbulent flows outside the porous media, the mean motion can be well described by Reynolds equations. The Reynolds stresses appearing in the Reynolds equations are modeled by quadratic forms of mean strain rates, which also relate Reynolds stresses to the turbulence energy k and its dissipation rate ϵ. The values of k and ϵ can be obtained by solving a set of transport equations for k and ϵ. The details of numerical implementations can be found in Lin et al. (1997) and Lin & Liu (1998).

2.2. Spatially Averaged Navier–Stokes equations

The flows in porous media can still be described by the NSE. However, due to the complex structures of porous materials, it is infeasible to solve the NSE directly.

In general, the NSE is averaged over a length scale which is larger than the typical pore size but smaller than the characteristic length scale of the physical problem. The resulting spatially averaged NSE are,

$$\frac{\partial \overline{U}_i}{\partial x_i} = 0 \tag{1}$$

$$\frac{1+c_A}{n}\frac{\partial \overline{U}_i}{\partial t} + \frac{1}{n^2}\overline{U}_j\frac{\partial \overline{U}_i}{\partial x_j} = -\frac{1}{\rho}\frac{\partial \overline{p_0}}{\partial x_i} + \frac{\nu}{n}\frac{\partial^2 \overline{U}_i}{\partial x_j \partial x_j} - \frac{1}{n^2}\frac{\partial \overline{U''_i U''_j}}{\partial x_j} \tag{2}$$

where n is the effective porosity and the overbar denotes the spatially averaged (from herein referred to as mean) quantities. Thus, $\overline{U}_i$ is the mean velocity vector in the i-th component, $\overline{p_0}$ the mean effective pressure ($p_0 = p + \rho g h_0$ with ρ the density, g gravitational acceleration, h_0 vertical distance from selected datum), ν the molecular viscosity, and U''_i the spatial fluctuation of the i-th velocity component. The coefficient c_A represents the added mass effect for accelerating fluid in porous media and it was suggested by van Gent (1995) that $c_A = \gamma\frac{1-n}{n}$ with $\gamma=0.34$ being an empirical coefficient. The correlation of spatial velocity fluctuations, the last term on the right hand side of (2), can be induced by both inhomogeneity of porous materials or turbulent fluctuations. This term is modeled in this paper by a combination of linear and nonlinear friction forces,

$$-\frac{1}{n^2}\frac{\partial \overline{U''_i U''_j}}{\partial x_j} = -ga\overline{U}_i - gb\overline{V}\,\overline{U}_i \tag{3}$$

where a are b are empirical coefficients and $\overline{V}$ is the characteristic velocity which can be estimated by $\overline{V} = \sqrt{\overline{U}_i \overline{U}_i}$. We currently adapt the same empirical formulae for a and b as in van Gent (1995),

$$a = \alpha\frac{(1-n)^2}{n^3}\frac{\nu}{gD_{50}^2} \tag{4}$$

$$b = \beta\left(1 + \frac{7.5}{KC}\right)\frac{1-n}{n^3}\frac{1}{gD_{50}} \tag{5}$$

where $\alpha = 1000$ and $\beta = 1.1$. KC is the Keulegan-Carpenter number, $KC = \frac{VT}{nD_{50}}$, T the typical wave period, and D_{50} characteristic diameter (mean size) of porous materials.

2.3. Boundary condition outside porous bed

The continuity of mean velocity and pressure is imposed across the interface of porous media and outside flow. The turbulence intensity outside the porous bed, however, needs special treatment. The modified log-law wall function for flow over porous bed was suggested by Ilegbusi (1989) as

$$\frac{u_t - \overline{U}_t}{(\tau_w/\rho)^{1/2}} = \frac{1}{\kappa}\ln\left\{\frac{Ex_n\left[(\tau_w + \tau_m)/\rho\right]^{1/2}}{\nu}\right\} \tag{6}$$

where u_t is the tangential velocity outside the boundary layer, $\overline{U}_t$ the tangential mean velocity along the porous bed, τ_w the wall shear stress, τ_m additional shear caused by suction or injection, κ the von Karman constant which equals to 0.41, E roughness coefficient, and x_n the normal distance from the wall. The added shear stress can be estimated by:

$$\tau_m = \frac{c_n \overline{U}_n u_t}{1 + (\tau_w/\rho)^{1/2}/\kappa u_t} \tag{7}$$

where $\overline{U}_n$ is the normal mean velocity across the porous bed and c_n is the empirical coefficient which is tentatively taken as 0.01 in this study as the result of numerical optimization. However, further verification is needed. From(6) and (7), we can use Newton Raphson method to calculate τ_w for given u_t, $\overline{U}_t$, and $\overline{U}_n$. The frictional velocity u_* can be found by $u_*^2 = \frac{|\tau_w|}{\rho}$. The value of u_* is then employed to define k and ϵ near the interface which is needed in the k-ϵ model outside porous media. The detailed information of boundary conditions for the k-ϵ equations can be found from Rodi (1980) and Liu & Lin (1997).

3. Model Calibration

The numerical model is first calibrated by a simple problem of flow passing through a porous dam. The experiments were conducted in a tank which is 89 cm long, 44 cm wide and 58 cm high. A porous dam which is 28 cm long and 27 cm high was placed at the center of the tank. The initial water level was 26.6 cm on one side of the porous dam and 1.0 cm on the other side. Gravels with mean diameter 0.625 inch (1.59 cm) and porosity 0.49 were used as porous material in the experiments. A gate located 2 cm upstream in front of the porous dam was used to hold the water as initial conditions. The gate was then suddenly pulled up in a short duration, 0.10 to 0.13 sec from close to fully open. A CCD camera with resolution 774 × 242 pixels was used to capture images. The camera was triggered by a PC at the frame rate of 10 fps. The uncertainties in the image-based free surface measurements are 2 mm in the region without porous media and 4 mm in the region filled with rocks.

In numerical computations, the entire domain is discretized by 178 × 112 grids with Δx=0.005m and Δy=0.0025m. We found that with coefficients suggested by van Gent (1995), i.e., (4) and (5), reasonable results can be obtained. However, in order to achieve better agreement between numerical results and experimental data, α and β have been modified to be 1100 and 1.17 (their originally suggested values are 1000 and 1.1), respectively. The comparisons of numerical results with modified coefficients and experimental measurement are shown in figure 1. Excellent agreement are obtained when flow passes through the porous dam (figure 1(b)). Even the details of returning breaking front is authentically simulated by the numerical model (figures 1(c) and (d)). It is our future plan

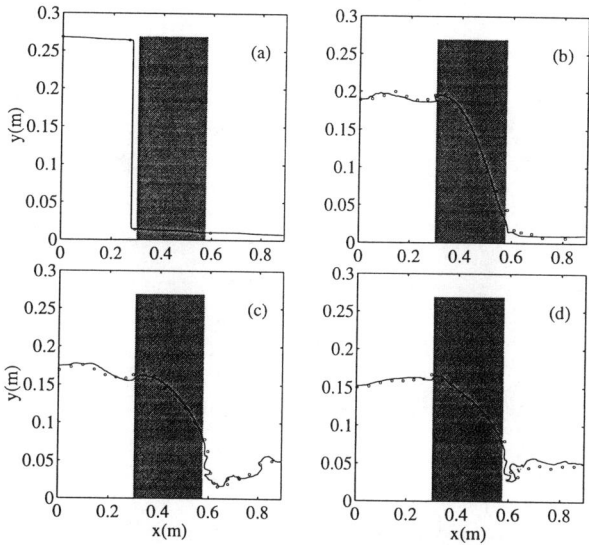

Figure 1: Comparison of calculated (—) and measured (o o o) free surfaces for flow passing through porous dam at (a) t=0.0s, (b) t=0.63s, (c) t=1.43s, and (d) t=1.83s.

to perform a series of experiments with different gravel sizes and different initial water levels to further calibrate the coefficients used in the model.

4. Simulation of breaking waves overtopping structures

The numerical model has also been used to study a practical problem: breaking waves overtopping a structure protected by an armour layer. Sakakiyama & Kayama (1997) has conducted a series of experiments to evaluate the overtopping rate of breaking waves over the protected coastal structures. The laboratory setup is shown in figure 2. The caisson breakwater is impermeable, while the armour layer in front of the caisson is made of concrete units called Tetrapod with the maximum height of 0.063m and porosity of 0.50. The origin of coordinates is at the intersection of still water level and the front wall of the caisson. In this study, we choose to simulate the case with the following wave parameters: wave period T=1.4s and wave height H=0.105m. The laboratory measurements were taken for more than 40s. The free surfaces were measured at x=-6.76m and x=-0.005m, respectively.

In the numerical computations, smaller computational domain and shorter simulation time are used to reduce computational time. The whole computational domain is discretized by 400 × 56 grids with Δx=0.02m and Δy=0.075m.

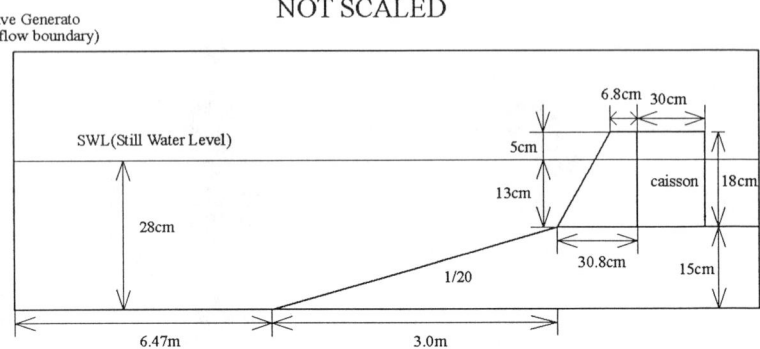

Figure 2: Schematic plot of experimental setup.

The simulation time is $t=9.8$s, which corresponds to 7 wave trains. In order to compare numerical results with experimental data, we have shifted the laboratory data 8.5s forward to match the phase of wave front at the section $x=-6.76$m (figure 3(a)). The same amount of shift is applied to section $x=-0.005$m (figure 3(b)). It is noted that at section $x=-0.005$m, the only meaningful comparisons are the last two waves, due to the short simulation time. Reasonably good agreements are obtained. The discrepancy in the free surfaces displacement in constant water depth (figure 3(a)) may be caused by the reflection from the vertical caisson, which appears after $t=4.5$s in experiments data ($t'=13.0$s in the actual experiment); while in the numerical computation, the reflection wave does not appear until the end of computation. The discrepancy in the free surface displacement near caisson ($x=-0.005$m) might be caused by the uncertainties of coefficients used in the numerical model for flow through porous media. This requires further verifications using more laboratory data.

The numerical simulation of wave overtopping is shown in figure 4. The series of wave profiles from $t=8.4$s to $t=9.8$s with the time intervals of $T/4$ is presented in five figures. The wave shoaling and steepening on the sloping beach is qualitatively simulated. The wave eventually breaks on the protection armour units (figure 4(c)) and overtopping occurs consequently (figure 4(d)). It is our plan to simulate cases with different type of armour units to study the stability of caisson with different porous media absorber.

4. Conclusion

This study shows preliminary results for two cases of flows passing through porous media. The agreement between numerical results and laboratory data is generally

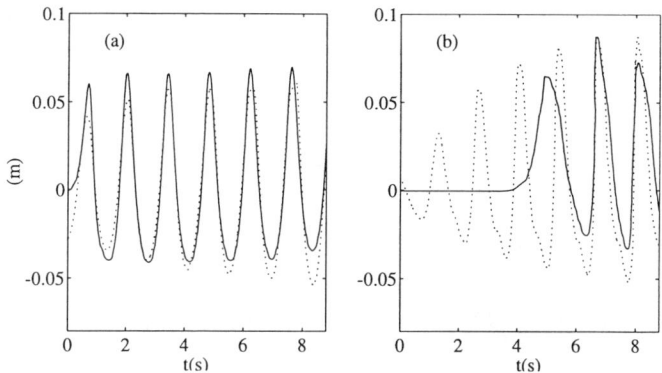

Figure 3: Comparison of calculated (—) and measured (- - -) free surface displacement at (a) $x=-6.76$m and (b) $x=-0.005$m for breaking wave overtopping coastal structures.

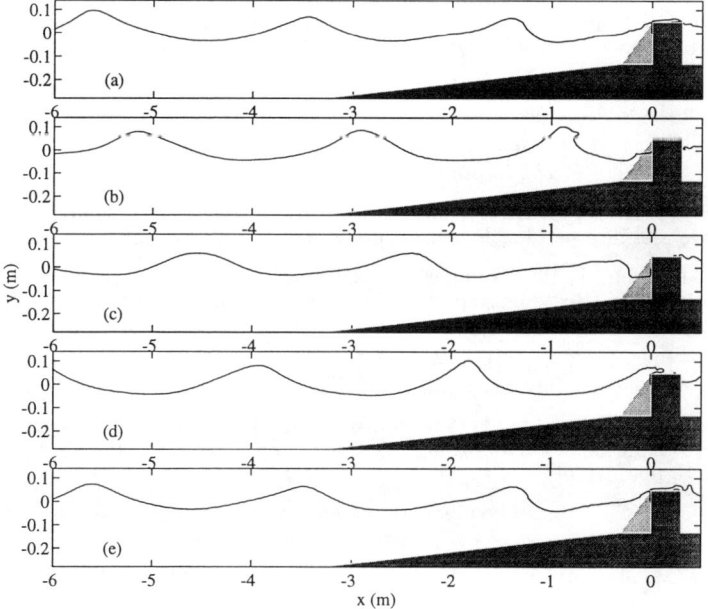

Figure 4: Simulation of breaking wave overtopping at (a) t=8.4s, (b) t=8.75s, (c) t=9.1s, (d) t=9.45s, and (e) t=9.8s.

good, indicating the model is applicable for studying wave-porous structures interaction. The empirical coefficients used in the model, however, needs further careful verifications by more laboratory data.

Acknowledgement

The research was supported, in part, by research grants from the National Science Foundation (CTS-9302203, CMS-9528013) and the Army Research Office (DAAL03-92-G-0116).

References

Aminti, P. & Lamberti, A. 1996 Interaction between main armour and toe berm damage. *Proc. 25th Int. Conf. Coastal Engng.*, pp. 1542-1555, ASCE.

Donnars, P. & Benoit, M. 1996 Interactions in the stability of toe-berm and main-armour for rubble mount breakwaters: an experimental study. *Proc. 25th Int. Conf. Coastal Engng.*, pp. 1617-1630, ASCE.

Ilegbusi O. J. 1989 Turbulent boundary layer on a porous flat plate with severe injection at various angles to the surface. *Int. J. Heat Mass Transfer*, **32**, n4, 761-765.

Kobayashi, N. & Wurjanto, A. 1990 Numerical model for waves on rough permeable slopes. *J. of Coastal Res.*, **7**, 149-166.

Lin, P., Liu, P. L.-F., & Chang, K.-A. 1997 Numerical Modeling of Deep-Water Wave Breaking. *Proc. Int. Conf. of Waves 97*, ASCE, in press.

Lin, P. & Liu, P. L.-F. 1998 A numerical study of breaking waves in the surf zone. *J. Fluid Mech.*, in press.

Liu, P. L.-F., Davis, M. H., & Downing S. 1996 Wave-induced boundary flows above and in a permeable bed. *J. Fluid Mech.*, **325**, 195-218.

Liu P. L.-F. & Lin, P. 1997 A numerical model for breaking wave: the volume of fluid method. Research Rep. No. CACR-97-02, Center for Applied Coastal Research, Ocean Eng. Lab., Univ. of Delaware, Newark, Delaware, 19716.

Rodi, W. 1980 Turbulence Models and Their Application in Hydraulics – A State-of-the-Art Review. I.A.H.R. publication.

Sakakiyama, T & Kayama, M. 1997 Numerical simulation of wave breaking and overtopping on wave absorbing revetment. *Proc. of Coastal Engng.*, **44**, JSCE, pp. 741-745.

van Gent, M. R. A. 1994 The modelling of wave action on and in coastal structures. *Coastal Engng.*, **22**, 311-339.

van Gent, M. R. A. 1995 Wave Interaction with Permeable Coastal Structures. Ph. D. thesis, Delft University.

Wurjanto, A. & Kobayashi, N. 1993 Irregular wave reflection and runup on permeable slopes. *J. Waterway, Port, Coastal, and Ocean Engng.*, **119**, 537-557.

Breaking Wave Statistics in Nearshore and Surf Zones

Ming-Yang Su,*J. C. Wesson, W. J. Teague, R.E. Burge

Abstract

During the Sandy Duck '97 Experiment at Duck, NC (Sept-Oct, 1997) we deployed bubble measuring arrays in the nearshore region (floating, 0.9-2.3km offshore, 8-14m water depths) and in the surf zone (bottom mounted, 4m water depth). Each array had acoustic resonators, to measure bubble density for bubbles with radii from 15μ to 1200μ, as well as void fraction sensors and an accelerometer. Outside the surf zone breaking waves are generated mostly by local winds. While shoaling effects, tides and swell contribute less to wave breaking outside the surf zone, they are most important within it. Bubbles generated by these waves and injected into the surface layer provide a measure of breaking intensity. The temporal and vertical response of the bubble density field to rapid changes in wind and waves, as measured by the acoustic resonator arrays, are presented here to analyze nearshore and surf zone wave breaking processes. The acoustic resonators resolve the roll-off in bubble density spectra at small radii, confirming previous optical determinations of this characteristic of the bubble density spectra.

1 Introduction

The Sandy Duck '97 Experiment (September -October, 1997) was a cooperative experiment hosted by the Army Corps of Engineers, Coastal Engineering Research Center (CERC), Field Research Facility (FRF), at Duck, NC. The major focus of the experiment is to improve the fundamental understanding of surf zone sediment transport (Birkemeier, 1997) and includes: 1)small and medium scale sediment transport and morphology, 2) wave shoaling, breaking and nearshore

---

*Naval Research Laboratory, Code 7330, Stennis Space Center, MS 39529

circulation and 3)swash processes, including sediment motion. The experiment location is primarily the surf zone in a box 700m along-shore and 550m across-

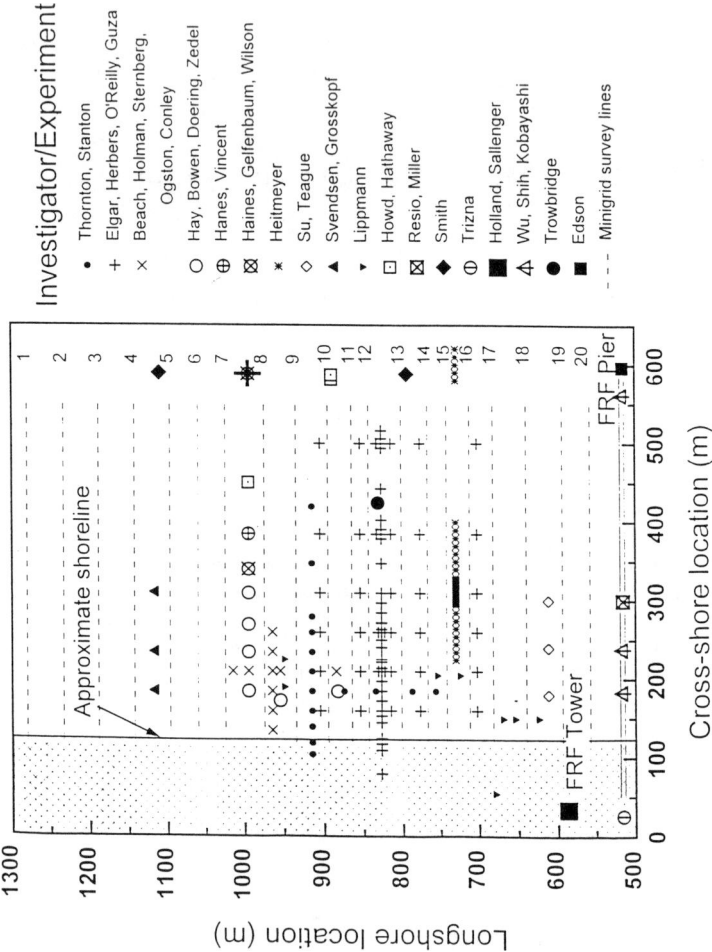

Figure 1: Sandy Duck '97 Surf Zone Layout: investigators and locations.

shore, on the north side of the 600m long FRF pier. More than 30 indivdiual experiments are included in the experiment (See Figure 1). As part of this experiment Naval Research Laboratory scientists based at Stennis Space Center (NRL-SSC) participated by measuring breaking wave statistics and bubble density fields at Duck, NC. This effort consisted of measuring void fraction in the surf zone using void fraction sensors mounted on swinging bars anchored at its bottom and held to the surface by floatation, and measuring bubble density distributions using acoustic resonators, both in the surf zone and outside it in the nearshore zone.

Bubbles due to wave breaking are significant to physical processes occurring in the near-surface boundary layers. In particular, bubbles play a major role in underwater acoustic propagation and scattering problems (Wu, 1987; Prosperetti, 1988) and in air-sea gas exchange. Measurement of bubbles in the surf zone is difficult due to the extreme physical conditions caused by breaking waves and currents.

The emphasis of this paper is on acoustic resonator measurements of bubble density distributions and is only part of the measurements made by NRL-SSC at Sandy Duck '97 which we report elsewhere. We describe the acoustic resonator systems that we used in section 2. We present results from the resonator arrays, gathered during a one of the storms we sampled, in section 3. Brief conclusions are in section 4.

2 Acoustic Resonator Arrays

The acoustic resonator (AR) is a sensor for measuring the bubble size spectrum and density, based on acoustical principles (See Cartmill and Su, 1993 and Su *et.al.* 1994 for details). It consists of two circular plate transducers, oriented parallel to each other and about 25cm apart to create a Fabry-Perot Interferometer. Thus, resonant modes of the system are determined by the plate separation and the speed of sound in water. The fundamental resonance is near 3kHz and all its harmonics are also resonances, so in a 200kHz bandwidth there are nearly 70 resonances. One energized plate acts as a sound source and the other as both reflector and receiver. The signal from the receiver plate is processed to produce a power spectrum from which the bubble size spectrum is derived. Bubbles are detected because they scatter and absorb acoustic energy, damping the resonant modes of the AR. Resonant frequencies of bubbles are inversely proportional to their radius, so the magnitude and frequency distribution of the damping are used to determine both the density and size distribution of bubbles. The acoustic resonators have been employed for deep water measurements and large wave tank measurements for several years (Su and Cartmill, 1993; Su et.al. 1994). Here we now employ them for nearshore investigations.

Figure 2a shows a resonator power spectrum generated in water without bubbles. Each peak with an asterisk is a resonant peak. Figure 2b shows the power spectrum when bubbles are present. The power spectrum in Figure 2b is much lower than that in Figure 2a. The difference is caused by the scattering and extinction due to bubbles. The two spectra, with and without bubbles, are used to derive the bubble size spectrum presented in Figure 2c. The bubble size spectrum (or density spectrum) is expressed as the number of bubbles per cubic meter per micron increment of bubble radius [$\#/m^3/\mu$]. It is effective for bubble densities

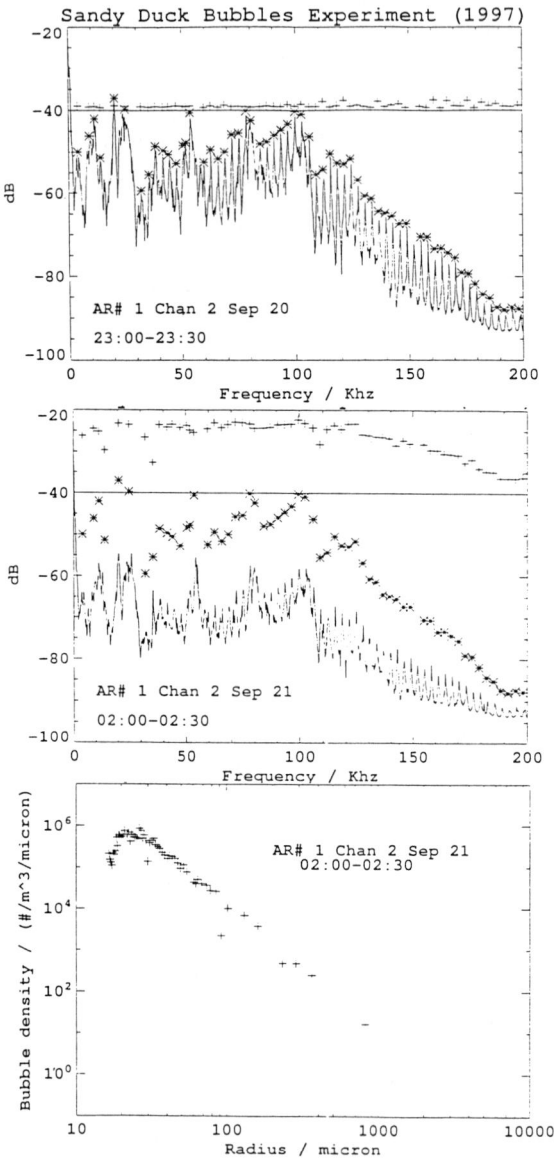

Figure 2: Acoustic resonator spectra(a and b) and bubble density spectrum(c). The minimal bubble condition (a) is a reference for the calculation of bubbles from (b) that is shown in (c).

up to 0.1% void fraction.

A floating resonator array consists of a 4m x 0.8m x 0.8m aluminum frame with instruments and floatation. Four acoustic resonators are installed on each of the four surface following arrays together with void fraction meters (also 4 per array), other sensors, a vertical accelerometer and an electronics box. For the surf zone array there are two acoustic resonators and two void fraction sensors on a bottom mounted frame. At each array the depth below sea surface of the four resonators differ. The purpose of this arrangement is to maximize the information content of the vertical distribution of bubble size densities. Table 1 lists the location and sensor information for the arrays. The void fraction sensors provide

System	Offshore(km)	Depth(m)	Chan1	Chan2	Chan3	Chan4
AR#1	2.3	14	0.39	0.85	1.31	2.39
AR#2	1.5	12	0.39	2.39	3.0	3.61
AR#3	1.0	10	0.39	0.85	1.31	3.0
AR#4	0.9	8	0.39	1.31	2.39	3.61
AR#5	0.2	4	2.0	2.3		

Table 1: Sensor locations for the acoustic resonator arrays, depths are in meters. Floating arrays were deployed directly east of the FRF Pier.

a coarser measurement of breaking waves, measuring void fraction from 1% to about 60%. Each array has floatation to keep it at the surface and is secured by three anchor lines. The electronics box in each array provides power to all sensors and collects data from all the sensors. The on-board 486 cpu Fourier transforms and ensemble averages the acoustic resonator signals before storing them. The on-board computer is also connected by a marine cable to shore. The sampling schedule and data are transmitted by modem to shore. Therefore, the status of each array can be monitored daily. The acoustic resonators on each array are sampled by a custom A/D board at 408kHz, giving spectra with a Nyquist frequency of 204kHz. Each spectrum is calculated from a 5ms sample for a fundamental frequency and interval spacing of 200Hz. We use the spectrum to 200kHz, averaging 200 spectra to produce each recorded ensemble average. The FFT processing time required for 4 channels, about 0.6 sec, produces a 2 minute ensemble time.

3 Statistics of Breaking Waves

Breaking waves are caused either by wind, primarily offshore, or by shoaling in shallow water. We expect essentially no bubbles in calm weather for the floating arrays because shoaling is not significant at their depths (8-14m). Instead, wind speed and direction, and short, choppy waves are most likely to produce measurable bubbles in the floating AR arrays. However, for the surf zone array, bubble generation by breaking waves is affected by shallow water shoaling, which in turn is determined by waves height and water depth (tides). Since the surf zone array is outside the main bar of the beach, moderate swell does not break over it, so we also expect very few bubbles there, except when the swell is sufficiently high to cause breaking near the surf zone array. Thus, our analysis of the acoustic resonator arrays will concentrate on storms. Here we will demonstrate the different response of the surf zone and floating arrays to one of the storm events we observed during Sandy Duck '97.

3.1 Bubble Generation

Over September 20-21, 1997 the wind and wave conditions changed rapidly. Figure 3 shows the change during the night in wind, waves, and spectra from the floating (3a), and surf zone arrays (3b). Wind speed and direction went from below 7 m/s to over 15 m/s and the wind direction rotated from west to north. These changes took place in less than an hour at midnight local time. Wave period also changed rapidly, from over 13 seconds for swells from offshore areas to less than 5 seconds for the locally generated sea The significant wave height grew more gradually, taking almost 8 hours to reach its peak value of 2.2m. The floating acoustic resonators measure an almost immediate change in the bubble conditions with the change in wind direction and wave period (3a). Here, a time series of the acoustic power (dB) of a single frequency component of the spectrum, shows the variability of the acoustic signal from one resonator, Channel 2 on AR#1. The frequency of 20kHz is a geometrical resonance that is also the resonance frequency for bubbles of 163μ radius. High values of acoustic power indicate a low density of bubbles of the relevant size, while the drop in the resonance amplitude indicates an increase in the bubble density. This change in resonance amplitude appears across the entire frequency range. Essentially the same rapid response to the wind was recorded by all four of the floating arrays. By contrast, the temporal change in a resonance amplitude for the surf zone array (Figure 3b) is much more gradual and is inversely related to wave height instead of following wind forcing or wave period.

3.2 Bubble Spectra

The bubble density spectra confirm a population of bubbles in the water, that is indicated by the step in the time series for a single frequency. Acoustic spectra at 2300 on Sep. 20 and at 0200 on Sep. 21 are shown in Figure 2a and 2b. The bubble density spectrum at 0200 (Figure 2c) results from bubbles generated by the onset of the storm: rising winds, change in wind direction and significant shortening of wave period. Bubble populations are high and the spectrum is well defined. This spectrum includes a roll-off in the bubble density with decreasing radius, starting between 25 and 30μ with a slope between r^4 and r^6. The ability

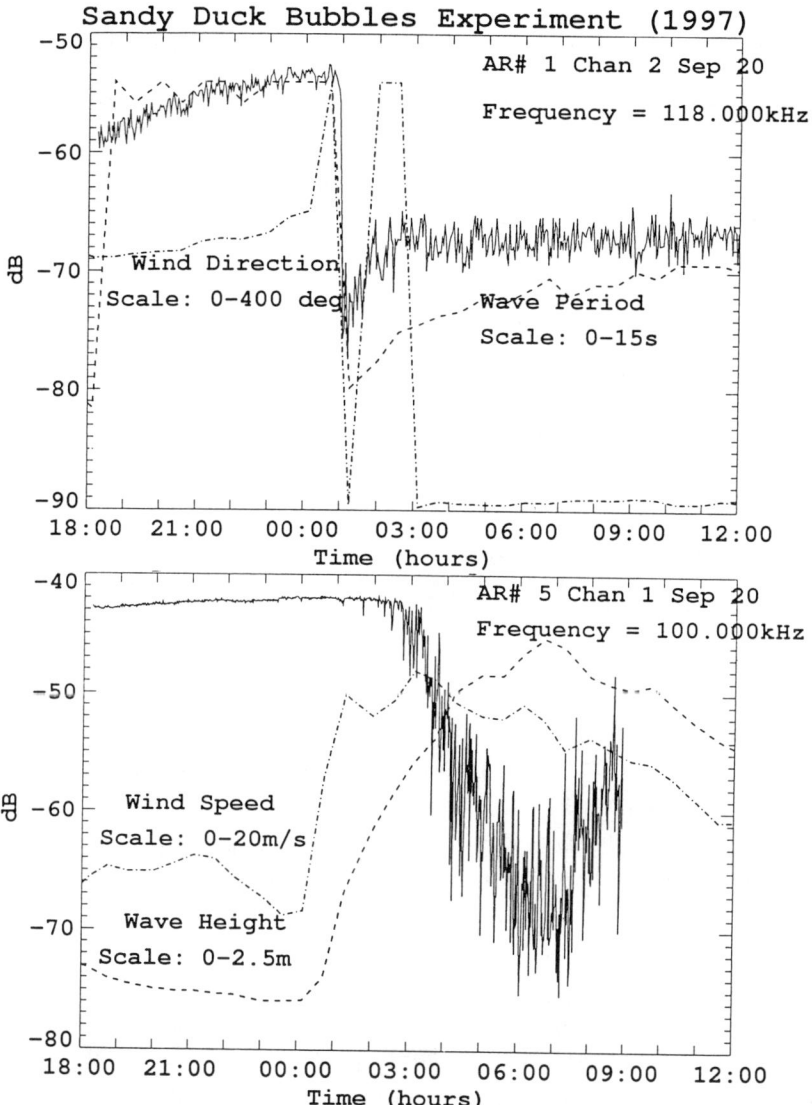

Figure 3: Storm variation during Sept 20-21. a) Wind direction, wave period and time series of the 118kHz resonance on Channel 2 of floating array #1. b) Wind speed, wave height and time series of the 100kHz resonance of Channel 1 of the surf zone bottom array.

of the acoustic resonator to produce distinctive peaks out to 200kHz resolves this size range of small bubbles. The bubble radius corresponding to the peak in bubble density around 25-30μ agrees well with our previous bubble measurements by an optical bubble sensor in the North Sea (Su et al., 1988). To our knowledge, this is the first time this bubble density peak has been measured directly by an acoustic technique. The fact that we observe the bubble density peaks around $r = 25 - 30\mu$, obtained by two totally different techniques, increases confidence in the validity of this particular important feature of the oceanic bubble size spectrum.

4 Conclusions

The floating acoustic resonator arrays we deployed during Sandy Duck '97 provide substantial new data on the vertical and horizontal distribution of bubble density due to wind-driven wave breaking during storms in the nearshore zone. Initial analysis shows that acoustic resonators detect bubble density changes quickly with storm events. Short period waves and winds produce observable bubbles offshore while wave height primarily determines bubble densities we observe in the surf zone. Furthermore, bubbles penetrate below the sea surface to depths of at least 3 or 4 meters at wind speeds of 15m/s. Perhap most importantly, we have confirmed the existence of a peak in the oceanic bubble density spectrum near $r = 25 - 30\mu$.

References

Berkemeier, W.A. 1997, "Sandy Duck '97", Coastal Engineering Research Center, Vol. CERC-97-2.

Cartmill, J. and M.Y. Su (1993). "Bubble Size Distribution Under Saltwater and Freshwater Breaking waves", *Dyn. Atmos. Oceans,* Vol 20, pp.25-31.

Prosperetti, A., 1988, "Bubble related Ambient Noise in the Ocean", *J. Acoust. Soc.* 84, pp.1042-1054.

Su, M.Y., D. Todoroff, and J. Cartmill (1994). "Laboratory Comparisons of Acoustic and Optical Sensors for Microbubble Measurement". *J. Atmos. Oceanic Tech.,* Vol 11, No 1, pp.170-181.

Su, M.Y., S.C. Ling and J. Cartmill (1988). "Optical Microbubble Measurements in the North Sea", Sea Surface Sound , B.R Kerman, Ed. Kluwer Academic Publishers, pp.211-224.

Wu, J., 1987, "Bubbles in the Near-Surface Ocean, A General Description", *J. Geophys. Res.,* 93, pp.587-590.

Discriminating Breaking in Deep Water Waves

Richard Seymour, F. ASCE[1], Charles-Alexandre Zimmermann[2]
and Jun Zhang[2], A. M.ASCE

Abstract

Records of deep water sea surface elevation alone, even when taken at high sampling rates, do not allow determination of which waves in the record are breaking. Although the white-capping and collapse of the front face of a breaking wave are clearly visible, they do not appear in any meaningful way in the output of an elevation gauge. Considerable research has been devoted to finding a method of detecting breaking events in a wave elevation record, but most of the models show little skill. Huang et al. [1992] suggested that an analytical method based on the Hilbert transform, which they called the Phase-Time Method, might have value. The present study evaluates the power of the Phase-Time method in detecting breaking wave in large scale laboratory experiments under a variety of deep water wave conditions. A useful model of deep water breaking wave detection is developed and its limits are stated.

Introduction

Direct observation remains a very important tool in the study of random wave breaking. Only the visual appearance of a breaker can, at present, define either the existence of breaking of an individual wave or its type. Statistics of breaking waves are critical to the study of ocean-atmosphere interactions such as gas exchange. Breaking is a dominant dissipation term in models for wave generation and propagation, yet there are few observations in deep water, and present models contain a high degree of uncertainty. In the past, much research has been devoted to detecting breaking in an elevation record. None of the criteria proposed proved to be efficient. Banner and Peregrine [1993] provide an overview of the deep water breaking wave detection. Modern video recording and image processing techniques have been particularly useful for whitecap cover measurements. Holthuijsen and Herbers [1986] demonstrated the

---
[1] Scripps Institution of Oceanography, University of California, San Diego, La Jolla, CA 92093
[2] Department of Civil Engineering, Texas A&M University, College Station, TX 77845

inadequacy of using a simple local wave slope criterion. Longuet-Higgins and Smith [1983] and Thorpe and Humphries [1980] developed a method relying on the rapid jump in surface elevation at the leading edge of the spilling region of a breaker. Narrow-beam Doppler radars are used to detect large-scale breaking events. These radars measure the significant increase in scatterer speed within breaking events [Keller et al., 1986]. Snyder et al. [1983] used the acoustic output from large-scale whitecaps to trigger a rapid sequence of photographs. Longuet-Higgins [1969] presented a simple statistical model for the loss of energy by wave breaking based on a crest downward acceleration threshold of 0.5g for the sharp-crested limiting Stokes wave. Finally, some recent studies have investigated the use of the local wave train properties derived from the Hilbert transform of the wave elevation signal [Melville, 1982; Hwang et al., 1989; Huang et al., 1992]. One method of particular interest is the Phase-Time Method (herein after referred to as PTM). Traditionally, research on the statistics of ocean waves has emphasized time domain properties. The work of Huang et al. [1992] presented a new approach using phase information to study frequency modulation, wave group structures, and wave breaking in the time domain. The PTM consists of using the Hilbert transform of the elevation time-series to obtain, after a few operations, a time series of the local phase and its derivative, a time-series of the variation of the local frequency from the mean frequency of the record. They implied that a rapid change in the phase function could signal wave breaking. Finally, Griffin et al. [1996] performed some experiments to study the kinematics and dynamic evolution of deterministically-forced laboratory deep water breaking waves, utilizing the PTM in their analyses. However, they did not develop any form of detection model based upon the PTM.

The present research was conducted to explore developing and validating a deep water breaking wave detection model based on the PTM. A laboratory-scale random-wave data set was acquired under a variety of deep water wave conditions. Each experiment contained a large number of breaking waves (up to 10%.). Video documentation was used to record the breaking events and to allow visual discrimination and subjective evaluation of the type of breaking. The PTM was applied to the data and explored as a detection method for breaking waves. A simple model for breaking detection was developed and validated.

The Phase-Time Method

Hilbert transforms are presently used in a variety of signal processing applications. A clear description of this transform is contained in Poularikas and Hahn [1996].

For a wave elevation time-series which can be written as

$$\eta(t) = \sum_{n=0}^{\infty} a_n \cos(n\sigma t) + b_n \sin(n\sigma t) \qquad [1]$$

The corresponding Hilbert transform of eq.[1] is

$$v(t) = \sum_{n=0}^{\infty} a_n \sin(n\sigma t) - b_n \cos(n\sigma t) \qquad [2]$$

The principal advantage of this transform is that if the original signal is a time domain signal, its transform will also be a time-domain signal. The PTM is based on the Hilbert transform, as follows. Using an elevation time-series record $\eta(t)$ and its Hilbert transform, $v(t)$

The phase function, $\Phi(t)$, is defined as

$$\Phi(t) = \arctan\left(\frac{v(t)}{\eta(t)}\right) \qquad [3$$

where $\Phi(t)$ can be decomposed into its mean value, which is a product of the mean frequency, n_0, and time, t, and a time-varying component, $\Theta(t)$

$$\Phi(t) = n_0 t + \Theta(t) \qquad [4]$$

the phase is first unwrapped and then detrended by subtracting the mean from eq.[4]. By definition, the time derivative of the phase function is the local frequency of the time-series

$$F = \frac{\partial \Phi}{\partial t} = n_0 + \frac{\partial \Theta}{\partial t} \qquad [5$$

Therefore, eq. 5 yields the variation of the frequency from its mean (in a local time domain). This approach, to be useful; requires a sampling rate fast enough so that the frequencies of interest are resolved. By deriving the Hilbert frequency directly, we have:

$$F = \frac{\frac{\partial v}{\partial t}\eta - \frac{\partial \eta}{\partial t}v}{\eta^2 + v^2} \qquad [6$$

From eq. 6 it is obvious that near mean sea level where η approaches zero, if v is also small (because of the 90° shift shown in eq. 2)., the denominator becomes very small and F very large. This is an artifact, but proved troublesome in the analysis of F(t) until it was removed.

Experimental Design

The experiment was designed to develop a model for detecting the breaking of random waves in deep water. Twelve data sets of JONSWAP model waves with peak frequencies ranging from 0.5 to 0.6 hz were obtained in the Offshore Technology Research Center Model Basin. A total of 11 wave staffs in two parallel arrays were employed to insure that a breaking event would actually be captured. Multiple video cameras were used to record breaking events. The arrangement is shown in Figure 1. The video images contained a clock signal to allow time synchronization. Each run had a duration of approximately 16 min with a sampling rate of 33 Hz (around 32700 data points for each gauge).

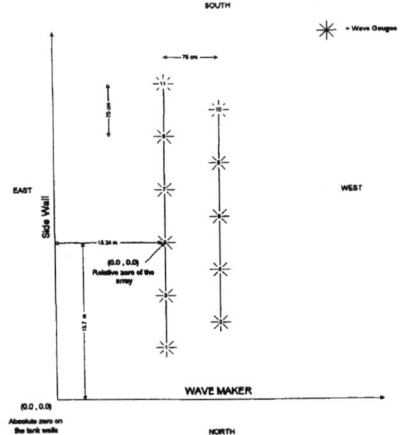

Figure 1 Experimental Setup

Detection Model

For every experimental run, the time for each breaking wave observed on the video record and the identification number of the gauge where the wave breaking occurred were recorded. In addition, a subjective evaluation was made of the intensity of the breaking The PTM procedure was applied to half of the experimental runs. The remaining runs were set aside to test the breaking detection model. The development of a detection model began with the elimination of extraneous large peaks in the frequency variation signal by excluding the events associated with near-zero wave elevations, since breaking never occurred at these points. A value of 1.5 standard deviations in the elevation record was found to be effective as a threshold for filtering the frequency variation record. Expressing this threshold in terms of a standard wave parameter yields $\eta_t = 0.38\ H_s$, where η_t is the value of the elevation record below which all frequency variation will be set to zero and where H_s is the significant wave height. The filtered frequency variation record for each of the gauges was then marked at the times of each breaking event allowing a search for characteristic features associated with breakers. There was a consistent frequency variation pattern corresponding to an incipient breaker -- a very high and sharp peak. An averaged shape of this pattern is shown in figure 2. A wave which will eventually be classified as a large breaker will exhibit this frequency signal shape at the gauge where the breaking initiates, and then this frequency variation signal will progressively decrease with the further breaking of the wave. The frequency signal at the peak of the breaking (i.e. when the white capping is very pronounced and when there is dramatic spilling down the front face) will, in most instances, be very reduced in magnitude compared to the incipient zone of breaking. Some waves in the record remain in the incipient stage without fully breaking. They exhibit a consistent and high frequency variation in every gauge crossed by them, whereas the largest breaking waves in the experiment show this high frequency variation for only one or two gages. This can lead to false detection because of the pattern produced by certain very high, steep, almost-breaking waves but which do not have quite enough energy to break. These waves may also exhibit a high frequency variation at one point during their evolution, and can misjudged as a real breaker.

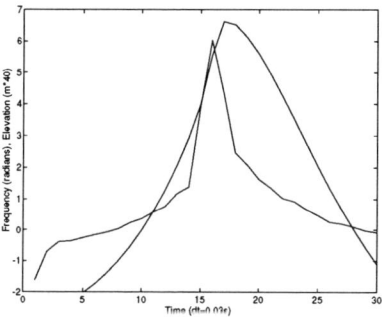

Figure 2. Local frequency variation peak and wave crest elevation for an average of 50 incipient breaking waves.

Another important characteristic pattern in the frequency variation signal is the appearance of double peaks in the record. This phenomenon occurs only with very large, almost-plunging, breaking waves, but is not present in every large breaker. In these cases, the frequency variation signal contains a double peak at the point of full breaking. The first peak corresponds to the point in time where the cascading front face of the wave passes the gauge, and the second one to the passage of the crest of the wave. An averaged shape of this double peak frequency is shown in Figure 3. The peak preceding the crest-associated peak is probably caused by the concentrations of high frequency energy in the spilling on the front face. At least one double peak event was found in every run and was independently identified as an unusually large breaker in the wave elevation video record prior to its discovery in the frequency variation record.

Griffen et al. found that the evolution of the Hilbert frequency of the packet traveling down the wave channel was nearly constant for non breaking waves, increased slightly for spilling breakers and grew rapidly for fully plunging breakers. This is in contradiction to the results found in the present experiments, where the higher frequency variation is most often found at the initiation of the breaking of the wave. The present study found a dissipation of the frequency variation for the large breakers instead of a progression, and finally many non-breaking waves are detected with quite high frequency variation but without constant behavior. The experiment of

Griffin et al. involved the coincident arrival of wave packets at one point within their gauge array in order to obtain breaking, whereas our experiment was based on random waves with JONSWAP type spectra which can break anywhere in the basin. The waves created and studied in the present experiments are closer to real ocean waves, and therefore it is more likely that the frequency variations obtained here will be found in the real ocean

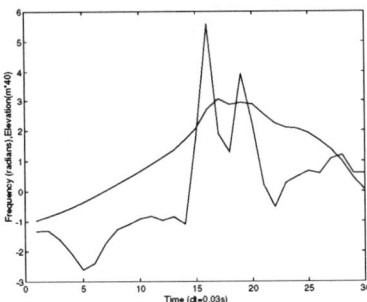

Figure 3. Local "double peak" frequency variation. and wave crest elevation records averaged for a number of very large breakers. Averages were accomplished by aligning the first (left) peak

The first step in the PTM analyses was to compare the peak values of frequency variation associated with each breaking event. The lowest frequency variation at breaking in each run was >=3.14 and <= 3.9 radians/s. The values were higher in the runs with higher peak spectral frequencies., suggesting that the frequency variation scales with the mean frequency of the wave train. A threshold of approximately 3 radians/s showed excellent skill in discriminating non-breaking and breaking waves. Scaling this threshold to the peak frequency of the spectrum yields $F_t = 1.2\ f_{peak}$, where F_t is the value of the frequency variation above which breaking is predicted. This threshold was very efficient in detecting breaking waves in those runs set aside for testing the model – predicting from 95% to 100% of real breaking waves and falsely detecting only about 5% As is typical in threshold-based models, raising the values sufficiently of either the frequency variation threshold, F_t, and/or the filtering threshold, η_t, will result in eliminating false positives. This is accomplished, of course, at the expense of skill in correctly predicting actual breakers.

Figure 4 shows an example of the detection model using the two parameters. The figure shows the filtered frequency variation and the surface elevation with the breaking waves marked. In this case, all the breakers were predicted and there were no false predictions. Because no parameter was found that could totally eliminate false detection, other parameters such as the frequency signal slope, Hilbert amplitude and a variety of wave elevation parameters such as the crest front steepness, the horizontal asymmetry, the vertical asymmetry were studied. None of these showed a great improvement in discrimination skill, largely because steep non-breaking waves have similar geometry to breakers.

One of the objectives of this research was to determine how rapidly the data must be sampled. The experimental sampling rate was 33 Hz. Scaling from the laboratory periods of 1.6 to 2 sec., to obtain the same accuracy the ocean sampling rate must be about 3 Hz (scale ratio on the order of 10). Ocean wave sampling is done commonly at 1 Hz and occasionally at 2 Hz. The time-series were sub-sampled at 16 Hz and 8 Hz and tested with the PTM. The results are not very satisfying. The 16 Hz sub-sampled signal showed some skill, but the 8 Hz sub-sampled signal is completely useless for detection. These results show that the model would be useful at a sampling rate of 25 times the peak frequency in the ocean, but accuracy would greatly improve with a sampling rate of 50 times the peak frequency.

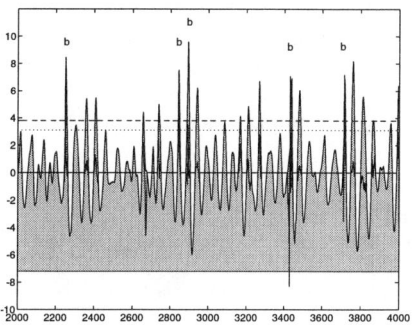

Figure 4. Wave elevation and filtered frequency variation for experiment 4, gauge 5. (b indicates a breaking wave). The elevation threshold is dashed and the frequency threshold is dotted. There was no false detection in this example.

Conclusions

This research effort, which was focused on exploring the feasibility of the PTM as a detection method for deep water breaking waves, resulted in the development of a detection model based on only two simple parameters. The model was tested against a quality data set, containing a wide range in heights and periods of realistic (although long-crested) waves, in which there was no uncertainty about the existence of breaking waves. The use of time-synchronized video as an observation tool was invaluable. The findings in the present work on the evolution of the frequency variation function were quite different from those in Griffen et al. This difference was most likely caused by the character of the waves used in the two experiments. The model developed here appears to have the capability of predicting breaking in deep water waves correctly at about the 95% skill level, with false detection limited to about 5%. The model requires sampling the surface elevation at rates only slightly greater than typically employed. The characteristic patterns in the frequency variation function appear to contain valuable information on the type and intensity of breaking.

References

Banner, M.L., and D.H. Peregrine, Wave breaking in deep water, *Annu. Rev. Fluid Mech.*, 25, 373-397, 1993.

Griffin, O.M, R.D. Peltzer, H.T. Wang, and W.W Schultz, Kinematic and dynamic evolution of deep water breaking waves, *J. Geophys. Res.*, 101, no. C7, 16,515-16,531, 1996.

Holthuijsen, L.H., T.H.C. Herbers, Statistics of wave breaking observed as whitecaps in the open sea, *J.Phys. Oceanography*, 16, 290-97, 1986.

Huang, N.E., S.R. Long, C.C. Tung, M.A Donelan, Y. Yuan, and R.J. Lai, The local properties of ocean waves by the phase-time method, *J. Geophys. Res. Lett.*, 19(7), 685-688, 1992.

Hwang, P.A., D. Xu, J. Wu, Breaking of wind generated waves: measurements and characteristics, *J.Fluid Mech.*, 202, 177-200, 1989.

Keller, W.C., W.J. Plant, G.R. Valenzuela, Observation of breaking ocean waves with coherent microwave radar, *Wave Dynamics and Radio Probing of the Sea Surface*, ed. O.M. Phillips, K. Hasselmann, 295-313, 1986.

Melville, W.K., The instability and breaking of deep-water waves, *J. Fluid Mech.*, 115, 165-185, 1982.

Longuet-Higgins, M.S., On wave breaking and the equilibrium spectrum of wind-generated waves, *Proceedings R. Soc. London Ser.*, 310, 151-159, 1969

Longuet-Higgins, M.S., N.D. Smith, Measurements of breaking by a surface jump meter, *J. Geophys.Res.*, 88, 9823-9831, 1983.

Poularikas, A.D., S.L. Hahn, The transforms and applications handbook, Boca Raton Fla, CRS Press, 1996.

Snyder, R.L., R.M. Kennedy, On the formation of whitecaps by a threshold mechanism, part 1: basic formation, *J. Phys. Oceanogr.*, 13, 1482-1492, 1983.

Thorpe, S.A., P.N. Humphries, Bubbles and breaking waves, Nature, 283, 463-465, 1980.

Wind Wave Prediction in Finite Depth Water

I.R. Young[1]

Abstract

The results of a field experiment aimed at investigating the evolution of fetch limited waves in water of finite depth are presented. In particular, non-dimensional growth curves, the parametric form of the one dimensional spectrum and the directional spreading of the spectrum are investigated.

Introduction

This paper presents the results of a fetch limited, finite depth growth experiment. The evolution of the wave spectrum has been measured at a number of points along a shallow lake. The relatively simple geometry and bathymetry of the lake and the absence of contaminating swell results in a near ideal test basin in which to investigate fetch limited growth in finite depth water.

The paper investigates three basic areas of finite depth fetch limited evolution: the development of non-dimensional energy and non-dimensional peak frequency with fetch, the one dimensional spectral form and directional spreading.

Description of Experiment

The site chosen for the present experiment was Lake George (see Figure 1). The lake is approximately 20 km long by 10 km wide and has a relatively uniform bathymetry with an approximate water depth of 2 m. A series of 8 measurement stations were established along the long North-South axis of the lake as shown in Figure 1. With the exception of Station 6, the sites consisted of minimum blockage space frame towers, designed to provide minimum contamination to either wind or wave measurements made from the towers. The central site, Station 6, consisted of a large platform with temporary accommodation for the research team. Each of the sites was instrumented with a surface piercing Zwarts poles. In addition, cup anemometers at a reference height of 10 m were located at Stations 2, 4, 6, 7 and 8. Air temperature and relative humidity and water temperature were also

---
[1]School of Civil Engineering, University College, UNSW, Canberra, ACT 2600, Australia

measured at Stations 2, 6 and 8. A spatial array consisting of 7 Zwarts poles was established adjacent to the platform at Station 6 to provide high resolution measurements of the directional wave spectrum.

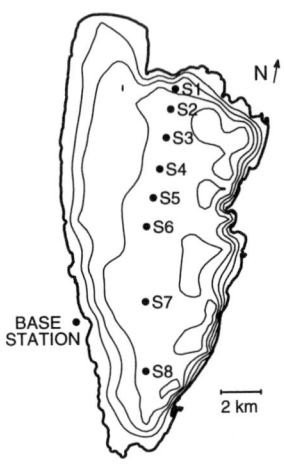

Figure 1: Map of the Lake George experimental site. The measurement locations are labeled S1 to S8. Data were transmitted to the Base Station on the western shore of the lake where it was logged under computer control. The contour interval is 0.5 m, with the maximum contour value 2 m.

Non-dimensional Growth Curves

It is convenient to represent the data in terms of the non-dimensional variables: non-dimensional energy, $\varepsilon = g^2 E/U_{10}^4$, non-dimensional frequency, $\nu = f_p U_{10}/g$, non-dimensional fetch $\chi = gx/U_{10}^2$ and non-dimensional depth $\delta = gd/U_{10}^2$ where g is gravitational acceleration, E is the total wave energy or variance of the wave record, f_p is the frequency of the spectral peak, d is the water depth, x is the fetch and U_{10} is wind velocity.

Figure 2 shows a scatter plot of ε verses χ. A similar plot could be presented for ν verses χ. The data have been partitioned into discrete intervals of δ. Figures 2 shows data in the range $\delta = 0.2 - 0.3$. A preliminary investigation of the north/south data indicated that the data which conformed to deep water conditions was consistent with previous deep water growth law formulations. In particular, ε was well modeled by the JONSWAP relationship (Hasselmann et al., 1973), $\varepsilon = 1.6 \times 10^{-7} \chi$ and ν by the Kahma and Calkoen (1992) form, $\nu = 2.18 \chi^{-0.27}$.

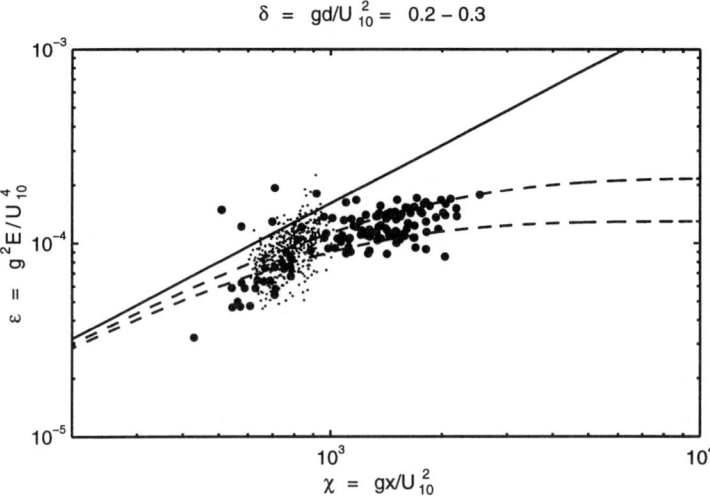

Figure 2: A scatter plot of non-dimensional energy, ε against non-dimensional fetch, χ. Only data with values of non-dimensional depth, δ between 0.2 and 0.3 are shown. The north/south data are shown as the large dots and the lower quality east/west data as the small dots. Equation (1) is shown for the two extremes of δ (ie. 0.2 and 0.3) by the two dashed lines. The deep water asymptotic form of Equation (1) is shown as the solid line.

Noting these deep water asymptotic limits to the data, a nonlinear least squares analysis of the data yielded the model

$$\varepsilon = 3.64 \times 10^{-3} \left\{ \tanh A_1 \tanh \left[\frac{B_1}{\tanh A_1} \right] \right\}^{1.74} \tag{1}$$

where

$$A_1 = 0.493 \delta^{0.75} \tag{2}$$
$$B_1 = 3.13 \times 10^{-3} \chi^{0.57} \tag{3}$$

and

$$\nu = 0.133 \left\{ \tanh A_2 \tanh \left[\frac{B_2}{\tanh A_2} \right] \right\}^{-0.37} \tag{4}$$

where

$$A_2 = 0.331 \delta^{1.01} \tag{5}$$
$$B_2 = 5.215 \times 10^{-4} \chi^{0.73} \tag{6}$$

Equation (1) is shown in Figures 2. As this figure contains data for a finite range of δ (ie. $\delta = 0.2 - 0.3$), two curves are shown, one for each of the extremes of δ for that figure. Generally, the proposed relationship (1) approximates the data well. As with all previous field measurements of this type there is some data scatter.

At short non-dimensional fetch the waves are in deep water and approach the deep water asymptotic limits which are consistent with the numerous previous deep water data sets. As the non-dimensional fetch increases, the effects of the finite water depth become more pronounced and the data progressively deviate from the deep water limit. At relatively large values of non-dimensional fetch, further spectral development ceases as shown by the "plateau" regions of the curves in Figure 2. Rather than there being a single universal growth curve, there are a family of curves, one for each value of non-dimensional depth.

Spectral Evolution

Bouws et al. (1985) proposed the TMA spectral form for the representation of wind generated waves in water of finite depth

$$E(f) = \alpha g^2 (2\pi)^{-4} f^{-5} \exp\left[\frac{-5}{4}\left(\frac{f}{f_p}\right)^{-4}\right] \cdot \gamma^{\exp\left[\frac{-(f-f_p)^2}{2\sigma^2 f_p^2}\right]} \cdot \Phi \qquad (7)$$

and

$$\Phi = \left\{ \frac{[k(f,d)]^{-3} \frac{\partial k(f,d)}{\partial f}}{[k(f,\infty)]^{-3} \frac{\partial k(f,\infty)}{\partial f}} \right\} \qquad (8)$$

Based on the wavenumber scaling arguments implicit in the TMA spectral form, Bouws et al. (1985, 1987) speculated that the spectral parameters should be functions of the non-dimensional wavenumber, $\kappa = U_{10}^2 k_p/g$, where k_p is the wavenumber of the spectral peak. To investigate this dependence with the present data set, the spectral parameter α is presented as a function of κ in Figure 3.

Within the data scatter, a relationship between α and κ is clear, with α an increasing function of κ (see Figure 3). A least squares fit to the data yields the power law relationship

$$\alpha = 0.0091 \kappa^{0.24} \qquad (9)$$

Equation (9) is shown on Figure 3, together with the TMA result, $\alpha_{TMA} = 0.0078 \kappa^{0.49}$. Both (9) and the TMA form are consistent with the data. In deep water the general TMA form reverts to that of JONSWAP. The deep water JONSWAP result scales α in terms of the non-dimensional frequency, ν, $\alpha_{JONSWAP} = 0.033 \nu^{0.67}$. Assuming a deep water linear dispersion relationship, this result can be converted to wavenumber space, $\alpha_{JONSWAP} = 0.01 \kappa^{0.33}$. This JONSWAP result is also shown in Figure 3 and is broadly consistent with the finite depth formulations.

All results are comparable and confirm that the trend towards decreasing values of α with increasing maturity of the waves, already observed in deep water, also holds in finite depth situations.

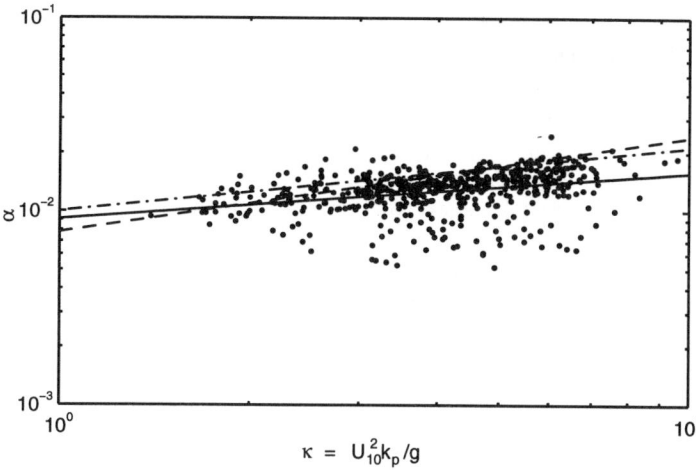

Figure 3: Values of the parameter α as a function of non-dimensional wavenumber, κ. The solid line is a least squares fit to the data [Equation (9)]. The TMA result is shown by the dashed line and the JONSWAP form, transformed from frequency to wavenumber space, by the dash-dot line.

In contrast to the observable trend in α with κ, no similar results were apparent for either γ or σ. This result is consistent with both TMA (Bouws et al., 1985, 1987) and JONSWAP (Hasselmann et al., 1973). There is, however, significant sampling variability associated with these parameters. The mean values of the data set yield $\gamma_{mean} = 2.70$ and $\sigma_{mean} = 0.12$.

Directional Spreading

Two common models have been proposed to represent the directional spreading of wind waves: $\cos^{2s}\theta/2$ and $\text{sech}^2\beta\theta$. A number of different techniques have been reported for the fitting of such analytical forms to the data. Mitsuyasu et al. (1975) and Hasselmann et al. (1980) analyzed their buoy data using the Fourier Expansion Method (Longuet-Higgins et al., 1963; Young, 1994) and determined the value of s from the first 2 components of the Fourier expansion. Donelan et al. (1985) compared this approach with one in which the directional form was matched to the half-power points of the measured spreading function. It was argued that matching the half-power points was more meaningful since interest is concentrated on the energetic region of the directional distribution. This technique appeared to reduce the scatter in the data.

In addition to these techniques, application of a nonlinear least squares fit

[Levenberg-Marquardt method, Press et al. (1986)] was also investigated with the present data. This least squares approach, was adopted here, together with a Maximum Likelihood Method analysis of the data from the directional array.

Both formulations of the spreading function yield qualitatively similar results, indicating the spectrum is narrowest at the spectral peak frequency and broadens at frequencies both larger and smaller than the peak value. The $\cos^{2s} \theta/2$ formulation exhibited marginally less scatter than the $\mathrm{sech}^2 \beta\theta$ form and was adopted for further analysis.

A least squares analysis of the directional data yields the following form for the present finite depth data

$$s = \begin{cases} 11 \left(\frac{f}{f_p}\right)^{2.7} & f < f_p \\ 11 \left(\frac{f}{f_p}\right)^{-2.4} & f \geq f_p \end{cases} \tag{10}$$

Equation (10) has been constrained to yield narrowest spreading at $f/f_p = 1$. The scatter in the data is such that the actual point of narrowest spreading cannot be determined with great accuracy. It is clearly, however, in the vicinity of the spectral peak frequency.

Equation (10) is shown in Figure 4 together with the deep water results of Mitsuyasu et al. (1975), Hasselmann et al. (1980) and Donelan et al. (1985). As the results of Mitsuyasu et al. (1975) and Hasselmann et al. (1980) are both wave age dependent, mean values typical of their respective data sets have been used to construct Figure 4 (Mitsuyasu et al., 1975 - $U_{10}/C_p = 1.1$; Hasselmann et al., 1980 - $U_{10}/C_p = 1.4$). As Donelan et al. (1985) provide no functional form expressed in terms of s, the result in Figure 4 was obtained from digitizing the result given in their Figure 30.

The present results are marginally narrower (higher s) than those of both Mitsuyasu et al. (1975) and Hasselmann et al. (1980), they are however significantly broader than the high resolution results of Donelan et al. (1985). As shown by Donelan et al. (1985), the results of both Mitsuyasu et al. (1975) and Hasselmann et al. (1980) are excessively broad due to the instrumentation and analysis technique utilized. The present result [Equation (10)] should have comparable resolving power to the result of Donelan et al. (1985). The inference is that finite depth wind wave spectra are broader than their deep water counterparts. Due to the relatively narrow range of $k_p d$ spanned by the present data set a more emphatic statement cannot be made. At present it is necessary to rely on these two independent data sets (Donelan et al., 1985 - deep water; Lake George - finite depth). There is always some possibility that, in addition to water depth, there are other unknown influences responsible for the different spreading.

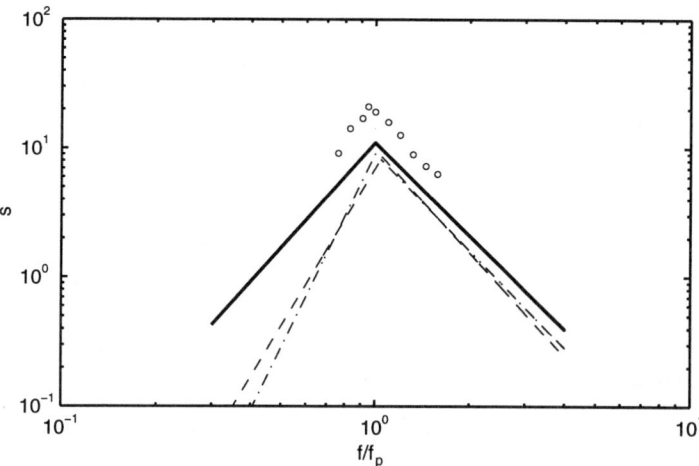

Figure 4: The dependence of the directional exponent s for the present finite depth data as a function of f/f_p [Equation (10)] (thick solid line). Also shown for comparative purposes are previous deep water data sets: Mitsuyasu et al. (1975) - 'dash dot line'; Hasselmann et al. (1980) - 'dashed line'; Donelan et al. (1985) - 'open circles'.

Conclusions

The data set presented in this paper represents the first comprehensive field study of the evolution of fetch limited waves in water of finite depth. The data set is sufficiently comprehensive to fully define non-dimensional growth curves for both energy and peak frequency. In contrast to deep water results, a single growth relationship does not exist for each quantity. Rather, a family of curves result, depending on the non-dimensional depth.

The one-dimensional spectrum has been shown to conform to the TMA form. The spectral parameter α is shown to be a function of the non-dimensional peak wavenumber, κ.

The directional spreading is qualitatively similar to deep water results. The spectra are narrowest at the spectral peak frequency and broaden at frequencies both above and below the peak. Compared with deep water results, finite depth spectra appear to exhibit broader spreading. This is possibly due to the different nonlinear coupling which will exist in finite depth cases.

The present data set is significantly more comprehensive than previous finite depth studies. The results of these previous studies are, however, consistent with the present results. This occurs, despite the fact that a wide range of bed materials exist for the various data sets. As a result, it can be speculated that bed material has little influence on spectral evolution under fetch limited conditions. This is markedly different to other finite depth cases, such as swell attenuation, where the rate of decay is very sensitive to the bed material.

References

Bouws, E., Günther, H., Rosenthal, W. and Vincent, C.L., 1985. Similarity of the wind wave spectrum in finite depth water, 1. Spectral form. J. Geophys. Res., 90: 975-986.

Bouws, E., Günther, H., Rosenthal, W. and Vincent, C.L., 1987. Similarity of the wind wave spectrum in finite depth water, 2. Statistical relationships between shape and growth stage parameters. Dtsch. Hydrogh. Z., 40: 1-24.

Donelan, M.A., Hamilton, J. and Hui, W.H., 1985. Directional spectra of wind-generated waves. Phil. Trans. R. Soc. Lond., A 315: 509-562.

Hasselmann, K. et al., 1973. Measurements of wind-wave growth and swell decay during the Joint North Sea Wave Project (JONSWAP). Dtsch. Hydrogh. Z., Suppl. A, 8, 12, 95pp.

Hasselmann, D.E., Dunckel, M. and Ewing, J.A., 1980. Directional wave spectra observed during JONSWAP 1973. J. Phys. Oceanogr., 10, 8: 1264-1280.

Kahma, K.K. and Calkoen, C.J., 1992. Reconciling discrepancies in the observed growth of wind-generated waves. J. Phys. Oceanogr., 22: 1389-1405.

Longuet-Higgins, M.S., Cartwright, D.E. and Smith, N.D., 1963. Observations of the directional spectrum of sea waves using the motions of a floating buoy. in 'Ocean wave spectra', Englewood Cliffs, Prentice Hall, Inc.: 111-136.

Mitsuyasu, H., Tasai, F., Suhara, T., Mizuno, S., Onkusu, M., Honda, T. and Rukiiski, K. , 1975. Observations of the directional spectrum of ocean waves using a cloverleaf buoy. J. Phys. Oceanogr., 5: 751-761.

Press, W.H., Flannery, B.P., Teukolsky, S.A. and Vetterling, W.T., 1986. Numerical Recipies. Cambridge University Press, 818pp.

Young, I.R., 1994. On the Measurement of Directional Wave Spectra. Applied Ocean Research, 16: 283-294.

Numerical Modeling of Storm Waves for Hurricanes Erin and Felix

Lihwa Lin[1]

Abstract

This paper compares numerically computed results of storm waves for Hurricanes Erin and Felix from the 1995 Atlantic hurricanes season. Two wave models, WISWAVE and HWAVE, were used for the computation of hurricane waves. Both models are based on discrete directional spectral method using a finite-difference, time-dependent scheme to calculate wave generation, propagation, and dissipation under wind and wave interactions. WISWAVE is a second generation model that has been used by the U.S. Army Corps of Engineers to generate wave hindcast information along all coastlines of the U.S. HWAVE was modified from WISWAVE with wind and wave interaction terms being recalibrated based upon data that are heavily influenced by tropical events. Comparison of model performance upon Hurricanes Erin and Felix showed that HWAVE is more efficient and accurate for prediction of hurricane waves than WISWAVE. This comparison showed that HWAVE generally uses less CPU time and computer memory than WISWAVE.

1. Introduction

Numerical modeling of hurricane waves has been traditionally carried out by using models which were calibrated based on data with no tropical events or with a limited number of tropical events. These traditional models are usually referred as either non-tropical or generic models. It is generally not known that if a non-tropical model is suitable for prediction of waves of a tropical event. Using non-tropical models for tropical events has shown to produce good result of prediction of waves if a tropical storm has small to moderate size and strength. However, applying non-tropical models to a bigger and stronger tropical storm often produces poor result of wave prediction. Therefore, it seems more appropriate to use a tropical-orientated model than a generic one for prediction of waves induced by a hurricane.

---

[1]Research Hydraulic Engineer. US Army Engineer Waterways Experiment Station, CEWES-CN-C, 3909 Halls Ferry Road, Vicksburg, Mississippi 39180.

Two wave models, WISWAVE and HWAVE, were used in the present study to simulate storm waves for Hurricanes Erin and Felix from the 1995 Atlantic hurricanes season. WISWAVE (Hubertz, 1992) is a second generation model that has been developed and used by the U.S. Army Corps of Engineers in the Wave Information Studies (WIS) program to generate a 40-year (1956-1995) hindcast wave database along all coastlines of the U.S. The model is a generic one which has been used for simulation of either tropical or non-tropical events. HWAVE was modified from WISWAVE with wind and wave interaction terms being recalibrated using data which are heavily influenced by tropical events. These two models share many similarities, both use discrete directional spectrum and time dependent scheme to simulate generation, dissipation, and propagation of ocean surface waves via wind and wave interactions.

Hurricanes Erin and Felix from the 1995 Atlantic hurricanes seasons were selected for the comparison of model performance. Erin is ranked as a Category 1 storm based on a Saffir/Simpson Scale (Simpson, et al., 1981) when it made landfall on the east coast of Florida (Rappaport, 1995). Erin is a relatively weak storm compared to many other hurricanes found in the Atlantic Ocean. The storm moved almost along a straight line in the Atlantic Ocean. However, it generated large offshore waves in the Atlantic Ocean according to the sea data measured by several NOAA buoys. Felix, on the other hands, is a Category 3 hurricane (Mayfield and Bevens, 1995), which never made landfall but created tremendous wave action in the western North Atlantic when it made a loop motion east of the North Carolina coast. These two hurricanes represent two very different tropical storm systems which generate large waves in the Atlantic Ocean.

2. Preparation of Wind Fields

Wind fields were prepared by a hurricane wind model HWIND. The HWIND is a standard 2-D gradient wind model (Harris, 1958). It assumes an ideal surface pressure distribution as

$$P = P_o + (P_n - P_o) \exp(-\frac{R}{r}) \quad (1)$$

and solves the surface winds from the following force balance equation:

$$U^2 + rfU = \frac{r}{\rho}\frac{\partial P}{\partial r} \quad (2)$$

where P and U are the surface pressure and wind speed, respectively, at a distance r from the storm center, P_o is the central pressure, P_n is the peripheral pressure, R is the radius of maximum wind, ρ is the density of air, $f = 2\omega\sin\lambda$ is the Coriolis parameter at a latitude λ, and ω is angular velocity of earth. The pressure and wind speed distributions described by Eqs.(1) and (2) are symmetric with respect to the storm center. The direction of the gradient winds is parallel to constant pressure line or an isobar. The wind speeds solved from Eq.(1) correspond to a stationary storm. In reality, a tropical storm will move and, therefore, the wind speeds solved from Eq.(1) need to be corrected by the forward motion of a storm. This correction due to storm's forward motion is computed in HWIND based on a simple form:

$$U_a = U_F [1 - \exp(-\frac{4R}{r})]$$

where U_a is the correction to be added to the gradient winds at each location of r, and U_F is the forward speed of the storm center.

Input information to HWIND including the center pressure, ambient pressure, forward speed of the storm center, radius of maximum winds, and storm track, are based on the North Atlantic Tropical Cyclone Climatology Data compiled at the National Hurricane Center (Hebert, et al., 1995) and Daily Weather Maps prepared by the National Meteorological Center, National Weather Service. In the present study, HWIND was used for computations of wind fields for Hurricanes Erin and Felix. The radius of maximum winds was selected to be equal to 65 km for both storms in the computations. It should be noted that to determine a precise radius of maximum wind is generally difficult. In reality, Erin may have a radius of maximum wind smaller than 65 km and Felix may have a radius of maximum wind greater than 65 km. The computations were carried out on a latitude-longitude parallel grid (0.25 degree spacing), which covers the area from 23°N to 45°N, and from 65°W to 81.5°W. Figure 1 shows the grid system for generation of wind fields. The computed wind fields were saved at every 3-hour interval for the entire length of a storm. Figures 2 and 3 show the storm tracks of Erin and Felix, respectively, in the Atlantic Ocean with a snap shot of surface wind fields computed from HWIND.

3. Computations of Wave Fields

Wave models WISWAVE and HWAVE were run for Hurricanes Erin and Felix on the same grid that was used by HWIND for preparation of wind fields. WISWAVE is capable of running deepwater to shallow water simulation whereas HWAVE is valid only for deepwater simulation. Both models were set for calculation of directional spectra with sixteen 22.5°-wide angle bands, centered on the compass directions of north, north-northeast, northeast, east-northeast, etc., and with twenty frequency bands, centered at frequencies corresponding to wave periods of 3, 4, 5, ..., 19, 20, 22, and 25 seconds. A time step of 720 seconds is used in the model simulation. WISWAVE and HWAVE compute discrete directional spectra using a finite-difference and time dependent scheme based on the wave energy transport equation:

$$\frac{\partial E(f,\theta)}{\partial t} + \frac{\partial E(f,\theta) C_g(f,\theta) \cos\theta}{\partial x} + \frac{\partial E(f,\theta) C_g(f,\theta) \sin\theta}{\partial y} = S(f,\theta)$$

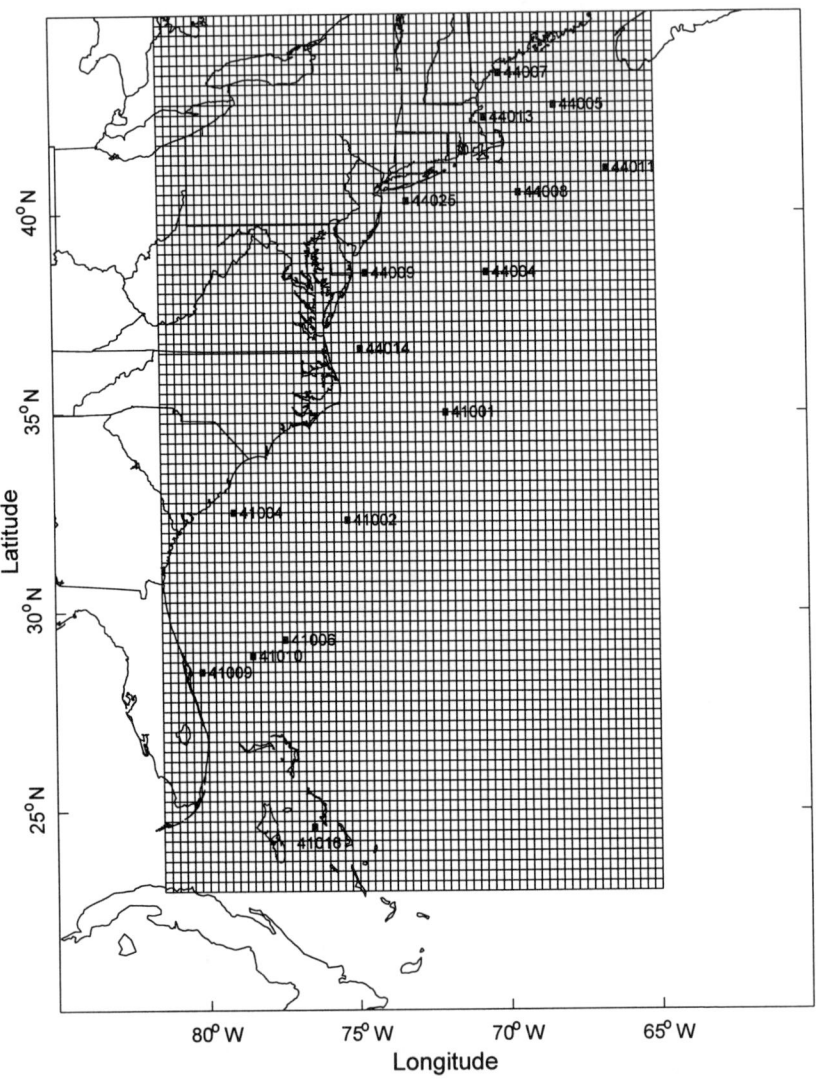

Figure 1: Numerical grid for wind and wave simulations.

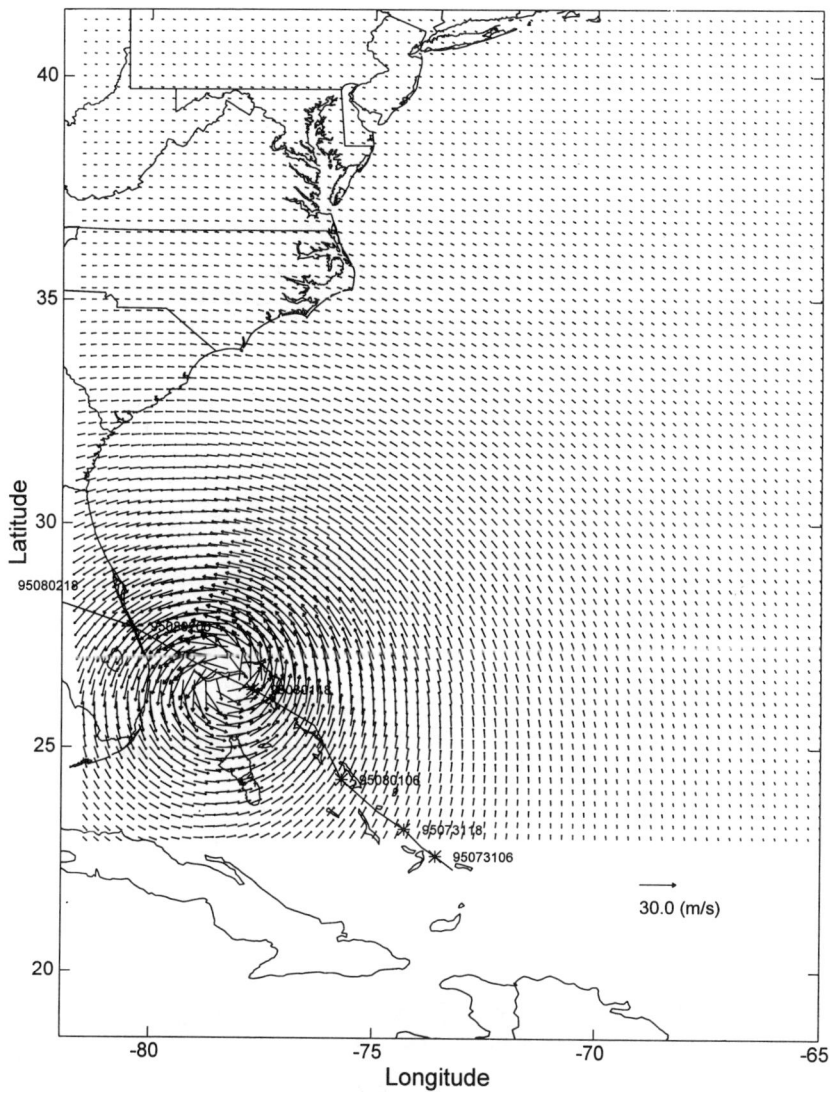

Figure 2: A snapshot of HWIND-simulated wind field for Erin.

326 OCEAN WAVE KINEMATICS, DYNAMICS AND LOADS ON STRUCTURES

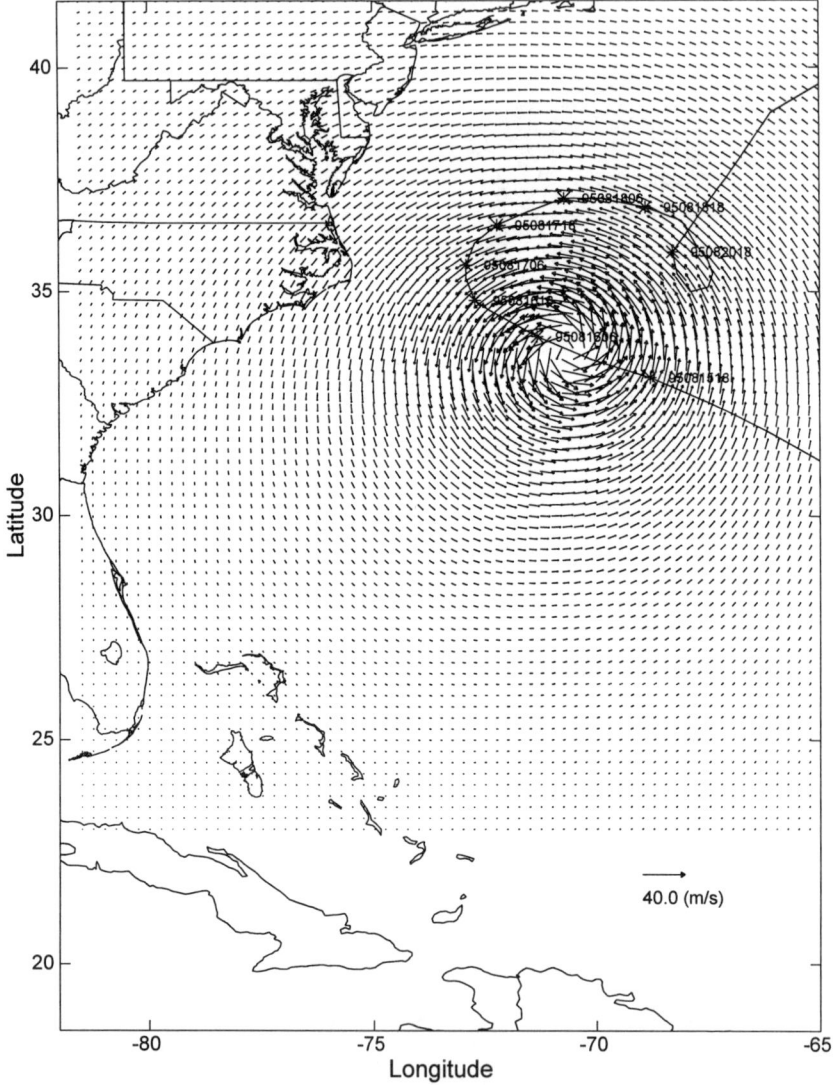

Figure 3: A snapshot of HWIND-simulated wind field for Felix.

where $E(f,\theta)$ is a directional spectrum on the (x, y) plane, f is frequency, θ is direction, $C_g(f,\theta)$ is the wave group velocity, t is the time, and $S(f,\theta)$ is a source function. The main difference between WISWAVE and HWAVE comes from the source function. The source functions, $S(f,\theta)$, given in WISWAVE and HWAVE are similar. However, they are calibrated based on quite different data sets. WISWAVE was calibrated based on data with a limited number of tropical cases whereas HWAVE was calibrated with data that emphasize tropical events. Figure 4 shows a snap shot of wave pattern computed for Hurricane Erin based on HWAVE.

4. Comparison of Model Performance

Model results from WISWAVE and HWAVE were saved and compared with ocean buoy data. Figures 5 to 7 show intercomparison of the computed and measured data for Erin at the Buoy 41006, 41009, and 41010 locations, respectively. Figure 8 shows the comparison for Felix at the Buoy 41001 location. For Erin, it is seen that the computed waves, from both WISWAVE and HWAVE, agree very well with the measured data. The waves computed from HWAVE were seen to have slightly greater height than those from WISWAVE. However, this larger wave height computed by HWAVE is clearly due to higher winds predicted by the HWIND model. For Felix, WISWAVE apparently overpredicts wave height whereas HWAVE yields reasonably good results as compared to the measured data.

Model efficiency was also compared for WISWAVE and HWAVE. Model computations were carried out by a 64-bits DEC ALPHA 400-MHZ Work Station computer. Table 1 presents the summary of CPU time and computer memory, in terms of mega words (MW), used in the model runs.

Table 1: Comparison of model efficiency.

Model	WISWAVE	HWAVE
Storm	Erin	Erin
Grid dimension	67x89	67x89
Grid spacing	0.25°	0.25°
Storm duration (days)	3.25	3.25
Simulation time step (sec)	720	720
Output time interval (hrs)	3	3
CPU (sec)	9	50
Computer memory (MW)	4.12	25.3

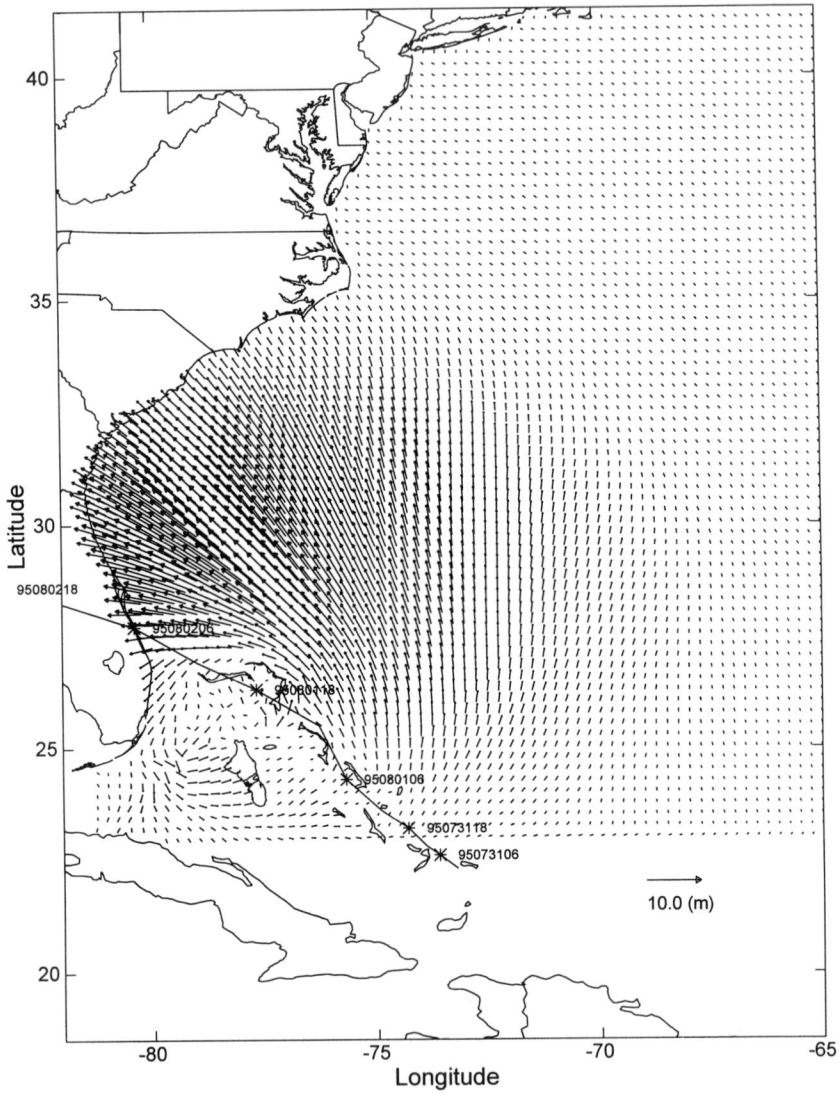

Figure 4: A snapshot of HWAVE-simulated wave field for Erin.

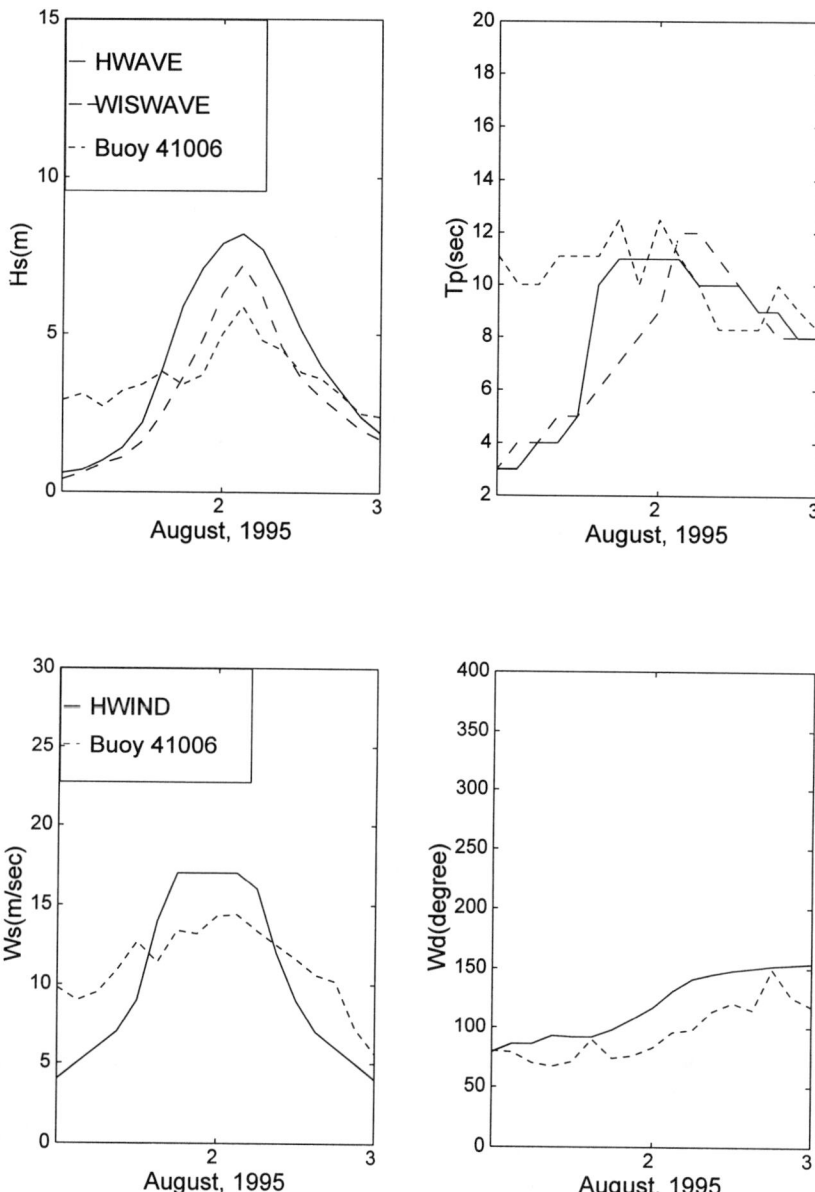

Figure 5: Comparison of model results and Buoy 41006 data.

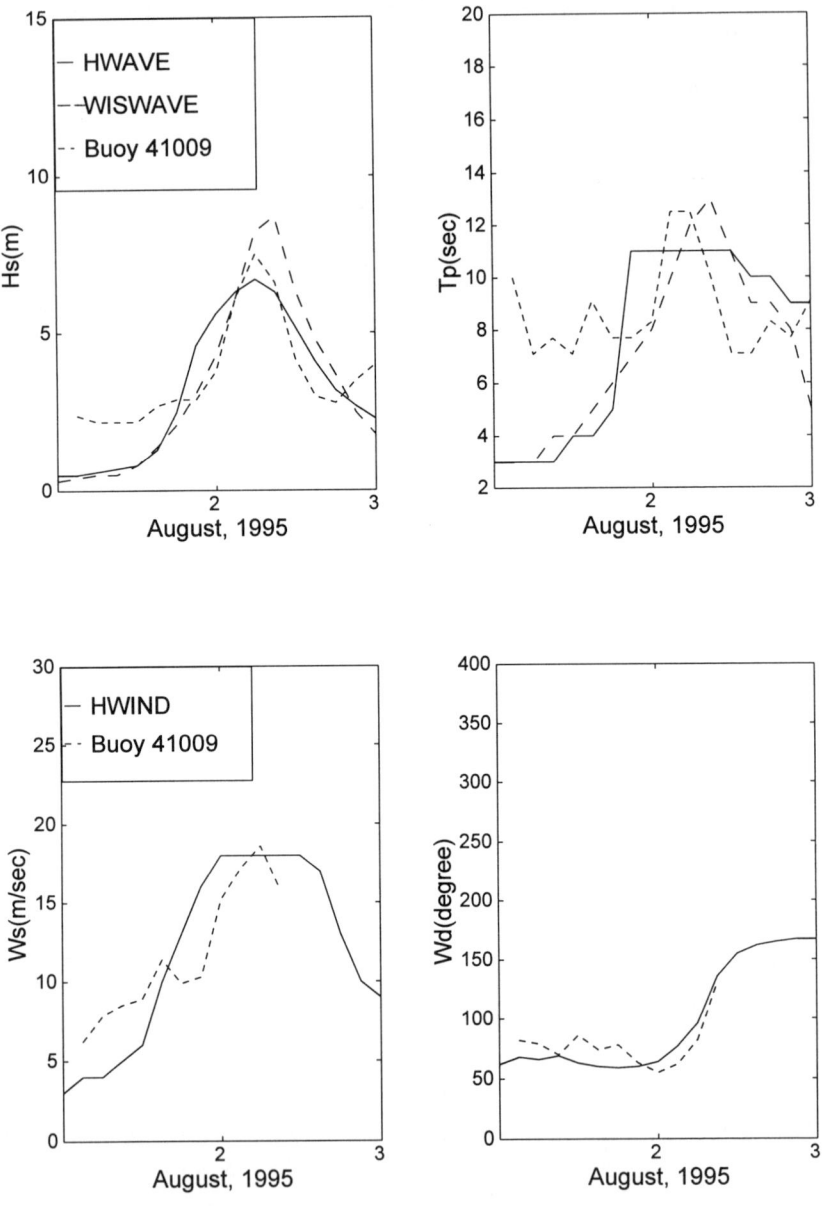

Figure 6: Comparison of model results and Buoy 41009 data.

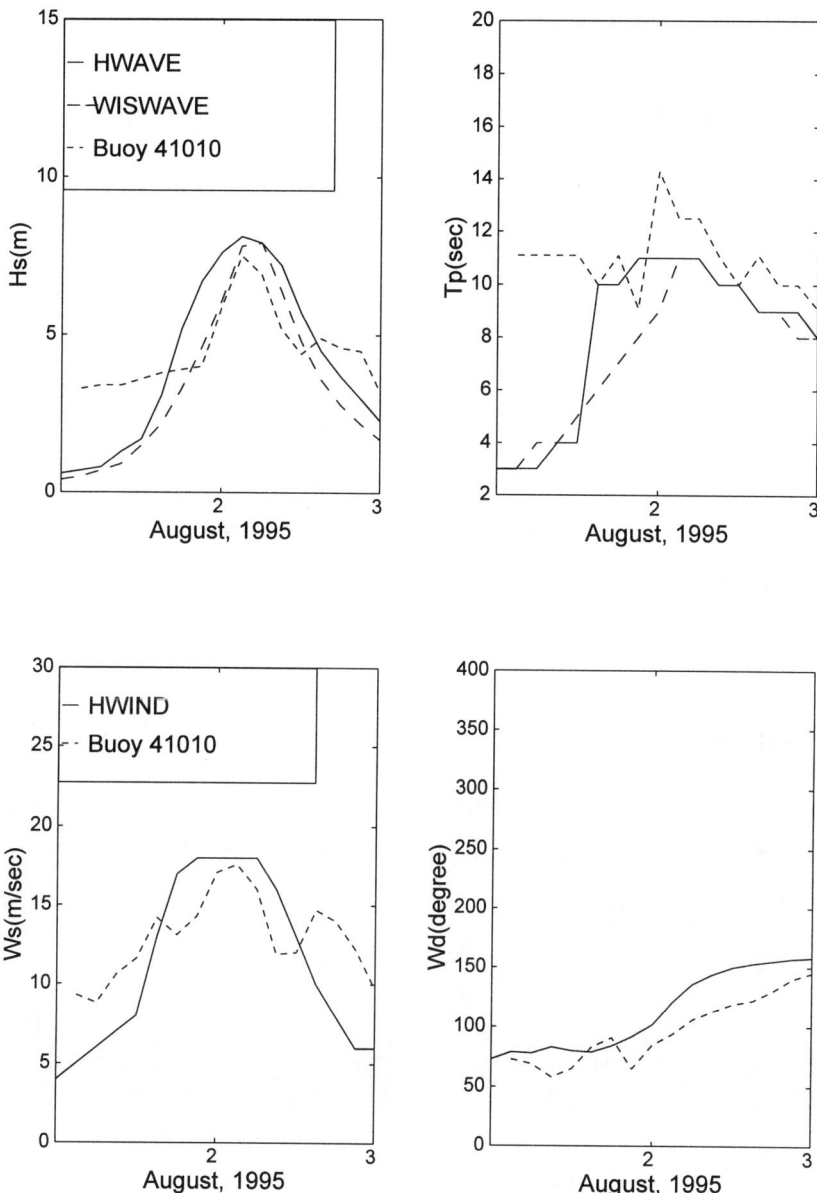

Figure 7: Comparison of model results and Buoy 41010 data.

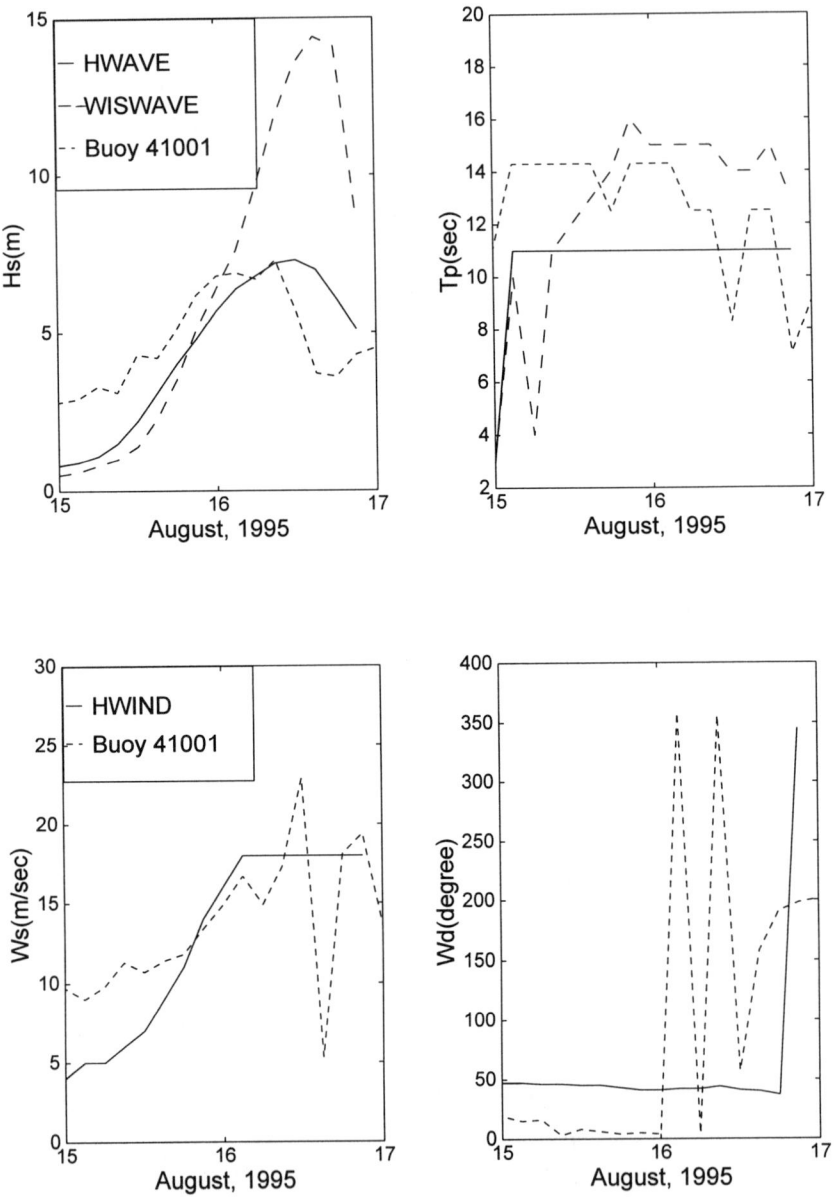

Figure 8: Comparison of model results and Buoy 41001 data.

5. Summary and Conclusions

Two computer models, WISWAVE and HWAVE, were used for the computation of ocean surface waves for Hurricanes Erin and Felix from the 1995 Atlantic hurricanes season. Both models use a discrete directional spectral method and a finite-difference, time-dependent scheme to calculate ocean wave fields under wind and wave interactions. WISWAVE is a second generation model that has been used by the U.S. Army Corps of Engineers to generate wave information along the U.S. coastlines. HWAVE was modified from WISWAVE with source terms being recalibrated based upon data that are heavily influenced by tropical storms. Computer modeling of waves for Hurricanes Erin and Felix presents an interesting problem since these two storms are very different in terms of storm intensity, duration and movement. Comparison of model results from WISWAVE and HWAVE with ocean buoy data showed that HWAVE is generally more efficient and accurate for prediction of hurricane waves than WISWAVE. It should be noted that the wind field data used as input to wave models were generated by a hurricane wind model HWIND. Therefore, there were no background winds that were used in the computer simulated wind fields. The wind field data simulated by HWIND were used solely as input information to WISWAVE and HWAVE in the present study.

Acknowledgments

The NOAA buoy data were received "on-line" through the National Buoy Data Center network at "http://seaboard.ndbc.noaa.gov". This research was supported by the Coastal Field Data Collection Program of the U.S. Army Engineer Waterways Experiment Station's Coastal and Hydraulic Laboratory. Permission was granted by the Chief of Engineers to publish this information.

References

Hebert, P.J., Jarrell, J.D., and Mayfield, M. (1995). 'The Deadliest, Costliest, and Most Intense United States Hurricanes of This Century,' Proceedings of the 17th Annual National Hurricane Conf., Atlantic City, N.J. April 11-14, 1995. pp12-50.

Hubertz, J.M. (1992). 'A users guide to the WIS wave model, Version 2.0,' WIS Report 27, U.S. Army Engineer Waterways Experiment Station, Vicksburg, MS.

Mayfield, M., and Beven, J. (1995). 'Preliminary Report, Hurricane Felix, 8 - 25 August 1995,' National Hurricane Center, Coral Gables, FL.

Rappaport, E.N. (1995). 'Preliminary Report, Hurricane Erin, 31 July - 6 August 1995,' National Hurricane Center, Coral Gables, FL.

Simpson, R.H. and Riehl, H. (1981). 'The Hurricane and Its Impact,' Basil Blackwell, Oxford.

WAM4 model/data comparison in the NW Gulf of Mexico

S. F. DiMarco and L. C. Bender, III[1]

Abstract

A model/data comparison of significant wave height and spectra at several locations on the Texas-Louisiana shelf in the northwest Gulf of Mexico show good agreement between observations and predictions using a third generation ocean wave prediction model (WAM Cycle 4). Our analysis shows a mean model bias magnitude of 0.09 m for five near-shore locations and 0.26 m for three deep water locations. The average root mean square difference is 0.23 m and 0.40 m for the near-shore and deep locations, respectively. Diurnal variation of modeled and observed wave height is also seen to be strongest during summer.

Introduction

We present a model/data comparison during a 10 month period (February 1994 through November 1994) between model output from cycle 4 of the third generation ocean wave prediction model (WAM) (WAMDI Group 1988, Bender 1996) and observations in the northwest Gulf of Mexico. Because of rapidly changing wind field conditions on the Texas-Louisiana shelf during frequent and numerous frontal passages during winter, the WAM model's inclusion of resonant, third-order, wave-wave interactions make it particularly suited to predict this region's wave spectra. Wind fields obtained from the National Center for Environmental Prediction (NCEP) were used for the Gulf of Mexico, while those generated by the Louisiana-Texas Shelf Physical Oceanography Program (LATEX) were used on the Texas-Louisiana shelf. These results were compared to significant wave heights from three National Data Buoy Center (NDBC) buoys and five LATEX bottom-mounted wave gauges. Figure 1 shows the northwest Gulf of Mexico, the model domain, and the locations of the NDBC buoys and LATEX wave gauges.

The model is run for the period of April 1992 through November 1994. The transport equation is solved on a two-level nested domain to incorporate the propagation of remotely generated swell from the deeper Gulf of Mexico onto the Texas-

---

[1] Department of Oceanography, Texas A&M University, College Station, Texas 77843-3146

Louisiana shelf. The outer level model (the Gulf domain) covers the Gulf of Mexico from 18°N to 31°N and 98°W to 80°W on a grid resolution of one degree and generates swell in the deep Gulf and propagates it to the nested domain of the Texas-Louisiana shelf. The nested level model (the shelf domain) covers the Texas-Louisiana shelf from 26°N to 30°N and from 98°W to 89°W on a finer grid resolution of one quarter degree.

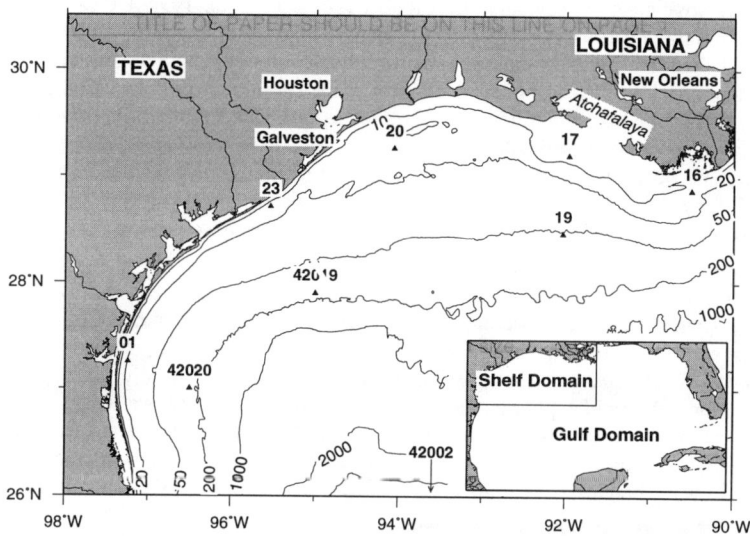

Figure 1. Map of Gulf of Mexico domain (inset) and shelf domain showing local bathymetry (m) and wave gauge and NDBC buoy locations.

The following section briefly describes the methodology used for this study, including descriptions of the wind fields, wave data, and model initialization. Next, we provide the results of the model/data comparison by presenting statistics during the model run period. We then provide specific analysis of shelf-wide wave events during April and June 1994. We conclude with a brief summary.

Methodology

In this section we provide a brief description of the observational data and computer parameters and methods used in this study.

a. Observational Data: Wind fields

The NCEP zonal and meridional wind fields are bilinearly interpolated from the spectral grid on which they were generated (a 18 x 23 point grid that spans the entire Gulf of Mexico) to the WAM-specified uniform 1° grid for the Gulf of Mexico. The wind fields for the Gulf domain are linearly interpolated in time to every hour by the

WAM model. We use LATEX hourly gridded zonal and meridional wind fields at 10-m height for the shelf domain on a half degree grid (Wang et al. 1998). These winds are supplied directly to the WAM model. No interpolation in time took place, however, LATEX winds were spatially interpolated to a quarter degree grid over the shelf domain.

b. Observational Data: Wave data

The observations used in this comparison come from two sources: five bottom-mounted wave gauges maintained by LATEX (DiMarco et al. 1994, DiMarco et al. 1995a) and three surface buoy records obtained from the National Data Buoy Center (NDBC). The five LATEX wave gauges were deployed near-shore along the Texas-Louisiana coast in depths ranging from 7 to 22 m. The instruments recorded hydrostatic pressure and current velocity at three hour intervals. Each of the wave gauge records was post-processed with a uniform high frequency cutoff of 0.222 Hz (4.5 s) to eliminate noise in the record attributed to hydrodynamic attenuation. The three NDBC buoys at 42002, 42019, and 42020 are 3-m discus buoys that ride the surface and estimate wave spectra from accelerations due to the wave motion. The NDBC buoys are in substantially deeper water and recorded hourly.

To obtain an estimate for the high-frequency wave energy lost by using the 0.222 Hz cutoff frequency, we apply an empirical formula for determining a wind-wave spectrum. The formula assumes a f^{-5} relationship between wave energy and frequency. Therefore, by using the highest frequency energy density bin from the observational data and applying this correction for higher energies, we obtain an analytical estimate for the lost high-frequency energy. As we will show, incorporating this energy into the significant wave height calculation substantially increases the estimated wave heights and significantly reduces the positive bias with the model.

c. WAM model initialization/parameters

The shelf domain is represented on a Cartesian grid, with shallow water propagation, bottom friction via the standard WAM parameter set, and depth refraction, but no current refraction. The source and propagation terms are integrated every five minutes to limit the Courant number to less than 0.3. Significant wave height and direction for the shelf and spectra at selected locations were saved after every six hours of integration.

The model is run from April 1992 through November 1994 to coincide with the full duration of the LATEX field program (Jochens and Nowlin 1994), however, the comparison is done only over the period when reliable wave data was consistently available from the five near-shore locations, i.e., from February 1994 through November 1994

Results

We begin with a statistical analysis of the model output as compared with the observational data measured at each wave location. Table 1 shows comparison statistics of the model with wave height estimates at each of the observation sites.

Table 1. WAM model/observation comparison of bias and root mean square difference

Site	Lon. °W	Lat. °N	type	No. obs.	no tail bias (m)	no tail rms (m)	w/tail bias (m)	w/tail rms (m)
01	97.25	27.26	bot	900	0.16	0.26	0.03	0.21
16	90.50	28.87	bot	382	0.19	0.28	0.09	0.24
17	91.97	29.20	bot	147	0.17	0.24	0.12	0.21
20	94.06	29.26	bot	924	0.06	0.19	-0.07	0.25
23	95.54	28.71	bot	162	0.01	0.16	-0.13	0.23
42002	93.60	25.90	buoy	1158	n/a	n/a	-0.14	0.35
42019	95.00	27.90	buoy	753	n/a	n/a	-0.34	0.45
42020	96.50	27.00	buoy	1160	n/a	n/a	-0.30	0.42

The five LATEX wave gauges were located in much shallower water near the Texas and Louisiana coasts than the NDBC buoys. As stated above, these bottom-mounted wave gauges did not provide accurate estimates of the wave energy at frequencies higher than 0.222 Hz (DiMarco et al. 1995a), so an empirical formula was used to estimate the wave energy for frequencies greater than 0.222 Hz. In general, wave heights are consistently highest in the winter and are higher on the western part of the shelf than they are on the eastern half of the shelf, during all seasons. Significant wave heights on the inner shelf are rarely greater than 2 m, except during hurricanes (DiMarco et al. 1995b). The minimum wave heights occur in the vicinity of the Atchafalaya Bay, where water depths are less than 10 m (DiMarco et al. 1995a). The largest significant wave heights are in nonsummer when frontal passage frequency is greatest (DiMego et al. 1974, Henry 1979, and A. E. Jochens private communication).

The statistics of Table 1 show the effects of incorporating the additional empirical high-frequency wave energy (i.e. a tail) in the observational estimates. In general, there is positive bias between the LATEX wave gauges and WAM output, indicating that the model tends to overestimate the significant wave height at these locations. The inclusion of the empirical high-frequency energy considerably improves the bias for 3 of the near-shore meters (01, 16, and 17). At location 20 the magnitude of the bias is virtually unchanged by including the additional energy; but the sign of the bias changes indicating that the model goes from slightly overpredicting to slightly underpredicting the wave energy. At location 23, the empirically derived energy substantially increases the magnitude of the bias. However, we note that location 23 is in a region of very energetic currents and that the model may be underpredicting the waves because wave-current interactions have been neglected in this comparison. The root mean square (rms) differences between the observations and model are similar to the bias in that the rms improves at location 01, 16, and 17, and becomes larger at locations 20 and 23 as the high-frequency energy is added to the observations. The mean wave height at the five near-shore stations during the comparison period was 59 cm and 58 cm for the model and observations, respectively.

The NDBC buoys as a group show a negative bias, indicating that the WAM model underpredicts wave heights in deep water. Buoy 42002 is a mid-Gulf deep water buoy. Buoys 42019 and 42020 are located on the western region of the Texas-Louisiana shelf near the 200 m isobath. In general, there is larger bias and rms for the NDBC buoys than the near-shore pressure gauges.

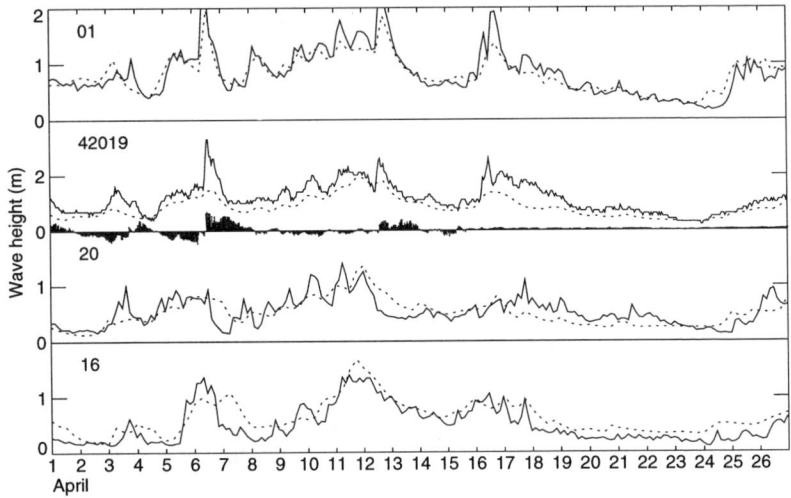

Figure 2. Observed (solid) and modeled (dot) significant wave height at locations 01 (top), 42019, 20, and 16 (bottom), during April 1994. Measured hourly wind vectors at location 19 are shown along center.

Figure 2 shows a comparison for a one month period (April 1994) of significant wave height estimated from the WAM model and the wave gauge at locations 01, 42019, 20 and 16 (reading west to east). Winds are shown in the meteorological convention (with direction from which winds are coming, north at top of page). The two time series compare quite well. The deep water buoy (42019) records a much greater mean wave height than the near-shore locations. Evident in this figure is an illustration of the tendency for the model to underpredict the wave height relative to the NDBC buoy locations and overpredict the wave height at the near-shore locations. Frontal passages are clearly seen in the wind field as impulsive events which have a corresponding increase in the significant wave height. Many of the gross scale features are seen shelf-wide, particularly the sudden increase in wave height during a frontal passage. The average spatial scale of winds over the Texas-Louisiana shelf is 270 km; however, this scale decreases somewhat during frontal passages (Wang et al. 1998). Visual comparison of modeled and observed unidirectional and directional wave spectra (not shown) at randomly selected times shows good agreement between 0.05 and 0.22 Hz.

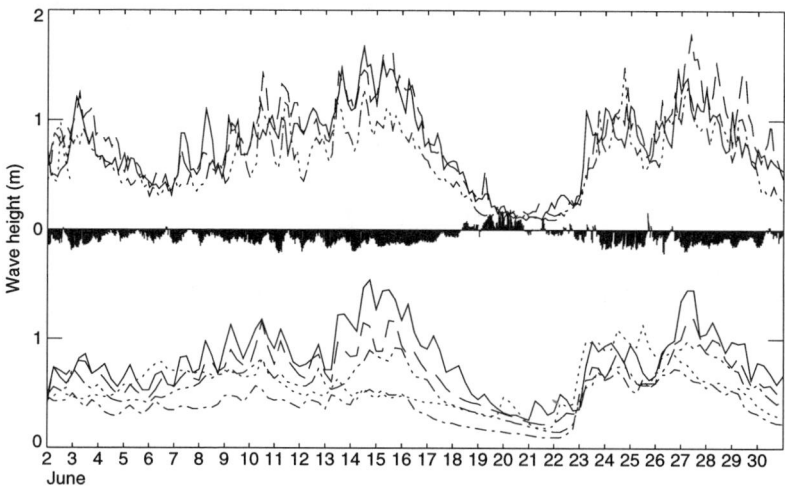

Figure 3. Observed (top) and model (bottom) significant wave height at locations 01 (solid), 23 (dash), 20 (dash-3 dot), 17 (dash-dot) and 16 (dot). Measured hourly wind vectors at location 19 are shown along center. Dates are UTC.

Figure 3 shows the modeled and observed time series of significant wave height estimated at all five near-shore locations from June 2, 1994 to June 30, 1994. This figure shows a distinct diurnal variation of the significant wave height in both the model and observations, particularly during the period of June 7 to June 16. The wind which is also superimposed in the center of this figure, also shows a diurnal variation during this period. We believe the diurnal variation of the wave height is related to fetch length, as the wave heights are reduced when the winds are seaward and increased when the winds are shoreward. In contrast to Figure 2, the large wave heights found in June 1994 are due to prolonged and steady winds from the central Gulf and not to strong and abrupt wind changes associated with a frontal passage. The significant wave height is dramatically reduced during June 19 to June 24 in both the model and observed data, when a week front turns the wind around from the north thereby decreasing the fetch length at the near-shore locations.

Howard et al. (1998) show the diurnal variation of the wind on the Texas-Louisiana shelf to be particularly strong during the summer months when frontal passage is infrequent. Figure 4 is a comparison by season of the spectral density of modeled and observed significant wave height at location 01. Here spring is defined to be March, April, May, while summer is defined to be June, July, and August, and fall is September, October, and November. The seasonal diurnal wind variation is clearly manifest in the significant wave height spectral density plots for both the model and observations as a prominent signal at 1.0 cpd during the summer. The diurnal variation of the observed wave height is not caused by tidal current-wave interaction because tidal current speeds on the shelf are small, typically less than 6 cm s^{-1} (DiMarco and Reid 1998).

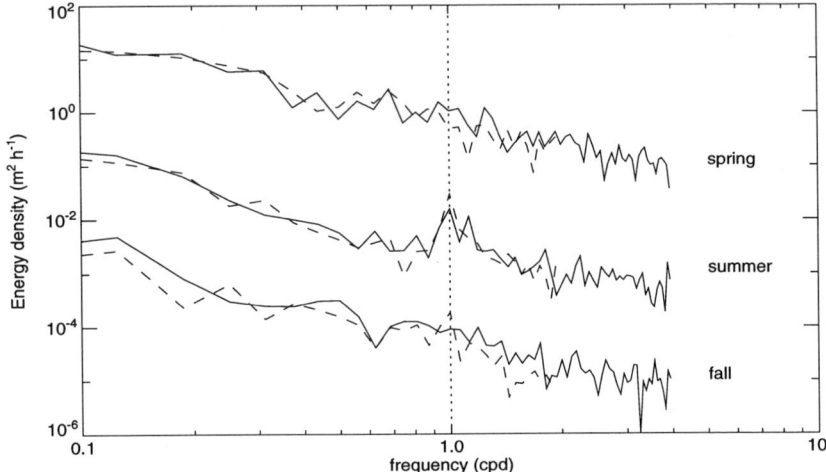

Figure 4. Spectral density of observed (solid) and modeled (broken) wave height by season during 1994 at location 01. Spectral estimates have 8-12 degrees of freedom.

Conclusions

We have shown that the WAM Cycle 4 model provides accurate estimates of the wave field over the Texas-Louisiana shelf based on comparison of significant wave height time series and spectra at five bottom mounted wave gauge and three floating buoy locations during February through November 1994. The bias (rms) difference is 0.09 m (0.23 m) and 0.26 m (0.40 m), at the near-shore and deep locations, respectively. The WAM model was also seen to accurately follow the diurnal variations in significant wave height observed during the summer months across the Texas-Louisiana shelf.

Acknowledgements

This study was funded by the U.S. Minerals Management Service under OCS contract No. 14-35-0001-30509. Additional funding has been provided by Texas A&M University, the Texas Engineering Experiment Station, and the Texas Institute of Oceanography. This report does not necessarily reflect the views or policies of the Minerals Management Service, and mention of trade names or commercial products does not constitute endorsement or recommendation by MMS.

References

Bender, L. C., Modification of the physics and numerics in a third generation ocean wave model. *J. Atm. Ocn. Tech.* *13*(3), 726-750, 1996.

DiMarco, S. F., F. J. Kelly, E. F. Childress, and D. Aubrey, Field comparison of two directional wave gauges, *MTS 94, Mar. Tech. Soc. Conf. Proc.*, Washington D.C. pp. 32-38, 1994.

DiMarco, S. F., F. J. Kelly, and N. L. Guinasso, Jr., LATEX Shelf Data Report: Mini-Spec Directional Wave Gauges. Volumes I and II. Texas A&M University, College Station, Texas. Technical Report 95-4-T. 565 pp. 1995a.

DiMarco, S. F., F. J. Kelly, J. Zhang, and N. L. Guinasso, Jr., Directional wave spectra on the Louisiana-Texas Shelf during Hurricane Andrew. *J. of Coastal Res.* Special Issue No. 21:217-233, 1995b.

DiMarco, S. F., and R. O. Reid, Characterization of the principal tidal current constituents on the Texas-Louisiana Shelf, 1998. *J. Geo. Res.* (in press).

DiMego, G. L., L. F. Bosart, and G. W. Endersen. An examination of the frequency and mean conditions surrounding frontal incursions into the Gulf of Mexico and Caribbean Sea. *Mon. Wea. Rev.*, 104:709-718, 1976.

Henry, W. K. Some aspects of the fate of cold fronts in the Gulf of Mexico. *Mon. Wea. Rev.*, 107:1078-1082, 1979.

Howard, M. K., S. F. DiMarco, and R. O. Reid, Seasonal variation of wind-driven current cycling on the Texas-Louisiana continental shelf, *EOS, Transactions, American Geophysical Union*, 79(1), OS188, 1998/Supplement.

Jochens, A. E. and W. D. Nowlin, Jr, eds. Texas-Louisiana Shelf Circulation and Transport Processes Study: Year 1 - Annual Report. Volume II: Technical Summary. OCS Study MMS-94-0030, U.S. Dept. of the Interior, Minerals Management Service, Gulf of Mexico OCS Region, New Orleans, LA. 207 pp. 1994.

WAMDI Group, the WAM Model - A third generation ocean wave prediction model, *J. Phys. Ocn.* *18*, 1775-1810, 1988.

Wang, W., W. D. Nowlin, and R. O. Reid, Analyzed surface meteorological fields over the northwestern Gulf of Mexico for 1992-1994; mean, seasonal, and monthly patterns, *Mon. Wea. Rev.* 1998. (in press)

Long-Term and Extreme Waves in the Gulf of Mexico

Chung-Chu Teng[1]

Abstract

Based on 17 years of wave data measured hourly from three National Data Buoy Center (NDBC) buoy stations in the Gulf of Mexico, long-term wave height and wave period distributions were studied. Extreme wave heights from the extratropical storms and tropical storms/hurricanes were analyzed. Relations between the extreme wave heights and corresponding wave periods were also examined.

Introduction

Long-term wave data, which are crucial for ocean engineering designs, marine operations, and wave and marine environmental studies, can be obtained from visual observations, instrumental measurements, or hindcast and forecast models. Among them, data from instrumental measurement are probably the most accurate and reliable. However, due to the complexity and high expense for setup, operation, and maintenance of measurement instruments, measured data usually have problems of insufficient coverage and long downtime for long-term purposes.

Since NDBC started wave measurement using its data buoys in 1978, a huge amount of wave data has been measured and archived. In this study, 17 years of hourly wave data measured at three NDBC buoy stations in the Gulf of Mexico were used. Distributions of long-term wave height and period were first studied. Then, extreme wave heights from the extratropical storms and tropical storms/hurricanes were examined. Finally, both the peak and average wave periods corresponding to the extreme wave heights were investigated.

Wave Data

Locations of the three buoy stations are shown in Figure 1. The western, center, and eastern buoys are designated as stations 42002, 42001, and 42003, respectively. The water depths for these three stations are 3,246, 3,200, and 3,164 m, respectively. Since these stations are in very deep water, wave data were not affected by any shallow-water effects. These buoy stations collected standard meteorological data (including wind speed/direction, air temperature, water temperature, and pressure) and wave data. These data were processed onboard the buoys, and the processed data were transmitted to the ground station through the Geostationary Operational Environmental Satellite (GOES) hourly. Once received on shore, the data were further processed and checked via both automatic and manual data quality control processes. Representative wave parameters used in the study were derived from wave spectra — significant wave height (H_s) is

---

[1]National Data Buoy Center, Building 1100, Stennis Space Center, MS 39529-6000

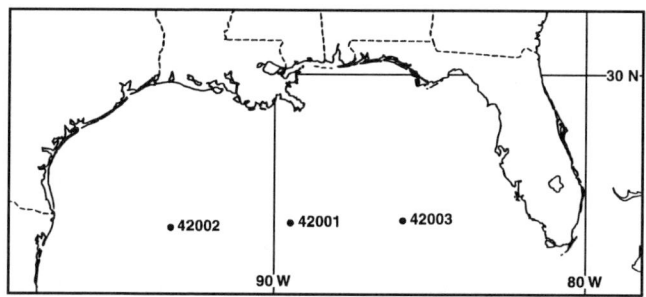

Figure 1. Locations of Buoy Stations

computed from the zeroth moment of the spectrum, peak wave period (T_p) is the period at the spectral peak, and average (or zero-crossing) wave period (T_z) is computed from the zeroth and second spectral moments.

Seventeen years (from 1980 to 1996) of hourly measured wave data were used. Table 1 presents the available data points and corresponding percentages for the stations. Due to many reasons (e.g., sensor failure, payload failure, periodic loss of data during transmission, or buoy not being on station due to refurbishment or mooring failure, etc.), data gaps or missing data exist in the data sets. For these stations, the data cover 84, 91, and 80 percent of the possible data in the 17 years. No single year had less than 60 percent of data for all three stations. Table 2 presents basic statistics of H_s, T_z, and T_p for the stations. The average wave heights and periods for the center and eastern stations (42001 and 42003) are approximately the same, and those for the western station (42002) are clearly higher and longer.

Monthly distributions of available data points in the 17 years are quite uniform for all three stations. No single month had data less than 8,600 data points (or 70 percent) in 17 years for the stations. Figure 2 shows the monthly averages and maxima of the significant wave heights for the 17 years for the three stations. Seasonal variations for the three stations are similar. Average wave heights were high between October and March,

Table 1. Available Data Points and Percentage in 17 Years

	Center Buoy 42001	Western Buoy 42002	Eastern Buoy 42003
Total data points	125,186	135,491	119,599
Percentage	84	91	80

Table 2. Statistics of Significant Wave Height and Periods

		Center Buoy 42001	Western Buoy 42002	Eastern Buoy 42003
H_s (m)	max	9.10	9.70	10.70
	mean	1.09	1.22	1.04
	s.d.	0.74	0.74	0.70
T_z (s)	max	12.20	12.40	9.70
	mean	4.74	4.92	4.79
	s.d.	0.86	0.83	0.80
T_p (s)	max	16.70	16.70	16.70
	mean	6.05	6.30	6.13
	s.d.	1.51	1.46	1.41

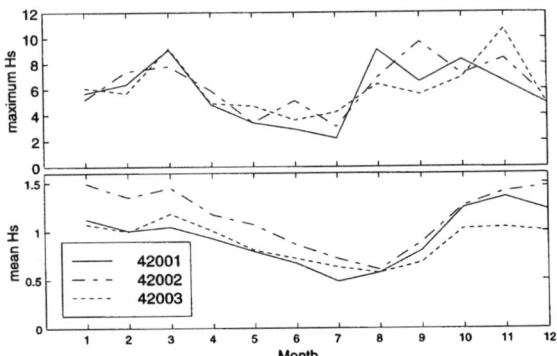

Figure 2. Monthly Distributions of (a) Maximum and (b) Mean Significant Wave Heights

and became very low in the summer. However, maximum wave heights were high in some summer months due to hurricanes. The average wave heights at the western station (42002) are always higher than those at the center and eastern stations (42001 and 42003).

Wave Height and Period Distributions

In this study, three probability distribution functions were used to fit the measured data: the log-normal, 3-parameter Weibull, and modified log-normal proposed by Fang and Hogben (1982).

- Log-Normal Distribution

$$p(x) = \frac{1}{\sqrt{2\pi}\,\sigma\,x} e^{-\frac{1}{2}\left(\frac{\ln(x) - \mu}{\sigma}\right)^2} \quad (1)$$

where μ and σ are parameters of the distribution. Based on the method of moments, these parameters were determined from the means and standard deviations (i.e., μ = mean and σ = standard deviation). Fang and Hogben (1982) proposed a modified version of the log-normal distribution that multiplies the original version by a skewness factor, a function of μ and σ.

- 3-Parameter Weibull Distribution

$$p(x) = \frac{C(x - A)^{C-1}}{B^C} e^{\left[-\left(\frac{x - A}{B}\right)^C\right]} \quad (2)$$

where A, B, and C are the location, scale, and shape parameters, respectively. The parameters can be estimated using the least square method (LSM) or the maximum likelihood method (MLM).

Figure 3 shows the comparisons between histograms of measured significant wave heights (based on 17 years of hourly wave data) and the fitted probability distributions for the three stations. For each station, the left side of the figure shows the fittings of the

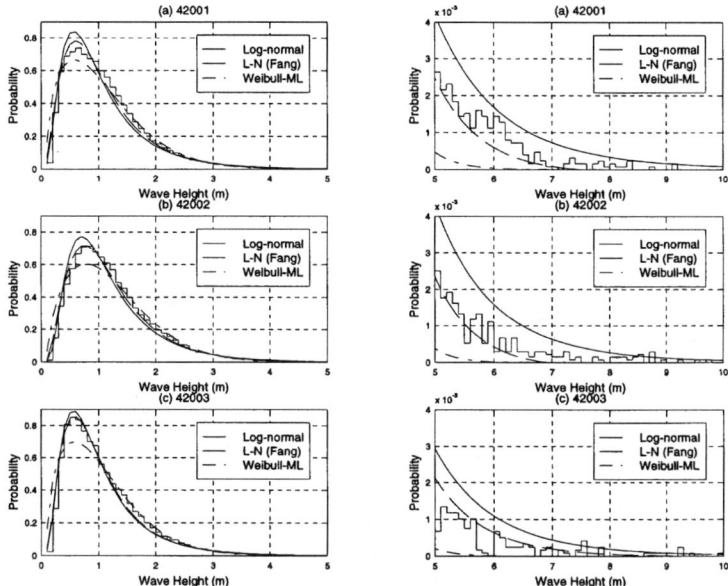

Figure 3. Comparisons Between Histograms of Measured Significant Wave Heights and the Fitted Probability Distributions for Stations (a) 42001, (b) 42002, and (c) 42003

distribution for H_s smaller than 5 m, and the right side shows a zoom-in view of the fittings for H_s greater than 5 m. From the plots on the left side, it is clear the modified log-normal distribution fits the data better than the other two distributions. The log-normal distribution always overpredicts the peak and underpredicts the higher end part right over the peak. The 3-parameter Weibull distribution underpredicts the peak, overpredicts the lower end, and predicts the higher end well. This result is similar to that presented by Teng and Palao (1996) for buoy stations in the northeastern Pacific Ocean. As shown in the plots on the right side, the distributions of the very high wave heights (i.e., extreme values) are very irregular so that none of the three distributions can fit the data. The extreme values of the wave height will be further examined in the following section.

Figure 4 shows the comparisons between histograms of measured average wave periods and the fitted probability distributions for station 42001. The results of fitting for the other two stations are the same and, thus, are not presented to save space. It was found that both the original and modified log-normal distributions fit the data well for all three stations. The modified log-normal distribution is a little better near the peak, but is slightly worse on the two sides of the peak. The difference between the two is very small and negligible. As shown in the figure, the fitting for the 3-parameter Weibull distribution is worse.

Extreme Waves

Extreme waves in the Gulf of Mexico can be generated by either extratropical storms or tropical storms/hurricanes. As recommended by the International Association for Hydraulic Research's (IAHR) Working Group on Extreme Wave Statistics (Mathiesen, et al., 1994) and Goda (1998), it is better to analyze extreme waves from these two meteorological systems separately because they may be from different populations.

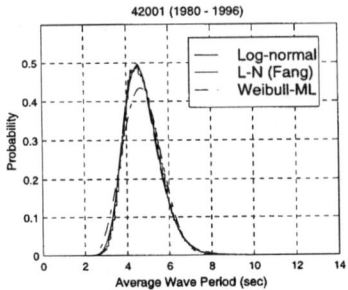

Figure 4. Histogram of Measured Average Wave Periods for Station 42001

Extreme values are identified based on the peak-over-threshold (POT) method using a threshold value of 4 m. To ensure statistical independence between extreme values, a 72-hour separation between two extreme values was imposed to make sure no two extreme values are from the same storm. After determining the extreme values based on the POT method, the extreme values generated by the hurricanes/tropical storms were identified based on the storm history and track maps officially published by National Oceanic and Atmospheric Administration's (NOAA) National Hurricane Center (that can be obtained from its Web site). Other extreme values are categorized as generated by extratropical storms. The top of Table 3 presents amounts of all extreme values, extreme values from hurricanes/tropical storms, and extreme values from extratropical storms for the three stations. Extreme wave height values, together with the date/time of occurrence, from hurricanes/tropical storms are listed in the middle of the table. Since too many

Table 3. Exteme Values From Both Extratropical Storms and Hurricanes/Tropical Storms for the Three Stations

	Center Buoy 42001			Western Buoy 42002			Eastern Buoy 42003		
Total Extremes	65			104			48		
Extremes From Hurricanes/ Tropical Storms	11			9			8		
Extremes From Extratropical Storms	54			95			40		
	Value (m)	Date	Time (UTC)	Value (m)	Date	Time (UTC)	Value (m)	Date	Time (UTC)
Extreme values and corresponding date/time from hurricanes and tropical storms	9.1	08/08/80	1700	9.7	09/16/88	0800	10.7	11/20/85	1700
	8.3	10/04/95	0400	8.4	11/13/80	1400	7.8	11/22/88	1300
	6.8	10/29/85	1300	7.2	10/29/85	0000	6.9	10/04/95	1300
	6.6	11/13/80	0100	6.9	08/08/80	2300	6.4	08/25/92	0100
	6.6	09/16/88	1500	6.1	10/15/95	0400	6.1	11/11/80	1400
	6.3	08/14/85	1700	6.1	10/06/96	2300	5.6	09/15/88	0200
	6.3	11/21/85	1900	5.1	06/19/93	0800	4.7	11/01/85	0000
	5.2	10/07/96	1900	5.1	10/15/95	0400	4.2	07/02/94	1800
	4.5	10/15/95	1400	4.7	10/15/89	0100			
	4.4	08/25/92	1200						
	4.2	11/22/88	1500						
Highest extreme values and corresponding date/time from extratropical storms	9.1	03/13/93	1100	7.8	03/13/93	0900	9.2	03/13/93	1800
	6.3	02/27/83	0200	7.4	02/27/83	0900	6.1	01/14/82	2000
	6.3	03/07/87	1200	6.1	02/27/84	2300	5.6	01/04/94	0200
	5.8	03/29/84	0600	5.9	04/14/80	0200	5.3	03/21/96	0400
	5.7	01/04/94	0100	5.7	02/11/81	1700	5.2	03/29/84	1100

extreme wave heights from extratropical storms existed, only the highest five extreme waves are listed in the bottom part of the table. As expected, most extreme values from extratropical storms identified here occurred between October and April. From this table, it can be seen that the very extreme waves generated by the tropical storms are higher than those generated by the extratropical storms. The highest waves from the extratropical storms for the three stations were all generated by the March 1993 Superstorm (also called Storm of the Century by some people). Details of this storm can be found in Gilhousen (1994).

Based on the study conducted by Teng (1997) and the recommendation by the IAHR group (Mathiesen, et al., 1994), the 3-parameter Weibull distribution (i.e., Eq. (2)) was used for the extreme wave analysis in this study. Parameters of the distribution were estimated by the LSM and the MLM. The three parameters in the Weibull distribution were determined simultaneously using an iterative numerical routine. This is different from the technique suggested by Goda (1998) that sets the shape parameter to a fixed value and solves for the other two parameters. To evaluate the fitting and parameters quantitatively, four goodness-of-fit statistics were used — coefficient of determination, mean square error, Cramer-von Mises, and Kolmogorov-Smirnov. Due to the data gaps and missing data, portions of the time series that are gaps or missing were not included in the effective data record length, as recommended by the IAHR group (Mathiesen, et al., 1994). Although extensive study on extreme wave analysis was conducted, only selected and concise results are presented in the paper due to the space limitation.

In this study, 100-year return wave heights were obtained from various extreme value data sets extreme values from extratropical storms, extreme values from tropical storms and hurricanes, all the extreme values (i.e., extratropical storms plus tropical storms/hurricanes), and extreme values from the annual maximum method. The results are summarized in Table 4.

For the extratropical storms, values of 100-year return wave heights for all three stations are about the same as or smaller than the highest measured wave height from the extratropical storms. Figure 5 shows the fittings of data to the 3-parameter Weibull distribution. Since the highest wave heights are much higher than the second highest wave

Table 4. 100-Year Return Wave Heights (Based on the Weibull Distribution)

	42001		42002		42003	
	MLM	LSM	MLM	LSM	MLM	LSM
Extratropical Storms	9.15	8.67	7.90	7.52	8.17	7.62
Tropical Storms/Hurricanes	11.11	12.90	11.15	12.22	11.56	12.26
All Extreme Data	11.86	11.94	10.09	9.63	10.93	10.26
Annual Maximum Method	10.63	9.99	10.64	10.84	10.93	10.38

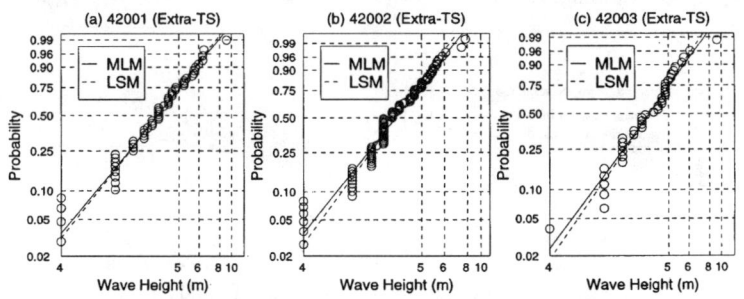

Figure 5. Weibull Distribution Fits for Exteme Values From Extratropical Storms for Stations (a) 42001, (b) 42002, and (c) 42003

heights (can also be seen in Table 3), the highest values are below the fitted lines, which may cause an underestimate of the return values. The highest values of the three stations look and act as outliers for the fittings. However, they are real (i.e., all were generated by the March 1993 Superstorm) and, statistically, are not outliers based on the outlier detection technique proposed by Goda (1998). Since 100-year return wave heights based on extreme values from extratropical storms are even smaller than the maximum measured waves, it is not appropriate or statistically reasonable to use them for engineering or operational applications.

Since tropical storms and hurricanes are rare events, 17 years of wave data still cannot provide enough extreme values (as shown in Table 3) to conduct a reliable extreme wave analysis. Although 100-year return wave heights look reasonable and consistent between stations, they can only be used as reference. Note the amounts of extreme values from the hurricanes/tropical storms only increase slightly when the threshold value was lowered from 4 to 3 m (i.e., from 11, 9, and 8 points to 13, 9, 13 points for the three stations), which are still not enough for a reliable analysis. Due to lack of long-term data on these rare events, alternative approaches of determining extreme waves, such as results from hindcasting (e.g., Haring and Heideman, 1980), should be considered. Despite that they may be from different extreme value populations, all of the extreme values from the POT method (including data from both extratropical storms and tropical storms/hurricanes) were also used to conduct an extreme wave analysis. Compared with the results from the extratropical storms, the fittings are a little better. However, the highest values are still below the fitted line (especially for stations 42002 and 42003), which may still cause an underestimate. Although the IAHR group (Mathiesen, et al., 1994) and other studies (e.g., Teng, 1997) show the annual maximum method is not appropriate for selecting data for extreme wave analysis, the IAHR group recommended the method be used when the fit based on the POT method is poor. Based on the recommendation, an extreme wave analysis using the annual maximum method was also conducted in this study. The fittings are similar to those shown in Teng (1997) for annual maxima from the northeastern Pacific Ocean, and the corresponding 100-year return wave heights are shown in Table 4 for reference.

For the wave periods corresponding to the extreme significant wave heights, Eq. 3 (in which T_i represents either T_p or T_z) with two empirical coefficients a and b is used.

$$T_i = a \cdot H_s^b \qquad (3)$$

Figure 6 shows scatterplots between the significant wave heights and wave periods (both T_p and T_z) for all H_s greater than 5 m for all three stations. The solid lines are the lower limits proposed by Teng, et al. (1994) (i.e., $a = 3.62$ and $b = 0.5$ for H_s versus T_p; $a = 3.28$

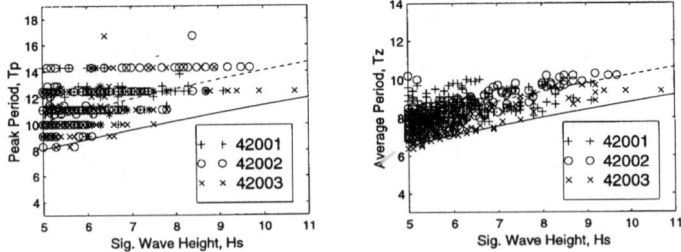

Figure 6. (a) Peak and (b) Average Wave Periods Versus Significant Wave Heights for the Three Stations

and $b = 0.43$ for H_s versus T_z). Evidently, these limits are also valid for extreme values in the Gulf of Mexico. The figure shows the wave periods at the eastern station (42002) are slightly higher than those at the center and western stations (42001 and 42003) for H_s greater than 8 m. In addition, for H_s greater than 8 m, the peak periods are always greater than 12 seconds for all three stations. The dotted lines in the plots are the regression lines based on the peak values from all three stations, with the coefficients $a = 4.85$ and $b = 0.46$ for T_p and $a = 4.04$ and $b = 0.40$ for T_z. Values of coefficient a are higher if only the western station is considered, and are lower if the center and eastern stations are considered.

Conclusions

The mean wave heights at the western location are higher, and the mean wave periods are longer than the center and eastern locations. The modified log-normal distribution fits the long-term significant wave heights better than the log-normal and Weibull distributions. Due to the rare extreme events in this area, the very high value end of the wave height distributions is irregular. Both the log-normal distribution and the modified version fit the average period well.

Both 100-year return wave heights and the highest measured wave heights (from the March 1993 Superstorm) from extratropical storms are smaller than the highest measured heights from hurricanes. In this area, extreme waves from hurricanes form the population to consider for design and engineering application. Relations between the wave periods and the extreme wave heights were proposed. Wave periods corresponding to extreme wave heights at the western station are longer than those at the center and eastern stations.

Appendix of References

Fang, Z.S., and Hogben, N., 1982, "Analysis and Prediction of Long-Term Probability Distributions of Wave Heights and Periods," Technical Report, National Maritime Institute, London.

Gilhousen, D.B., 1994, "The Value of NDBC Observations during March 1993's Storm of the Century," Weather and Forecasting, Vol. 9, No. 2, pp. 255-264.

Goda, Y., 1998, "Statistical Analysis of Extreme Waves," Chapter 11, Random Seas and Design of Maritime Structures (Second Edition), World Scientific (in press).

Haring, R.E., and Heideman, J.C., 1980, "Gulf of Mexico Rare Wave Return Period," Journal of Petroleum Technology, January, pp. 35-47.

Mathiesen, M., Goda, Y., Hawkes, P., Mansard, E., Martin, J., Peltier, E., Thompson, E., and Van Vledder, G., 1994, "Recommended Practice for Extreme Wave Analysis," Journal of Hydraulic Research, Vol. 32, No. 6, pp. 803-814.

Teng, C.C., 1997, "Design Wave Heights Estimated from Long-Term Measurements," Third International Symposium on Ocean Wave Measurement and Analysis, Virginia Beach, VA, November (in press).

Teng, C.C., and Palao, I., 1996, "Wave Height and Period Distributions from Long-term Wave Measurement," 25th International Conference on Coastal Engineering, Orlando, FL, September, pp. 368-379.

Teng, C.C., Timpe, G., and Palao, I., 1994, "The Development of Design Waves and Wave Spectra for Use in Ocean Structure Design," Transactions, Society of Naval Architects and Marine Engineers, Vol. 102, pp. 475-499.

Parametric Characterization of Surface Wave Data

J.M. Niedzwecki[1], F. ASCE
J.W. van de Lindt[2], S.M. ASCE

Abstract

There has been increasing interest on the part of practitioners and engineering researchers to develop better parametric characterizations of surface waves and their kinematics in an effort to reconcile field measurements and waves generated in model basin studies. The many nonlinear features of individual waves and the wave fields in localized regions observed in field studies, have raised further questions regarding our collective understanding of this natural phenomena. Better understanding and description of aspects, such as, wave forms, dispersion and general behavior in long and short crested seas are of great interest for science and engineering studies. Presently there is a variety of parameters that have been developed to characterize both the global and more detailed local scale phenomena. In this study the primary emphasis is placed on global scale characterization of the wave field. A dimensionless spectral peakedness parameter that relates measured data to a spectrum model selected by the designer is investigated. This parameter provides one measure of the appropriateness of the model selected to represent the data. Single and double peaked wave elevation spectrum models are used to illustrate the behavior of the dimensionless spectral peakedness parameter.

Discussion

There have been numerous research contributions made over the years by investigators studying the basic phenomena leading to the engineering characterization of ocean surface waves. To a large extent these contributions are contained in technical books on the subject, which include for example those by Kinsman (1984), Goda (1975), Ochi (1990) and, Dean and Dalrymple (1991). In

---

[1] Wofford Cain'13 Professor, Department of Civil Engineering, Texas A&M University College Station, TX 77843-3136
[2] Graduate Student, , Department of Civil Engineering, Texas A&M University College Station, TX 77843-3136

the process of designing deepwater platforms, design waves and linear spectral analysis techniques have been used extensively. Methods have been proposed and used that allow one to estimate the velocity and acceleration components in wave crests by effectively extending linear theory. Gudmestad (1993) presented an extensive investigation of deepwater wave kinematics modeling and provides a practitioners perspective on the effectiveness of current modeling procedures. The introduction of a platform into the wave field leads to a wide range of interesting wave-platform interaction problems including the platform dynamic response behavior. The coupled response behavior of offshore platforms with the extreme behavior of ocean surface waves and their kinematics was the topic of a recent OTRC sponsored industry workshop (Seymour 1996). In that workshop an effort to relate platform response events of interest, to offshore engineers, with key parameters characterizing the wave field was discussed. A more general discussion that includes many of the parameters discussed at the workshop can be found in the article by Niedzwecki, van de Lindt, and Sandt (1998). In that study a dimensionless form of Goda's (1985) spectral peakedness parameter was introduced. It was defined as the ratio of spectral peakedness of measured data to that of a spectrum model selected to model the seas for design purposes. A plot of the dimensionless spectral peakedness parameter for single-peaked JONSWAP and a two-peaked Ochi-Hubble spectrum is presented in Figure 1. When the model selected directly corresponds to the data the dimensionless spectral peakedness parameter has a value of unity.

The data sets used in this study were generated using a JONSWAP wave spectrum model with peakedness parameters varying between the limits shown on the figure. First consider a case when the JONSWAP spectral peakedness parameter has a value of unity and the model selected is the Pierson-Moskowitz wave spectrum. This case corresponds to the middle curve in Figure 1. In this case, a value of unity only occurs when the data corresponds to the Pierson-Moskowitz wave spectrum. The dimensionless spectral peakedness increases in value rapidly as the intensity of the simulated storm data increases. Next, suppose that one selects the average or mean JONSWAP wave spectrum model identified by the numeric value of 3.3 in Figure 1. Again it can be observed that only when the data corresponds precisely to the mean JONSWAP parameters does the dimensionless ratio take on a value of unity. Note that the slope of this curve is much steeper, indicating that the general model has features similar to the data. This is in sharp contrast to the previous case, where the slope is much less steep. Finally, consider the case when the data is single-peaked but the designer selects a double-peaked Ochi-Hubble model. The resulting variation of the spectral peakedness parameter is presented as the rightmost curve in Figure 1. As expected the data and the model never correspond to one another and so the parameter never actually becomes unity. However, for the limit when the data approaches the Pierson-Moskowitz wave spectrum limit, the secondary peak prevents the dimensionless spectral peakedness parameter from becoming unity. Further, the slope of the curve indicates a general mismatch between the data and the model selected.

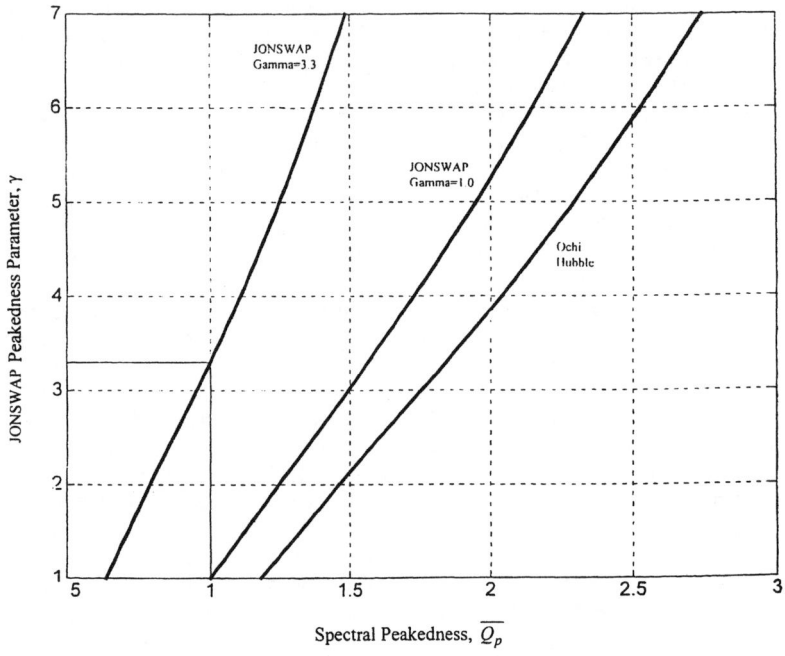

Figure 1: Variation of the dimensionless spectral peakedness parameter, $\overline{Q_p}$

Conclusion

The use of the dimensionless spectral peakedness parameter for differentiating between data and pre-selected spectral models was presented. It was found that a general mismatch between the data and a design model selected could be inferred from the slope of the curves. This parameter seems reasonably promising but it should be used in conjunction with other information about the data. Some other useful parameters have been developed and will be discussed at the meeting.

Acknowledgements

The writers gratefully acknowledge the partial support of the Offshore Technology Research Center, National Science Foundation Engineering Research Centers Program Grant Number CDR--8721512 during this study. The senior writer would also like to acknowledge the partial support of ONR Grant No. N00014-93-1-0620.

References

Dean, R.G. and Dalrymple, R.A. (1991). *Water Wave Mechanics for Engineers and Scientists*, World Scientific, Singapore.

Goda, Y. (1985). *Random Seas and Design of Maritime Structures*, University of Tokyo Press, Japan.

Gudmestad O.T., (1993). Measured and Predicted Deep Water Wave Kinematics in Regular and Irregular Seas, *Marine Structures*, Vol. 6, 1-73.

Kinsman, B. (1984). *Wind Wave: Their Generation and Propagation on the Ocean Surface*, Dover Press, New York.

Ochi, M.K. (1990). *Applied Probability & Stochastic Processes*, Wiley Interscience, New York.

Extreme Ocean Wave Characteristics Workshop, ed. Seymour, R.J. (1996). Offshore Technology Research Center Workshop Report, Houston, Texas, Nov. 1996.

Niedzwecki, J.M., van de Lindt, J.W. and Sandt, E.W. (1998). "Characterizing Random Wave Surface Elevation Data," *Ocean Engineering*, in press.

A Global Data Base for Wind Speed and Wave Height

I.R. Young[1]

Abstract

Mean monthly values of altimeter wind speed and wave height are compared with data from NDBC buoys. As a result of these comparisons, corrections are made to the raw data products available from these satellites. Data from the GEOSAT, TOPEX and ERS1 missions, corrected in this fashion are used to show that there have been no measurable changes in the global wind and wave climate during the ten years spanned by these various missions. It is proposed that the corrected values of wind speed and wave height provide the basis for the formation of a long term global data base which spans the periods of these multiple missions.

Introduction

Global values of wind speed and wave height can be obtained from satellite based radar altimeters. Such data are invaluable for applications including long term climate studies and engineering design. As satellites generally have a relatively short life (less than 5 years), the long term data set must be obtained from a series of satellites. As a result, it is important to ensure that each satellite altimeter is accurately calibrated. This study follows the approach of Cotton and Carter (1994) and compares mean monthly altimeter wave height and wind speed values with buoy data.

Satellite Data

Satellite data were compiled for the GEOSAT, TOPEX and ERS1 altimeters. Data from the following periods were utilized:

- GEOSAT (Geophysical Data Records): November 1986 - January 1990
- TOPEX (Merged Geophysical Data Records): September 1992 - October 1995

---
[1]School of Civil Engineering, University College, UNSW, Canberra, ACT 2600, Australia

- ERS1 (CERSAT OPR Data Products): August 1991 - August 1995

Buoy Data

Data from a total of 16 NDBC buoys were analyzed. Buoys were selected such that neither the buoy data nor the satellite data, from a corresponding 4° × 4° square, would be influenced by proximity to shore. The buoy data were partitioned into periods corresponding to each of the three satellite periods and monthly means determined for each of these periods. All buoys had sufficiently long records to produce statistically reliable estimates of mean monthly values of U_{10} and H_s corresponding to each of the satellite periods.

Satellite-Buoy Data Comparisons

Cotton and Carter (1994) have shown that the comparison of buoy and satellite mean monthly values is a valid method for the calibration of satellite altimeter values of H_s. The use of mean monthly values greatly increases the available data as colocated measurements in both space and time are not required.

Figure 1 shows a comparison between buoy mean month values of H_s and the GEOSAT mean monthly values. A total of 203 points were available from the 16 buoys. A least squares linear regression yields the result

$$H_s(\text{buoy}) = 1.144 H_s(\text{GEOSAT}) - 0.148 \qquad (1)$$

This result is shown in Figure 1. Carter et al. (1992) compared GEOSAT values of H_s with data from 13 NDBC buoys and concluded that GEOSAT values were consistently low by 13% (i.e. $H_s(\text{buoy}) = 1.13 H_s(\text{GEOSAT})$). This result is also shown in Figure 1 and is clearly in close agreement with the present data as represented by (1). This agreement adds confidence that the use of mean month values, together with a satellite averaging region of 4° × 4° yield reliable results.

Figure 2 shows the comparison between buoy mean monthly values of H_s and the TOPEX mean monthly values. Based on the 192 available values a linear regression yields

$$H_s(\text{buoy}) = 1.067 H_s(\text{TOPEX}) - 0.079 \qquad (2)$$

As shown in Figure 2 this result is in excellent agreement with the result obtained by Cotton and Carter (1994), based on a smaller data set (i.e. $H_s(\text{buoy}) = 1.089 H_s(\text{TOPEX}) - 0.172$).

The ERS1 intercomparison is shown in Figure 3. Based on the 192 points in the intercomparison a linear regression yields

$$H_s(\text{buoy}) = 1.243 H_s(\text{ERS1}) + 0.040 \qquad (3)$$

This results is again in very good agreement with the result obtained by Cotton and Carter (1994) (i.e. $H_s(\text{buoy}) = 1.267 H_s(\text{ERS1}) + 0.107$).

Figure 1 shows the comparison between buoy mean monthly values of U_{10} and GEOSAT values. A linear regression yields the result (196 points)

$$U_{10}(\text{buoy}) = 0.874 U_{10}(\text{GEOSAT}) + 0.337 \qquad (4)$$

Forcing the regression through zero yields $U_{10}(\text{buoy}) = 0.913 U_{10}(\text{GEOSAT})$, apparently indicating that buoy data is approximately 9% lower than the GEOSAT observations. Such a result appears at variance with the large number of validations performed for GEOSAT wind speed data. Indeed, the relationship between altimeter σ_0 and U_{10} utilized in this study (Chelton and Wentz, 1986) was developed for GEOSAT data.

The vast majority of the buoy data available during the GEOSAT mission were obtained with the GSBP buoy sensor package. The GSBP package determined wind speeds using a vector average, whilst later sensor packages (i.e. those available during the TOPEX and ERS1 missions) used a simple scalar average. Gilhousen (1987) has reported that NDBC buoy wind speeds calculated using the GSBP vector averaging method are approximately 7% low. A similar result has been reported by Gower (1996). Hence, (4) should be treated with caution, the apparent difference between GEOSAT values of U_{10} and buoy results probably being due to the manner in which the buoy wind speed was determined.

Figure 2 shows the comparison between TOPEX mean monthly values of U_{10} and buoy values (190 points). The TOPEX values of U_{10} are clearly lower than the buoy data, consistent with the recommendation by Callahan et al. (1994) that TOPEX values of σ_0 should be reduced by 0.7dB. A linear regression to the data of Figure 2 gives

$$U_{10}(\text{buoy}) = 0.943 U_{10}(\text{TOPEX}) + 1.847 \qquad (5)$$

Figure 3 shows the intercomparison between mean monthly buoy values of U_{10} and ERS1 values. A linear regression based on 192 points gives

$$U_{10}(\text{buoy}) = 0.849 U_{10}(\text{ERS1}) + 1.217 \qquad (6)$$

Close inspection of the data shows that the large zero offset in (6) is largely a result of the scattered values above approximately 10 m/s. If these values are excluded from the fit and the regression is forced to pass through zero, an acceptable fit to the data can be achieved with the result

$$U_{10}(\text{buoy}) = 0.998 U_{10}(\text{ERS1}) \qquad (7)$$

Equation (7) indicates that the ERS1 values of U_{10} are unbiased.

Satellite-Satellite Data Comparisons

Adopting the calibration relationships developed above, it is possible to carry out cross-validations between the satellite missions. Examining all the 4° × 4°

squares on the Earth's surface, a total of 15,095 were identified which had sufficient passes of the respective satellites to yield reliable mean monthly values of H_s. Equations (1), (2) and (3) were applied to the respective raw values of the satellite H_s and the mean monthly values determined. The results are shown in Figure 4 (GEOSAT Vs. TOPEX), Figure 5 (GEOSAT Vs. ERS1) and Figure 6 (ERS1 Vs. TOPEX). Linear regression for these data sets yield the results: $H_s(\text{TOPEX}) = 0.978 H_s(\text{GEOSAT}) + 0.158$, $H_s(\text{ERS1}) = 0.998 H_s(\text{GEOSAT}) + 0.046$ and $H_s(\text{TOPEX}) = 0.980 H_s(\text{ERS1}) + 0.111$. These results indicate that for all practical purposes the three satellite missions have recorded the same global average wave conditions. This is not surprising for TOPEX and ERS1 as the mission periods analyzed largely overlap. There is, however, no overlap between these satellites and GEOSAT. Hence, the present results indicate that there has been no measurable change in the global wave field over the period of time spanned by the satellite missions. This is consistent with the conclusions of Cotton and Carter (1994). Not surprisingly, the comparisons between GEOSAT and TOPEX, and GEOSAT and ERS1 have greater scatter than between ERS1 and TOPEX since the latter satellites largely overlap.

A similar intercomparison can be carried out between satellite values of U_{10}. As the global mean monthly wave fields measured by the various satellites are essentially the same, the global mean monthly wind fields could also be expected to be the same. As altimeter measurements of U_{10} are less reliable than altimeter measurements of H_s (Young and Holland, 1996), the global 4° × 4° data set is smaller, consisting of 11,457 points. The GEOSAT U_{10} data were assumed to be unbiased. Equations (5) and (7) were used to "correct" the TOPEX and ERS1 U_{10} data respective, and global mean monthly values were determined. As expected, ERS1 and GEOSAT U_{10} data agreed well. Correlations of both ERS1 and GEOSAT values of U_{10} with TOPEX values showed a clear difference. As TOPEX and ERS1 overlapped, it would be surprising if such a difference really existed. Also, the more reliable H_s data showed no such trend. This leads to the possible conclusion that the correction to the TOPEX values of U_{10} represented by (5) may be in error. An alternative fit between the buoy and TOPEX mean monthly values of U_{10} was investigated which would remove this bias between the TOPEX values of U_{10} and the other satellites. A relationship which achieves this is

$$U_{10}(\text{buoy}) = 0.99 U_{10}(\text{TOPEX}) + 1.61 \tag{8}$$

Equation (8) is shown in Figure 2 and provides a fit which is visually equal to (5).

The final intercomparisons between satellite values of U_{10} using (7) and (8) are shown in Figure 4 (GEOSAT Vs. TOPEX), Figure 5 (GEOSAT Vs. ERS1) and Figure 6 (ERS1 Vs. TOPEX). Linear regression for these data yield the results: $U_{10}(\text{TOPEX}) = 1.004 U_{10}(\text{GEOSAT}) - 0.117$, $U_{10}(\text{ERS1}) = 0.999 U_{10}(\text{GEOSAT}) - 0.233$ and $U_{10}(\text{TOPEX}) = 1.006 U_{10}(\text{ERS1}) + 0.111$.

Conclusions

Mean monthly values of H_s and U_{10} obtained from altimeters on GEOSAT, TOPEX and ERS1 have been compared with NDBC buoy data. As a result of these comparisons and the requirement that intercomparisons of H_s and U_{10} between the satellites must yield consistent results, calibration relationships have been developed for each of the satellites. The recommended equations are:

$$H_s = 1.144 H_s(\text{GEOSAT}) - 0.148 \qquad (9)$$
$$H_s = 1.067 H_s(\text{TOPEX}) - 0.079 \qquad (10)$$
$$H_s = 1.243 H_s(\text{ERS1}) + 0.040 \qquad (11)$$
$$U_{10} = U_{10}(\text{GEOSAT}) \qquad (12)$$
$$U_{10} = 0.99 U_{10}(\text{TOPEX}) + 1.61 \qquad (13)$$
$$U_{10} = U_{10}(\text{ERS1}) \qquad (14)$$

These results extend and refine previous comparisons between the altimeter data sets and buoy data. They provide the basis for the development of long term data series of both wind speed and wave height from the multiple altimeter missions.

References

Callahan, P.S., C.S. Morris, and S.V. Hsiao, 1994, "Comparison of TOPEX/POSEIDON σ_0 and significant wave height distributions to Geosat", *J. Geophys. Res., 99*, 25,015-25,024.

Carter, D.J.T., 1992, "Wave height and wind speed from satellite radar altimeters", in *Guide to Satellite Remote Sensing of the Marine Environment, IOC Manual and Guides*, 24, 33-81, UNESCO, Paris.

Chelton, D.B., and F.J. Wentz, 1986, "Further development of an improved altimeter wind speed algorithm", *J. Geophys. Res., 91*, 14,250-14,260.

Cotton, P.D., and D.J.T. Carter, 1994, "Cross calibration of TOPEX, ERS1, and Geosat wave heights", *J. Geophys. Res., 99*, 25,025-25,033.

Gilhousen, D.B., 1987, "A field evaluation of NDBC moored buoy winds", *J. Atmos. and Ocean Tech., 4*, 94-104.

Gower, J.F.R., 1996, "Intercalibration of wave and wind data from TOPEX/POSEIDON and moored buoys off the west coast of Canada", *J. Geophys. Res., 101*, 3817-3829.

Young, I.R., 1994, "Global ocean wave statistics obtained from satellite observations", *Applied Ocean Research, 16*, 235-248.

Young, I.R. and Holland, G.J., 1996, "Atlas of the oceans: wind and wave climate", *Pergamon Press*, ISBN 0-08-042519-4, 241pp.

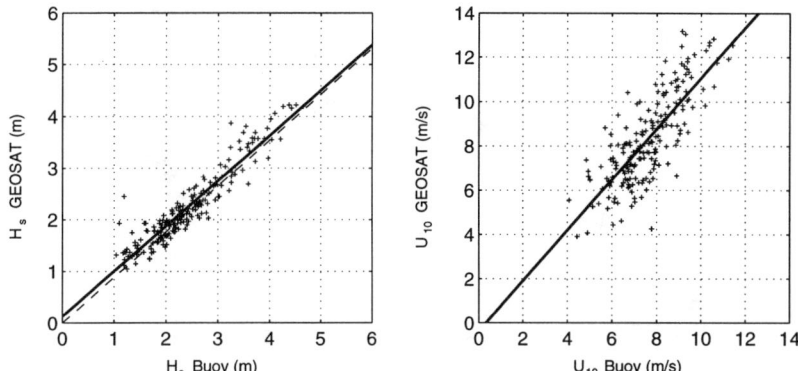

Figure 1: [Left] Mean monthly values of buoy H_s compared with mean month values of GEOSAT altimeter H_s. The solid line is (1) and the dashed is the relationship proposed by Carter et al. (1992). [Right] Mean monthly values of buoy U_{10} compared with mean month values of GEOSAT altimeter U_{10}. The solid line is (4).

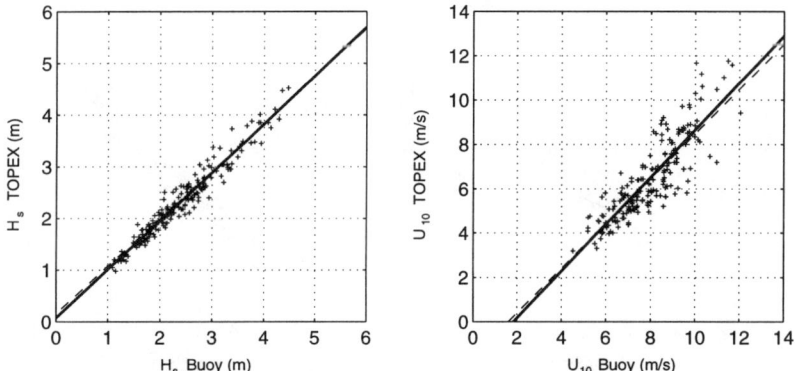

Figure 2: [Left] Mean monthly values of buoy H_s compared with mean month values of TOPEX altimeter H_s. The solid line is (2) and the dashed line is the relationship proposed by Cotton and Carter (1994). [Right] Mean monthly values of buoy U_{10} compared with mean month values of TOPEX altimeter U_{10}. The solid line is (5) and the dashed line is (8).

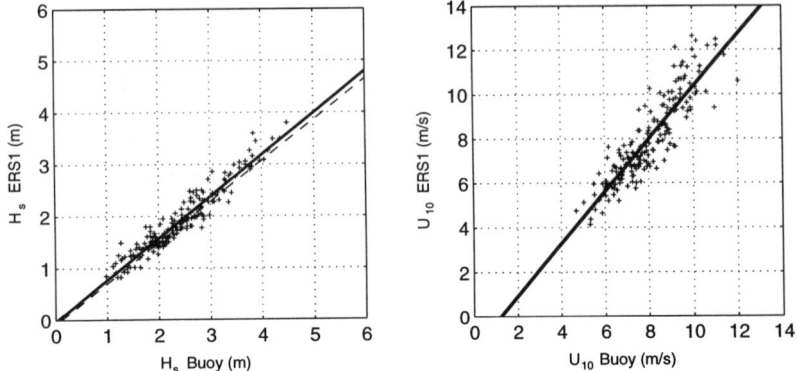

Figure 3: [Left] Mean monthly values of buoy H_s compared with mean month values of ERS1 altimeter H_s. The solid line is (3) and the dashed line is the relationship proposed by Cotton and Carter (1994). [Right] Mean monthly values of buoy U_{10} compared with mean month values of ERS1 altimeter U_{10}. The solid line is (6).

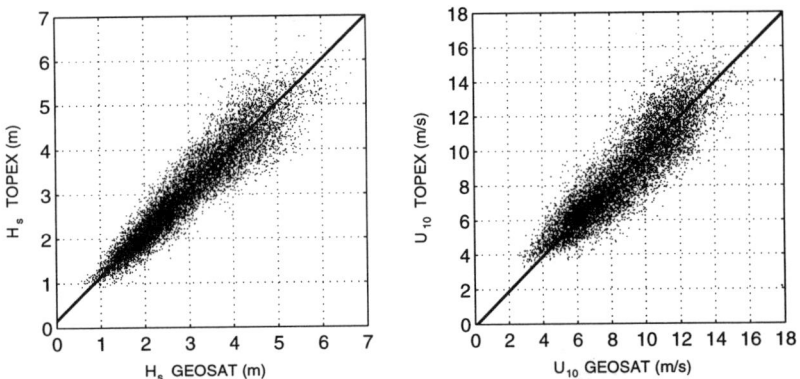

Figure 4: [Left] Global mean monthly values of corrected GEOSAT altimeter H_s compared with mean monthly values of corrected TOPEX altimeter H_s. The solid line is the regression result $H_s(\text{TOPEX}) = 0.978 H_s(\text{GEOSAT}) + 0.158$. [Right] Global mean monthly values of corrected GEOSAT altimeter U_{10} compared with mean monthly values of corrected TOPEX altimeter U_{10}. The solid line is the regression result $U_{10}(\text{TOPEX}) = 1.004 U_{10}(\text{GEOSAT}) - 0.117$.

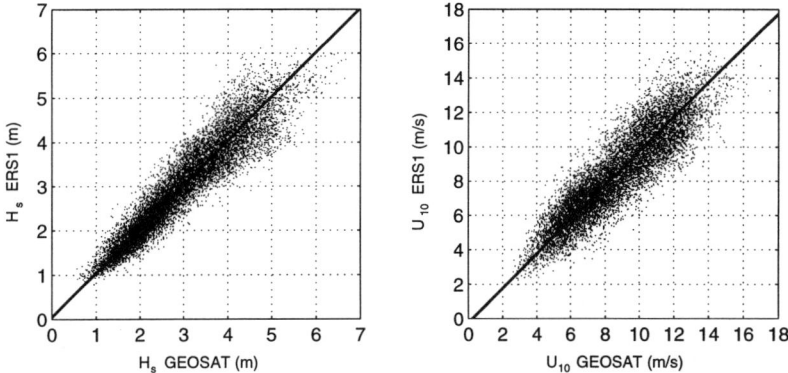

Figure 5: [Left] Global mean monthly values of corrected GEOSAT altimeter H_s compared with mean monthly values of corrected ERS1 altimeter H_s. The solid line is the regression result $H_s(\text{ERS1}) = 0.998 H_s(\text{GEOSAT}) + 0.046$. [Right] Global mean monthly values of corrected GEOSAT altimeter U_{10} compared with mean monthly values of corrected ERS1 altimeter U_{10}. The solid line is the regression result $U_{10}(\text{ERS1}) = 0.999 U_{10}(\text{GEOSAT}) - 0.233$.

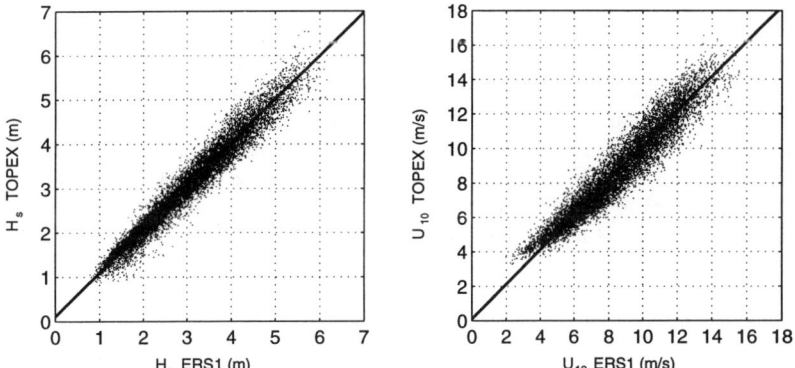

Figure 6: [Left] Global mean monthly values of corrected ERS1 altimeter H_s compared with mean monthly values of corrected TOPEX altimeter H_s. The solid line is the regression result $H_s(\text{TOPEX}) = 0.980 H_s(\text{ERS1}) + 0.112$. [Right] Global mean monthly values of corrected ERS1 altimeter U_{10} compared with mean monthly values of corrected TOPEX altimeter U_{10}. The solid line is the regression result $U_{10}(\text{TOPEX}) = 1.006 U_{10}(\text{ERS1}) + 0.111$.

Joint statistics of waves and water levels

David P. Hurdle[1] and Gerbrant Ph. van Vledder[1]

Abstract

Two studies have been performed to analyse the sensitivity of design parameters on different statistical analysis techniques. In the first study two methods for synthetic data generation of hurricane wind fields in the Gulf of Khambhat have been applied to derive the probability of exceedance of wind speed at a location in the North of the Gulf of Khambhat. The first method is based on random track replacement of historical hurricanes. The second method is an extension to the first one, in the sense that the peak wind speeds are taken from a logarithmic distribution. The simulations indicate that the inclusion of a random hurricane intensity can have significant effects on the design wind speed. In the second study an analysis was made of the sensitivity of the design deck level on various approaches of taking the joint occurrence of waves and water levels into account. In addition, the directionally sorted peaks method has been applied. The results indicate that a separate analysis of waves and water levels gives the highest results of the design deck level. Inclusion of information about the joint occurrence of waves and water levels decreases the design deck level. Besides, inclusion of directional information may lead to a further decrease of the design deck level.

Introduction

The statistical analysis of extremes of a single parameter e.g. wave height, has made much progress in recent years and standard techniques have been accepted (e.g. Van Vledder and Zitman, 1992; Van Vledder et al., 1993). For the analysis of joint extremes of several variables the situation is more complicated. There are two main obstacles to deriving the joint statistics of the extremes of

---

[1] Research engineers, ALKYON Hydraulic Consultancy & Research, PO Box 248, 8300 AE Emmeloord, The Netherlands. E-mail: hurdle@alkyon.nl & vledder@alkyon.nl.

several variables: the availability of sufficient historical data to allow analysis and the absence of any established statistical analysis methods which can generally be applied.

The lack of sufficient data can be overcome by using synthetic data generation techniques. For hurricanes such techniques have been developed by e.g. Chouinard and Liu (1993). The usual assumption with respect to hurricanes is to assume that their occurrence is random in time, independent of each other and that is they follow a Poisson process. Further, their intensity is often described by a specific probability function which is independent of space. For severe storms in the North Sea, synthetic data generation techniques have been developed by e.g. Ciesliekiwicz and Graff (1996). These techniques make use of parametric descriptions of pressure fields, such as Empirical Orthogonal Functions. Techniques to transform joint distributions have been developed by e.g. de Valk (1991). These techniques have been further developed in the NEPTUNE project.

In this paper, various of the above methods will be discussed and elaborated for a particular situation. To this end, readily available data and numerical model simulations have been applied to generate wind fields of tropical cyclones in the Gulf of Khambhat, located along the north-west coast of India. Statistical techniques are not yet generally available to derive joint statistics for engineering applications. Therefore, it is common practise to use approaches based on conservative assumptions. This paper explores such approaches and contrasts the results in a number of experiments based on measured North Sea data.

Synthetic data generation

In the NEPTUNE project methods were developed to generate large amounts of data for the purposes of statistical analysis without having to carry out large amounts of numerical modelling. In addition, methods were developed for parameterizing wind fields and a method linking the sea state conditions to the parameterised wind fields to be developed (see Ciesliekiwicz and Graff, 1996). This method allows generation of large amounts of sea state data in situations where wind and pressure data are available for an extensive and measured or hindcast sea state parameters are available for a limited period. Similarly, a parametric link was developed between the offshore and nearshore wave parameters (Van Vledder et al., 1997).

The availability of data can be considered at various levels. In shallow water, the hydraulic design conditions can vary significantly over a distance of a few kilometres because of differences in the nearshore depth profile and the presence of bank systems. Thus measurements or other data for one location can not directly be used for a nearby neighbouring location. In many situations (including that in the Netherlands), statistics of offshore conditions in water of about 20 m depth generally vary on a length scale of order of tens of kilometres (e.g. changes in the fetch length for generation of waves). This means that measurements or statistics at a limited number of offshore stations are sufficient to cover the whole offshore coastal length. However, in situations where fetch varies rapidly with location (e.g. the Gulf of Khambhat) it should be expected that offshore design wave and water level conditions vary more rapidly with position.

The statistics of storm intensity (atmospheric pressure and wind speed) and typical storm tracks over sea tend to vary on a length scale of order of hundreds of kilometres. In situations where too little data is available on a local level, the obvious approach is to look for it at a more fundamental level and to use meteorological or hydraulic modelling to obtain the required results.

In the following paragraphs the example of the situation in the Gulf of Khambhat on the north west coast of India is further developed. In particular, the design wind speed at the site is examined to illustrate the effect of taking a limited number of parameters chosen for the purposes of generating synthetic storms. The following activities have to be carried out to generate synthetic (metocean) data:

- The identification of critical parameters;
- derivation of distributions for the critical parameters. e.g. the peak wind speed U_{MAX} may follow the logarithmic distribution;.
- the identification of the interdependency of parameters and if necessary to develop multivariate distributions for them, e.g. for tropical storms there may be some relation between U_{MAX} ; the characteristic radius of the maximum wind speed R_{MAX}, and the pressure difference ΔP;
- obtain a number of samples from the parent distribution critical for the area of interest;
- for these samples use numerical models to simulate the required coastal parameters;
- carry out (single or multivariate) statistical analysis to obtain extreme values for the design parameters.

Hurricanes in the Gulf of Khambhat

Tropical storms which struck the Indian Coast between 1960 and 1996 are shown in Figure 1 (taken from the U.S. Navy Dept. of Commerce CD ROM: Global Tropical / Extratropical Cyclone Climatic Atlas). An analysis of the storm tracks has shown that only two storms relevant for wave and water level conditions passed over the Gulf of Khambhat in this period. This is certainly not enough for the purposes of a reliable statistical analysis.

The critical parameters for tropical storms are the maximum wind (U_{MAX}), storm radius (R_{MAX}, the radius of the maximum wind), the pressure difference (ΔP) and the path of the storm. The description of the path of the storm may be split into two parts: the location of the coastal crossing point; and the track movement relative to this. For design wind speeds and wave heights, the coastal

crossing point of the storms and the value of U_{MAX} are probably the two most important parameters. For the present study, synthetic storms were then generated in two ways:
- displacing the tracks of a limited number of existing storms for which information was available;
- displacing the tracks of these storms and scaling their intensity.

The distribution of the coastal crossing points for the tracks is shown in Figure 2, which also shows the theoretical distribution of storms along the coast under the hypothesis that they are uniformly distributed. This agreement is sufficiently good to accept the hypothesis. For the Gulf of Mexico, however, there are strong indications that the storm intensity is affected by local oceanographic features (Chouinard et al., 1997). An analysis of the distribution of the maximum wind speeds in the 9 storms for which these are available, shows that a logarithmic distribution is sufficiently accurate to represent the distribution of maximum wind speeds. Further, as usual, it has been assumed that the maximum wind speed and the coastal crossing point are not correlated.

Monte-Carlo simulations have been performed to generate 30 storms crossing the line CD shown in Figure 1 using the two methods mentioned above. According to the distribution of coastal crossing points this number of storms is to be expected in a period of 540 years. In the first method, a storm has been selected at random from 5 basic storms for which all required parameters are known. A coastal crossing point was selected at random and the track of the storm translated to pass over this. In the second method the same procedure has been followed, but, in addition, a value for U_{MAX} was selected at random from the logarithmic distribution for this parameter and the wind speeds in the basic storm was scaled to give this value. The wind speed at the site in the north of the Gulf of Khambhat was computed in the course of the storm using a parametric model for tropical storms as developed by Cooper (1988).

The results of these two types of simulations is summarised in Figure 3 in terms of the exceedance probability of wind speed at a location in the North of the Gulf of Khambhat. Also shown in Figure 3 is the probability of exceedance corresponding to the '100-year' wind speed. For lower values of the probability of exceedance, or similarly for higher return periods, the results of both techniques deviate considerably, with much higher values of the design wind speed for the case that the peak wind speeds have been randomised. For the present study, however, no verification data are available to confirm or contradict the present results. On the basis of studies published in literature (e.g. Chouinard and Liu, 1997), it is found that a random storm intensity should be included in a statistical parametric description of hurricanes.

Joint Statistics of waves and water levels

Various statistical techniques used to describe the combined hydraulic conditions will be demonstrated by considering the design level for a jetty deck which depends on both the extreme wave height and water level. In particular, results of the following four methods will be compared:

A. Separate computation of the n-year return value of the design wave height and water level by applying statistical extrapolation techniques, followed by a linear summation of the results;
B. application of the simultaneous peak value method in which the value of e.g. the storm surge occurring simultaneously with the peak value of the wave height, is taken as the value to be extrapolated. This technique has been applied in the NESS project (Shaw, 1992 and Van Vledder and Zitman, 1992);
C. computation of the total wave crest level above MSL through each storm (obtained by numerical modelling or parametric relationships) and statistical extrapolation of the results;
D. the directionally sorted peaks method, in which the values of the critical values are first sorted according to the wind direction at the time they are predicted before a statistical analysis is carried out. Combination of extreme values for different wind direction sectors will be carried out after statistical extrapolation.

Method A is the most conservative approach since it assumes that the peak wave height occurs simultaneously with the peak water level.

Method B involves a measure of uncertainty since it is not certain that the peak load during a storm occurs at the moment when one of the design parameters is at a maximum (i.e. it may give results that are too low). On the other hand, it may give results which are too high since the ranking of the severity of the storms is not necessarily the same for all design parameters.

Method C above requires the least assumptions to carry out statistical analysis and therefore gives a good benchmark for the other techniques. An advantage of this method is that only one environmental parameter is defined, for which a standard uni-variate statistical analysis can be applied. It has, however, the considerable disadvantage that it is not only site specific but also specific to the particular item being designed.

Method D has the disadvantage that it requires the consideration of design scenarios for several wind direction sectors. Further, the essential problems of Method B are also inherent in method D, although it is expected that by sorting the conditions into types (according to the directional sector of the wind) the problem of the ranking of the severity of the storms will be reduced.

The above four methods have been applied to a continuous time series of winds, waves and water levels of 14 year duration, measured every 3 hours near the K13 offshore platform located in the southern North Sea. These data are routinely collected by the Ministry of Transport and Public Works in the

Netherlands. For the present study these data have been statistically analysed to obtain the design deck level (DL). The deck level is defined as:

$$DL = h + 1.15 H_s \qquad (1)$$

In this equation the factor 1.15 is the product of the factors 1.9 and 0.6, in which 1.9 is the assumed ratio of the maximum individual wave height H_{max} and the significant wave height H_s, and were 0.6 is the assumed ratio of the crest level and the wave height. The factor 0.6 is a measure of the non-linearity for the design wave (for a linear sinusoidal wave this factor is 0.5). The factors 1.9 and 0.6 are not based on sophisticated wave theories, such as used by e.g. Haring and Heideman (1976). They are needed as realistic, although conservative, estimates to be able to illustrate the differences between the various analysis techniques.

The extremal analysis of the waves and water levels has been performed using the Peak over Threshold method. The three-parameter Weibull distribution has been fitted to various sets of maxima using the Least Squares Method (c.f. Van Vledder et al., 1993). The data threshold factor B has been kept constant in each fit procedure, such that only the shape factor k and the scale factor A have been estimated. By varying the data threshold factor B, the quality of the fit can be improved.

The main results of the statistical analysis are summarised in Table 1. In this table the results are shown for two return periods, 5 year and 1000 year, for the omni-directional case and for two direction sectors of 30° width and mean directions of 270° and 330°. The associated design water level (i.e. the one occurring simultaneously with the peak significant wave height) is indicated as h[Hs]. As an example, the fit of the Weibull distribution for the selected set of peak Deck Levels for the direction sector 270° is shown in Figure 4. It can be seen that the quality of the fit is quite good.

Method	Parameter	5 year			1000 year		
		270°	330°	Omni-directional	270°	330°	Omni-directional
A	Hs (m)	5.13	5.51	5.67	7.36	9.60	8.14
	h (m)	1.31	1.20	1.56	1.81	2.00	2.46
	DL (m)	7.21	7.54	8.08	10.27	13.04	11.82
B	h[Hs] (m)	1.15	1.10	1.38	1.56	2.08	2.30
	DL (m)	7.05	7.44	7.90	10.02	13.12	11.66
C	DL (m)	7.02	7.32	7.66	8.68	12.21	11.11

Table 1: Design wave height, water level and deck level computed with the methods A, B and C, for the omni-directional case and computed with the directionally sorted method, shown in the columns with the headers 270° and 330°.

For the omni-directional case as well as for the two direction sectors the results indicate that the results of method A produce the highest values of the design deck level, whereas the results of method C give the lowest design deck levels. This decrease is found for both the 5-year as for the 1000-year return period. The decrease in design deck level is expected since method C includes the maximum amount of information on the joint occurrence of waves and water levels. The relative decrease in design deck level is not very large, but this may be different when data from another location are analysed.

The results of the directionally sorted peaks method show that a further decrease of the design deck level may be obtained when the analysis is performed per direction sector. This further decrease was found for most cases, but not for the 1000-year deck level corresponding to the 330° direction sector. Inspection of the basic data leading to this result, have shown that the number of measured data used in the fit procedure is rather small and that the quality of the fit is rather poor.

Conclusions

The analysis of two methods for generating synthetic wind fields of hurricanes indicate that the inclusion of a random peak wind speed can have significant effects on the design wind speed, especially for high values of the return period. Second, the analysis of different methods of including information of the joint occurrence of waves and water levels indicate that a separate statistical analysis of waves and water levels gives the highest values for the design deck level. These values can be considered as conservative. Application of information on the joint occurrence of waves and water levels reduces the design deck level. Finally, application of the directionally sorted peaks method, may yield a further decrease of the design deck level, at least when sufficient data is taken into account.

Acknowledgements

The wind, water level and wave height data for the offshore platform K13 have been provided by the Ministry of Transport and Public Works in the Netherlands.

References

Chouinard, L.E. and C. Liu., 1993: probability of severe hurricanes across the Gulf of Mexico. Proc. 24th Offshore Technological Conf., OTC 6832, OTC, Houston, 223-234.

Chouinard, L.E., and C. Liu, 1997: Model for recurrence rate of hurricanes in Gulf of Mexico. Journal of Waterway, Port, Coastal, and Ocean Engineering, Vol. 123, No. 3. 113-119.

Chouinard, L.E., C. Liu, and C.K. Cooper, 1997: Model for severity of hurricanes in Gulf of Mexico, Journal of Waterway, Port, Coastal, and Ocean Engineering, Vol. 123, No. 3., 120-129.

Ciesliekiwicz, W. and J. Graff, 1996: Sea state parameterization using empirical orthogonal functions, Proc. 25th Int. Conf on Coastal Engineering.

Cooper, C.K., 1988: Parametric Models of Hurricane Generated Winds, Waves and Currents in Deep Water. Offshore Technology Conference, Houston, Texas. OTC 5838, May 1988.

Graff, J., and W. Ciesliekiwicz, 1996: NEPTUNE - An integrated approach to determining NW European coastal extremes. Proc. 25th Int. Conf. on Coastal Engineering. 1375-1388.

Haring, R. C., and J.C. Heideman, 1978: Gulf of Mexico rare wave return periods. Proc. 10th Annual Offshore Technology. Conf., OTC 3229, OTC, Houston, 1523-1536.

Shaw, C.J., 1992: "The North European Storm Study (NESS)". Proc. Conf. on Impacts of technical developments on safety cases, March 19, 1992, London.

Van Vledder, G.Ph., and T. Zitman, 1992: Design waves: statistics and engineering practice. Proc. ISOPE 1992, San Francisco.

Van Vledder, G.Ph., Goda, Y., Hawkes, P., Mansard, E., Martin, M.J., Mathiesen, M., Peltier, E., and Thompson, E., 1993: Case studies of extreme wave analysis: A comparative analysis, Proc. WAVES'93, New Orleans, USA, 978-992.

Van Vledder, G.Ph., and D.P. Hurdle, 1997: parameterisation of numerical wave transformations to derive joint coastal extremes. Proc. 3rd International Symposium on Ocean wave Measurement and Analysis, Nov. 3-7, Virginia Beach, Virginia.

Valk, C.F. de, 1991: Definition of joint extreme condition statistics of hydraulic parameters off the coast of the Netherlands, DELFT HYDRAULICS report H957.

370   OCEAN WAVE KINEMATICS, DYNAMICS AND LOADS ON STRUCTURES

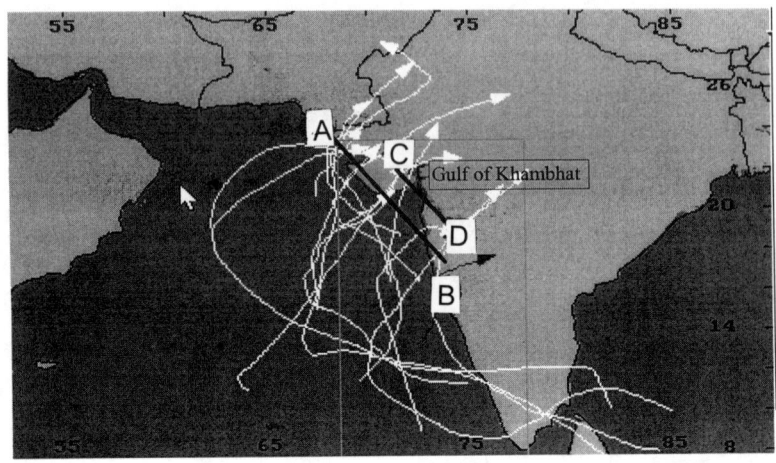

Figure 1   Hurricane tracks which have struck the west Indian coast in the period between 1960 and 1996.

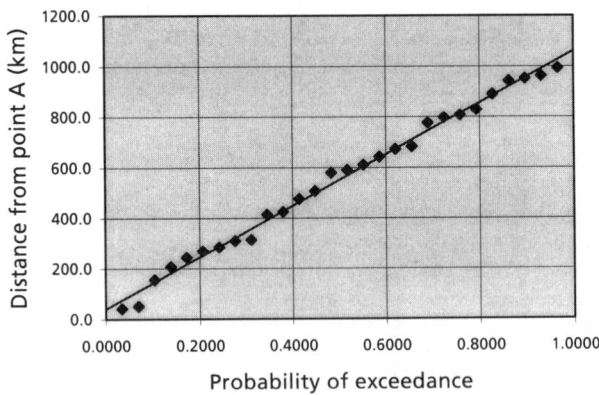

Figure 2:   Distribution of coastal crossing points along the line AB for the hurricanes shown in Figure 1.

OCEAN WAVE KINEMATICS, DYNAMICS AND LOADS ON STRUCTURES   371

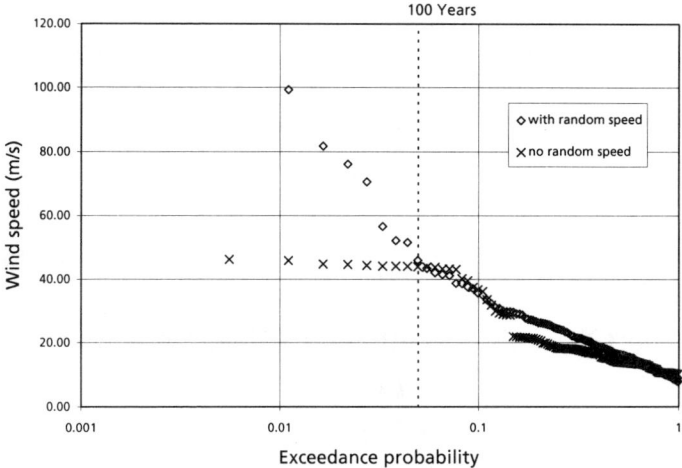

Figure 3: Probability of exceedance of wind speeds for a location in the Gulf of Khambhat based on two type of synthetics data generation techniques, with and without a random peak wind speed.

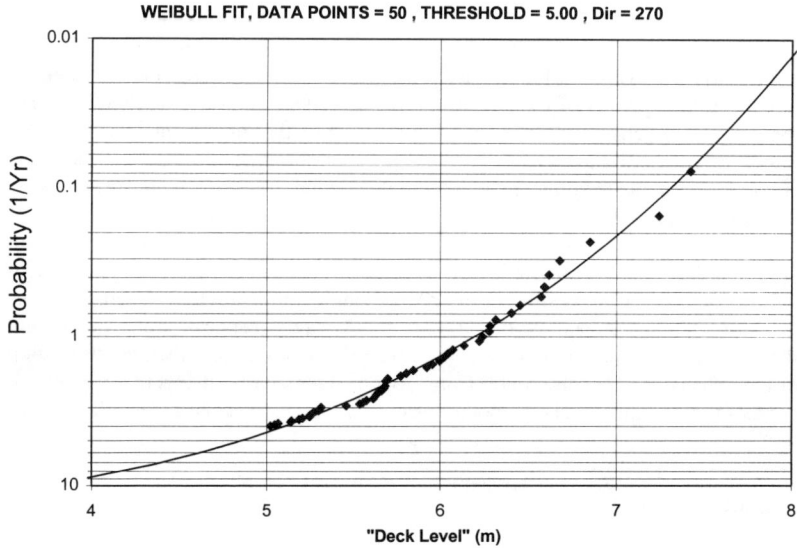

Figure 4: Fitted Weibull distribution to the selected set of points for the Deck Level for the case of a data threshold of 5 m and the direction sector 270°.

Wave Crest Distributions: Observations and Second Order Theory

George Z. Forristall[1]

Abstract

Many empirical and heuristic distribution functions for wave crest heights have been proposed, but their predictions differ considerably. Part of the lack of agreement is due to the difficulty of making measurements that accurately record the true height of the wave crests. Surface following buoys effectively cancel out the second order nonlinearity by making a Lagrangian measurement. Pressure transducers filter the nonlinear components of the signal in complicated ways. Wave staffs have varying degrees of sensitivity to spray. The location of the instruments also plays an important role. We have clear evidence from measurements in the North Sea that spurious crests due to spray are a problem downwind even from mounting supports that appear transparent.

Much of the theoretical nonlinearity can be captured by calculations correct to second order. Explicit calculation of the interactions of each pair of components in a directional spectrum is straightforward although computationally intensive. This technique has the advantage that the effects of wave steepness, water depth, and directional spreading are included with no approximation other than the truncation of the expansion at second order. Comparisons with measurements that we believe to be of the best quality show good agreement with these second order calculations.

---

[1] Shell International Deepwater Services B.V., P.O. Box 3006, 2280 MH Rijswijk, The Netherlands, gforrist@euronet.nl

Introduction

An E&P Forum (1995) workshop recently identified the definition of a suitable air gap for setting the minimum deck elevation of structures as one of the most important uncertainties in the design process for fixed structures offshore. Most of this uncertainty comes from our poor knowledge of the distribution of crest heights.

The problem is to calculate the statistical distribution of crest heights given the directional spectrum of the waves and the water depth. By crest height, we mean the highest point on a wave trace between the time it crosses above mean water level and the time it crosses below mean water level, the 'zero-crossing crest height.' The alternate definition of a crest as a local maxima is not useful for engineering purposes since wave records can show many small maxima between zero crossings.

To first order, the water surface can be represented as Gaussian noise with a reasonably narrow frequency band. The crest heights then have the same distribution as the envelope of the noise, which has the Rayleigh distribution

$$P(\eta_c > \eta) = \exp\left[-8\frac{\eta^2}{H_s^2}\right] \tag{1}$$

where η_c is the crest height, $H_s = 4m_o^{1/2}$ is the significant wave height and m_o is the variance of the wave spectrum.

Real waves show a small but easily noticed departure from a Gaussian surface. The crests are higher and sharper than expected from a summation of sinusoidal waves with random phase, and the troughs are shallower and flatter. It is easy to tell by inspection whether a wave record is right side up.

The shape of regular progressive waves can now be calculated to a very high degree of accuracy, but no generally accepted theory for random waves exists. A theory for the second order interactions of waves in a random directional sea has been available for some time, but there does not seem to have been a concerted effort to use it to develop wave crest distributions, perhaps because it is not clear how to derive a distribution directly from the theory. We proceed by simulating long time series of waves including the second order interaction terms and find the sample distributions of the simulations. Jha (1997) has recently performed similar simulations and compared them to laboratory and field measurements. We extend such calculations to simulate directional spectra and compare the results to several sets of field measurements in extreme conditions.

We begin with a review of some of the methods which have previously been proposed for calculating wave crest heights for engineering design. The next section discusses the principles of operation of some instruments commonly used to measure waves, and the problems they may have in accurately measuring crests. Then we present the equations for the second order interactions between wavelets in a random directional sea in intermediate water depth. The results of using those equations to simulate wave crests are compared to measurements and to previous methods. Simulations for a set of JONSWAP spectra of varying steepness are also shown. We close with conclusions and recommendations for future work.

Previous estimates of extreme crests

For design purposes, a crest height is often estimated by taking the height and period of the design wave and applying a high order regular wave theory such as Stokes fifth order. Since such regular waves are often used as input to calculate forces on a structure, this method has the advantage that the crest height used to set the deck elevation is consistent with the wave used in the force calculations. It has the disadvantage of neglecting the random and directionally spread nature of the real sea.

A popular empirical crest height distribution was presented by Haring et al. (1976) and is given in equation 2.

$$P(\eta_c > \eta) = \exp\left\{-\frac{1}{2}\left(\frac{\eta^2}{m_o}\right)\left[1 - 4.37\frac{\eta}{d}\left(0.57 - \frac{\eta}{d}\right)\right]\right\} \qquad (2)$$

This equation was derived by empirical fitting to 376 hours of storm wave records including measurements with Baylor wave staffs in the Gulf of Mexico and Waverider buoys in the North Sea and the Gulf of Alaska. The distribution is a function of the variance in the spectrum, m_o, divided by water depth, d. It gives a higher probability of high crests for shallow water, as would be expected theoretically from the increased nonlinearity of waves in shallow water. It does not have any dependence on wave steepness and it reduces to the Rayleigh distribution in very deep water, which cannot be strictly correct, since deep water waves are also nonlinear. Equation 2 is, however, significantly different from the Rayleigh distribution for large waves anywhere on the continental shelf.

It is reasonable to suppose that the nonlinear wave surface could be approximated by an amplitude modulated Stokes waves. Tayfun (1980) and Huang et al.(1986) produced crest height distributions using this assumption, and Kriebel and Dawson (1990) found a simple form for the resulting equation. According to Kriebel and Dawson,

$$P(\eta_c > \eta) = \exp\left[-8\frac{\eta^2}{H_s^2}\right]\exp\left[8R\frac{\eta^3}{H_s^3}\right] \qquad (3)$$

where $R = kH_s$ is the wave steepness, and k is the wavenumber associated with the mean wave frequency. Kriebel (personal communication, 1998) actually used the mean period of the highest waves, estimated as $0.95/f_p$ where f_p is the peak frequency of the spectrum, in his calculations. This choice seems reasonable since we are trying to estimate the distribution of the highest crests.

If the wave steepness becomes large, the probability density function for equation 3 can become negative, and Kriebel and Dawson (1993) used the same assumptions to derive the slightly different formula

$$P(\eta_c > \eta) = \exp\left[-8\frac{\eta^2}{H_s^2}\left(1 - \frac{1}{2}R\frac{\eta}{H_s}\right)^2\right] \qquad (4)$$

Kriebel and Dawson (1993) also extended their distribution to shallow water using the depth dependent terms from the Stokes second order expansion. The resulting distribution is the same as equation 4 with R replaced by R_*, an effective steepness given by

$$R_* = kH_s f_2(kd) \qquad (5)$$

where

$$f_2(kd) = \frac{\cosh kd(2 + \cosh 2kd)}{2\sinh^3 kd} - \frac{1}{\sinh 2kd} \qquad (6)$$

Instrumentation

Verification of crest height distributions has been hampered by the difficulty of accurately measuring wave crests in extreme sea states. Measurements from one type of sensor often disagree with those from another type, and there is no agreement on which is correct. The basic problem is the lack of any absolute standard against which the accuracy of the sensors can be judged. The notes below illustrate the problems for some of the sensors in common use. They are not meant to be exhaustive.

Buoys

Buoys are the most popular instruments for collecting information on wave climate. Instrument comparisons, most notably those in WADIC (Allender et al., 1989), have demonstrated that the popular models can accurately measure integral properties of the wave field. Crests measured by buoys are, however, generally smaller than those measured by other instruments.

This underestimation of crest heights is often thought to be due to the buoy partially submerging in a crest or sliding sideways away from the highest point on a high crest. These mechanisms may play a role, but even a perfect surface following buoy will underestimate wave crests. A buoy that acts as a particle on the surface will move forward in the direction of wave propagation in the crest and backward in the trough. It will therefore spend more time than a wave staff at a fixed location in the crest, and less time in the trough.

The orbital motion of the buoy distorts the shape of the wave profile, but does not by itself make the crest measurements lower. However, almost all buoys measure wave elevation through double integration of the vertical acceleration. The absolute value of the still water elevation is thus not known, and crests heights are measured from the mean of the elevation measurements. Since the buoy spends extra time in the crest, the mean water level will be slightly higher than the true still water level, and the crest height above mean water level will be slightly too small.

The motion of the buoy is a finite amplitude effect. James (1986), Srokosz and Longuet-Higgins (1986) and Longuet-Higgins (1986) have all considered aspects of this problem and show that to second order, the vertical displacement of the mean surface is equal to the amount that the crest is raised. Therefore, the Lagrangian motion of the buoy cancels out the second order non-linearity of the wave crest. The details of the buoy motion are greatly complicated by its mooring line and the random nature of the real sea, and have not yet been worked out completely. Nevertheless, the main features of the argument must still be important, so that accelerometer buoys cannot be considered as a real choice for measuring the distribution of wave crests.

Pressure transducers

Pressure transducers are useful for measuring waves at shallow water sites or on platforms where the sensor can be mounted relatively close to the sea surface. The signal must be corrected for the attenuation of the pressure fluctuations, which increases with increasing depth and increasing wavenumber. These corrections are reasonably accurate for at least the low frequency part of the wave spectrum, but it is hard to see how the results could be useful for estimating the crest distribution. If the corrections are made with first order theory, as is usual, the resulting surface will be Gaussian. If the difficult problem of making higher order corrections is faced, the result will only be an expression of the theory used rather than an independent check of it.

The Baylor wave staff

The Baylor wave staff consists of a pair of stainless steel wire ropes separated by insulators about 20 cm long. The transducer measures the natural frequency of the inductive loop made by the two wires and the sea surface, from which the length of the loop is found. The instrument is robust and relatively immune to fouling. It has been particularly popular for wave measurements from platforms in the Gulf of Mexico.

Tests have shown that it has a linearity better than 1% and a response time of better than 300 m/sec. Our experience from calibrating Baylor staffs is that quite a firm short is necessary before the sensor responds. It thus seems unlikely that it would be affected by spray.

EMI laser

The EMI laser is a pulsed range finder operating in the near infrared region. Narrow pulses of light are produced by a laser diode, and the radiation from the target is used to stop a time interval measurement. The time of travel is converted to an analog voltage which is proportional to the distance to the reflector. The response of the optical unit is 10 to 15 Hz, but the output is usually filtered by a 2 Hz Butterworth filter to eliminate high frequency noise. We are aware of some tests which showed that the instrument responded to artificial spray, but we do not have details of the tests.

Marex radar

The Marex wave radar is a ranging device derived from a radar altimeter. The radar operates in the microwave (J) band with a beamwidth of $\pm 6°$. It must be positioned on a structure so that sidelobes of the beam do not reflect from members of the structure. We do not know of any tests of its response to spray.

Second order wave profiles

A second order expansion of the sea surface can capture the effects of wave steepness, water depth and directional spreading with no approximations other than the truncation of the expansion at second order. The second order wave interactions for infinite water depth were calculated by Longuet-Higgins (1963), and the calculations were extended to intermediate water depth by Sharma and Dean (1979). We reproduce the latter result for completeness. Let the first order water surface be given by

$$\eta = \sum_{n=1}^{N} a_n \cos\left(\mathbf{k}_n x - \sigma_n t + \epsilon_n\right) \qquad (7)$$

where t is time, x is the position vector in the plane, σ_n, ϵ_n and $\mathbf{k}_n$ are, respectively, the radian frequency, phase, and vector wavenumber of wavelet n, and a_n is its amplitude. The frequencies and wavenumbers are related by the linear dispersion equation

$$\sigma_n^2 = g|\mathbf{k}_n| \tanh\left(|\mathbf{k}_n|d\right) \qquad (8)$$

where g is the acceleration of gravity and d is the water depth. The second order correction to the wave surface given by Sharma and Dean (1979) is then

$$\eta^{(2)} = 1/4 \sum_{i=1}^{N} \sum_{j=1}^{N} a_i a_j \left\{ K^- \cos(\psi_i - \psi_j) + K^+ \cos(\psi_i + \psi_j) \right\} \qquad (9)$$

where

$$K^- = \left[D_{ij}^- - (\mathbf{k}_i \bullet \mathbf{k}_j + R_i R_j)\right] (R_i R_j)^{-1/2} + (R_i + R_j) \qquad (10)$$

$$K^+ = \left[D_{ij}^+ - (\mathbf{k}_i \bullet \mathbf{k}_j - R_i R_j)\right] (R_i R_j)^{-1/2} + (R_i + R_j) \qquad (11)$$

$$D_{ij}^- = \frac{\left(\sqrt{R_i} - \sqrt{R_j}\right)\left\{\sqrt{R_j}\left(\mathbf{k}_i^2 - R_i^2\right) - \sqrt{R_i}\left(\mathbf{k}_j^2 - R_j^2\right)\right\}}{\left(\sqrt{R_i} - \sqrt{R_j}\right)^2 - k_{ij}^- \tanh k_{ij}^- d}$$
$$+ \frac{2\left(\sqrt{R_i} - \sqrt{R_j}\right)^2 \left(\mathbf{k}_i \bullet \mathbf{k}_j + R_i R_j\right)}{\left(\sqrt{R_i} - \sqrt{R_j}\right)^2 - k_{ij}^- \tanh k_{ij}^- d} \qquad (12)$$

$$D_{ij}^+ = \frac{2\left(\sqrt{R_i} + \sqrt{R_j}\right)^2 \left(\mathbf{k}_i \bullet \mathbf{k}_j - R_i R_j\right)}{\left(\sqrt{R_i} + \sqrt{R_j}\right)^2 - k_{ij}^+ \tanh k_{ij}^+ d}$$
$$+ \frac{\left(\sqrt{R_i} + \sqrt{R_j}\right)\left\{\sqrt{R_i}\left(\mathbf{k}_j^2 - R_j^2\right) + \sqrt{R_j}\left(\mathbf{k}_i^2 - R_i^2\right)\right\}}{\left(\sqrt{R_i} + \sqrt{R_j}\right)^2 - k_{ij}^+ \tanh k_{ij}^+ d} \qquad (13)$$

$$k_{ij}^- = |\mathbf{k}_i - \mathbf{k}_j| \qquad (14)$$

$$k_{ij}^+ = |\mathbf{k}_i + \mathbf{k}_j| \qquad (15)$$

$$R_i = |\mathbf{k}_i|\tanh(|\mathbf{k}_n|d) = \sigma_i^2/g \qquad (16)$$

and

$$\psi_i = \mathbf{k}_i x - \sigma_i t + \epsilon_i. \qquad (17)$$

For infinite water depth, equation (7) reduces to equation (3.7) of Longuet-Higgins (1963) except that the latter equation is missing a factor of $1/2$.

Form of the interaction kernel

The positive interaction terms given by equation 13 occur at the sum of the frequencies of the interacting wavelets. They produce the sharpening of the crests and flattening of the troughs that we associate with second order Stokes waves. The negative interaction terms given by equation 12 occur at the difference of the frequencies of the first order wavelets. These interactions give the setdown of the water level under wave groups.

The skewness kernel is defined as $(K^- + K^+)/4$ and is a measure of the strength of the second order interaction. It is a function of the frequencies of the two interaction wavelets as well as their angular separation and the water depth. The strength of the interaction is much greater in shallow water, matching the observation that the wave profile is more skewed in shallow water.

Figure 1 shows an example of the interaction kernel for relatively shallow water. The depth is 1 meter and the $f_1 = 0.5859$. The ratio of the frequencies of the two wavelets is shown on the x-axis and the difference in the direction of travel of the wavelets is on the y-axis, with $0°$ in the center of the scale. In this shallow water the peak of the interaction is not for co-linear wavelets, but for wavelets separated by a small angle. The kernel has a depression for waves that are nearly in the same direction which is due to the large negative value of the negative interaction term for wavelets with nearly the same direction and frequency in shallow water.. This depression is particularly deep and sharp near $f_1/f_2 = 1$. It does not appear in deep water. Figure 1 shows that wave setdown effects will be much more important for unidirectional waves produced in a wave tank than they are for natural waves in the ocean, which are always spread directionally to some extent.

$$d = 1 \text{ m}, \quad f_1 = 0.5859$$

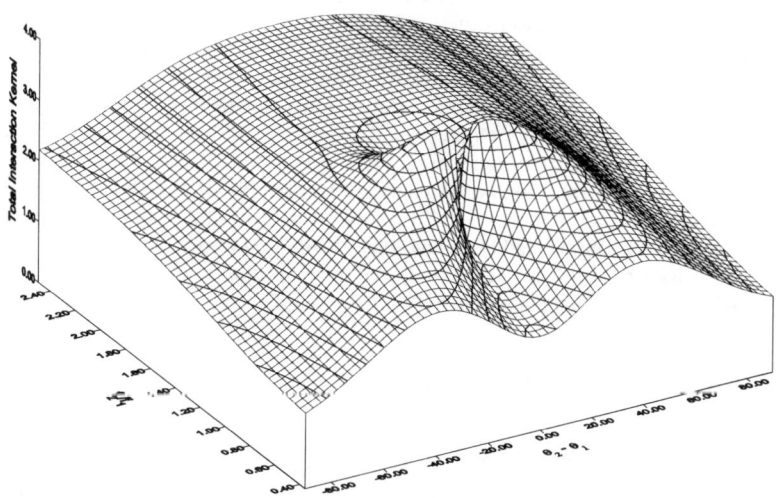

Figure 1: Second order interaction kernel for waves in shallow water

Numerical implementation

The calculations indicated by equations 9–17 are easy to implement, but represent a fair amount of computation since there are very many interactions to calculated, particularly in the three dimensional case. A typical simulation would have 4096 time steps at 4 Hz. A first order simulation of a given wave spectrum is produced from equation 7 by choosing the phases ϵ_n from a uniform random distribution and calculating the wave height using an inverse fast Fourier transform.

The interactions do not have to be calculated for all of the 2048 frequencies in the Fourier transform of the time series, partly because the energy in the first order spectrum is very small at the higher frequencies, but also because the energy in the measured spectrum at these high frequencies appears to be mostly due to nonlinear interactions. It is possible to use an iterative scheme to produce a first order spectrum from the measured spectrum, but it is much simpler to simply truncate the second order calculations at a small multiple of the peak frequency of the spectrum. This simplification has little effect on the crest height distributions. The calculations described in this paper were typically truncated for $f_1 + f_2$ greater than 4 or 5 times the peak frequency.

We generally want to perform many simulations using the same spectrum in order to produce stable statistics for rare crest heights. Since the skewness kernel is the same for all of the repetitions, it is efficient to calculate it once for each spectrum and store it. The calculations also exploit the fact that the kernel is symmetric in the two frequencies and depends only on the difference in the two angles. We use an angular resolution of $12°$ for the three dimensional simulations.

Repeating a simulation with precisely the same spectral variance at each Fourier line does not include all of the natural variability of waves. The variances are actually random variables with a Chi-squared distribution. If a directional spectrum is simulated, the addition of several wavelets with different directions and the same frequency will automatically produce this distribution in the frequency spectrum. For consistency, we therefore multiply the input spectral lines in two dimensional simulations by a Chi-squared random variate.

Comparisons with measurements

Storms at Tern in the North Sea

The Tern platform is located in the northern North Sea between the Shetland Islands and Norway. It is about 150 km from the nearest shoreline and in 167 m water depth. It is a fairly standard 8 legged steel oil production platform that was installed in 1988. It was equipped with a structural monitoring system including strain gauges, two wave height sensors, and an electromagnetic current meter. Jonathan et al. (1994) and Jonathan and Taylor (1995) give descriptions of the measurement system and the oceanographic conditions in the storms considered here.

Figure 2 shows an outline plan of the platform with the sensor locations marked. The rectangles in the figure show the outline of the structural members of the platform at mean sea level (m-s-l), at the 41 m depth of the Marsh McBirney electromagnetic current meter, and at the mud line. A Marex wave radar was mounted under the southeast corner of the platform deck and an EMI laser wave sensor was mounted under the deck on the southwest corner. Both of these sensors measure the distance from the instrument to the instantaneous water surface.

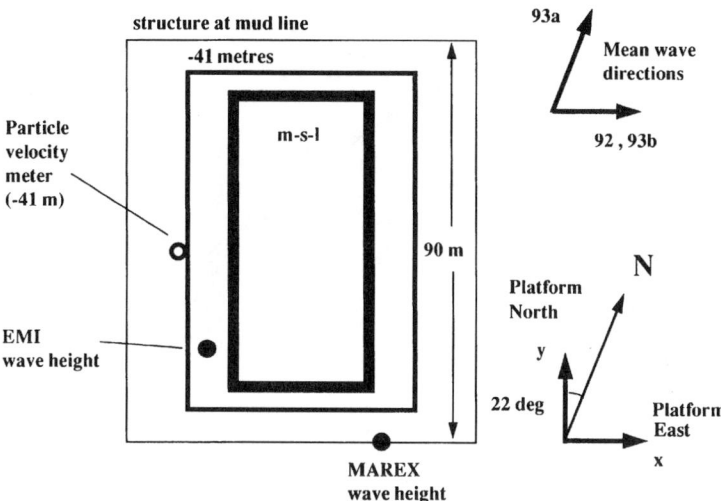

Figure 2: Outline plan of the Tern platform with locations of the wave height sensors and the mean wave directions for the storms.

Three storms at Tern have been studied in detail. One of the storms occurred in January 1992 and two in January 1993. The storm on 4 January 1993 is referred to as 93a and the one on 17 January 1993 is referred to as 93b. Conditions in all three storms were extreme, with peak significant wave heights above 12 m. Storms 93a and 93b are particularly useful for wave process studies since the conditions remained stationary, nearly within the limits of sampling variability, for 8 or 9 hours. Storm 92 built quickly to a peak H_s of 13.8 m and then declined.

Figures 3 and 4 show the probability distributions of the wave crest heights for storms 93a and 92 respectively. The crest heights are normalized by the significant wave height during the hour the crest was measured. If the waves were linear, so that the surface elevation had a Gaussian distribution, the crest heights between zero crossings would have a Rayleigh distribution. This theoretical distribution is shown by the solid line in the figures. As expected, the sample distributions from the measurements show an excess of high crest heights above the Rayleigh curve.

The sample distributions also show a significant disagreement between the results from the two wave sensors. In storm 93a, the crests measured by the Marex radar are about 10% higher than the crests expected from linear theory, but the highest crests from the EMI laser measurements are 20-30% higher than linear theory.

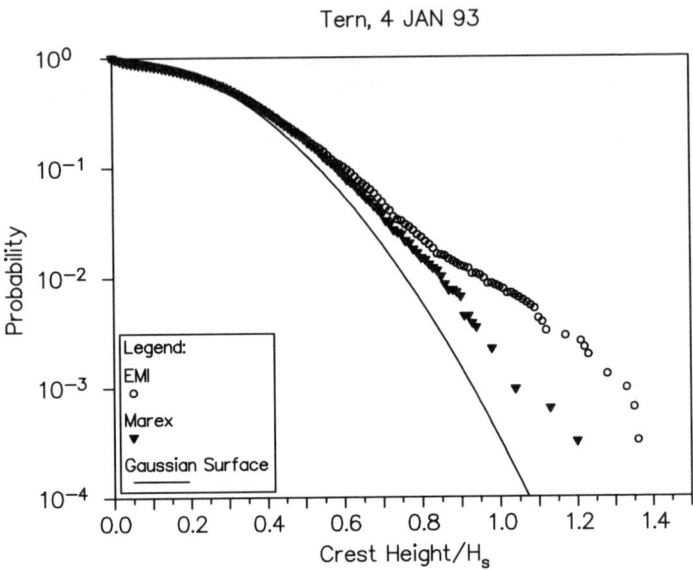

Figure 3: Probability distribution of normalized crest heights measured at Tern during the storm on 4 Jan 1993. The crest heights are normalized by the significant wave height during each hour of measurements. Nine hours of measurements with an average significant wave height of about 12 m were combined to produce the observed distributions.

In storm 92, however, the situation is reversed. The measurements from the EMI laser are about 10% above the linear curve while those from the Marex radar are much higher. It seems quite likely that the difference between the storms can be explained by the difference in wave directions, as shown in Figure 2. In storm 93a, the waves were propagating to the north, so that their crests passed the Marex radar before encountering any structural elements on the platform. On the other hand, the leg on the southwest corner of the platform is upwave from the EMI sensor, and it is quite likely that spray caused by a wave crest hitting that leg would sometimes pass under the EMI laser. The response of wave sensors in the presence of spray is not known with any certainty, but it is reasonable to suppose that spray from structural interference could cause a laser gauge to record higher crests than those in the ambient waves.

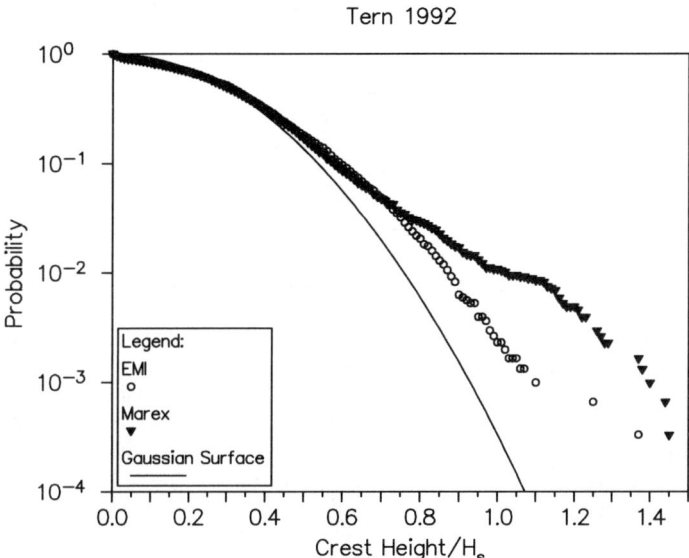

Figure 4: Probability distribution of normalized crest heights measured at Tern during the storm on 1 Jan 1992. The crest heights are normalized by the significant wave height during each hour of measurements. Eight hours of measurements with an average significant wave height of about 11 m were combined to produce the observed distributions.

This hypothesis is supported by the fact that in storm 92 the wave were propagating to the east, so that the EMI laser was on the windward side of the platform while the Marex radar was in the lee of structural members. Apparently, both of the sensors measured crests higher than the ambient waves due to spray from structural members, although there is no obvious evidence of spray in the measured time series. The location of a wave sensor with respect to the platform it is mounted on thus may be at least as important as the response characteristics of the sensor itself.

Given the likely platform interference, we compare the second order simulations with the Marex measurements during storm 93a and the EMI measurements during storm 92. Figure 5 shows the comparison for 93a. The ordinate in this figure gives the ratio of the crest height to the height predicted by the Rayleigh distribution at the probability level of the abscissa. It thus shows that the measured wave crests were about 10% higher than expected from linear theory, and that the ratio increases slightly at lower probability levels. The measurements are for 9 hours of the storm, which included about 3000 crests.

The simulations were based on spectra calculated from the measurements for each hour. For the two dimensional simulations, 200 repetitions for each of the 9 spectra were made. The simulations were made at a 5.12 Hz sample rate, the same as the measurements, and each simulation was 4096 points long. There were therefore about 44 hours of simulation for each hour of measurements, so the statistics of the simulations are much more stable than those of the measurements. The three dimensional simulations were based on directional spectra calculated from the wave radar and the electromagnetic current meter. Since the three dimensional simulations demand much more computer time, only 30 repetitions were made for each spectrum, giving 6.67 hours of simulation for each hour of measurement.

The statistics of the simulations are very similar to those of the measurements, about 10% higher than linear theory and increasing slightly at lower probability levels. The three dimensional simulations are about 2% lower than the two dimensional simulations. The difference is caused by the slightly lower values of the second order interaction kernel for wavelets which are not co-linear. The two dimensional simulations appear to be slightly higher than the measurements while the three dimensional simulations appear to be slightly lower.

Figure 6 shows the measured and simulated statistics for storm 92. Both the measurements and the simulations are for 8 hours at the peak of the storm when the significant wave height ranged from 7.85 to 13.78 m. The 2D and 3D simulations are very similar to those for storm 93a, but the data points in the top 10% of the crests are consistently higher than the simulations. There is no obvious difference in the spectra of the two storms that would explain the difference in their crest statistics. It is possible that the response of the EMI and Marex sensors is slightly different even when they have good exposure to the ambient waves.

Hurricane Opal

This storm entered the Bay of Campeche in the Gulf of Mexico on September 30, 1995 as a tropical depression. It intensified rapidly as it accelerated north-northwestward on October 3 and 4 on its way to a landfall in the panhandle of Florida. Its maximum intensity was observed early on October 4 as the eye passed 27.3^o N, 88.5^o W with a 916 mb central pressure and a very small eye 10 nm in diameter.

Waves from Hurricane Opal were measured at the Bullwinkle oil production platform located at 27.9^o N, 90.9^o W in 410 m of water. A comprehensive instrumentation system was installed on Bullwinkle in order to measure the environment and the response of the structure. Swanson and Baxter (1989) give a good description of this system. A Baylor wave staff was used to measure wave elevations and three

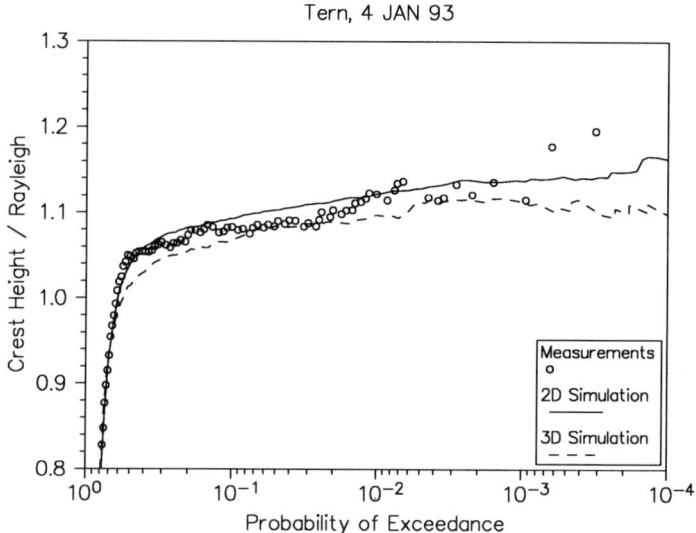

Figure 5: Distribution of crest heights in storm 93a at Tern. The ordinate show the crest height normalized by the height predicted at that probability level by the Rayleigh distribution. The measurements are from the Marex wave radar.

Marsh-McBirney electromagnetic current meters measured the wave particle velocities.

Since the eye of the tight storm passed 125 miles to the east of Bullwinkle, the waves there were never extremely high. We used the 14 hours of measurements from 0500Z to 1800Z on October 4, 1995 in our analysis, and the significant wave height during that time ranged from 4.78 to 6.14 m. The measurements were recorded at 4 Hz. Figure 7 show the observed and simulated distributions of crest heights for Hurricane Opal. The simulations are about 2.5% lower than the simulations for the Tern storms since the waves were not as steep. The three dimensional simulations produce crests about 1% smaller than the two dimensional simulations, but they are still 2% larger than the bulk of the observations.

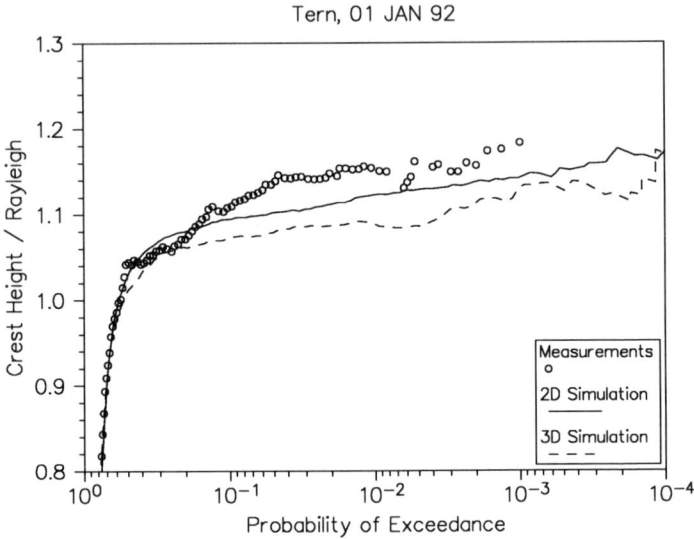

Figure 6: Distribution of crest heights during storm 92. For an explanation of the curves, see Figure 5.

Hurricane Camille

Camille was one of the most intense and destructive storms to strike the United States this century. Waves were recorded by a Baylor wave staff at station 1 of the ODGP (Hamilton, 1976) which was on the South Pass 62A oil production platform located at 29° 04'50" N, 88° 44'30" W. Camille passed almost directly over the station, and the large waves that were measured had a great influence on setting the standard criteria for platform design in the Gulf. The measurements were recorded on an analog tape recorder, and unfortunately were only digitized at a 1 Hz sampling rate. The wave staff broke at 1630 CDT on August 17, 1969, which hindcasting studies show to be very close to the peak of the storm.

Measurements and simulations of the crest heights from 1200 to 1630 CDT are shown in Figure 8. The significant wave height grew from 9.96 to 13.40 m during this period. Only two dimensional simulations are shown, since directional spectra

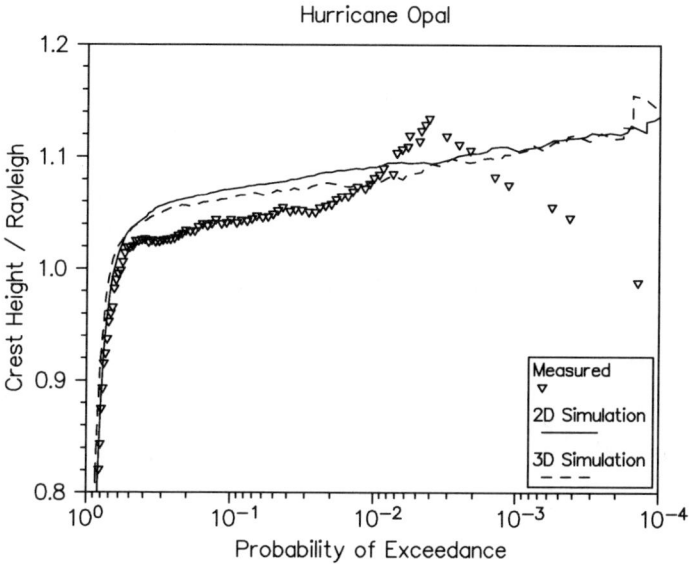

Figure 7: Distribution of crest heights during Hurricane Opal at Bullwinkel. For an explanation of the curves, see Figure 5

were not measured during the storm. The simulations made at a one second time step agree reasonably well with the measurements, although there appears to be considerable scatter in the measurements which cause the distribution to deviate from a smooth curve. We also made simulations with a 1/4 second time step, and they are significantly different, about 2.5–3% higher. The difference is not due to the energy content of the sea at frequencies above 1 Hz, which is very small, but to the fact that a one Hz sampling frequency is likely to miss the very peak of longer waves. This effect was investigated in some detail by Tayfun (1993), who found that the higher waves would be underestimated by approximately $(\pi^2/6)\left(\Delta/\bar{T}\right)^2$ where Δ is the sampling interval and $\bar{T}$ is the average wave period. For the 10 second average wave periods in this part of Camille, the formula gives an error of 1.6%, which is close to the difference in the simulations. The relatively long sampling interval in the digitized record of the Camille waves clearly causes an underestimate of the true crest heights.

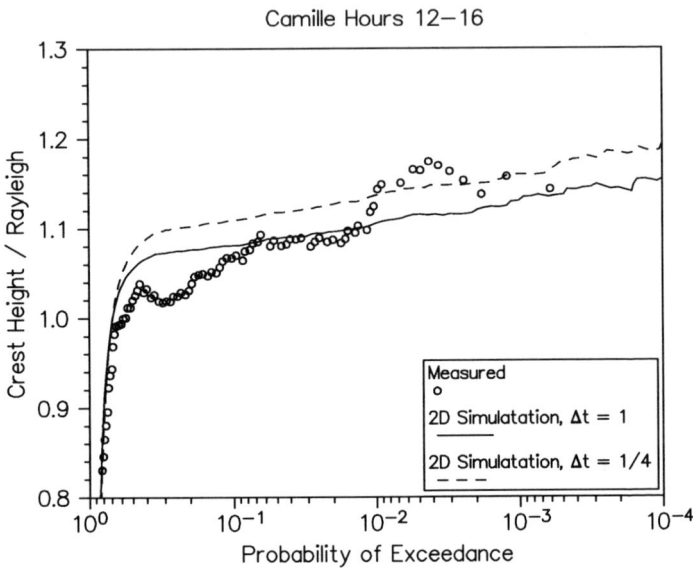

Figure 8: Distribution of crest heights during Hurricane Camille at SP62 for 1200-1630 on August 17, 1969. For an explanation of the curves, see Figure 5.

Shallow water waves in Lake Ontario

The Canada Centre for Inland Waters maintains a research tower at the western end of Lake Ontario. Many interesting studies of wave processes have been conducted there, including the definitive measurements of directional spectra by Donelan et al. (1985).. The tower is in 12 m of water. An easterly a storm on January 10, 1977 produced waves that reached a significant height of 3 m at the tower reached. During that storm, 14 wave staffs on the tower and extensions from it were operational, and we have examined measurements from them during one hour at the peak of the storm. The measurements were recorded at 5 Hz.

Despite the fact that the structure appears to be very transparent, a few unusually high crests were observed in the lee of the platform but not by the wave staffs on the windward side of the platform. Therefore in Figure 9 we only included data from the 6 wave staffs with the clearest upwind exposure. The two dimensional simulation agrees very well with the measurements, while the three dimensional simulation appears to be a bit high, at least at low probabilities. In shallow water, waves with a narrow directional spreading can be more non-linear than unidirectional waves because the interaction kernel reaches its maximum for wavelets slightly separated in direction, as shown in Figure 1.

Figure 9 also shows the results of a two dimensional simulation made with the water depth increased to 1000 m, labelled 'Deep Water.' The crests in this simulation are about 3% lower than those for the true water depth of 12 m, showing the increased non-linearity of the waves due to the shallow water at the tower.

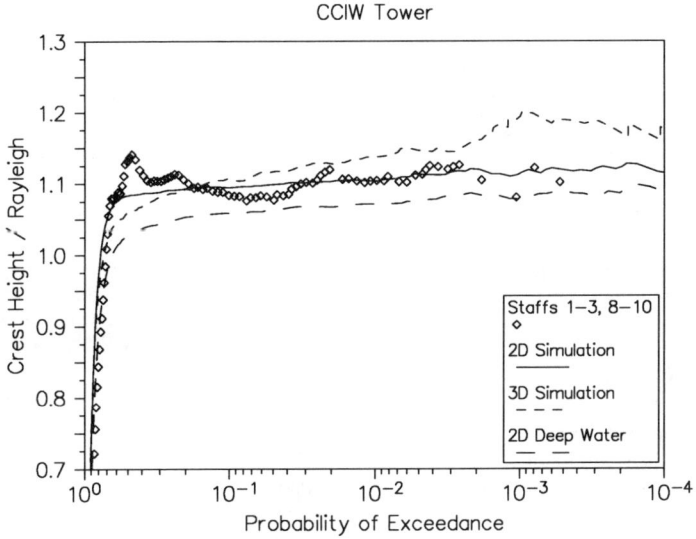

Figure 9: Distribution of crest heights at the CCIW tower on January 10, 1977.

Comparisons with previous estimates

Figures 10 and 11 compare our two dimension, second order wave simulations with previous methods of estimating crest heights. Figure 10 is for conditions during the peak of storm 93a at Tern, with for $D = 167$ m, $T_p = 14.3$ sec, $H_s = 12.0$ m. Using $T_s = 0.95T_p$, we have $T_s = 13.6$ sec for use in equation 4. The steepness, $R = 0.2628$, and the effective shallow water steepness R_* are virtually identical for this relatively deep water.

Our two dimensional second order simulation is shown as the solid line in the figure. Kriebel and Dawson's (1993) distribution from equation 3 agrees quite well with the simulations, although it seems to extrapolate to higher crests at very low probabilities. The distribution of Haring et al.(1976) lies below the simulations, because the water depth at Tern is deep enough that the nonlinear adjustment in equation 2 is too small. We also tested equation 2 for the 410 m water depth at Bullwinkle, and the resulting distribution (not shown) was very close to the Rayleigh distribution.

Crest heights calculated from Stokes fifth order waves are shown as small circles in Figure 10. The wave heights at several probabilities of exceedence were calculated from the Rayleigh distribution and the crest height of a Stokes fifth wave with that height and a period of 13.6 sec was found. These crests are just slightly lower than those from the second order simulations. The steps in this calculation are standard practice, but they are somewhat inconsistent since actual trough to crest wave heights are lower than given by the Rayleigh distribution. If an empirical distribution such as the one due to Forristall (1978) were used to estimate the wave heights, the Stokes crest heights for this example would be about the same as given directly by a Rayleigh distribution of crest heights.

Figure 11 is for conditions during the peak of storm in Lake Ontario, with for $D = 12$ m, $T_p = 8.36$ sec, $T_s = 7.94$ sec, and $H_s = 3.0$ m. The steepness is $R = 0.251$, and the effective shallow water steepness $R_* = 0.613$. The Haring et al. (1976) distribution is close to the Stokes fifth order wave crests, but both are considerably larger than the simulations. The Kriebel and Dawson (1993) shallow water distribution unrealistically higher than any of the others, probably because the very high value of effective steepness invalidates the mathematical assumptions that were made in the derivation of the distribution by reversion of series. The Stokes crest heights may be too high because the regular wave method effectively concentrates all of the energy of the spectrum at one frequency, and the positive interaction term is largest for self interaction. The method also does not include the negative interaction which produces a set down under high wave groups.

The Haring et al. (1976) distribution is too low for very deep water and too high for very shallow water. These differences should not be too surprising since the water depths in these examples are outside the range of water depths in the data from which this empirical distribution was developed.

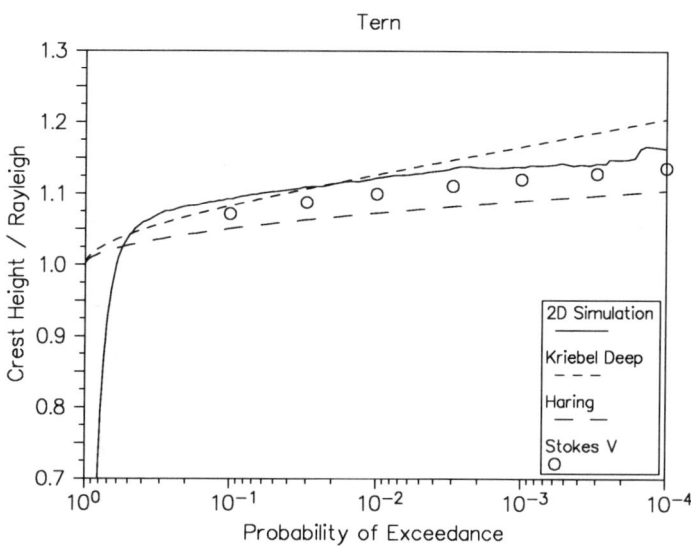

Figure 10: Comparison of crest height distributions for $D = 167$ m, $T_s = 13.6$ sec, $H_s = 12.0$ m.

Crest heights for JONSWAP spectra

Figure 12 shows simulated crest height distributions for a sequence of pseudo-JONSWAP spectra with regularly increasing steepness. The spectra were constructed using the equations of Goda (1985) which give spectra similar to the JONSWAP form but with a specified peak period and significant wave height. All of the spectra had a peak period of 8.98 seconds and a peak enhancement factor $\gamma = 3.3$. The significant wave height was increased in increments of 1 m from 1 m to 12 m, giving increments of steepness, R, of 0.05.

All of the simulations consisted of 1000 repetitions of a 4096 sample time series with a sampling interval of 4 Hz. The simulations were two dimensional. Since the simulation of each spectrum began with the same random number seed, and the input spectra were proportional to each other, the detailed behavior of the simulations, including the high values at very low probabilities, are also proportional to each

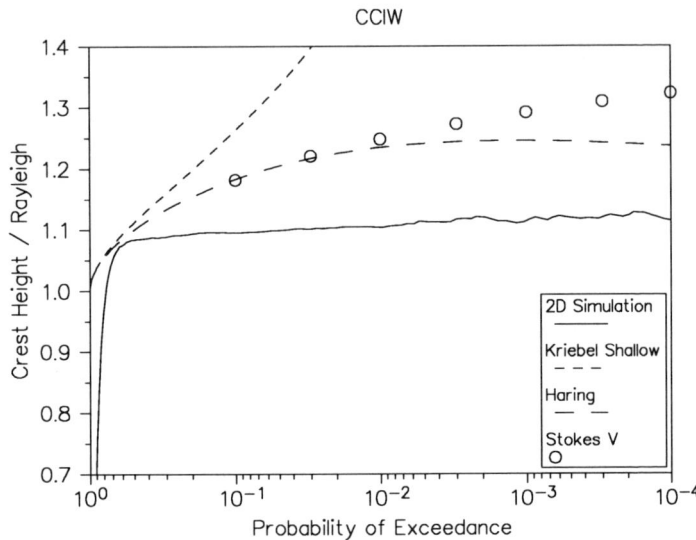

Figure 11: Comparison of crest height distributions for $D = 12$ m, $T_s = 7.9$ sec, $H_s = 3.0$ m.

other. These high values are no doubt due to sampling variability which persists despite the long simulations.

The simulations show that, at least for spectra which are similar to each other, the crest height distribution is almost exactly proportional to the wave steepness. Many more simulations will be needed to establish the effects of spectral shape and directional spreading, but it seems likely that it will be possible to parameterized these distributions in a way that will permit easy use in engineering design.

Conclusions and Recommendations

Second order simulations of wave crests agree well with measurements of high wave crests made in both deep and shallow water. Three dimensional simulations that account for the directional spreading of waves produce crests that are about 2% lower than two dimensional simulations in deep water. Shallow water makes the

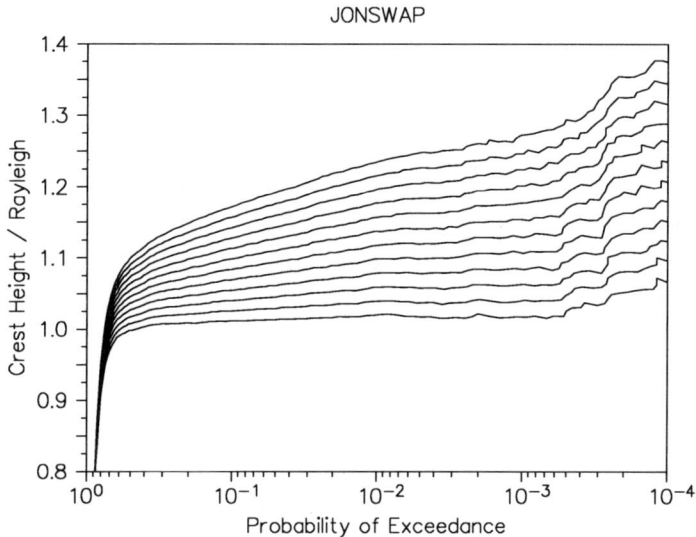

Figure 12: Crest height distributions for psuedo-JONSWAP spectra. The spectra all have a peak frequency of 8.98 sec, and the significant wave height increases in increments of 1 m, giving steepness increments of 0.05.

simulated waves more nonlinear and the crests higher as expected. All of these features appear to agree with measurements, but we cannot be too definite about the accuracy of the simulations because there is still doubt about the accuracy of the measurements.

The measured height of wave crests is apparently influenced by both the type of sensor used and the location of the sensor on a platform. The two influences have been difficult to sort out because there have been few comparisons of different instruments placed near each other. To remedy this situation, we have begun the Wave Crest Sensor Intercomparison Study (WACSIS) as a Joint Industry Project on the Meetpost Noordwijk research platform in the North Sea. Through the winter of 1997-1998, waves will be continuously recorded using a Baylor wave staff, Marex radar, Saab radar, and EMI laser along with other sensors.

The second order simulations appear to have a greater range of applicability than previous methods which have been proposed for estimating crest height distributions. Simulations of pseudo-JONSWAP spectra indicate that the crest height distribution increases almost linearly with wave steepness. A systematic investigation of simulations for various input spectra and water depths should enable us to find a parametric distribution that matches the simulations and is accurate enough for engineering use.

Acknowledgments

Mark Donelan provided the data from Lake Ontario and Mike Vogel sent us the data from Hurricane Opal. The Tern data were provided by Peter Tromans and Paul Taylor, with whom I have had many stimulating discussions on the subject of wave statistics and dynamics. Dave Kriebel was very helpful in discussing the use of his distribution functions with me.

References

[1] J. Allender, T. Auduson, S. F. Barstow, S. Bjerken, H. E. Krogstad, P. Steinbakke, L. Vartdal, L. E. Borgman, and C. Graham. The WADIC project: A comprehensive field evaluation of directional wave instrumentation. *Ocean Engineering*, 16:505–536, 1989.

[2] M. A. Donelan, J. Hamilton, and W. H. Hui. Directional spectra of wind-generated waves. *Philosophical Trans. Roy. Soc. London*, A315:509–562, 1985.

[3] E&P Forum. *Uncertainties in the Design Process, Report 3.15/229*. The Oil Industry International Exploration & Production Forum, London, 1995.

[4] G. Z. Forristall. On the statistical distribution of wave heights in a storm. *Journal of Geophysical Research*, 83:2353–2358, 1978.

[5] Y. Goda. *Random Sea and Design of Maritime Structures*. University of Tokyo Press, Tokyo, 1985.

[6] R. C. Hamilton and E. G. Ward. Ocean Data Gathering Program: Quality and reduction of data. *Journal of Petroleum Technology*, 28:337–344, 1976.

[7] N. E. Huang, L. F. Bliven, S. R. Long, and C. C. Tung. An analytical model for oceanic whitecap coverage. *Journal of Physical Oceanography*, 16:1597–1604, 1986.

[8] I. D. James. A note on the theoretical comparison of wave staff and wave rider buoys in steep gravity waves. *Ocean Engineering*, 13:209–214, 1986.

[9] A. K. Jha. *Nonlinear stochastic models for loads and responses of offshore structures and vessels*. PhD dissertation, Stanford University, 1997.

[10] P. Jonathan and P. H. Taylor. On irregular, nonlinear waves in a spread sea. *Journal of Offshore Mechanics and Arctic Engineering*, 119:37–41, 1995.

[11] P. Jonathan, P. H. Taylor, and P. S. Tromans. Storm waves in the northern North

Sea. In *Proc. 7th International Conference on the Behaviour of Offshore Structures (BOSS)*, Boston, 1994.
[12] D. L. Kriebel and T. H. Dawson. Nonlinear effects on wave groups in random seas. 113:142–147, 1991.
[13] D. L. Kriebel and T. H. Dawson. Nonlinearity in wave crest statistics. In *Proceeding of the 2nd International Symposium on Ocean Wave Measurement and Analysis, ASCE*, pages 61–75, New Orleans, 1993.
[14] M. S. Longuet-Higgins. The effect of non-linearities on statistical distrbutions in the theory of sea waves. *Journal of Fluid Mechanics*, 17:459–480, 1963.
[15] M. S. Longuet-Higgins. Eulerian and lagrangian aspects of surface waves. *Journal of Fluid Mechanics*, 173:683–707, 1986.
[16] J. N. Sharma and R. G. Dean. Development and evaluation of a procedure for simulating a random directional second order sea surface and associated wave forces. *Ocean Engineering Report 20*, 1979. University of Delaware, Newark.
[17] M. A. Srokosz and M. S. Longuet-Higgins. On the skewness of sea surface elevation. *Journal of Fluid Mechanics*, 164:487–497, 1986.
[18] R. C. Swanson and G. D. Baxter. The Bullwinkle platform instrumentation system. In *Proceedings of the 21st Annual Offshore Technology Conference*, page OTC 6052, Houston, 1989.
[19] M. A. Tayfun. Narrow-band nonlinear sea waves. *Journal of Geophysical Research*, 85:1548–1552, 1980.
[20] M. A. Tayfun. Sampling-rate errors in statistics of wave heights and periods. *Journal of Waterways, Ports, Coastal and Ocean Engineering*, 119:172–192, 1993.

Wave Climate Parameterization of the Distribution of Maximum Wave Heights and Crest Elevations

Marc Prevosto[1], Raymond Nerzic[2]

Abstract

Standard models for the distribution of wave heights and crest elevations are useful for computing maximum wave and crest heights, on the basis of significant wave heights and mean wave periods, for the purpose of designing offshore structures. However, there can be differences of 10-20% in the results obtained with different models, which are unsuitable in terms of design criteria. The discrepancies could be explained by the problems of accurately representing wave properties in a given sea state, particularly for strong and steep waves, because of the non-linearity in the wave kinematics which is not suitably taken into account in the common models.

A Corrected Weibull-Stokes (CWS) model was thus proposed by Nerzic and Prevosto (1997) on the basis of standard Weibull models, modified using a third order Stokes expansion. This model was validated with North Sea wave data, but the analysis shown that the model might be site dependent, in relation with water depth and sea state properties such as directionality, and spectral shape.

The present paper describes a procedure which allowed to derive the parameters of the CWS model from satellite based directional wave spectra, using a second order non-linear wave model. The procedure was validated for the same North Sea area and the developments indicated that the methodology could be an effective tool to parameterize the distribution of maximum wave and crest heights for design purposes in any oceanic area.

Introduction

The Corrected Weibull-Stokes (CWS) model proposed by Nerzic and Prevosto (1997) for the distribution of wave and crest heights is based on standard Rayleigh and Weibull models, modified using a third order Stokes expansion of the wave en-

---

1.IFREMER, DITI/GO/COM, BP 70, 29280 PLOUZANE, FRANCE
2.OPTIMER, 3 rue Jean Monnet, District de Montpellier, 34830 CLAPIERS, FRANCE

velope.

In the model, the Stokes expansion allows to take into account the non-linearity in the wave kinematics, in relation with wave steepness. The effect of the steepness on wave crest elevation is particularly clear from site measurements, as shown on figure 1 from wave data at Frigg field in the North Sea. It can be observed also that the effect of steepness on wave height is small. And this is consistent with the Stokes expansion model.

However, other wave properties can influence the distribution of wave heights and crest elevations, for example the directionality and the spectral shape as well as the water depth or the wave length-to-depth ratio. And this tends to indicate that the distributions of wave heights and crest elevations might be site dependent, in relation with wave climate and water depth. Unfortunately, there are only few oceanic sites with a wave data base that can allow analysing the wave distribution. And this particularly because of the limited accuracy of measurement of maximum wave crests by wave buoys.

Wave observations from satellite give access to second order statistical information: the significant wave heights from the altimeter and the wave directionality and the wave spectral shape from the synthetic aperture radar (SAR). These data do not permit to describe the non-linear characteristics of the wave field but they provide information about the crestedness and the frequency peak narrowness of the sea state.

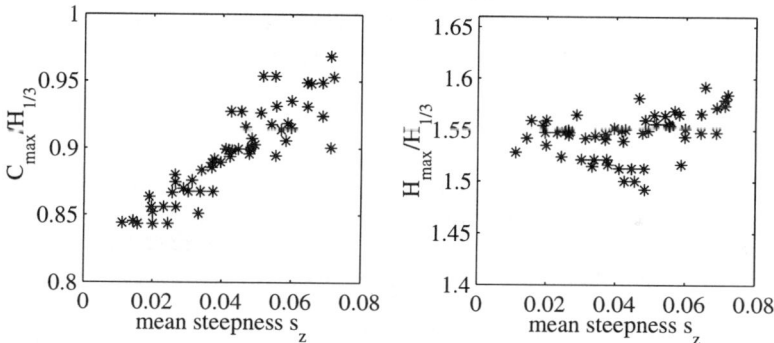

Figure 1. $C_{max}/H_{1/3}$ mode (left) and $H_{max}/H_{1/3}$ mode (right) vs. Steepness – Frigg field data

Second order non-linear wave models allow computing time series of non-linear waves from directional spectra. This permits to supplement the spectral information given by satellites in order to generate data bases of wave elevation time series on a specific site and then the maximum wave heights and crest elevations. So, we can avoid the difficult theoretical transformation of the statistics of the wave spectral data set to the wave statistics through the non-linear transfer of the wave kinematics. Then, the wave data base can be processed as actual data and thus, they can provide the fitted parameters of the distribution of the maxima in any oceanic area.

This methodology was applied to directional wave spectra from satellite observations around Frigg field and the results were compared to the parameters derived from the analysis of site measurements. The analysis allowed stating the ability of the proposed methodology to replace the processing of in-situ wave data.

Weibull-Stokes model

Weibull model with Stokes expansion. The Weibull-Stokes model is based on the Weibull (or Rayleigh) distribution and on the assumption that the sea surface process is a distortion of a first-order Gaussian process due to non-linear effects. The amplitudes of the first order process are distributed as a Weibull distribution (or a Rayleigh distribution for narrow banded process) and the non-linear effects can be introduced with a second- or third-order Stokes expansion.

Thus, after Vinje (1989), the following relations between crest and wave heights (C and H) and wave amplitude (a) of the first order process can be considered at infinite water depth:

$$H = 2a(1 + b_3(k_m a)^2) \text{ and } C = a(1 + b_2(k_m a) + b_3(k_m a)^2) \qquad (1)$$

where $b_2 = 1/2$ and $b_3 = 3/8$ are the parameters of the third-order Stokes expansion, in infinite water depth and k_m is a mean wave number, related to the mean wave period T_m, defined either from the spectral moments ($T_m = (m_0/m_2)^{1/2}$) or as the mean zero down-crossing period T_z.

Then, the maximum wave heights and crest elevations are asymptotically distributed as a Gumbel distribution with parameters derived from the parameters of the Gumbel distribution associated to the first order process. For example, with a_N the mode parameter of the maxima distribution of the first order process, it gives for the mode parameter of the maximum crest elevations:

$$a_{CN} = a_N(1 + b_2(k_m a_N) + b_3(k_m a_N)^2) \qquad (2)$$

Corrected Weibull-Stokes model. After the analysis of wave data from the Frigg field in the North Sea (cf. Nerzic and Prevosto (1997)), a Corrected Weibull-Stokes (CWF) model was proposed, with two correction factors. The first correction factor, α_k, was for the mean wave steepness or wave number and the second factor, α_a, for the wave energy term or the wave amplitude. This resulted in a simple asymptotic parameterized Gumbel law for the non-normalized maxima of wave height and crest elevation with the mode and the scale parameters given by relations like (2).

The parameters θ, β (resp. scale and shape parameter of the Weibull law) and α_k were as follows (Table 1), with the two different definitions of the couple (H_s, T_m), first with (H_{m0}, T_{02}) and then with ($H_{1/3}, T_z$). The correction factor α_k applies to the wave number k_m given by T_m transformed by the dispersion relation and the correction factor α_a is entered in θ.

Table 1. Parameters of the Weibull-Stokes model – Frigg field data

Weibull-Stokes Model	H_{max}/H_s			C_{max}/H_s		
	θ	β	α_k	θ	β	α_k
$H_s = H_{m0} / T_m = T_{02}$	0.73	2.38	0.7	0.71	2.08	0.6
$H_s = H_{1/3} / T_m = T_z$	0.77	2.38	0.7	0.78	2.18	0.7

The differences resulting from the selected definition of significant wave height and mean wave period are mainly on the scale parameter, about 5% to 10% larger for $H_{1/3}$ than for H_{m0}, reflecting the bias between the two significant wave height parameters. The main results are presented in figure 2, with the data-to-model ratios from the mode parameters, for both H_{max}/H_{m0} and C_{max}/H_{m0}. The bias are nil and the standard deviations are less than 2 or 3%.

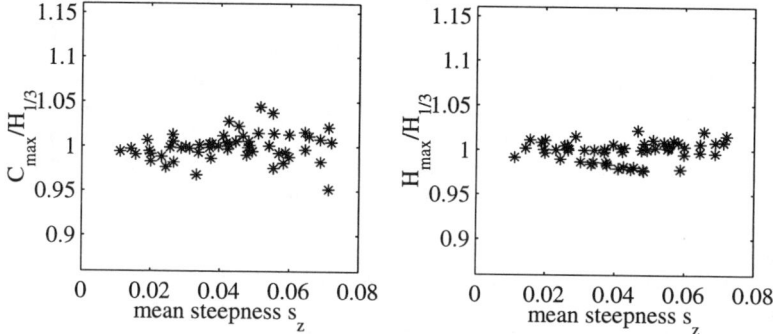

Figure 2. $C_{max}/H_{1/3}$ (left) and $H_{max}/H_{1/3}$ (right) Model-Data mode ratio vs. Steepness – Frigg field data / Weibull-Stokes model

PARAMETERIZATION OF THE WEIBULL-STOKES MODEL

If the CWS model of distribution of maximum wave and crest heights is a very general model, the parameters of the model are site dependent because the wave properties vary from one oceanic area to an other. The main site characteristics which can influence the shape of maximum waves are the water depth, because it is clear that offshore waves are distorted when propagating in shallow waters, and also the meteorological and oceanographic features which act on the wave directionality and on the wave spectral shape. The directional and spectral properties of waves act on the steepness of waves and on the wave pattern, and can lead to very different sea states, from long unidirectional swell to mixed wind seas. Thus, the site dependent wave climate has a main influence on the distribution of maximum wave and crest heights.

However, there are very few site data to analyse the influence of wave climate on the distribution of maximum wave and crest heights. There are few extensive wave surveys available for such analysis, excepted for the North Sea and the Gulf of Mexico. And often the available wave data come from buoy measurements, therefore with limited accuracy of wave crest data.

But the developments of second-order non-linear wave simulation models allow to generate wave elevation time series from directional wave spectra. This allows to generate synthetic data bases of wave statistics from a climatic description of wave conditions in any oceanic area, provided that the requested sea state statistics are available, that means statistics of significant wave height, directionality and spectral shape. And the only extended source for such wave statistics are from satellite observations.

Wave statistics from satellite observations. Satellite observations provide statistics of offshore sea states in two different ways: significant wave heights from altimetric data and directional spectra from SAR images. The main advantage of satellite observations is their availability for any oceanic area, with only a limitation at distances less than 5 to 10 km from shore.

The satellite data have some limitations, particularly the SAR directional spectra are limited to long waves, with a cut-off frequency about 0.14 Hz. But a procedure described in (Hajji, 1993) allows to complement the spectra in the high frequency band with wind wave spectra generated from scatterometer data.

Thus, the directional wave spectra derived from satellite observations are provided with a resolution of 24 angular bands (15° angle) and 22 frequency bands (from 1/25 Hz to 1/4 Hz).The frequency resolution is quite low, with a cut-off frequency about 0.25 Hz, and the angular resolution is 15°. This can lead to some limitations in the use of the spectra, which are discussed in Bitner-Gregersen & al. (1996). The limitations affect mainly the accuracy of wave period parameters (peak frequency, mean period, etc. and therefore the steepness parameters) rather than the accuracy of wave height parameters, because the wave energy outside the frequency band remains very low. Thus, the results of second order wave simulations are only slightly affected, with mainly some bias on wave period parameters.

Second-order wave simulation. The wave simulation model applied for the generation of non-linear irregular waves is a second-order non-linear wave model for multidirectional waves developed by Prevosto (1998) on the basis of the model described by Ding & al. (1994). Note that 25 seconds were necessary to simulate a wave time series of 4096 time step by a Personal Computer with a 133 MHz processor and 32 Mb RAM.

Procedure for parameterization. The basis for the parameterization of the CWS model is simply to generate a data set of maximum wave and crest heights, H_{max} and C_{max}, from a data base of directional wave spectra derived from satellite observations, using the second-order wave simulation model.

To be adequate for the purpose of deriving the distribution of maximum wave and crest heights at a given oceanic site, the wave spectra data base must be representative of the wave climate at the site. In particular, wave spectra for storm conditions must be present to analyse the distribution of extreme wave and crest heights. Thus, several wave time series have to be simulated to get a sufficient data set of H_{max} and C_{max} to allow the analysis of their distributions.

Then, the H_{max} and C_{max} data have to be grouped by class intervals of H_s and T_m and the H_{max}/H_s and C_{max}/H_s ratio distributions in each class interval have to be fitted to Gumbel distributions. The mode and scale parameters of the fitted distributions can then be analysed in relation with the proposed Corrected Weibull-Stokes model, in order to determine the parameters θ, β and α_k.

This procedure can be applied for any oceanic area, provided that satellite based directional wave spectra are available. And with the procedure, a model of distribution of maximum wave and crest can therefore be derived for any oceanic area, but the procedure requests that a representative data base of wave spectra be built.

Site case analysis

The methodology was applied to a wave spectra data base over 3 years from satellite observations in an area of 200km×200km centred on the Frigg field in the North Sea. The data base consisted of 118 directional wave spectra. The wave height – period and the wave height – steepness scatter diagrams of this data base are presented at figure 3, with $H_s = H_{m0}$, the significant wave height, $T_m = T_{02}$, the spectral mean period, $s_m = \frac{k_m}{2\pi} H_s$, the mean steepness and k_m, the mean wave number, related to T_m with the dispersion relation.

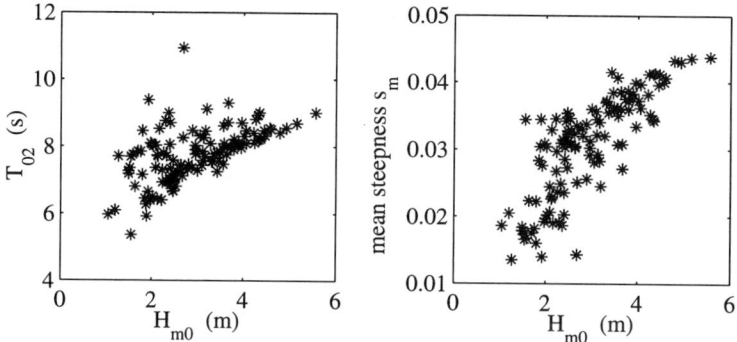

Figure 3. Wave height – period (left) and wave height – steepness (right) scatter diagrams – Satellite data base

The data base was limited to 118 spectra, with a maximum significant wave height of 5.6 m and a maximum steepness of 0.05. This can be compared to the maximum significant wave height of 12.4 m and to the maximum steepness of 0.08, in the Frigg wave data base used for the validation of the Weibull-Stokes model. However, the data base was sufficient to test the methodology, because several synthetic wave time series can be generated from each directional wave spectra. 100 wave time series were simulated for each wave spectra, i.e. a total of 11800 wave time series.

The H_{max} and C_{max} data were grouped by class intervals of H_s and T_m, with bandwidth of resp. 0.5 m and 1.0 s. Then, the H_{max}/H_s and C_{max}/H_s ratio distributions in each class interval were fitted to Gumbel distributions, using the maximum likelihood method. And the parameters of the fitted distributions were analysed in relation with the proposed Corrected Weibull-Stokes model.

The procedure was applied for the two different definitions of the couple (H_s,T_m) considered in the study. The parameters θ, β and α_k derived from the analysis are presented hereafter:

Table 2. Parameters of the Weibull-Stokes model – Satellite data

Weibull-Stokes Model	H_{max} / H_s			C_{max} / H_s		
	θ	β	α_k	θ	β	α_k
$H_s = H_{m0} / T_m = T_{02}$	0.67	2.11	0.6	0.70	2.00	0.5
$H_s = H_{1/3} / T_m = T_z$	0.80	2.48	0.6	0.82	2.32	0.5

The main results are presented in figure 4, with the data-to-model ratios of the mode parameters, for both H_{max}/H_s and C_{max}/H_s. They show a good agreement between the model and the data. The bias are nil and the standard deviations are less than 2 to 4%.

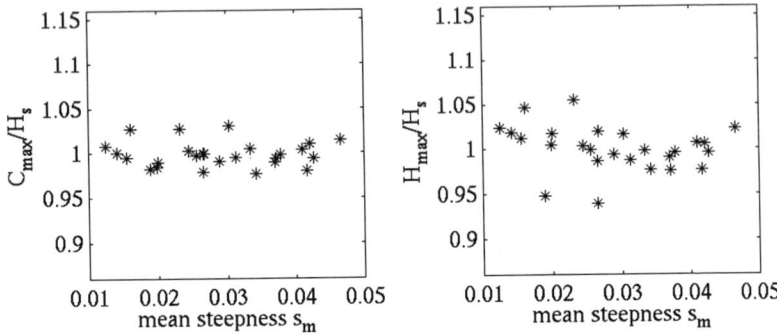

Figure 4. C_{max}/H_s (left) and H_{max}/H_s (right) Model-Data mode ratio vs. Steepness – Satellite data / Weibull-Stokes model

There are some differences between the parameters θ and β calculated for (H_{m0}, T_{02}) and for $(H_{1/3}, T_z)$, which cannot be explained only by the bias between H_{m0} and $H_{1/3}$. In fact, several pairs of parameters θ and β would be suitable for the model, that means with low bias and standard deviations between the mode and the scale parameters from the models and from the data.

The parameters θ and β show also some differences with the parameters derived from field measurements and presented in table 1, particularly the parameters for H_{m0} and T_{02}. However, the predictions of actual maximum wave and crest heights by the present models are good, as indicated by the ratios between the field data and the predictions with the model based on satellite data. The data-to-model ratios of the mode parameters are presented in figure 5 and they show a good agreement. The bias are less than 1% and the standard deviations are less than 3%.

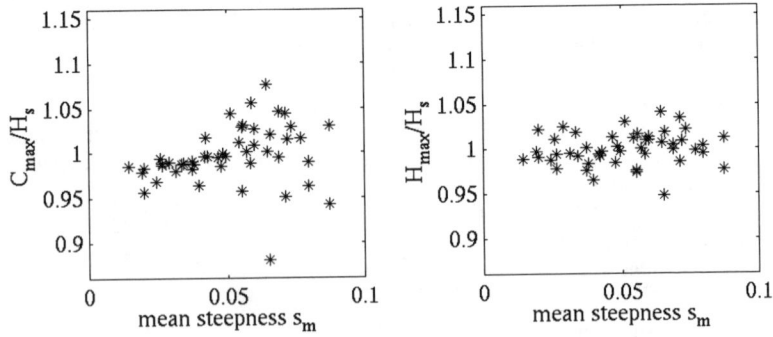

Figure 5. C_{max}/H_s (left) and H_{max}/H_s (right) Model-Data mode ratio vs. Steepness – Frigg field data / Weibull-Stokes model from Satellite data

Conclusions

The methodology developed for the parameterization of the distribution of maximum wave and crest heights from satellite based directional wave spectra, using second-order wave simulations, has proven its efficiency for the tested site case. This gives a complementary evidence of the suitability of the Weibull-Stokes model to represent the distribution of maximum wave and crest heights and thus to derive the extreme wave height and crest elevation parameters for design purposes.

The results indicated also the ability of the procedure to derive the distribution of maximum wave and crest heights from sea observations by satellites, without the help of site measurements, even if site data can provide a better accuracy. This important result came probably for two reasons. First, the satellite observations allow deriving the directional wave spectra with sufficient accuracy for the purpose, in spite of the low accuracy and resolution of the data. Secondly, the second-order wave simulation models are adequate to represent the non-linear characteristics of ocean waves which concern the wave heights and crest elevations.

The procedure was tested for a site of the North Sea in 100 metres of water, in almost deep-water conditions. In shallower waters, the non-linearities are affected also by the water depth and mainly by the wavelength-to-depth ratio. The ability of the proposed procedure to parameterize the distribution of maximum wave and crest heights in shallow waters still has to be demonstrated.Acknowledgments

The authors are grateful to Elf for permission of using Frigg Field data, and Météomer for providing the data set of inverted SAR spectra

References

Bitner-Gregersen E. & al. (1996). "World-wide characteristics of Hs and Tz for long-term load responses of ships and offshore structures", *Proc. 6th ISOPE Conf.*, vol. III, pp. 95-102.

Ding, P.-X., Sun, F., Yu, Z.-W. (1994). "Study of the second-order nonlinear characteristics of ocean waves (1) -- theoretical derivation", *Sci. China Ser. B*, vol. 37, no. 5, pp. 625-633.

Hajji, H., Bonicel, D., Chaperon, B., Lasnier, P., Loeuil, S. (1993). "Use of SAR wave mode, altimeter and wind scatterometer data in operational wave forecasting system", *Proc. 2nd ERS-1 Symposium*, E.S.A., Hamburg.

Nerzic, R., Prevosto, M. (1997). "A Weibull-Stokes Model for the Distribution of Maximum Wave and Crest Heights", *Proc. of the 7th ISOPE Conf.*, vol. III, pp. 367-377.

Prevosto M. (1998) "Effect of directional spreading and spectral bandwidth on the nonlinearity of the irregular waves", *Proc. ISOPE 98*, to be published.

Vinje T. (1989). "The statistical distribution of wave heights in a random seaway", *Applied Ocean Res.*, vol. 11, no. 3.

Numerical forecasting of infra-gravity waves near harbour entrances

Alexandru Sheremet[1]

Abstract

We present a numerical model for the forecasting of the infra-gravity waves generated during the shoaling of arbitrary gravity-wave directional spectra, developed as a part of the Sea21 Project for forecasting operability conditions of marine installations. The aim of the model is to provide information about the energy distribution in the infra-gravity wave band (with frequency less than 0.05Hz) of the spectrum, at water depths that are typical of harbour entrances (between 10 to 20m). The model accounts for wave refraction and nonlinear quadratic interactions, and is designed to accommodate relatively large spatial scales (several hundreds of spectral peak wave-lengths).

1. Introduction

Information about the essentially nonlinear evolution of shoaling infra-gravity waves plays an important role in the prediction of low frequency harbour oscillations and the evaluation of operability conditions of marine installations. At open sea, the infra-gravity waves are nearly absent from the spectrum; they are generated during the shoaling process through nonlinear interaction among wind waves, a mechanism that becomes significant as the water depth decreases. Most of the existing operative numerical models are either linear or use some nonlinear interactions parameterizations that mainly transfer energy to the second harmonic of the spectral peak, and are therefore inadequate for the prediction of long waves. The present model forecasts the spectral density in long wave band, and is designed to accommodate relatively large spatial scales (several hundreds spectral peak wave-lengths).

The quadratic nonlinear mild slope equation (NEMS) derived by Sheremet [6] provides a simple and consistent description of wave refraction, diffraction

---

[1] Center for Coastal Studies, Scripps Institution of Oceanography, UCSD, 9500 Gilman Drive La Jolla, CA, 92093-0209

and nonlinear wave interactions over a fully three-dimensional bathymetry and has the advantage that no a-priori assumptions are made about the shape of the free surface; it deals with the directionality of the wave spectrum in a natural way and all directions of propagations are taken into account. However, NEMS integration over large spatial and frequency domains is a heavy numerical task; it also requires adequate boundary conditions set along a sometimes geometrically complicated shore-line; to describe correctly topographically trapped waves, it needs to be extended to include essential higher order cubic terms, making the problem even more difficult to solve.

Two steps are taken in this work to simplify the problem and preserve the essential physical mechanisms responsible for the long waves generation: A WKB expansion that transforms NEMS to a more tractable initial value problem, and a formal decoupling of the equations, based on the observation that in most of the shoaling region the long waves energy is significantly smaller than the spectral peak energy.

In Section 2 we present the mathematical formulation of the problem and give a short description of the algorithms used for the calculations of the wind- and infra-gravity waves respectively. Numerical simulations and comparison with measured data are presented in Section 3. Section 4 summarizes the results.

2. Formulation of the problem

The steady state NEMS (see the derivation in Sheremet [6]) has the form:

$$\nabla \cdot (cC_g \nabla \varphi) + \omega^2 \frac{C_g}{c}\varphi = \frac{i\omega}{4\pi}\int_{-\infty}^{\infty}\left[\frac{\omega_1\omega_2}{g^2}\left(\omega^2 - \omega_1\omega_2\right)\varphi_1\varphi_2 + \right.$$
$$\left. + 2\nabla\varphi_1\cdot\nabla\varphi_2 + \frac{\omega_1}{\omega}\varphi_1\nabla^2\varphi_2 + \frac{\omega_2}{\omega}\varphi_2\nabla^2\varphi_1\right]\delta(\omega - \omega_1 - \omega_2)d\omega_1 d\omega_2, \quad (1)$$

where C_g and c are the group and phase velocities, $\varphi_1 = \varphi(...,\omega_1)$ and $\varphi = ig/\omega\,\hat{\eta}$ where $\hat{\eta}$ is he Fourier transform of the free surface displacement. Equation (1) is valid under the basic assumptions that the gradient of the water depth is small (the mild slope assumption), the varying bathymetry affects the waves evolution on the same scale as the leading order nonlinear wave interactions, and the effect of background currents and wave setup/set down is negligible.

Other processes not taken into account in the present implementation are wave breaking, energy dissipation by bottom friction, energy input from the wind. The dependency of the deep-water directional spectrum on the spatial coordinates is also neglected.

The WKB expansion amounts to representing the free surface as a superposition of slowly modulated plane wave of amplitude A and phase θ, resulting in the evolution equation:

$$\frac{1}{2}A\nabla\cdot\mathbf{C}_g + \mathbf{C}_g\cdot\nabla A = \int_{-\infty}^{\infty}\mathbf{W}_{0:1,2}A_1 A_2 e^{-i\Delta_{0:1,2}^{\theta}}\delta(\omega - \omega_1 - \omega_2)d\omega_1 d\omega_2 \quad (2)$$

where $\mathbf{W}_{(0,1,2)}$ is an interaction coefficient depending on ω, ω_1 and ω_2 and $\Delta^\theta_{0;1,2} = \theta_0 - \theta_1 - \theta_2$.

Given the deep-water wave spectrum, Equation (2) may be solved as an initial value problem. Due to the nonlinearity, however, the information about the modal amplitudes and phases at a given point may be obtained only if all the upstream dependencies related to that point are resolved, that is, if the history all the waves reaching that point is known. This may become a major obstacle when the shoaling domain spans several hundreds wave-lengths.

One may argue, however, that in most of the shoaling region the long waves energy is significantly smaller than the spectral peak energy and therefore the energy feedback from the long waves to wind waves should be negligible within the leading order 'difference' coupling (while the 'sum' interaction is a higher order process, much less efficient). Neglecting the energy feedback from the long waves to the wind waves, the system may be formally written as:

$$\text{long waves}: \quad \mathcal{L}(A_L) = \mathcal{N}(A_W, A_W), \qquad (3)$$
$$\text{wind waves}: \quad \mathcal{L}(A_W) = 0, \qquad (4)$$

where $\mathcal{L}$ and $\mathcal{N}$ denote the linear and nonlinear operators appearing in Equation (2) and the indices W and L stand for wind- and long waves.

Equation (4) for the wind-waves is linear and may be integrated by fast standard ray methods to obtain the wave field at any point. The present numerical model implements the method due to Longuet-Higgins [4], (see also Graber et al. [3]).

After bringing Equations (4) and (3) to a discrete form, defining the wave ray as a curve everywhere tangent to the group velocity vector and performing some standard transformations, one may write for the infra-gravity wave spectral mode "j" the equation:

$$\frac{dB_j}{ds_j} = i \left(\frac{\sigma_j}{C_{g,j}}\right)^{1/2} \sum_{\alpha,\beta \in \mathcal{W}} 2\mathbf{W}_{(j,-\alpha,\beta)} A^*_\alpha A_\beta e^{i\Delta^\theta_{\beta;j,\alpha}} \delta(\omega_\beta - \omega_j - \omega_\alpha), \qquad (5)$$

where $\mathcal{W}$ is the set of indices defining the wind-wave band, s_j is the coordinate along the ray corresponding to the spectral mode "j", σ_j refraction coefficient and

$$B_j = A_j \left(C_{g,j}\sigma_j\right)^{1/2}. \qquad (6)$$

Equation (5) is implemented in the present model to describe the long waves evolution. To solve this new initial-value problem, one needs to know the geometry of the infra-gravity wave ray and the wind-wave field along it. Note that Equation (5) does not provide any information about the geometry of the ray and that this information cannot be extracted from the deep water spectral data, since the long waves are usually absent from the deep water spectra (the second order long waves are bound waves forced by pairs of wind waves). In deep water a directional wind-wave spectrum will generate a complex, directionally spread spectrum of bound infra-gravity waves; however, if the wind-wave spectrum has a well defined peak and small angular spread, the main part of the energy in the locked infra-gravity wave band is concentrated on the direction of the peak. For the purpose of the present model, which is intended to simulate the infra-gravity wave generation up to the entrance of a harbor, at 15m-20m depth, the simplification adopted was to identify the long waves rays with the spectral peak ray.

3. Numerical implementation

3.1 The numerical model

The approach described in the previous sections has been implemented as the infra-gravity waves forecasting module of the operative model for forecasting the operability conditions of marine installations, 'Sea21', in pilot stage since September 1997.

The integrations may be divided into two distinct steps: the geometry-related calculations, requiring only geometrical data such as the bathymetry matrix, the direction and the wavelength of the spectral peak in deep water, and the actual infra-gravity waves calculations. At the first stage the spectral peak propagation ray is computed up to the target point (harbor entrance) together with the transfer functions for the wind-wave refraction problem, for all the points that describe the spectral peak ray. Any incoming deep water spectrum having the given direction and peak wave-length will be refracted and shoaled using the data produced by this module, so that these data are, in a certain sense, of general use. If the possible incoming directions are discretized, the full structure of the actual directional spectrum is not important at this stage in the calculations; two spectra having the same peak parameters will be treated by the first package of programs in the model as the same spectrum, irrespective, for example, of the significant wave-heights. The first module needs then to be run only once for a given direction and peak frequency, and the computed rays and transfer functions may be used directly by the long waves evolution integrations.

At the second stage Equation (5) is solved using as initial data the full deep-water directional wind-wave spectrum, evolving along the ray and using the transfer functions computed at the first stage. The phases of the wind-wave modes along the ray are computed with respect to the first point of the ray, generating a set of ready-made phases for all the wind-wave frequencies. A "realization" is obtained by translating these phase arrays by random numbers uniformly distributed between 0 and 2π. For each such realization Equations 5 are integrated to yield the infra-gravity wave spectrum at the target point. The final gravity-wave spectrum is the average of a number of such realizations.

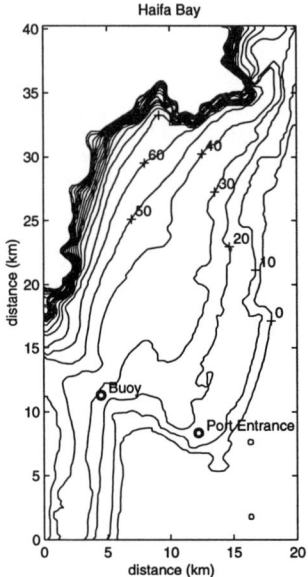

Figure 1: The bathymetry used in the numerical simulations, with the locations of the Datawell buoy and the Haifa Port entrance (contour levels in meters).

3.2 Haifa Bay simulations

Haifa Bay provides a good setting for testing the performance of the model, due to its dimensions and shape – the relevant refraction/shoaling domain spans an average of 20 to 30km over a slowly varying bathymetry of less than 50m depth. We present here samples from a series of preliminary tests, carried out using the data provided by the directional Datawell buoy and pressure sensors (IOLR Report No. H06/96 [5]) located at the breakwater head – Haifa Port entrance. A sketch of the bathymetry used and the locations of the two gauges is shown in Figure 1.

	Datawell 95-02071826	Bw. Head 02-07-18	Num.	Datawell 96-03071147	Bw.Head 03-07-12	Num.
H_{m_0}(m)	4.23	1.18	1.6	4.52	1.73	2.0
T_p(s)	9.1	10.3	9.1	12.3	11.6	11.4
θ(deg)	251	332		268	354	
H_{04}(m)		0.18	0.21		0.27	0.30

Table 1: Simulated and measured data for two storms (see text for explanations)

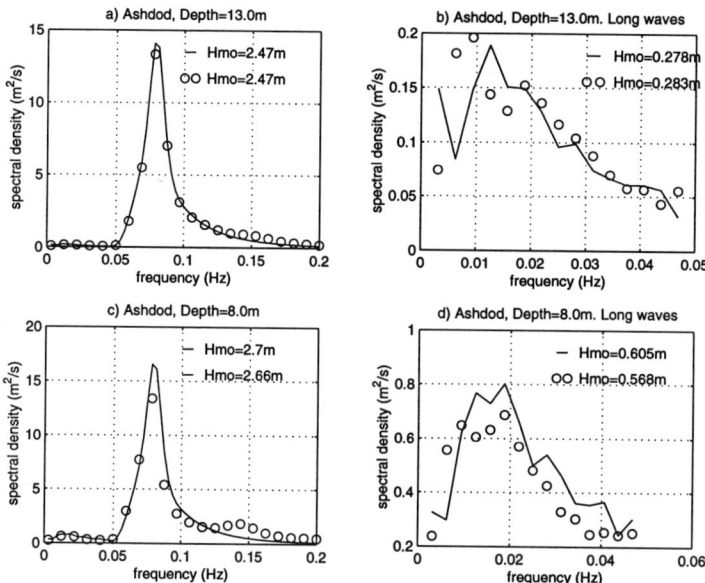

Figure 2: Numerical results of the present model (full line) compared to the fully nonlinear unidirectional model (circles) for Ashdod Beach. Deep water spectral parameters: Direction WNW, significant heigh 2.33m, peak period 12.5s. a,b) the shoaled spectrum, 13m depth; c,d) shoaled spectrum, 8m depth.

Table 1 summarizes the numerical results for storms 7 February 1995, (18:26 hrs.) and 7 March 1995, (11:47 hrs). Three columns are displayed for each storm: The Datawell buoy records, the breakwater head records and the numerical results. Each column lists the total significant wave height (H_{m_0}), the spectral peak period (T_p), the propagation direction (θ) and the significant wave height corresponding to the domain from 20s to 4min (H_{04}). The long wave band in the numerical simulations, used to compute the 'equivalent' H_{04} is from 20s to 320s. The numerical model is seen to provide a rather good estimation for the energy accumulated in the infra-gravity wave band.

An consistency check was performed on the model by comparing its results to those yielded by the integration of the unidirectional, fully nonlinear evolution equation (see Agnon et al. [1]) for the Ashdod beach, which is approximately cylindrical. The directional model shoaled the deep-water spectrum along the ray corresponding to a period 12.5s wave propagating almost normally to the shore (incoming direction WNW). The spectrum shoaled by the unidirectional model is of a JONSWAP type:

$$S(\xi) = \frac{H_s^2 T_p}{16N} \xi^{-5} \exp\left(-\frac{5}{4}\xi^{-\varpi}\right)\gamma^{\exp[-\frac{1}{2\sigma^2}(\xi-1)^2]}, \tag{7}$$

with S is the spectral density, $\xi = f/f_p$, f and f_p the frequency and peak frequency respectively ($\gamma = 2.0$, ϖ=-4). N is a normalization constant, defined so as to have total energy=$H_s^2/16$. An equivalent directional spectrum,

$$S(f, \theta, \theta_0) = S(f_p \xi) D(\theta, \theta_0), \quad \text{with} \quad D(\theta, \theta_0) = \cos^{2s}\left(\frac{\theta - \theta_0}{2}\right), \tag{8}$$

($s = 10$) was used as the deep water spectrum for the directional simulation. A comparison of the results of the two models for the infra-gravity wave band spectral density is shown in Figure 2. The present model behaves consistently up to a depth of about 7m.

Wherever a typical storm can be defined for a harbour, characterized by a small number of parameters, the infra-gravity wave band at the harbour entrance can be pre-calculated for all the storms of interest, and the real time operation of the model reduces to a database search, with dramatic decrease of runtime. Such a database was built for Haifa Bay, using model spectra from Definitions (8) and (7). The three relevant parameters that characterize the evolution of the deep-water wind-wave spectrum are the propagation direction, the peak frequency and the significant wave-height, which organizes the database corresponding to a 3-dimensional array consisting of discretized values of these parameters. Figure 2 shows the contour map of the infra-gravity wave band 'significant height' at the port entrance as a function of the direction and peak period of the incoming spectrum, for two total significant heights, 3.0m and 5.0m respectively. Note the large waves generated along rays coming around the Cape Carmel, which might be explained by the relatively shallow water and the longer span the waves have to develop along. Although storms of 5.0m height from SW in the 8s range are not realistic, the diagrams serve well to illustrate the general behaviour of the model and some characteristic features of topography.

4. Conclusions

A numerical model was derived for the infra-gravity waves generation in the domain ranging from deep water up to 15-20m water depth, based on the WKB expansion of the nonlinear elliptic mild slope evolution equations (NEMS) and the un-coupling of the wind- and infra-gravity waves, resulting in a pair of linear systems of equations. The infra-gravity waves evolution equation contains a forcing terms due to wind-waves. The approach preserves the essential mechanisms of infra-gravity wave generation. Numerical simulations compare well to the available measured data. Additional consistency checks were carried out for an almost cylindrical beach, in which the results of the present model compared well also to that of a unidirectional fully nonlinear model.

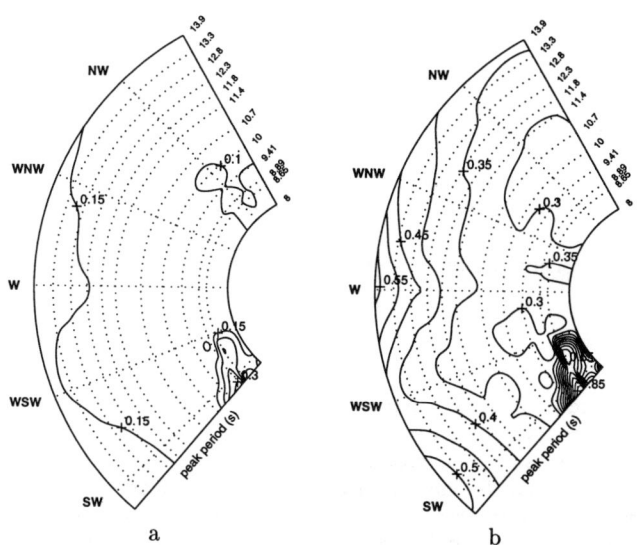

Figure 2: The infra-gravity wave band significant height (contour levels in meters) at the Haifa Port entrance, as a function of the direction and period of the deep-water spectral peak, for deep-water wave heights of a) 3.0m, b) 5.0m.

References

1. Y. Agnon, A. Sheremet, Gonsalves, J., and M. Stiassnie. A unidirectional model for shoaling gravity waves. *Coastal Engineering*, **20**, pp. 29-58, 1993.

2. J.C.W. Berkhoff. Computation of combined refraction diffraction. *Proc. 13th Conf. Coastal Eng. ASCE*, **1**, pp.471-490, 1972.

3. H.C Graber, M. Byman, and H Gunther. Numerical simulation of surface wave refraction in the North Sea, part 1: Kinematics and part 2: Dynamics. *Dt. hydrogr. Z.*, **43 - 44** pp.1-17/1-15, 1990.

4. M.S. Longuet-Higgins. On the transformation of a continuous spectrum by refraction. *Proc. Camb. Phil. Soc.*, **53** p. 226-229, 1957.

5. D.S. Rosen and Raskin L. Long waves monitoring in Haifa Port. Technical Report H06/96, IOLR,Israel Oceanographic and Limnological Research, Israel, 1996.

6. A. Sheremet. *Wave interactions in shallow water*. PhD thesis, Technion – Israel Institute of Technology, Haifa 32000, Israel, 1996.

7. M. Stiassnie and M. Glozman. Sea21 - forecasting operability of marine installations. *Submitted to the 5-th International Workshop on Wave Hindcasting and Forecasting, Melbourne, Florida*, 1998.

Development and Study of Green's Function for Water Waves Over Variable Bathymetry Domains

Gerassimos A. Athanassoulis[1] and Konstadinos A. Belibassakis[2]

Abstract

An enhanced local-mode representation of the wave potential in variable bathymetry regions is applied to the problem of propagation of water waves, emitted from a monochromatic point source. The representation includes, except of the propagating mode and the infinite system of evanescent modes, also an additional mode, which takes into account the effects of the bottom slope. Using this representation in conjunction with a variational principle, a forced system of horizontal coupled-mode equations is derived for the determination of the complex modal-amplitude functions. This system contains an additional equation associated with the sloping-bottom mode and produces solutions consistent with the bottom boundary condition. The present method provides the means for the efficient treatment of complex wave-body interaction problems, in the case when the structures are large and thus the local bathymetry variation effects are important.

Introduction

The interaction of floating and/or immersed bodies with the uneven bottom topography in the coastal environment is a mathematically difficult problem that, recently, has also become practically important, as the dimensions of large floating obstacles increase. That means that the characteristic lengths: horizontal dimension of the body L, wave length λ, bottom variation length λ_b, and average depth h may be all comparable and, thus, the ratio h/λ may be neither large enough (> 0.5) nor small enough (< 0.07) so that the deep or the shallow water models to be applicable. Furthermore, the majority of the existing shallow water models, developed for coastal

---

[1] Prof. Dept. of Naval Architecture and Marine Engrg., National Technical Univ. of Athens P.O. Box 64070, Zografos 15710, Athens, Greece

[2] Res. Engr., Dept. of Naval Architecture and Marine Engrg., National Technical Univ. of Athens P.O. Box 64070, Zografos 15710, Athens, Greece

environment applications, are based on the assumption of small or mild bottom slope. See, e.g., the relevant surveys by Porter and Chamberlain (1997) and Dingemans (1997). Thus, it seems timely to investigate water wave problems in a marine environment such that λ, λ_b and h are all of the same order of magnitude, the latter being also non-constant because of a bottom topography variation. Although the non-linearity of water waves is of great importance, a consistent linear solution is still very useful, providing a great deal of information concerning the wave field.

Consider the marine environment illustrated in Fig. 1. This environment consists of a water layer D_{3D} bounded above by the free surface $\partial D_{F,3D}$ and below by a rigid bottom $\partial D_{\Pi,3D}$. It is assumed that the bottom slope exhibits an arbitrary one dimensional variation in a subdomain of finite length, i.e. the bathymetry is characterised by parallel, straight bottom contours lying between two regions of constant but (possibly) different depth. The liquid is assumed homogeneous, inviscid and incompressible, and the flow is excited by a monochromatic point source of angular frequency ω, located at a point in the variable bathymetry region. Our main concern in formulating and studying this problem is to investigate the far-field wave pattern which, in this case, is azimuthally strongly anisotropic.

Differential formulation of the problem

A Cartesian coordinate system is introduced, with its origin on the mean water level, the z-axis pointing upwards and the y-axis being parallel to the bottom contours, Fig.1. The liquid domain is represented by $D_{3D} = D \times R$, where $D = \{(x,z): x \in R, -h(x) < z < 0\}$ is the intersection of D_{3D} by a vertical plane perpendicular to the bottom contours, $R = (-\infty, +\infty)$ is a copy of the real line and $h(x)$ represents the local depth. The depth function is assumed sufficiently smooth, and $h(x) = h(a) = h_1$, for $x \le a$, $h(x) = h(b) = h_3$, for $x \ge b$. The domain D_{3D} is decomposed in three subdomains $D_{3D}^{(i)} = D^{(i)} \times R$, $i = 1,2,3$, defined as follows: $D_{3D}^{(1)}$ is the constant-depth subdomain characterised by $x < a$, $D_{3D}^{(3)}$ is the constant-depth subdomain characterised by $x > b$, and $D_{3D}^{(2)}$ is the variable bathymetry subdomain lying between $D_{3D}^{(1)}$ and $D_{3D}^{(3)}$. Finally, we define the vertical interfaces $\partial D_{I,3D}^{(12)} = \partial D_I^{(12)} \times R$ and $\partial D_{I,3D}^{(23)} = \partial D_I^{(23)} \times R$, which are the common vertical boundaries of subdomains $D_{3D}^{(1)}$ and $D_{3D}^{(2)}$, and $D_{3D}^{(2)}$ and $D_{3D}^{(3)}$, respectively. Clearly, $\partial D_I^{(12)}$ and $\partial D_I^{(23)}$ are vertical segments at $x = a$ and $x = b$, respectively.

Assuming that the surface elevation and the velocity field are small, the linearised water wave equations can be used. See, e.g., Wehausen and Laitone (1960). Then, the velocity field is time harmonic with angular frequency ω, the same as the pulsating source, and can be represented by a velocity potential of the form $\mathscr{P}(x,y,z;t) = Re\{\Phi(x,y,z;\mu)\exp(-i\omega t)\}$, where $\mu = \omega^2/g$ and g is the acceleration due to gravity and $i = \sqrt{-1}$. The wave problem $\mathscr{P}_{3D}(D_{3D}, \mu, \bar{x}_0)$ is formulated in terms of the frequency-dependent potential $\Phi = \Phi(x,y,z;\mu)$ and consists of the Laplace equation, the free-surface boundary condition, the bottom boundary condition and

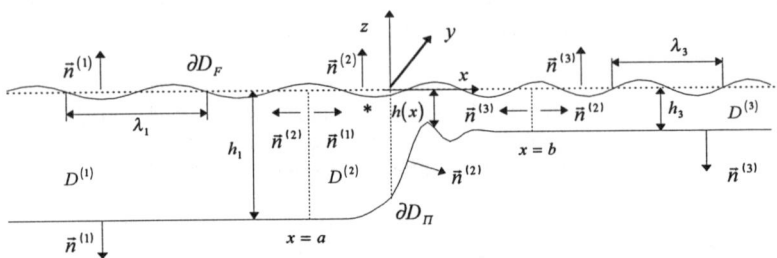

Figure 1. Domain decomposition and notation. The source is denoted by (*).

of the radiation condition, requiring:

$$\Phi \sim \text{outgoing waves as } r = \sqrt{(x-x_0)^2 + (y-y_0)^2} \to \infty, \tag{1}$$

where $\vec{x}_0 = (x_0, 0, z_0)$ is the location of the point source.

Since the geometry of the domain is of "cylindrical" character, i.e. $D_{3D} = D \times R$, it is possible to reduce the dimensionality of the problem by taking the Fourier transform with respect to y. The transformed potential is denoted by $\varphi(x, z; \xi, \mu) = \varphi(x, z; \xi)$, (the μ-dependence being suppressed), and is given as follows:

$$\varphi(x, z; \xi) = \frac{1}{2\pi} \int_{-\infty}^{+\infty} \Phi(x, y, z) e^{-iy\xi} dy, \quad \Phi(x, y, z) = \int_{+\infty}^{-\infty} \varphi(x, z; \xi) e^{+iy\xi} d\xi. \tag{2.a,b}$$

Applying the Fourier transform to the problem $\mathscr{P}_{3D}(D_{3D}, \mu, \vec{x}_0)$, we obtain the following two-dimensional wave problem:

$$\nabla^2 \varphi - \xi^2 \varphi = -\frac{1}{2\pi} \delta(x - x_0) \delta(z - z_0), \quad \text{in } D, \tag{3a}$$

$$\frac{\partial \varphi}{\partial z} - \mu \varphi = 0 \quad \text{on } \partial D_F, \quad \frac{\partial \varphi}{\partial n} = 0 \quad \text{on } \partial D_\Pi, \tag{3b,c}$$

$$\varphi \sim \text{outgoing waves as } |x| \to \infty. \tag{3d}$$

The boundary value problem (3a,b,c,d) will be referred to as the *transformed wave problem* $\mathscr{P}_\xi(D, \mu, \vec{x}_0)$. This problem can be reformulated as a *transmission problem* in the bounded subdomain $D^{(2)}$ with the aid of appropriate representation of the wave potentials $\varphi^{(i)}(x, z; \xi)$ in the subdomains $D^{(i)}$, $i = 1,2,3$, respectively, and by requiring matching of the potential and its normal derivative the at the common vertical interfaces. In the case of the left half-strip $D^{(1)}$ ($x \le a$) this representation reads :

$$\varphi^{(1)}(x,z;\xi) = C_0^{(1)}(\xi)\exp\left(-xK_0^{(1)}(\xi)\right)Z_0^{(1)}(z) + \sum_{n=1}^{\infty} C_n^{(1)}(\xi) Z_n^{(1)}(z) \exp\left((x-a)K_n^{(1)}(\xi)\right), \quad (4)$$

where $K_0^{(1)}(\xi) \equiv K_0^{(1)} = \sqrt{\xi^2 - (k_0^{(1)})^2} \, \mathrm{sign}(k_0^{(1)} - |\xi|)$. A similar expansion is also derived for the right half-strip $D^{(3)}$; see Athanassoulis and Belibassakis (1997b). In the above expansions $C_n^{(1)}(\xi) \equiv C_n^{(1)}$, $n = 0,1,2,..$, $i = 1,3$, are undefined coefficients to be determined by means of the appropriate matching conditions at the interfaces $\partial D_I^{(12)}$ and $\partial D_I^{(23)}$. Also, the sets of numbers $\{ik_0^{(i)}, k_n^{(i)}, n=1,2,...\}$, $i = 1,3$, and the sets of vertical functions $\{Z_n^{(i)}(z), n=0,1,2,...\}$, $i = 1,3$, are, respectively, the eigenvalues and the corresponding eigenfunctions of regular Sturm-Liouville problems obtained by separation of variables in the half-strips $D^{(i)}$, $i = 1,3$. The eigenvalues are given as the roots of the dispersion relation $\mu h_i = -k^{(i)} h_i \tan(k^{(i)} h_i)$, $i = 1,3$, and the eigenfunctions $\{Z_n^{(i)}(z), n = 0,1,2,...\}$ are given by

$$Z_0^{(i)}(z) = \frac{\cosh\left(k_0^{(i)}(z+h_i)\right)}{\cosh\left(k_0^{(i)} h_i\right)}, \quad Z_n^{(i)}(z) = \frac{\cos\left(k_n^{(i)}(z+h_i)\right)}{\cos\left(k_n^{(i)} h_i\right)}, \, n=1,2,..., \quad i=1,3. \quad (5a,b)$$

The correctness (completeness) of the expansions (4) follows by the standard theory of regular eigenvalue problems. We only recall here that the series expansion in terms of the basis $Z_n^{(i)}(z)$, $n = 0,1,2,...$, of any $C^2(-h_i, 0)$-function is uniformly convergent in $[-h_i, 0]$, only if it satisfies the same (as the basis) boundary condition.

Variational formulation and the coupled-mode system of equations

The transmission problem admits a variational formulation in accordance with Bai's construction (Bai and Yeung 1974; Mei 1978). Consider the functional $\mathcal{F} = \mathcal{F}\left(\varphi^{(2)}, \{C_n^{(1)}\}_{n=0,1,...}, \{C_n^{(3)}\}_{n=0,1,...}\right)$ defined by

$$\mathcal{F} = \frac{1}{2}\int_{D^{(2)}}\left[(\nabla\varphi^{(2)})^2 + \xi^2(\varphi^{(2)})^2 - \frac{1}{\pi}\delta(x-x_0)\delta(z-z_0)\varphi^{(2)}\right]dV - \frac{1}{2}\mu\int_{\partial D_F^{(2)}}(\varphi^{(2)})^2 dS +$$
$$+ \int_{\partial D_I^{(12)}}\left(\varphi^{(2)} - \frac{1}{2}\varphi^{(1)}\right)\frac{\partial \varphi^{(1)}}{\partial n^{(1)}}dS + \int_{\partial D_I^{(23)}}\left(\varphi^{(2)} - \frac{1}{2}\varphi^{(3)}\right)\frac{\partial \varphi^{(3)}}{\partial n^{(3)}}dS, \quad (6)$$

where $\varphi^{(1)}, \varphi^{(3)}$ are considered only on the interfaces $\partial D_I^{(12)}$, $\partial D_I^{(23)}$, respectively. The variational formulation of the transmission problem is stated in the following

<u>Theorem A</u>: The functions $\varphi^{(2)}(x,z;\xi)$ in $D^{(2)}$, and $\varphi\left(x,z;\{C_n(\xi)\}_{n=0,1,...},\xi\right)$ in $D^{(i)}$, $i = 1,3$, form a solution of the transmission problem iff they render the functional $\mathcal{F}$ stationary, i.e. iff they satisfy the variational equation

$$\delta\mathcal{F}\left(\varphi^{(2)}, \{C_n^{(1)}\}_{n=0,1,...}, \{C_n^{(3)}\}_{n=0,1,...}\right) = 0. \quad (7)$$

Details about the proof of Theorem A can be found in Athanassoulis and Belibassakis (1997a). We simply note here that he appearance of the δ-function in the functional is just a matter of conciseness. It can be avoided if we introduce a decomposition of $\varphi^{(2)}$ of the form $\varphi^{(2)} = \varphi^{(2)}_{\text{reg}} + G_\xi(\bar{x}; \bar{x}_0)$, where $G_\xi(\bar{x}; \bar{x}_0)$ is the free-space Green function of the operator $\Delta - \xi^2$. In this case the functional $\mathscr{F}$ is defined on $\varphi^{(2)}_{\text{reg}}$ instead of $\varphi^{(2)}$ and, thus, no singular terms appear, at the expense of some additional (known) boundary terms, coming from the contribution of $G_\xi(\bar{x}; \bar{x}_0)$. In fact, this approach, which is also beneficial from the numerical point of view, is followed for deriving numerical results.

The above variational principle can be used to obtain an alternative, semi-discrete (Kantorovich) formulation in terms of *local modes*. This family of basis functions is obtained by formulating and solving a local vertical Sturm-Liouville problem in the interval $[-h(x), 0]$. The form of functions $Z_n(z; x)$, $n = 0,1,2,...$, is similar to (5a,b), with h_i and $k_n^{(i)}$, $n = 0,1,2,...$ being replaced by the local depth $h(x)$ and the local eigenvalues $k_n(x)$, $n = 0,1,2,...$. However, all these functions are incompatible with the bottom boundary condition (3c) whenever $dh(x)/dx \neq 0$. To remedy this inconsistency we introduce an additional mode, denoted by $\varphi_{-1}(x; \xi) \cdot Z_{-1}(z; x)$ and called the *sloping-bottom mode*, so that the expansion of the $\varphi^{(2)}(x, z; \xi)$ in the variable-bathymetry domain is (Athanassoulis and Belibassakis 1997a,b)

$$\varphi^{(2)}(x, z; \xi) = \varphi_{-1}(x; \xi) \cdot Z_{-1}(z; x) + \varphi_0(x; \xi) \cdot Z_0(z; x) + \sum_{n=1}^{\infty} \varphi_n(x; \xi) \cdot Z_n(z; x), \qquad (8)$$

where $Z_{-1}(z; x)$ is satisfying the conditions $Z_{-1}(0; x) = 0$, $dZ_{-1}(0; x)/dz = 0$ and $Z_{-1}(-h(x); x) = 0$, $dZ_{-1}(-h(x); x)/dz = 1$. Each term in the expansion (8) satisfies the free surface condition (3b) individually. Thus, representation (8) renders the free surface condition an *essential condition* in relation to the variational formulation (7). The term, $\varphi_0(x; \xi) Z_0(z; x)$ is the *propagation mode*, and the terms $\varphi_n(x; \xi) Z_n(z; x)$, $n = 1, 2, ..$, are the *evanescent modes*. The sloping-bottom mode $\varphi_{-1}(x; \xi) \cdot Z_{-1}(z; x)$ is not needed when the bottom is flat. Thus, the following end conditions can be imposed on the amplitude of the sloping bottom mode: $\varphi_{-1}(a; \xi) = \varphi_{-1}(b; \xi) = 0$.

Having constructed a complete local-mode (semi-discrete) representation of the function $\varphi^{(2)}(x, z; \xi)$, we proceed by using it in order to change arguments of the functional $\mathscr{F}$ and obtain an alternative formulation of our problem. Thus, after substituting (8) in (7), making the necessary algebraic manipulation and using the standard arguments of the Calculus of Variations with respect to the new arguments (degrees of freedom of the system), which are now: $\varphi_n(x; \xi)$, $a < x < b$, $n = -1, 0, 1, 2, ...$, $\varphi_n(a; \xi)$, $\varphi_n(b; \xi)$, $n = 0, 1, 2, ...$, and $C_n^{(i)}(\xi)$, $i = 1, 3$, $n = 0, 1, 2, ...$, we obtain the following result (a prime denotes differentiation with respect to x):

<u>Theorem B</u>: The variational equation (7) and, thus, the transformed wave problem (3), are equivalent to the following system of second-order ordinary differential equations

$$\sum_{n=-1}^{\infty} a_{mn}(x)\varphi_n''(x;\xi) + b_{mn}(x)\varphi_n'(x;\xi) + \left(c_{mn}(x) - \xi^2 a_{mn}(x)\right)\varphi_n(x;\xi) = -\frac{1}{2\pi}\delta(x-x_0)Z_m(z_0), \quad (9a)$$

in $a < x < b$, $m = -1, 0, 1, \ldots$, supplemented by the boundary conditions

$$\varphi_{-1}'(a;\xi) = 0, \qquad \varphi_{-1}'(b;\xi) = 0, \tag{9b}$$

$$\varphi_0'(a;\xi) + K_0^{(1)}(\xi)\varphi_0(a;\xi) = 0, \quad \varphi_n'(a;\xi) - K_n^{(1)}(\xi)\varphi_n(a;\xi) = 0, \quad n = 1,2,3,\ldots \tag{9c}$$

$$\varphi_0'(b;\xi) - K_0^{(3)}(\xi)\varphi_0(b;\xi) = 0, \quad \varphi_n'(b;\xi) + K_n^{(3)}(\xi)\varphi_n(b;\xi) = 0, \quad n = 1,2,3 \tag{9d}$$

The coefficients $C_n^{(i)}(\xi)$ are then obtained by the equations:

$$C_0^{(1)}(\xi) = \varphi_0(a;\xi)\exp\left(a K_0^{(1)}(\xi)\right), \quad C_n^{(1)}(\xi) = \varphi_n(a;\xi), \quad n = 1,2,3,\ldots \tag{10a}$$

$$C_0^{(3)}(\xi) = \varphi_0(b;\xi)\exp\left(-b K_0^{(3)}(\xi)\right), \quad C_n^{(3)}(\xi) = \varphi_n(b;\xi), \quad n = 1,2,3,\ldots \tag{10b}$$

The x-dependent coefficients $a_{mn}(x)$, $b_{mn}(x)$ and $c_{mn}(x)$ are specified as functions of the frequency and the local depth and can be found in Athanassoulis and Belibassakis (1997a,b).

Solution of the 3D problem by Fourier inversion and far field structure

As regards the numerical solution, the infinite system (9) is solved by truncation, retaining a sufficient number N_m of evanescent modes, the total number of equations being thus reduced to $N_m + 2$. Being able to solve the boundary value problem (9) for any value of the Fourier parameter ξ, we can reconstruct the 3D wave potential $\Phi(x,y,z)$ by applying the inverse Fourier transform (2b). Taking into account the rapid (exponential) decay of $\varphi_{reg}(x,z;\xi)$, for large values of ξ, the Fourier inversion is effectively performed by FFT in a bounded interval of the form $[0, \Xi]$. The Fourier inversion of the singular part $G_\xi(\bar{x};\bar{x}_0)$ is performed analytically. Furthermore, by applying the method of stationary phase to the Fourier integrals (2b) in the left $(x \ll a)$ and right $(x \gg b)$ half-strips corresponding to the subdomains $D_{3D}^{(i)}$, $i = 1,3$, respectively, we obtain the following leading-order asymptotic estimates

$$\Phi(r,\vartheta,z) \sim c^{(i)} C_0^{(i)}\left(k_0^{(i)}\sin(\vartheta)\right)\left(k_0^{(i)}r\right)^{-1/2}\left(1 + \frac{\tan(\vartheta)}{k_0^{(i)}\cos(\vartheta)}\right)^{-1/2}\exp\left(ik_0^{(i)}r\right)Z_0^{(i)}(z) + O(r^{-1}), \tag{11}$$

where r is the horizontal distance between the source and the field point, $\vartheta = \arctan((y-y_0)/(x-x_0))$, $0 \leq \vartheta \leq \pi$, is the azimuthal angle and $c^{(i)}$, $i = 1,3$, are complex constants. We observe in Eq. (11) the geometrical spreading law, the appearence of a shadow zone along the bottom irregularity $(\vartheta = \pi/2)$, and that the far field approaches an outgoing wave pattern with wavenumbers corresponding to the constant depths h_i in the subdomains $D_{3D}^{(i)}$, $i = 1,3$, respectively.

Presentation of numerical results

Numerical results are presented for the environment shown in Fig. 2, which models a smooth but steep underwater shoaling, with maximum bottom slope 94% and mean bottom slope 20%. The pulsating source is located at $(x_0, y_0, z_0) = (0, 0, -1\text{m})$ and its angular frequency is $\omega = 2\,\text{rad/sec}$. In this case, the ratios $h_1/\lambda_1 = 0.4$ and $h_3/\lambda_3 = 0.17$, fall well outside the deep or shallow water

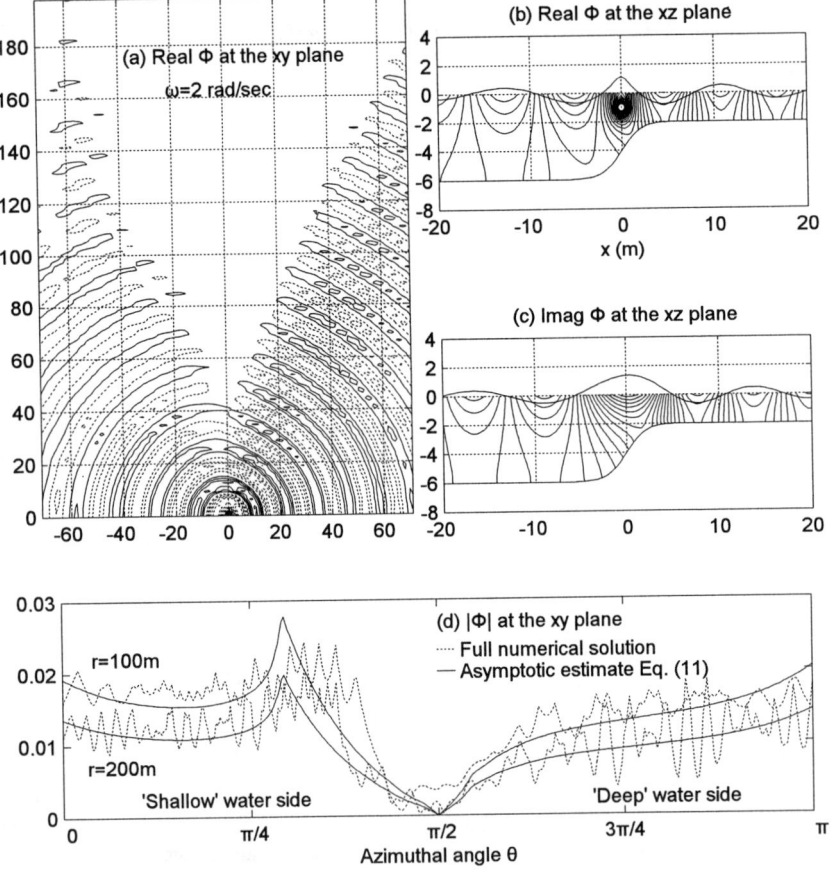

Figure 2. The case of a smooth underwater shoaling. (a) Real part of the potential at the free surface. (b,c) The wave field at the xz-plane. (d) The modulus of the wave potential at intermediate distances from the source on the free-surface, as calculated by direct Fourier inversion (2b) and by using the asymptotic estimate (11).

limits. The horizontal structure of the wave potential (and also of the free-surface elevation η, since $\eta = i\omega\Phi/g$), is shown in Fig. 2(a). From this figure we observe the shadow zone, which is formed along the bottom irregularity. In Figs. 2(b,c), the vertical structure of the wave potential in the xz-plane containing the point source is visualised with the aid of equipotential contours. We are able to observe the correct representation of the singularity, which is located at $z_0 = -1\text{m}$, and the satisfaction of the bottom boundary condition $\partial\varphi/\partial n = 0$, which is equivalent to the orthogonality between the equipotential contours and the bottom. Finally in Fig. 2(d) a comparison is given between the results obtained by direct Fourier inversion and the application of the asymptotic estimate (11) at intermediate distances from the source.

Conclusions

The water wave field generated by a pulsating point source which is located in a region with variable bottom bathymetry has been studied, using full linear theory, without any simplifications concerning the depth or the geometry (slope and curvature) of the bottom. The main results of this work are: (a) The problem is reformulated as a boundary value problem for a system of second-order ODEs in a bounded interval $(a \leq x \leq b)$. (b) The far-field wave pattern exhibits a strong azimuthal anisotropy, its main features being a shadow zone centered at the area of variable bathymetry. As $x \to \pm\infty$, in sectors not including the variable bathymetry region, the asymptotic behaviour of the far field approaches the standard one, corresponding to outgoing waves propagating on the constant depth h_1 or h_3, respectively.

References

Athanassoulis, G.A. and Belibassakis, K.A. (1997a). "Enhanced coupled-mode theory for the propagation of small-amplitude water waves over variable bathymetry regions", submitted.
Athanassoulis, G.A. and Belibassakis, K.A. (1997b). "Water wave Green's function for a 3-D uneven bottom problem with different depths at $x \to +\infty$ and $x \to -\infty$", *Proc. IUTAM Symposium on Computational Methods in Unbounded Domains*, University of Colorado, Boulder, July 1997, Intern, Union of Theoretical and Applied Mechanics.
Bai, K.J. and Yeung, R. (1974). "Numerical solutions of free-surface and flow problems", *Proc. 10th Symp. Naval Hydrodyn.*, Office of Naval Research, 609-641.
Dingemans M.W. (1997) "Water wave propagation over uneven bottoms", World Scientific.
Mei, C.C. (1978). "Numerical methods in water wave diffraction and radiation", *Annual Rev. Fluid Mech.*, 10, 393-416.
Porter, D. and Chamberlain, P.G. (1997). "Linear wave scattering by two-dimensional topography", Ch. 2 in: *"Gravity Waves of Finite Depth"*. Edited by Hunt, J.N. Computational Mechanics Publications, Southampton.
Wehausen, J.V., and Laitone, E.V. (1960). "Surface Waves". In *Handbuch der Physik*, Vol. IV/3. Edited by Flugge, W., Springer-Verlag, Berlin, 446-778.

FOCUSSED WAVE GROUPS ON DEEP AND SHALLOW WATER

By Paul H. Taylor [1] and Erwin Vijfvinkel [2]

Abstract

The evolution of steep uni-directional wave groups with realistic spectra is studied using an efficient and robust numerical scheme due to Craig and Sulem. The size of the wave group at the instant of focus, defined as the occurrence of the highest (or lowest) surface elevation at any time, is strongly affected by the water depth. In deep water (Kd~4), the highest waves are considerably higher and the groups more compact than predicted by linear or 2nd order wave theory. As the depth is reduced to Kd~1, the predominately 3rd order wave-wave interactions responsible for the non-linear evolution of the group are much weaker – in contrast to the mostly 2nd order vertical asymmetry of the wave group, which is much stronger in shallower water. The wave kinematics for these groups is also discussed. For these isolated groups particularly on shallow water, the return flow associated with the localised set-down beneath the group is very large.

Introduction

For the design and re-assessment of offshore platforms in areas affected by severe weather, accurate models are required for the height, shape and internal flow kinematics of extreme waves. Waves in the open sea are broad-banded. Thus, most of the properties of a random sea-state can be estimated using the statistics of a random Gaussian process. However, the largest waves in a random sea are also the most non-linear. In this paper we investigate the effect of wave non-linearity on the focussing of localised groups of extreme waves using a fully non-linear numerical scheme.

---

[1] Dept. of Eng. Science, University of Oxford, Parks Road, Oxford OX1 3PJ, U.K. (author for correspondence). Formally at Shell International Exploration and Production B.V., Rijswijk, The Netherlands.
[2] Rijks*universiteit* Groningen, The Netherlands.

Numerical method

We use an improved version of a pseudo-spectral method by Craig and Sulem (1993) to simulate uni-directional wave evolution on water of constant water depth. The Craig and Sulem method is applicable so long as the surface elevation is a single-valued function of a horizontal co-ordinate (i.e. for breaking waves up to the instant when the surface becomes vertical). It proves to be both efficient and numerically robust.

Like most methods, the free surface boundary conditions are used to advance the solution in time, using the surface profile and the velocity potential on this surface. The heart of the Craig and Sulem approach is the Dirichlet-Neumann operator. This operator, applied to the free surface velocity potential, gives the velocity component of a fluid particle at the free surface normal to the free surface. Craig and Sulem obtain a rapidly convergent Taylor expansion of this operator solely in terms of the shape of the surface and the surface potential. This Taylor series can be calculated quickly by switching between physical and spectral spaces - products are performed in physical space and differentiation in spectral space via FFTs, which then also become products. We have also rearranged the terms in the Taylor expansion; our scheme is 4x faster the original Craig and Sulem algorithm for the cases reported here. We use their explicit two-step Adams-Bashford time marching scheme. The total computational effort is almost linear in the number of collocation points along the surface (scaling roughly as $O(M^2 N \log N)$) per time-step, where M is the order of truncation of the Taylor series and N is the number of points). Thus, problems with a wide range of length–scales can be modelled easily.

Focussed wave groups – the initial condition

Taking the surface of the open sea as a linear random Gaussian process greatly simplifies the statistics. Then, the most probable shape of the ocean surface in the vicinity of an extreme crest tends to a scaled version of the auto-correlation function (see for example Jonathan et al. 1994), denoted previously as NewWave. Here we are interested in changes in this behaviour due to wave non-linearity. Thus, waves in the vicinity of an extreme wave event are simulated numerically using a defocussed NewWave, using the profile 20 periods before the extreme crest occurs as the initial condition. The linear NewWave group is constructed from a fixed wave number spectrum for all simulations irrespective of the water depth. This wave number spectrum is obtained using the deep water dispersion relation from the frequency spectrum of a severe storm measured on the 4th of January 1993 at the Tern platform in the northern North Sea (Jonathan et al. 1994). The measured frequency spectrum has a high JONSWAP-like peak and a close to ω^{-4} tail.

The numerical simulations last 40 wave periods (based on the peak of the wave spectrum), starting 20 periods before linear focus, using 512 Fourier components.

All the numerical simulations used the same input surface profile. We use values for the surface elevation and velocity potential from linear wave theory as the initial conditions, making no attempt to include the mostly 2nd order bound wave structure associated with real waves.

Focussed wave groups – the structure around the focus point

Figure 1 shows a weakly non-linear wave group focussing up. On the left is the initial spatial profile. In the centre is the profile at the instant the maximum crest elevation occurs - this profile is very close to being front-back symmetric. On the right the final profile is shown 40 periods after the start. This is virtually identical in form to the initial profile apart from being front-back reversed - the short waves in the group move more slowly than the longer waves so they are overtaken.

Figure 2 shows steep wave focussing on both shallow and deep water. The initial surface profiles for the two cases are identical. If the focus of each was linear, both peak crest elevations at the instant of focus would have a value of AK=0.26, based on the size of the crest and the wave number of the peak of the amplitude spectrum. For the deep-water case, this is the steepest group, which can be run through focus without the calculation breaking down. The length and time scales of all the calculations reported here are scaled such that the water depth of the Kd=4 case is $d=1m$, giving $T_p \sim 1s$.

The maximum elevation obtained by focussing to an extreme crest is much larger in deep water than in shallow water. In deep water, the focussed wave crest can be 30% larger than the linear value. This extra elevation is partly due to the 2nd order bound harmonics, but the rest is due to 3rd order wave-wave interactions. In shallow water the extreme elevation can even be slightly smaller than the linear value. This reduction of elevation at the focus point is due to the sag of the free surface beneath the main group, and to the virtual elimination of 3rd order wave-wave interactions due to the reduced depth.

In all cases the focussed profiles obtained by crest focussing and trough focussing show almost perfect vertical symmetry - the position in space and time of the highest crest is virtually identical to that of the deepest trough, when the initial condition is inverted. Thus, wave group focussing is an envelope property even for these realistically broad-banded groups. Each element of figure 2 shows two wave profiles at the instant of focus, one for a group focussed to an extreme crest, and the second for a group focussed to an extreme trough. The main group is virtually front-back symmetric for the shallow water case. In contrast, the deep-water group is strongly asymmetric. Similar predictions have been made for focussed groups by Lo and Mei (1985), based on a non-linear envelope equation.

In shallow water, the 2nd order wave components bound to a steep group are strong. However, a linear initial condition was used for the simulations. Thus, the bound wave structure, which should be present, is cancelled by destructive interference from an identical set of free-wave components. These free waves are predominately of two forms – a 'hump' cancelling out the bound set-down beneath the group, and free-waves centred on a wave number 2x the peak in the linear wave spectrum cancelling the 2nd order trough-crest asymmetry. Due to linear dispersion, these free wave 'errors' propagate away from the main group. In the shallow water simulations, these components can be recognised as a hump ahead of the main group and a high-wave number train behind. Fortunately, neither significantly affects the behaviour of the main group.

Figure 3 shows the whole focussing event as a set of (x,t) diagrams. The surface elevation profiles at each time run horizontally, with the initial profile at the bottom of the figure. The profiles for subsequent times are shown further up the figure. The wave focus is clearly visible halfway up (the profiles are shown in a frame of reference moving with the centre of the main group). Also visible are the free-wave systems arising from errors in the initial condition on either side of the main group.

As well as the height of the focussed group being affected by non-linear interactions, so is the position and time of the highest wave. For steep waves on shallow water, the focus point is only slightly shifted. For deeper water, one or more upstream jumps in the position of the focus point, each of about two periods, occurs as the amplitude is increased. Then, the focussing event is mainly determined by the focussing of components from around the peak of the wave spectrum, the high wave number components from the spectral tail focus up later.

We also present results for the horizontal velocity field beneath the focussed waves groups using the Fourier solver of Fenton and Rienencker (1981). The velocity profiles beneath the extreme crests and troughs for the cases in figure 2 are shown in Figure 4. The groups studied here are physically compact. Thus, the return flow beneath the group is an important velocity component. For both water depths, the fluid velocity backward beneath the deepest troughs is substantially larger than the equivalent forward velocity beneath a focussed crest at a fixed level.

Conclusions

For deep-water uni-directional waves with realistic spectra, the wave-wave interactions identified by Lo and Mei and observed experimentally by Baldock, Swan and Taylor (1996) can lead to important, slow but cumulative changes in the focussing of steep wave groups. Thus, linear theory, even allowing for the bound wave components, does not describe wave focussing adequately for groups on deep water. In contrast, for wave groups in shallower (intermediate) water depth the focussing is effectively a linear process, despite the increased strength of the bound-wave components.

Acknowledgement - The authors are grateful to Peter Tromans and George Forristall of Shell International Deepwater Services B.V. for enlightening discussions.

References

Baldock, T.E., Swan, C., and Taylor, P.H. *A laboratory study of non-linear surface waves on water.* Phil. Trans. Soc. Lond. Ser. A **345** (1996), 1-28.

Craig, W. and Sulem, C. *Numerical simulation of gravity waves.* J. Comp. Phys. **108** (1993), 73-83.

Fenton, J.D., and Rienecker, M.M.A., *A Fourier method for solving non-linear water-wave problems: application to solitary-wave interactions.* J. Fluid Mech. **118** (1982), 411-443.

Jonathan, P., Taylor, P.H. and Tromans, P.S. *Storm waves in the northern North Sea.* Proc. 7th Internal Conference on the Behaviour of Offshore Structures (BOSS) **2** (1994), 481-494.

Lo, E. and Mei, C.C. *A numerical study of water-wave modulation based on a Higher order non-linear Schrodinger equation.* J. Fluid Mech. **150** (1985), 395-416.

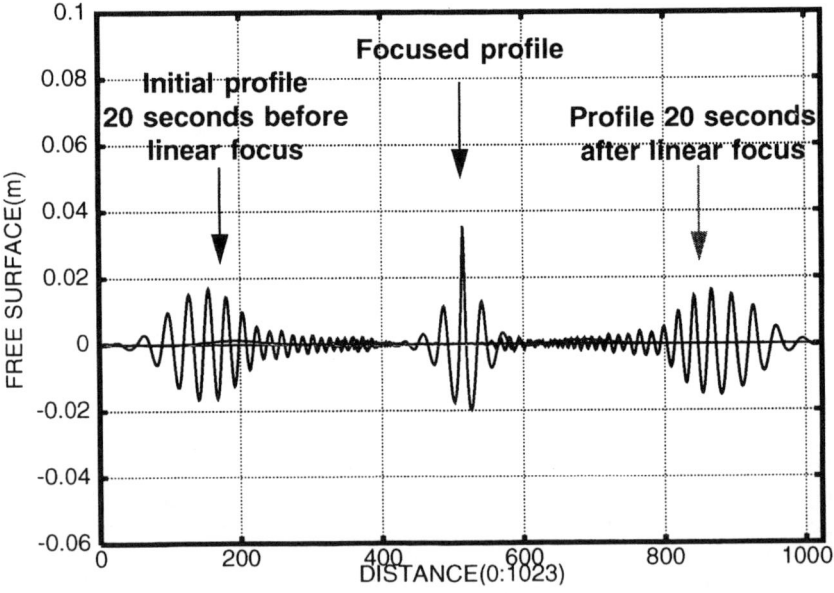

Figure 1. Example of an approximately linear focus.

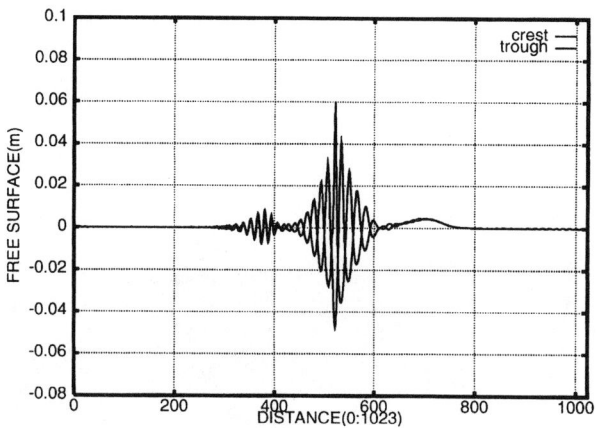

(a) Shallow water (Kd=1, KA=0.26)

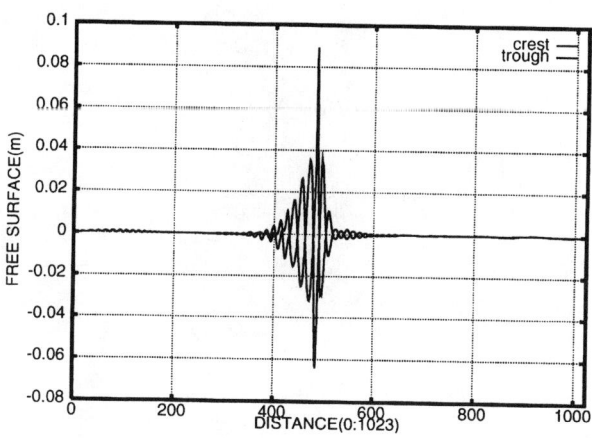

(b) Deep water (Kd=4, KA=0.26)

Figure 2. Spatial profiles of steep wave groups at focus, showing extreme crest and trough focussing.

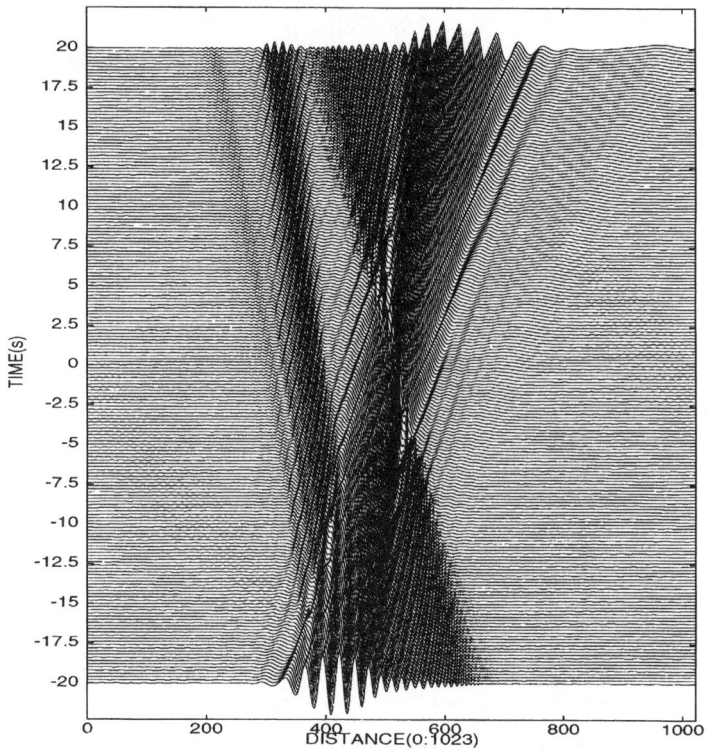

Figure 3.
Evolution of the free surface for a steep wave group on shallow water (Kd=1, KA=0.26).

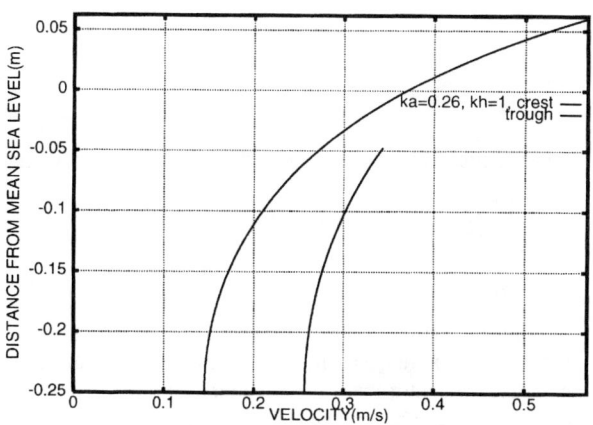

(a) Shallow water

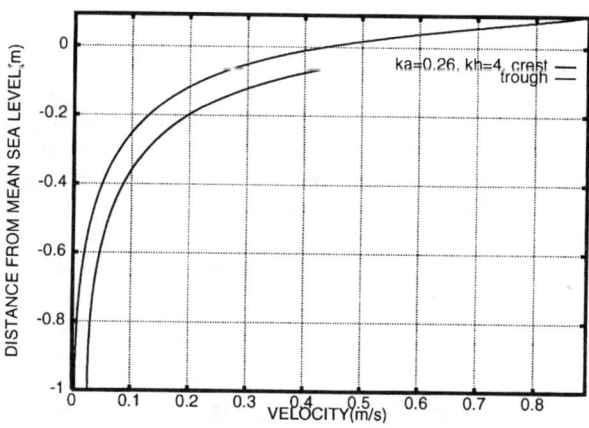

(b) Deep water

Figure 4. Horizontal velocity profiles beneath steep wave groups at the instant of focus.

Turbulence Measurement In a Towing Tank Using Hot-Film Anemometry

Shukai Wu[1]

Abstract

Hot-film anemometry as a tool for measuring the velocity of a fluid flow is described. Compared with pitot tubes conventionally used for flow measurement, hot-film anemometry has the advantage of being highly accurate and sensitive to low-speed flows. It also has much higher spatial and frequency resolutions. Results of sample turbulence measurement done behind a ship model in a towing tank is presented, indicate its applicability for offshore model testing.

Introduction

In the model testing of many offshore structures in calm water or uniform current, it is necessary to keep the turbulence level as low as possible. For certain types of model testing, it is well known that presence of turbulence can significantly affect the test results. An example of this is in the experimental study of the onset of vortex shedding and the occurrence of vortex induced vibrations of a cylindrical structure subject to a uniform current. Yet, due to the various practical difficulties, the turbulence level in the current generated in a model basin or towing tank has rarely been quantified. One of the difficulties involved is in the measurement of turbulent velocity. Conventional wave gages and velocity measuring devices, such as pitot tubes, are limited in term of accuracy and frequency resolution as well as the ability to measure the directionality of the flow. Furthermore, pitot tubes are insensitive to low-speed flows (pressure drop is proportional to speed squared).

This paper describes experience gained in using hot-film anemometry for measuring highly turbulent flows, such as the wake flow behind a ship model in a towing tank. It is hope that this technique would enable the turbulence level in offshore model tests to be quantified accurately, if desired.

---

[1] Staff Engineering, Aker Engineering Inc, 11757 katy Freeway, Suite. 1300, Houston, TX 77079

Hot-Film Anemometry

Hot-film anemometry is an instrument used for measuring the velocity of fluid flows. In its simplest form, it uses a tiny hot-film sensor placed at the tip of the sensor holder known as the hot-film probe. In the normal operation, the sensor is heated electrically and is kept at a constant temperature. When placed in a flow field perpendicular to direction of the flow, the probe loses heat at a rate which is directly related to the velocity of the flow. As a result, the velocity of the flow can be measured by measuring the electrical power needed to keep the temperature of the probe sensor constant.

Typically, the size of the hot-film sensor is in micrometers and that of the probe is in millimeters. As a result, hot-film probes has a very high level of accuracy and spatial and frequency resolutions, and are ideal for measuring highly turbulent flows. Moreover, hot-film probes are extremely sensitive in low-speed flows which is in contrast with pitot tubes. By using a probe with two such hot-film sensors arranged in a V configuration, the directionality of the flow velocity as well as the magnitude can be measured (Wu & Bose 1991).

Calibration

For a constant-temperature hot-film anemometry, the power supply to the probe to keep the sensor temperature constant (therefore, the sensor resistance is constant) is proportional to the voltage squared. The calibration curve of such hot-film anemometry typically follows a modified form of King's Law (Wu & Bose 1994):

$$V^2 = \Sigma_k A_k U^{kn}$$

Where V is the anemometer output voltage; U is the flow velocity, A's and n are calibration constants. Generally, use of k≤3 is sufficient to obtain an calibration accuracy of 0.2%. Figure 1 shows a typical calibration curve done in water. It is worth noting that at low speeds, the hot-film probe is extremely sensitive, indicating its capability for measurement in low-speed flows.

Example Measurement

Measurement of the turbulent flow velocity behind a ship model was made in a nominal wake survey experiment. The model was towed at a constant speed of 1.45m/s. The axial velocity component was measured at various locations of the propeller disk (Wu & Bose 1992). Figure 2 shows the energy spectral density of the turbulent velocity component (=total velocity - mean velocity). The data sampling frequency used was 300Hz. Details of the experimental setup and procedure are referred to the above reference.

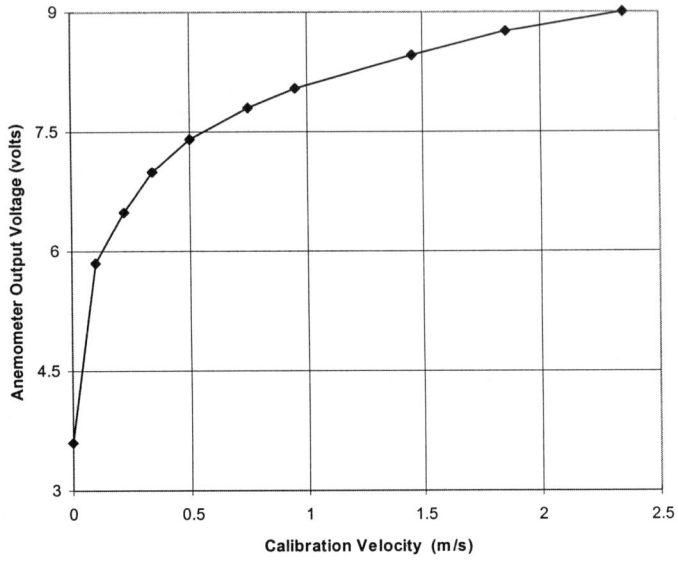

Figure 1   Typical Calibration Curve for Hot-Film Anemometry

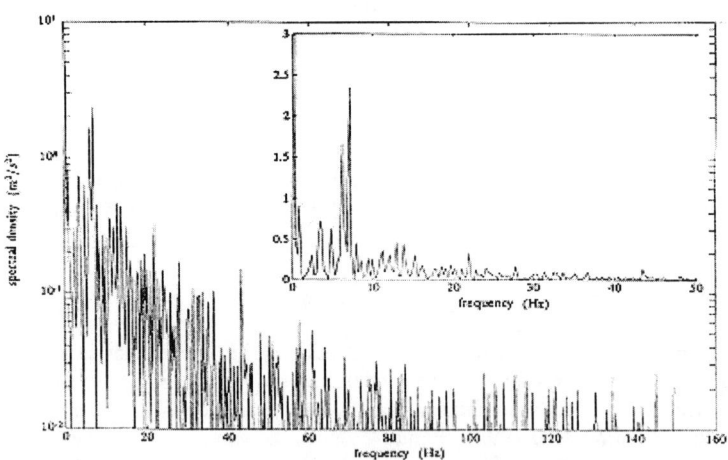

Figure 2   Energy Spectral density Distribution of a Typical Converted Velocity Recording Made Behind a Ship Model

It is interesting to note that the major portion of the turbulence has frequencies below 10Hz. However, there is also a significant level of turbulence between 10 and 20Hz. The results also serve as an indication of the level of frequency resolution achievable with hot-film probes, which is far beyond the capabilities of pitot tubes.

Concluding Remarks

Constant temperature hot-film anemometry has been successfully used to measure highly turbulent flows behind ship models. Besides its high accuracy, the instrument has a high spatial and frequency resolution and should be very useful, when used for offshore model tests, in providing quantitative information on the turbulence level present in the model testing. It is hoped that the information presented in the paper will serve as an introduction to the instrument, and will lead to its more widespread use in the model testing of offshore structures.

References

1. Wu, S. and Bose, N. (1991). Calibration of a Wedge-Shaped Vee Hot-Film Probe for Towing Tank Measurements. Ocean Engineering Research Centre Report: OERC91-WTT-TR004, Memorial University of Newfoundland.

2. Wu, S. and Bose, N. (1992). Axial Wake Survey behind a Fishing Vessel Model by Using a Wedge-Shaped Hot-Film Probe. J. Kansai Soc. N. A. Japan, no. 218, pp29-40.

3. Wu, S. and Bose, N. (1994). An Extended Power Law Model for the Calibration of Hot-Wire/Hot-Film Constant Temperature Probes. Int. J. Heat mass Transfer, vol. 37, no.3, pp437-442.

Turbulent Wave Boundary Layers
in the Surf- and Swash Zone -
Analysis of Small Scale Field Measurements

Volker Müller[1]

Abstract

The evolution of the bathymetry of a natural beach is determined by the bottom stress and its temporal variation during the tide. Up to now, the complexity of the interaction processes between shallow water waves and related flow acting on sedimentary processes during different tidal phases on beaches and tidal flats is not fully understood. Using data of a monitoring measuring programme, funded by the Federal Ministry of Science and Technology and the Technical University Hamburg-Harburg, an attempt for verification of shallow water wave theories at different tidal phases for calculating bottom stress or wave friction factor on basis of laminar or turbulent wave boundary layer theory was made. As the main result it could be shown, that there exist mainly three different regions of validity for the various theoretical approximations and assumptions and these regions of validity can clearly be separated by the local value of wave height to water depth ratio.

Introduction

Today shore protection becomes more and more important. This is because of the rising sea level and frequency of storms. Different methods for shore protection are known. At the coast of the German Wadden Sea the method of repeated beach nourishment is the mostly used method. But erosion of the protected coast can not be stopped by this method. Rates of sediment erosion up to $250 m^3/m/year$ on sandy beaches could be observed. Also measuring a cumulative erosion depth of 12 cm and

---

[1]Senior Engineer, Technical University Hamburg - Harburg, Dept. of Ocean Engineering 1, Lauenbruch Ost 1, 21079 Hamburg, Germany

a cumulative sedimentation height at nearly the same range went on under calm wave conditions for one tide, too. These dynamics of the combined erosion- and sedimentation processes at tidal beaches is not fully understood until now..

The main hydrodynamic parameter which determines the probability of erosion and sedimentation and also the net sediment transport is the bottom shear stress τ_b or the friction velocity u_*. At a research programmes funded by the Federal Ministry of Research and Technology or the Technical University Hamburg-Harburg the tangential velocity profile close to the bottom at tidal beaches and tidal flats was measured with a high temporal resolution. Also the dynamic pressure was measured. To get a correlation of the dynamics of sediment and hydrodynamic processes these data will be systematised here.

Experiments

The measuring data were sampled during various measuring programmes at the beach of the island Norderney in the Frisian Wadden Sea (part of the North Sea) and at tidal flats at the estuary of the river Elbe in the years 1993 till 1995 (see Tab. 1).

Tab. 1: Table of experiments used at this systematics

time	location	Nb. of measuring locations	d [m]	f_s [Hz]
Oct. 1993	Beach (Norderney)	2	0.0 .. 1,5	5
Aug. 1994	Tidal flat (Elbe)	1	0.0 .. 1.2	5
Oct. 1994	Beach (Norderney)	1	0.0 .. 1,8	5
March 1995	Tidal flat (Elbe)	4	0.0 .. 1.5	5
Aug. 1995	Tidal flat (Elbe)	6	0.0 .. 1,5	20
Oct. 1995	Beach (Norderney)	1	0.2 .. 2.2	100

A hot film anemometer system was used for measuring the profiles of the tangential velocity close to the bottom (6 single sensors, distance of the single hot film sensors to the bottom surface ranges from -5 to 12 cm) (GUST; 1988). Hot - film anemometry also provides the opportunity to measure the bottom shear stress, but unfortunately it was not possible to locate a hot film sensor on the surface of the sandy beach directly for the whole measuring period. Another problem of the hot film anemometry is the strong dependence of the measuring principle (forced heat convection) to the ambient temperature. A software - related, a posteriori temperature compensation scheme for hot film anemometer signals in non-isothermal water flows was used (HENSSE et.al.; 1997). Beside the tangential velocity dynamic pressure under waves and ambient temperature were sampled with the same sampling frequency f_s like the hot film sensor signals (see tab. 1). All signals were synchronised, so it was possible to correlate the pressure data to the velocity data (fig. 1).

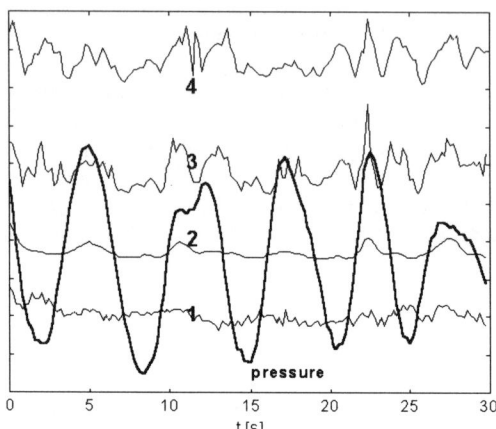

Fig. 1: Example of synchronised signals of pressure sensor and hot film sensors at different heights above the bottom: 1: 0.5 cm under the surface (amplification factor = 400); 2: 0.5 cm above the surface (amplification factor = 10); 3: 3.5 cm above the surface (amplification factor = 1); 4: 7.0 cm above the surface (amplification factor = 1);

Erosion and sedimentation could be recognised by changing the characteristics of the hot film signals due to a rapid change of heat convection characteristics. The pressure data were used to estimate the mean water depth d, wave angular frequency ω, wave number k and wave amplitude H/2. If more than one measuring system was used, it was possible to determine the celerity c and the direction of wave propagation directly.

Systematics of the results

A comparison of all measured data shows, that there exist always the same correlation characteristics between the mean water depth depending on the tidal phase, the wave amplitude and the probability of sediment erosion processes (fig. 2). At a time period, marked with (C) at fig. 2, the probability of sediment erosion processes is much more higher than at all the other periods at a whole tide. This period distinguish the periods with the strong correlation between wave amplitude and water depth (periods marked with (T) at fig. 2) from the periods with only a weak correlation between these both parameters (periods marked with (W) at fig. 2).

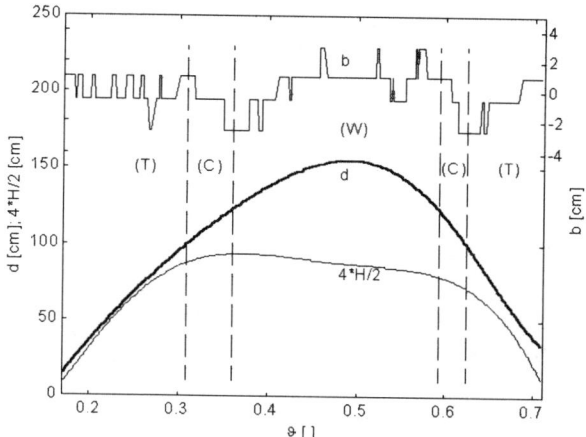

Fig. 2: Characteristic example of relative changes of the bottom surface b related to the water depth d and the mean wave amplitude H/2 depending on the tidal phase $\vartheta = t/T_{tide}$

The values for the so-called tidal-dominated range (T), the wave-dominated range (W) and the interchanging range (C) are

$$(T): \quad \frac{d}{H/2} < 3.2$$

$$(C): \quad 3.2 < \frac{d}{H/2} < 4.8$$

$$(W): \quad \frac{d}{H/2} \geq 4.8$$

Values of hydrodynamic parameters, especially higher moments of the statistical distributions of tangential velocity close to the bottom and shear stress (wall shear stress) are significantly different at these ranges. At the tidal-dominated range the wave amplitude is correlated with the mean water depth like

$$H/2 = (0.41...0.45)d$$

The factor is slightly different at the different measuring locations and measuring times and is in agreement with data from literature (THORNTON, GUZA; 1989).

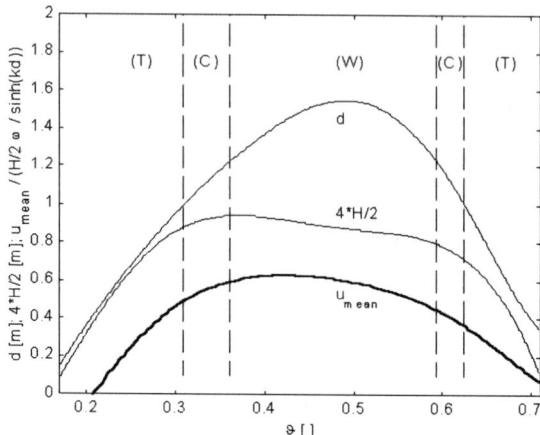

Fig. 3: Spatial arithmetic mean of the dimensionless tangential velocity at the range from 0 cm to 6 cm above the bottom (the comparative velocity was calculated from the pressure data)

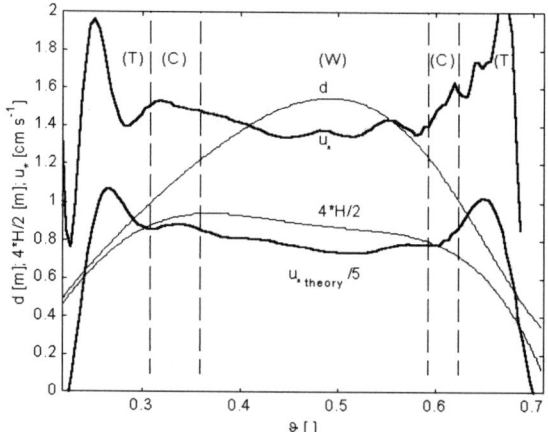

Fig. 4: Values of mean wall shear stress calculated a) from the measured tangential velocity profile at the range from 0 to 12 cm above the bottom (curve marked with u∗) and b) calculated by turbulent wave boundary layer theory using pressure data (curve marked with u∗ theory)

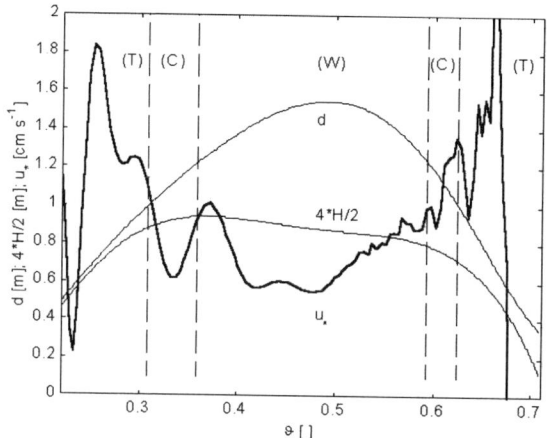

Fig. 5: Values of mean shear stress in the boundary layer at a range from 2 cm to 8 cm above the bottom calculated from the measured tangential velocity profile at the range from 0 to 12 cm above the bottom (curve marked with u∗)

Tidal – dominated range (T): The correlation of pressure and near-bottom velocity in this range is weak. The distribution of velocity is skewed to the right and very leptokurtic compared with the pressure distribution which is nearly symmetric and mesokurtic at this range. Calculated data of bottom shear stress using turbulent or laminar boundary layer at this range are not correct. But direct measurement of wall shear stress was not possible.

Wave – dominated range (W): At this range there was found a high correlation between pressure and near-bottom velocity. The statistical parameters of the distributions of pressure and velocity are nearly the same if the measuring principle of hot film anemometry and its influence on the statistical parameters is taken into account (JUNGLEWITZ et.al.; 1997). The higher moments of the pressure distribution are nearly the same as in the tidal-dominated range. Calculating the bottom shear stress by turbulent wave boundary layer (from the parameters of the flow regime this theory has to be used) gives values which are much more higher than expected (the critical friction velocity is in the region of 1.1 cm s^{-1} for the sediment at the measuring locations (see tab. 1)). But calculating the bottom stress by laminar linear wave boundary layer theory the values are nearly the same if the bottom stress is calculated from the measured velocity profile normal to the bottom.

Interchange range (C): At this range the statistical parameters of the velocity distribution change. A comparison of measured data with calculated ones gives no indication for the high erosion probability at this range. Only the inner shear stress in the boundary layer (fig. 5) has a local minimum at this range. So higher values of bottom shear stress can be expected, but the spatial resolution of the hot film measurements was not high enough.

Conclusions

Three different hydrodynamic effective regions can be distinguished in the surf and swash zone by the water depth – wave amplitude ratio. The comparison between measured and calculated data for the near bottom hydrodynamics gives different results at this regions. Laminar and turbulent wave boundary layer theory can be used to calculate the bottom shear stress at a range $d/(H/2) > 4.8$ successfully. Also it is not necessary to use higher orders of wave theory if the aim is to calculate the erosion or deposition probability. At a range of $3.2 < d/(H/2) < 4.8$ high erosion probabilities were found at all experiments (tab. 1). This couldn't be calculated from measured data also from the experimental values too. Further investigations are necessary.

References

Gust, G.: 1988, "Skin Friction Probes for Field Applications", Journal of Geophysical Research, Vol. 93, No. C11, 1988, pp. 14,121-14,132

Henße, J., Mueller, V. Gust, G.: "Dynamic temperature compensation for hot film anemometry in turbulent flows - necessity and realisation, Proc. 3.rd Int. Conf. On Fluid Measurements and its Application, Beijing, 1997

Junglewitz, A., Mueller, V., Gust, G., Calibration of a hot film sensor for application in turbulent ship model testing, ASME, FEDSM97-3468, Vancouver, (1997)

Thornton, E. B., Guza, R. T.: "Wind Wave Transformation", in R. J. Seymour (Ed.), "Nearshore Sediment Transport", Plenum Press, New York, 1989

WIND EFFECTS ON WAVE TRANSFORMATION AND UNDERTOW

Joshua Carter[1], William Hobensack[1], Douglas Kennedy[1], and Daniel Cox[1]

ABSTRACT: A detailed laboratory study of the wind effects on cross-shore wave transformation and undertow was made using a 2D wind-wave flume. The test program consisted of cases with swell waves only, wind waves only, and combinations of swell and wind waves to quantify the relative importance of the two mechanisms. The free surface elevations were measured at fifteen locations to provide estimates of the cross-shore wave height and setup variations. The undertow profile was measured at four locations. Comparisons indicate that (1) the wave heights of the combined wind and swell waves were much less than the sum of the wind waves and swell waves generated separately due to breaking by the wind, (2) an effect of the swell waves was to lower the peak frequency of the wind waves, and (3) the undertow induced by the wind waves was roughly 20-50% larger than that due to the swell waves even though the wind wave heights were smaller. This paper shows that although the wind does not have a large impact on the cross shore variation of the root-mean square wave height, the wind effects the undertow and may be an important process to consider when modeling undertow. Further research is needed to better understand the interactions of wind, swell waves, and undertow.

INTRODUCTION

Many coastal regions experience sea breezes (onshore winds roughly 5 to 10 m/s) due to the heating of the inland areas. Recent field investigations show that even a modest sea breeze can induce a cross-shore current three times larger than that generated by the swell waves alone and can significantly effect the longshore current and suspended sediment concentration (Pattiaratchi and Masselink, 1996). The potential increase due to storm and hurricane winds is even greater. Although wind-induced currents in the open ocean have been well-studied, the role of the wind on nearshore wave transformation and undertow has received little attention. Furthermore, many coastal models have been developed without considering wind effects even though it is the experience of a casual beach comber that the wind can have a dramatic influence on coastal conditions. In light of this, a preliminary study

---

[1] Ocean Engineering Program, Dept. of Civil Engineering, Texas A&M Univ., College Station, TX 77843-3136, USA

was made of the undertow induced by swell waves alone, wind waves alone and a combination of the two to quantify the relative importance of these two mechanisms in the nearshore region. This investigation was conducted in a small-scale laboratory flume since this flume offers precise control over the waves and wind relative to field experiments.

EXPERIMENTAL SETUP

A 36 m long by 0.67 m wide by 0.91 m deep glass-walled wind wave flume was used as shown in Figure 1. A programmable wavemaker was used to generate irregular swell waves based on a JONSWAP spectrum. A rigid lid and fan system was used to generate the wind waves. The water depth of the constant-depth section was $d = 0.65$ m, and the space from the still water level (SWL) to the rigid lid was $b = 0.26$ m. An impermeable 1:20 slope was placed with the toe at $x = 17.3$ m from the wavemaker ($x = 0$ m).

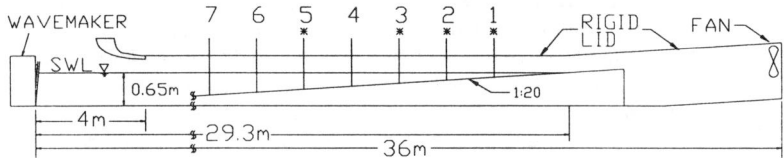

Figure 1: Wind-wave flume and measuring lines (Figure not to scale).

Seven measuring lines were established as shown in Figure 1 with the locations L7 to L1 given as $x = 20.8, 21.5, 22.2, 22.9, 23.6, 24.3$, and 25.0 m, respectively. The free surface elevation was measured at each line using an array of 3 resistance type wave gages with Gage 2 centered over the line, Gage 1 seaward of Gage 2 at $\Delta x = -0.5$ m, and Gage 3 landward of Gage 2 at $\Delta x = 0.2$ m. The horizontal velocities were measured using a two-component acoustic-Doppler velocimeter (ADV) at L5, L3, L2, and L1 (as indicated in Figure 1 by '*') located below Gage 2. The ADV probe was aligned to minimize disturbances to the flow in the direction of wave propagation. The horizontal velocity components measured were u in the direction of wave propagation and v in the cross-tank direction. The vertical velocity w was not measured. At a given measuring line, the vertical spacing of the velocity measurements was on the order of 1 cm. Wind speed was measured using a cup anemometer at $x = 12.8$ m and approximately midway between the SWL and the rigid lid. The fan was started two minutes before data acquisition so the wind waves were fetch-limited. The wavemaker and data acquisition system were synchronized with a digital trigger to assure repeatability of the runs. The free surface elevations, $\eta(t)$, water particle velocity, $u(t)$, and wind speed, $V(t)$, were recorded at 25 Hz for 330 s. The first 30 s and last 20 s of the recorded signal were removed to eliminate transitional effects.

Five cases were run as listed in Table 1 using swell waves only (Cases 1 and 2), wind waves only (Case 3), and combinations of swell and wind waves (Cases 4 and 5). The peak frequency was chosen to give the swell waves a lower peak frequency ($f_p = 0.53$ Hz) than the wind waves ($f_p = 2.08$ Hz). The amplitude of the input time series was scaled down to produce small swell (H_{rms} = 2.75 cm), and scaled up to produce large swell (H_{rms} = 5.75 cm). The distinction between 'large' and 'small' is relative to the energy of wind waves as will be discussed in the next section. The values of the root-mean-square wave height, H_{rms}, and peak frequency, f_p, listed in Table 1 were measured at Gage 2, L7. The mean wind speed, $\overline{V}$, was measured by the anemometer. The setting for the fan speed was not changed for Cases 3, 4, and 5 although the measured mean wind speed changed slightly for Cases 4 and 5 due to the presence of the swell waves. Although the surface gravity waves can be scaled from laboratory to field using Froude similitude, it is possible that the scaling of the wind does not follow Froude similitude. Nevertheless, given a characteristic length scale of 1:25, wave heights presented in Table 1 would scale to 0.7 m $\leq H_{rms} \leq$ 1.8 m and the frequency would scale to 0.09 Hz $\leq f_p \leq$ 0.42 Hz (2.4 s $\leq T_p \leq$ 11 s) which is realistic for ocean waves. The 6.75 m/s wind speed would scale to 34 m/s which corresponds to a severe gale.

Table 1: Five experimental cases.

Case	Input Conditions		Measured Values		
	Swell	Wind	H_{rms} [cm]	f_p [Hz]	$\overline{V}$ [m/s]
1	Small	None	2.75	0.53	0.00
2	Large	None	5.75	0.53	0.00
3	None	Constant	3.66	2.08	6.75
4	Small	Constant	4.48	0.49	7.00
5	Large	Constant	7.02	0.45	6.77

ANALYSIS OF WAVE TRANSFORMATION AND UNDERTOW

Figure 2a shows the measured wave spectra S(f) at L7 outside the surf zone for Case 1 (small swell only), Case 3 (wind only), and Case 4 (small swell with wind). This figure shows a single peak frequency for Cases 1 and 3 and a bimodal spectrum for Case 4, as is characteristic of a locally generated sea (e.g., Pattiaratchi and Masselink, 1996). Figure 2b shows the measured wave spectra at the same location for Case 2 (large swell only), Case 3, and Case 5 (large swell with wind). Figures 2a and 2b reveal that the effect of the swell wave is to shift the peak frequency of the wind wave energy. As the swell wave energy increases, the shift becomes more pronounced. The mechanism for this shift is not clear. However, for both figures it is clear that spectra measured for the combined swell and wind cases (Cases 4 and 5) are not simply the algebraic sum of the two cases measured separately.

Figure 3a shows the wave spectra at L1 inside the surf zone for Cases 1, 3, and 4. Overall, there is a decrease in wave energy due to breaking. It is interesting to note that the energy at $f_p \approx$ 0.5 Hz is slightly larger for Case 4 than for Case 1 even though the wind energy is input at much higher frequencies. A similar effect can be observed for Cases 2 and 5 in Figure 3b.

Figure 4a shows the cross-shore variation of H_{rms} and undertow, $\overline{u}$, and Figure 4b shows the setup, $\overline{\eta}$, for Cases 2, 3, and 5. For Case 2, most wave breaking begins between L3 and L2, and for Cases 4 and 5, wave breaking begins before L3. The uncertainty of $\overline{u}$ is approximately ±0.1 cm/s for Case 2 and ±0.5 cm/s for Cases 3 and 5 based on earlier measurements of maximum and minimum mean velocities for several runs at mid-depth. The uncertainty in H_{rms} and $\overline{\eta}$ can be inferred from the locations where the time series were measured twice, shown on Figures 4a and 4b. The uncertainty in $\overline{\eta}$ is large and has an unrealistic setdown for Case 2. Nevertheless, the relative setup and, more importantly, the setup gradient can be discussed quantitatively among Cases 2, 3 and 5.

In this figure there are several important observations. The cross-shore variation of H_{rms} for Case 3 (wind only) is qualitatively similar to the variation of H_{rms} for Case 2 (large swell). This suggests that the fetch length is too short in the nearshore region for the wind to greatly affect $H_{rms}(x)$. It was observed, however, that the wind greatly increased breaking when added to the swell waves (Case 5). The effect of the increased breaking is that H_{rms} of the combined swell and wind was is much less than that generated by the two mechanisms separately.

In comparing the magnitude of the local wave height to the magnitude of the undertow at a given measuring line, it can be observed that the smaller wave heights for the wind (Case 3) produce larger undertow than the swell waves (Case 2). This has significant implications for nearshore models in which the depth-averaged undertow is coupled to the local water depth and wave height (e.g., Smith et al., 1992; Kennedy et al., 1998). This relation can be expressed

$$\overline{U} = -\frac{C_u \sqrt{g\overline{h}}}{8}\left(\frac{H_{rms}}{\overline{h}}\right)^2 \qquad (1)$$

where C_u is an empirical coefficient. Earlier work suggests that this coefficient is on the order of 1 but that it increases in the surf zone due to effects of the surface roller. The work presented here shows that surface currents induced by the wind also effects this coefficient.

Figure 4b shows the cross-shore variation of $\overline{\eta}$ for the three cases. Because $\overline{\eta}$ is a difficult quantity to measure, there is a large scatter in the data. For $\overline{\eta}$ in Case 2, the error appears to be a systematic. Nevertheless, it is interesting to compare the setup gradients $\partial \overline{\eta}/\partial x$ inside the surf zone for the three cases. Drawing a best fit line by eye through the points landward of L3 for each of the three cases and estimating the gradient gives $\partial \overline{\eta}/\partial x \approx 3.6 \times 10^{-3}, 7.5 \times 10^{-3}$, and 8.6×10^{-3} for Cases 2, 3, and 5 respectively. These rough estimates indicate that the setup gradient is 2.1 times larger for the wind waves alone versus swell waves alone and is 2.4 times larger for the wind and swell waves versus the swell waves alone. This increase in the setup gradient is roughly consistent with the increased undertow seen at L2 in Figure 4a. Future work entails a quantitative analysis of the cross-shore momentum balance including the wind shear stress.

SUMMARY

This qualitative study of the wind effects on wave transformation and undertow suggest the following

1) The effect of the swell waves is to shift the peak frequency of the wind waves toward that of the swell wave. The larger the swell energy, the larger the shift. The peak frequency of the swell wave remains unchanged.

2) The cross-shore variation of H_{rms} for the wind waves is qualitatively similar to that of the swell waves. The cross-shore variation of $\overline{\eta}$ is significantly different.

3) For strong winds, the undertow cannot be related to the local water depth and wave height through Equation 1 unless effects of the surface current are considered.

4) Increases in the measured undertow due to the wind are consistent with the measured increases in the setup gradient.

Additional tests, including those with lower wind speeds, are required to assess the generalities of these conclusions. Direct measurements of the wind stress will allow for a detailed analysis of the cross-shore momentum balance.

ACKNOWLEDGMENTS

The authors thank John Reed for his technical support and Dr. Jun Zhang for his comments on this work. Joshua Carter was supported by the NSF Research Experience for Undergraduates Program and funded by the National Science Foundation, the Offshore Technology Research Center, and the Texas Engineering Experiment Station.

REFERENCES

Kennedy, D.L., Cox, D.T., and Kobayashi (1998) "Application of a kinematic undertow model to irregular waves on barred beaches and reflective coastal structures." *Proc 26th Int. Conf. on Coastal Engrg.* (abstracted accepted).

Pattiaratchi, C., and Masselink, G. (1996) "Sea breeze effects on nearshore coastal processes." *Proc. 25th Int. Conf. on Coastal Engrg.*, ASCE, 4200-4213.

Smith, J.M., Svendsen, I.A., Putrevu, U. (1992) "Vertical structure of the nearshore current at Delilah: measured and modeled." *Proc. 23 Int. Conf. on Coastal Engrg.*, ASCE, 2825-2838.

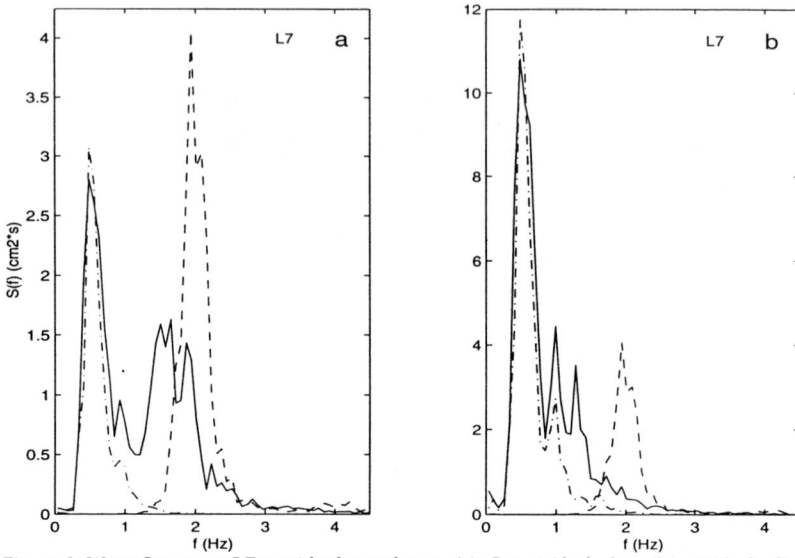

Figure 2: Wave Spectra at L7 outside the surf zone. (a) Case 1 (dash-dot), Case 3 (dashed), and Case 4 (solid); (b) Case 2 (dash-dot), Case 3 (dashed), and Case 5 (solid).

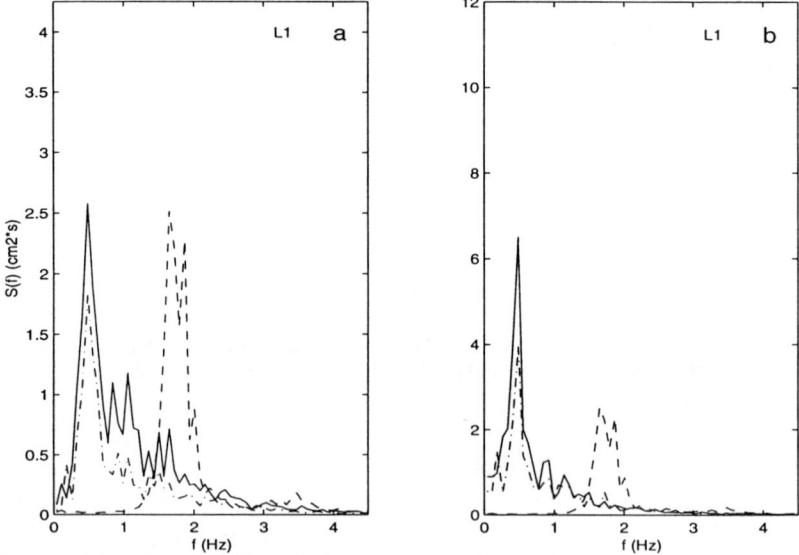

Figure 3: Wave Spectra at L1 inside the surf zone. (a) Case 1 (dash-dot), Case 3 (dashed), and Case 4 (solid); (b) Case 2 (dash-dot), Case 3 (dashed), and Case 5 (solid)..

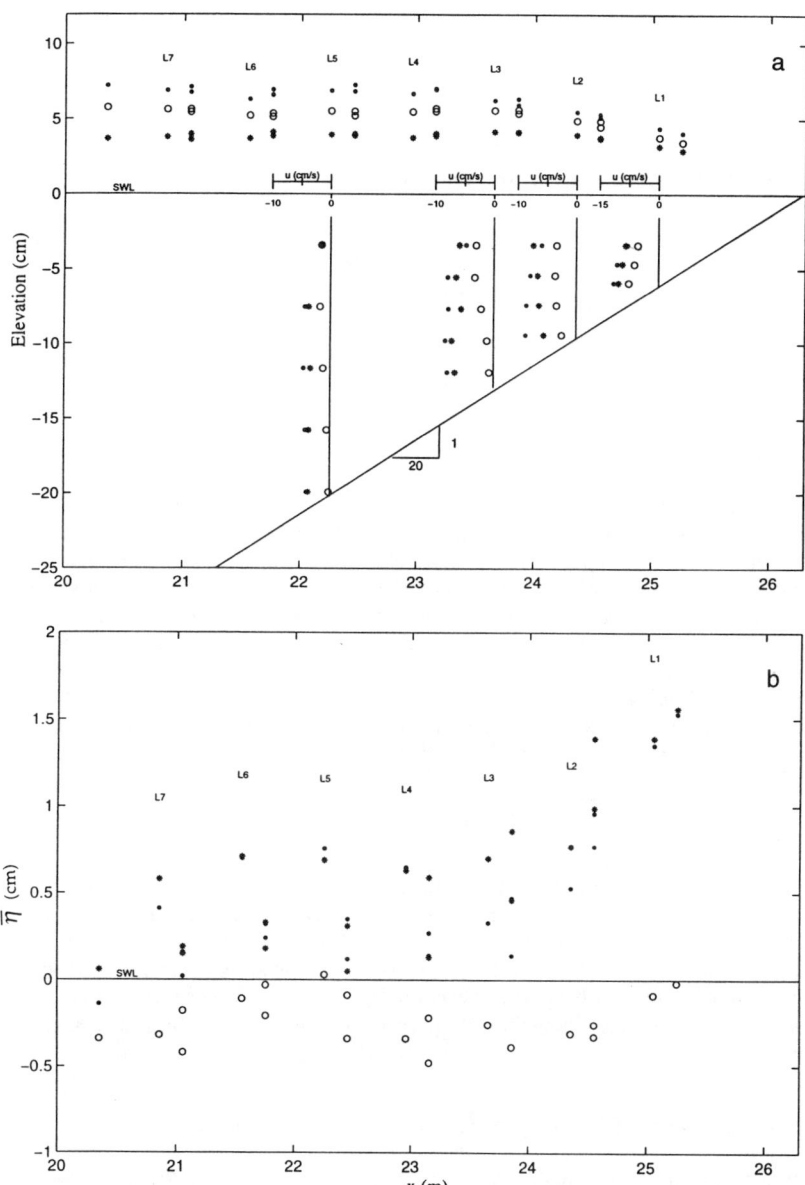

Figure 4: (a) Cross-shore variation of H_{rms}, $\bar{u}$ for Case 2 (o), Case 3 (*), and Case 5 (●); (b) cross-shore variation of $\bar{\eta}$.

Numerical Simulation of Bed-Load Sediment Transport by Long-shore Currents

M. Di Natale[1], Member ASCE, G. Piccirillo[2]

Abstract

In the paper a mathematical model to value the bed sediment transport due to a long-shore current is presented. This model, differentiating from the previous empirical formulas proposed by the other Authors, is founded on a hydrodynamic approach, which requests, at first time, the determination of the hydrodynamic flow-field (hydrodynamic module) and, then, the sediment transport rate evaluation (sediment transport module), founded on the Meyer and Peter-Muller theory. Some applicative examples, referred to the scheme of a rectilinear coastal line and uniform bed slope, are reported to determine the influence of the different characteristic parameters. These results are compared with those obtained by the application of the other empirical formulas.

1. INTRODUCTION

The net sediment transport rate, moved along a coastal line by a long-shore currents (due to oblique incidence of wave motion), is one of the first causes of many erosion or landfill process in a sand beach or near a marine structure. The determination of the sediment rate, Q_s, which is generated by a wave motion with a predefined values of deep water wave height, H_0, deep water wave length, L_0, and wave direction, α_0, constitutes a problem that has already been analized by many Authors.

---

[1] Professor, Dept. of Civ. Eng., Seconda Università degli Studi di Napoli, Aversa (CE), Italy
[2] Post-Grad. St., Dept. of Civ. Eng., Seconda Università degli Studi di Napoli, Aversa (CE), Italy

The empirical expression most widely used in the technical applications is the CERC formula (SPM, 1984):

$$Q_s = K P_{ls} \qquad (1)$$

where:
P_{ls} = long-shore energy flux factor; K = dimensionless coefficient.

Considering Q_s as a flow-rate expressed in m^3/yr, the K coefficient is equal to 1290 about [m^4/W yr] (Bruno et al., 1981). Regarding the energy flux factor, among the different empirical expressions the most used is the following relation (SPM, 1984):

$$P_{ls} = 0.0884 \rho\, g^{3/2}\, H_{sb}^{5/2}\, \text{sen}\, \alpha_b \qquad (2)$$

being
ρ = water density; α_b = angle of incidence in the surf-zone; H_{sb} = significant breaking wave height.

The Q_s expression previously reported has been the object of many considerations regarding to the absence of some important parameters like the bed slope, i, in the surf-zone and the sediment characteristic diameter, commonly identified as D_{50} diameter. A correlation study between the different available data (Swart, 1976) indicates the following relationship between the K coefficient and the diameter D_{50}:

$$K = 1876\, \log(0.00146 / D_{50}) \qquad (3)$$

The equation (3), in which D_{50} is expressed in m., can't be used for a D_{50} values greater then 1.46 mm, so the most frequent technical cases are excluded.

The K dipendence on bed slope, i, in the surf-zone has been obtained by experimental studies (Kamphuis and Readshaw, 1978; Vitale, 1981). In particular, it has been discovered that:

$$K = 708 \text{ for } 0.4 < \xi_b < 1.4 \qquad (4a)$$
$$K = 1254 \text{ for } \xi_b > 1.4 \qquad (4b)$$

Being $\xi_b = i/(H_0/L_0)^{0.5}$ the Irribarren number and $H_0 = H_{bs}/\sqrt{2}$.

By the relations (4a-b) it is possible to emphasize that the variability of the sediment transport rate, Q_s, grows with bed slope, i, until a value equal to $0.7\xi_b$, then it becomes constant even if the bed slope increases. The reported expressions are clearly incomplete; in fact, in the (3) only the grain size influence is considered, neglecting the bed slope dipendence, and vice-versa in the (4a-b).

By means of best fit analysis of experimental and empirical formula results, it has been possible to obtain (Schoones and Theron, 1994) that for the fine sand materials (D_{50}<1.00 mm) the K coefficient, in the CERC formula, is equal to 1379, assuming a value for the coefficient of linear correlation equal to 0.77.

Recently, during an experimental study in a 3D-basin, a relationship has been proposed (Kamphuis, 1991) to value the net transport rate Q_s:

$$Q_s = 6.4 \times 10^4\, H_{sb}^2\, T_p^{1.5}\, i^{0.75}\, D_{50}^{0.25}\, \text{sen}^{\,0.6}(2\alpha_b) \qquad (5)$$

It is necessary to observe that the equation (5), even though if it considers the scale effects of the model and the effects of the grain size, D_{50}, and bed slope, i, has been obtained by a limited data-set (i=0.10 and D_{50} = 0.105 ÷ 0.18 mm).

In this work the valuation of sediment transport rate, Q_s, has been done by a hydrodynamic approach. The reliability of the model results, comparing with the experimental data founded in literature, is proved quite encouraged. Moreover using this numerical model it is possible to value the sediment transport rate, Q_s, for a generic condition of bed morphology and coastal line geometry, for every bed slope in the surf zone and sediment diameter, D_{50}.

The limits of the proposed model are represented by the utilisation of the regular wave theory, the expression of the components of radiation stress tensor by the second order Stokes theory, the negligence of the suspended sediment transport.

The applicative example are referred to the scheme of a constant bed slope and regarding this cases there are reported the comparison between the models results and those obtained by application of the empirical formula. However, to give a more realistic example it is reported the results, concerning the bed sediment transport along Terracina beach, located in Italy along the Tyrrhenian Sea coasts.

2. HYDRODYNAMIC AND SEDIMENT TRANSPORT MODEL

The problem has been solved by a realisation of two different modules:
- hydrodynamic module;
- sediment transport module.

Hydrodinamic module

Regarding the first module, the 2DH continuity and momentum equations, where the velocity components averaged in the wave period, T, are used; the constitutive equations are reported below (Di Natale, 1997):

$$\frac{\partial \eta}{\partial t} + \frac{\partial UH}{\partial x} + \frac{\partial VH}{\partial y} = 0 \qquad (6)$$

$$\frac{dU}{dt} = -g\frac{\partial \eta}{\partial x} - \frac{\tau_{bx}}{\rho H} - \frac{1}{\rho H}\left(\frac{\partial S_{xx}}{\partial x} + \frac{\partial S_{xy}}{\partial y}\right) \qquad (7)$$

$$\frac{dU}{dt} = -g\frac{\partial \eta}{\partial y} - \frac{\tau_{by}}{\rho H} - \frac{1}{\rho H}\left(\frac{\partial S_{yx}}{\partial x} + \frac{\partial S_{yy}}{\partial y}\right) \qquad (8)$$

where U and V are the depth integrated velocity components, S_{ij} the components of radiation stress tensor and τ_b the bed shear stress.

To examine the refraction phenomena, the irrotationality equation of the wave number $\nabla \times \mathbf{K}$ and the energy flux consevation $\nabla C_g\ E=0$ are considered; moreover, the wave height in the surf zone, H_b, is computed as equal 0.78 h_b, where h_b is the corresponding local water depth.

On the basis of the equations previously described it is possible to determine the U and V components of the steady flow field in accordance with the lay-out reported in Fig. 1.

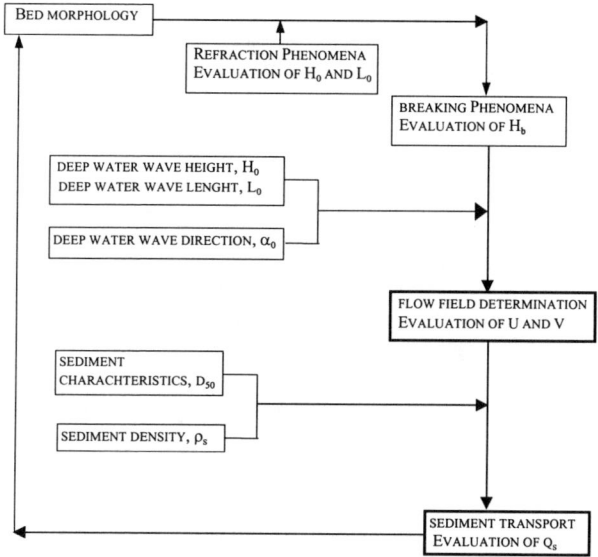

Figure 1. Model Lay-out

The numerical solution of the *hydrodynamic module* has been carried out by a difference finite scheme of the (6), (7) and (8) in which the radiation stresses terms are computed by the solution of the refraction equations (Chieffi and Di Natale, 1996).

Sediment transport module

Concerning the bed sediment transport module, assuming that the analysed non-steady flow field can be sketched as a steady flow fields succession, the Meyer and Peter-Muller theory is considered. Referring to the scheme of Fig. 2, the valuation of sediment transport rate Q_s in a generic transversal section, j, in the surf zone has been done as a whole amount of the sediment rate of a single cell of the calculation domain, $q_{sj}\Delta y$, when the U_{ij} and V_{ij} values are determined.
The expression of bed sediment transport rate per unit width, q_{sj}, is provided by the following equation:

$$q_{sj} = 8(\frac{1}{\Psi} - 0.0047)^{3/2} / (\frac{\rho_s - \rho}{\rho} g D_{50}^2)^{0.5} \qquad (9)$$

where: ρ_s and ρ = sediment and water density; $1/\Psi = \rho v^* / gD_{50}(\rho_s - \rho)$ (Shields parameter); $v^* = (\tau_0/\rho)^{0.5}$ (friction velocity); $\tau_b = A_f U \sqrt{U^2 + V^2}$ (bed shear stress); A_f = bottom friction coefficient.

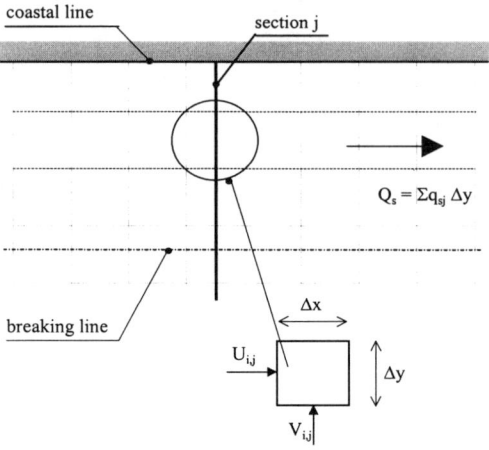

Figure 2. Calculation grid

3. NUMERICAL RESULTS

Some numerical simulations have been done to value the Q_s bed transport with the variations of the different characteristic parameters, relatively to a schematic case of uniform bed slope and linear coastal line. In the Figs. 3-4-5 the computed values of Q_s are reported versus the sediment grain size (characterised by D_{50} diameter) for predetermined values of α_0 and i (Fig. 3), for a fixed values of H_0 and α_0 (Fig. 4) and, then, for fixed values of H_0 and i (Fig. 5).

The results interpolation lines show that the transport rate, Q_s, decreases, in all the considered cases, with the growth of the grain size, resulting more sensitive to the variation of the α_0 parameter than H_0 and i. Fig. 5 shows that Q_s increases, for a fixed D_{50}, with the angle of incidence, assuming a maximum value in correspondence of $\alpha_0 = 50°$.

Regarding the influence of the bed slope parameter, in the Fig. 6 it is reported the interpolation curves of Q_s versus i for a fixed value of H_0 and α_0 ($H_0 = 1.0$ m, $\alpha_0 = 50°$) and for some D_{50} values. It is possible to note that Q_s grows with the bed slope, assuming the maximum in correspondence of a bed slope i equal to 0.03; then, the sediment transport rate decrease for greater values of i.

To validate the results obtained by the numerical model proposed, a comparison between other different model results has been done. In the Fig. 7 it is reported an explicative case ($H_0 = 1.00$ m, $\alpha_0 = 50°$, $i = 0.04$) from which it has been obtained: i) the CERC formula (Q_s constant for different D_{50}) over-estimates the sediment transport rate; ii) the Kamphuis results (Eqn. 5) are in agreement to those calculated; iii) in the Swart eqn. (K coefficient related only to sediment diameter D_{50}) the transport rate is strongly dependent from D_{50} and Q_s values are

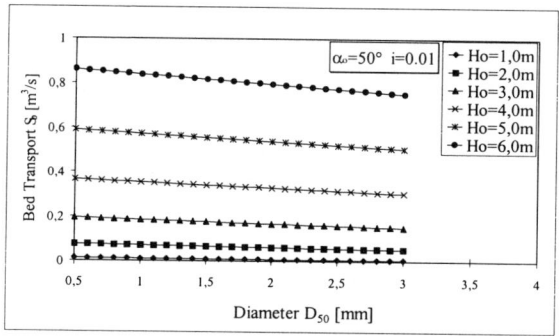

Figure 3. Variation of sediment flow-rate versus diameter D_{50} and wave height H_0

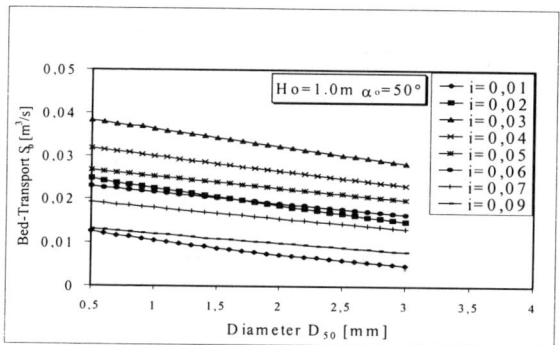

Figure 4. Variation of sediment flow-rate versus diameter D_{50} and bed slope i

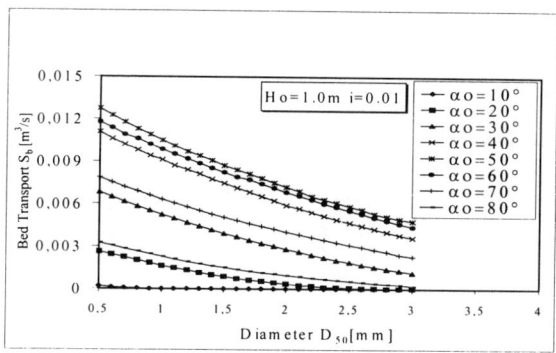

Figure 5. Variation of sediment flow-rate versus diameter D_{50} and wave angle α_0

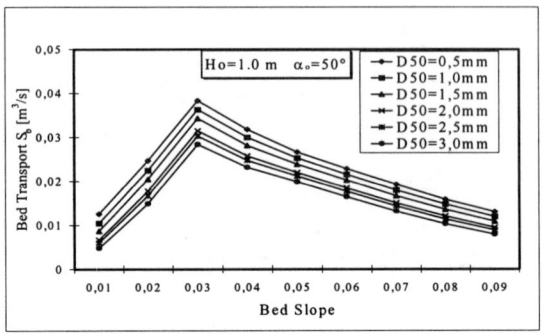

Figure 6. Variation of sediment flow-rate versus bed slope

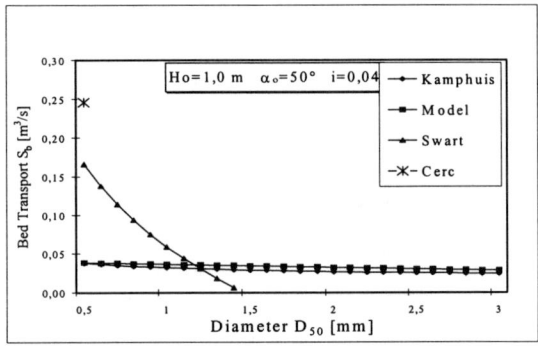

Figure 7. Comparison between different model results

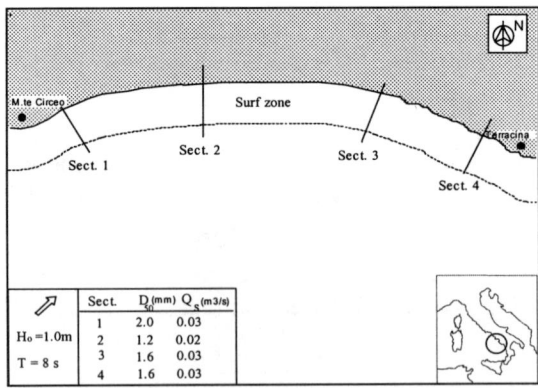

Figure 8. Applicative example (Terracina beach, Italy)

quite different from those obtained by the proposed model. Furthermore, as reported in previous papers (Schoonees and Theron, 1994), the Swart and CERC eqns. yeld an over-estimation of Q_s.

In conclusion, in the Fig. 7 an applicative example is reported for Terracina beach in the central coast of Tyrrhenian Sea. This coast, characterised by an oblique incidence wave motion is subjected to a long-shore sediment transport. The study shows how it is possible, using the proposed model, to calculate the bed sediment transport rate in the particular transversal sections of the coastal line.

4. CONCLUSIONS

The presented numerical model permits to value the bed long-shore sediment transport along a coastal line, with prefixed morphologic characteristics, due to a determined wave motion. The simulations have showed that the sediment transport rate is related to H_0, L_0 and α_0, as reported in the CERC formula, and to grain size, D_{50}, and bed slope, i, in the surf zone.

Regarding the last parameter, it is necessary to underline that the bed slope, considering fixed values of the other characteristics parameters, has an important role in the bed sediment transport. In particular, it has been noticed that, for a bed slope $i \approx 0.03$, there is the maximum value of sediment transport rate.

The comparison between the proposed model and the empirical expressions founded in literature shows a good agreement with Kamphuis relation while the Swart expression and the CERC formula provide Q_s values greater than model results.

REFERENCES

Bruno, R.O., Dean, R.G., Gable C.G. and Walton, T.L. (1981), *"Longshore sand transport study at Channel Islands Harbour, California"*, Techn. Paper, CERC, U.S. pp. 48

Chieffi, Di Natale (1996), *"Sul trasporto superficiale di inquinanti effluenti da condotte sottomarine in presenza di correnti longshore"*, AIOM, Padova (taly)

Di Natale, M. (1998), *"Non Linear Hydrodynamic Effects of Opposing Jet-current on Waves"*, 8th International Offshore and Polar Enginneering Conference, ISOPE.

Kamphuis, J W and Readshaw, J S (1978), *"A model study of alongshore sediment transport rate"* 16th International Conf. On Coastal Eng, Hamburg Vol. 2, pp. 1656-1674.

Kamphuis, J W (1991), *"Alongshore Sediment transport rate"*, ASCE Journ. of Water, Port, Coastal Ocean Eng.

Schoones, J.S. and Theron, A. K. (1994), *"Accuracy and Applicability of the SPM Longshore Transport Formula"*, Coastal Engineering pp. 2595 – 2609.

Swart, D H (1976), *"Predictive equations regarding coastal transports"*, 15th International Conf On Coastal Eng, Honolulu, Hawaii, Vol. 2, pp. 1113 – 1132.

U.S. Army, Corps of Engineers (1984), *"Shore Protection Manual"*.

Vitale, P. (1981), *"Movable-bed laboratory experiments comparing radiation stress and energy flux factor as predictors of longshore trasport rate"*, U.S. Army Coastal Eng. Res. Center, Fort Belvoir, Miscellaneous Report No. 81-4.94 pp.

Deposition of Cohesive Sediments under Waves
S. Abdel-Mawla[1]   A. Matheja[2]   C. Zimmermann[3]

Abstract

The deposition of cohesive sediment depends on a combination of different factors, including size, settling velocity and strength of the settling units. It may be hypothesised, that deposition of flocs is controlled by stochastic turbulent processes in a zone near the bed. A large difference exists in the basic characteristics of the response of water-mud system to wave-induced motion. In this paper, a physical model is studied for deposition of natural cohesive sediments under regular waves. Measuring velocities and dynamic pressure on the bottom, the bottom shear stress can be estimated. It is found, that for nearly identical bottom shear stress values, the rate of deposition is changed according to the pressure steepness (P/gT^2) on the bottom.

Introduction

Siltation of harbours and coastal zones is often undesirable as it diminishes navigability. Besides, deposited mud is often polluted, because it specifically absorbs contaminants from water. For managers of harbours and coastal zones it is therefore important to know prior to execution whether planned projects will reduce or increase siltation. The performance of the current cohesive sediment transport models, which are being used as a tool for prediction, needs considerable improvements. The shortcomings are caused to a great extent from insufficient knowledge of mechanisms driving cohesive sediment transport.

Cohesive sediment mainly consist of clay particles, which are single or aggregations of flocs. Mechanism for deposition and re-entrainment of mud particles are given in Fig. 1. For more comprehensive discussions the reader is refered to Mehta (1993, 1996) and Teisson et al. (1993).

---

[1] M.Sc., Research Eng., FRANZIUS-INSTITUTE for Hydraulic, Waterways and Coastal Engineering, Nienburger Str. 4, 30167 Hannover, Germany
[2] Dr.-Ing., Senior Eng., FRANZIUS-INSTITUTE for Hydraulic, Waterways and Coastal Engineering, Nienburger Str. 4, 30167 Hannover, Germany
[3] Prof. Dr.-Ing., Director, FRANZIUS-INSTITUTE for Hydraulic, Waterways and Coastal Engineering, Nienburger Str. 4, 30167 Hannover, Germany

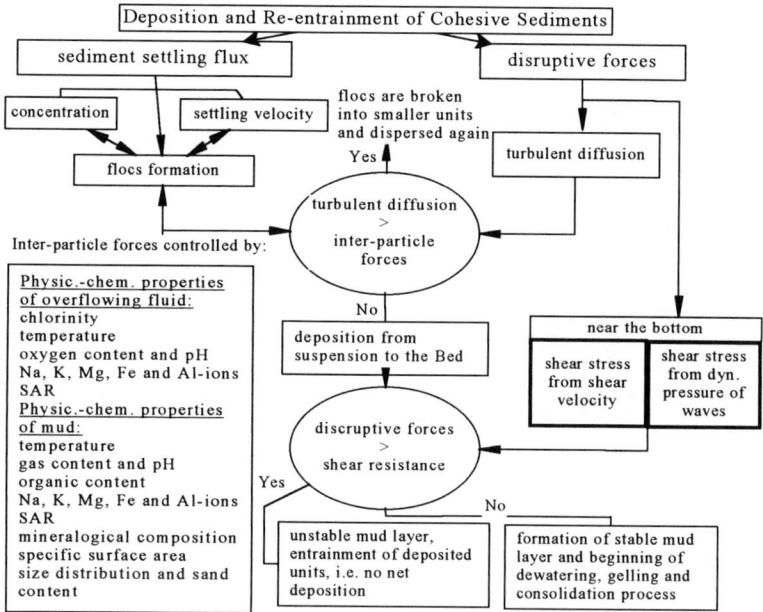

Fig. 1. Deposition and Re-entrainment Mechanism of Cohesive Sediments

Review of Previous Work

In most cohesive sediment deposition models, as summarized by Teisson et al. (1993) or Mehta (1996), deposition was modelled according to Krone (1962):

$$dm/dt = (1 - \tau_b/\tau_d) w_s c \tag{1}$$

This empirical formula is developed for unidirectional stream flows. It is applied also to predict deposition under waves as the deposition process is considered similar. Hawang (1989) developed a formula for settling velocities w_s in flocculation and hindered settling regions:

$$w_s = \frac{a c^n}{(c^2 + b^2)^m} \tag{2}$$

Mimura (1993) studied the rates of erosion and deposition of cohesive sediments under wave action. Bottom shear stress was calculated using a wave friction factor. He found, that deposition and erosion coexist in some range of bottom shear stress.

De Wit (1995) studied the effect of pore water pressure change in the mud layer on the liquefaction process. The author found the mud to be liquefied when pressure-induced shear stresses in the bed exceed the yield strength of the mud.

Some of previous experimental studies in the field of cohesive sediments are summarized in Tab. 1.

Table 1. Laboratory Facilities for Selected Studies
* t=bed sample thickness ** d=water depth over the bed ***T,H =wave period and height

Authors	Nature of Soil Samples	Lab. Facility	Flow Condition
Otsubo (1988) erosion	artificial kaolinite, bentonite, nat. mud (t = 2 cm)	pipe with fresh water (d = 5 cm)	steady currents
Horikawa (1986) erosion	artif. kaolinite mixed with water (diff. w_c), (t = 9.5 cm)	wave flume (d = 10 cm)	reg. waves, T = 1 s, H = 1.5-1.7 cm
de Wit (1995) erosion	artificial china and westwald clay, (t = 20 cm)	wave flume	reg. waves, current, T = 1.5s, H=1.5-9.3 cm
Feng (1992) bed fluidization	kaolinite clay (t = 16 cm)	wave flume (d = 35 cm)	reg. waves, T = 1.0 s, diff. H
van Kessel (1997) erosion/liquefaction	china clay, natural mud (t = 10 cm)	wave flume (d = 30 cm)	reg. waves, T = 1.65s, H = 0.4-5.5 cm

In this study, an attempt was made to estimate the bottom-shear stress using the measured velocity components to define the role of dynamic wave pressure on deposition of clay.

Experimental set-up and procedure

The wave flume is shown in Fig. 2. The water depth was kept constant at 40 cm with a fixed bed.

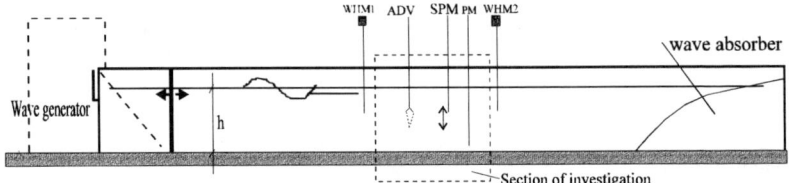

Fig. 2. Sketch of the Wave Flume

To measure the local deposition Acoustic Doppler Velocimeter (ADV), Spectro Photo Meter (SPM), Pressure Meter (PM) and two Wave Height Meters (WHM) were used to record velocity components, sediment concentration, pressure fluctuation on the bed and wave height.

Experiments covered wave periods of 0.74, 087, 1.06, 1.40 s, wave heights from 0.88 cm to 10.5 cm and three initial concentration ratios (3.5, 8.0 and 12.0 g/l). Concentrations of 8.0 and 12.0 g/l were tested for wave periods of 0.74 s and 1.40 s.

The cohesive material was natural clay (Tab. 2 to Tab. 5). Settling velocity formula of Hawang (1989) from eq. (4) was calibrated by an experiment using settling column (a = 0.14, b = 49, n = 1.3 and m = 2.3). Starting with the initial concentration c_0 ADV was used to measure velocity fluctuations from bottom to surface. SPM was used to record the variations of concentration with time, while deposition was related to wave characteristics.

Table 2. Grain Size

Soil classification	Grain size	Percent %
clay	<0.002	55
very fine silt	0.002-0.006	25
fine silt	0.006-0.02	15
silt	0.02-0.06	2
very fine sand	0.06-0.2	3

Table 3. Characteristic Diameters

d_{10}	d_{30}	d_{50}	d_{60}
-	-	0.0017	0.0028

Table 4. Atterberg Limits

liquid limit (W_L)	63.1%
plastic limit (W_P)	21.2%
plasticity index (I_P)	41.9%
volumetric density (g/cm³)	2.679

Table 5. Chemical Properties

pH of water	8.3
organic content	5.49 %
$CaCO^3$ content	4.5 %

For each point, more than 50 waves were recorded, while sampling frequency of ADV (0.1 to 25 Hz) was adjusted according to wave period.

Data Analysis

Turbulent and wave-induced components of the free surface elevations and velocities were separated by phase averaging over 50 waves, showing that setdown and setup were negligible. Phase-averaged horizontal velocities are used to estimate the shear velocity. Wave-induced current $\bar{u}_a$ is neglected. Near the bottom, a logarithmic velocity profile is assumed at each phase over the wave period:

$$u_a(Z_b) = \frac{u_*}{\kappa} \ln(Z_b - d_* / Z_0) \quad \text{for } Z_b \geq (d_* + Z_0) \quad (3)$$

Cox et al. (1996) showed, that the expression of $d_* = 0.7 Z_0$ was a adequate estimate in many commonly encountered types of roughness. Eq. (3) can be reduced to a linear equation:

$$Y_i = \beta X_i + \alpha \quad \text{for } i = 1, 2, \ldots, n \quad (4)$$

$$X_i = [u_a]_i \quad \text{and} \quad Y_i = \ln\left[(Z_b)_i - d_*\right] \quad (5)$$

$$\beta = \kappa / u_* \quad \text{and} \quad \alpha = \ln(Z_0) \quad (6)$$

Maximum shear velocity $u_{*,max}$ and bottom roughness Z_0 ($\kappa = 0.4$, $d_* \cong 0.7 d_{50}$ with $d_{50} = 0.90$ mm) were computed using least squares. The actual value of d_* is selected as the best fit from $d_* = 0.04, 0.05, 0.07$ and 0.08 (elevation Z_b includes the uncertainty of the initial elevation of the ADV measuring volume). Values for U_* and Z_0 were computed three times for n = 3, 4 and 5.

Following Jonsson and Carlsen (1976) variation of bottom shear stress with time for sinusoidal flow is given by:

$$\tau_b(t) = \rho |u_*| u_* \quad (7)$$

Assuming the maximum bottom shear stress within a short part of the wave period, the use of time-averaged bed shear stress (over half a wave cycle) is more realistic as shown in eq. (8). The rate of deposition R_d [kg/(s m^2)] is calculated from the concentration profiles using eq. (9).

$$\tau_b = \frac{1}{2}\rho U_*^2 \tag{8}$$

$$R_d = \frac{1}{t}\int_{t=i}^{t=i+1}\int_{z=0}^{z=h} c(z,t)\,dz\,dt \tag{9}$$

Results

Experimental time for total deposition increased according to the used wave characteristics (experiment stopped, when $\Delta c \cong 0$, where Δc is the difference between two sucessive concentration profiles). The runing time reached up to 24 hours. Measuring the concentration profiles each time and according to eq. (9), total deposited sediments relationships are drawn for all wave characteristics and initial sediment concentrations. Deposition under still water (H/gT² = 0) was taken as reference for deposition under waves.

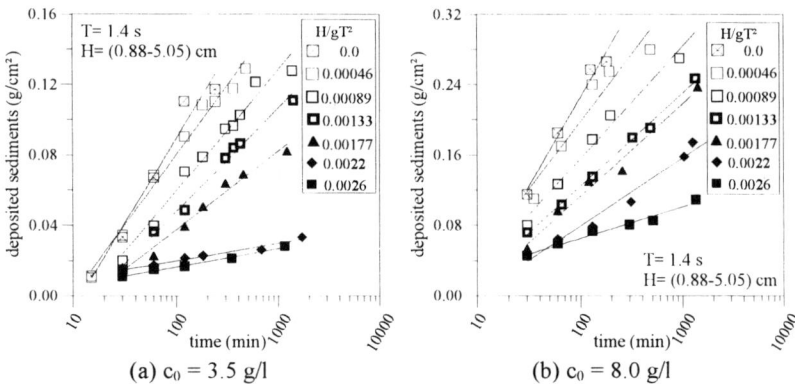

(a) $c_0 = 3.5$ g/l   (b) $c_0 = 8.0$ g/l
Fig. 3. Sediment Deposition with Time as Function of Wave Height

Logarithmic behaviour of deposition may be explained by sorting and flocculation mechanism of cohesive sediments mixture. Fig. 3 is an examples for this group of relationships. As shown, the rate of deposition decreases as the wave height increase at the same wave period. Comparing Fig. 3 (a) and (b), it is found that the rate of deposition for $c_0 = 8.0$ g/l is more than for 4.0 g/l. This can be explained by increasing settling velocities due to increasing concentration up to a certain limit (max. settling velocity in flocculation zone from eq. (2) is at about 12 g/l).

Fig. 4 conclude all used wave conditions, expressed as dimensionless parameter (H/gT² and T), against deposited sediments for initial concentration 3.5 g/l. Comparing the behaviour of deposition it is found that as wave period decreases, the effect of wave steepness (H/gT^2) decreases and deposition increases at the same wave steepness.

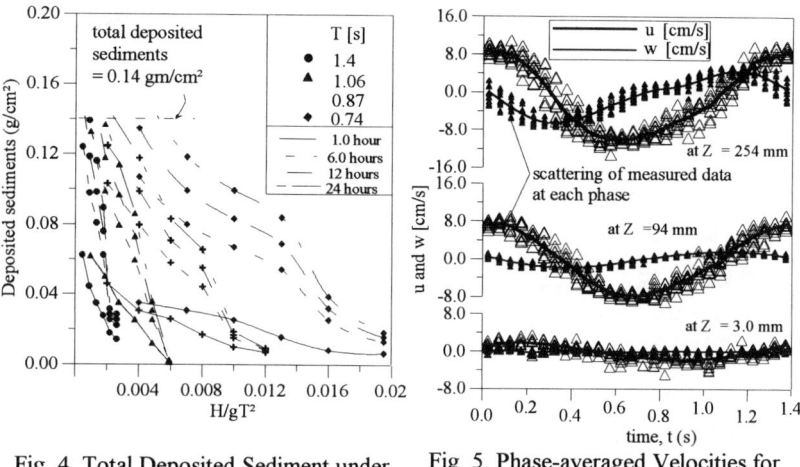

Fig. 4. Total Deposited Sediment under variing Wave Height and Period

Fig. 5. Phase-averaged Velocities for Measured Data ($H/gT^2 = 0.00263$, T = 1.4 s)

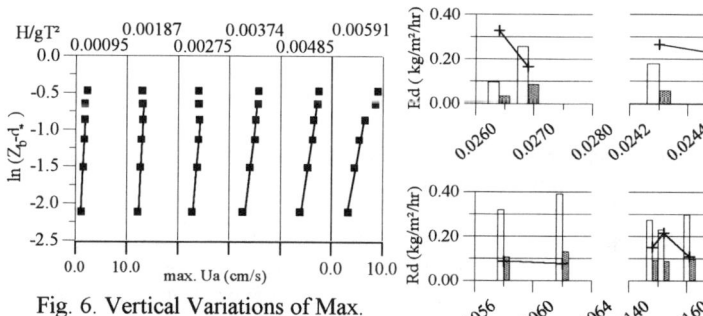

Fig. 6. Vertical Variations of Max. Phase-averaged Horizontal Velocity $u_{a,\,max}$ (T = 1.06 s)

Fig. 7. Rate of Deposition for Nearly Identical Shear Stress Values ($c_0 = 3.5$ g/l)

The phase averaged velocities are calculated at each measuring point. Fig. 5 shows an example including data scattering. Fig. 6 shows an example for the results of application the algorithm explained through eq. (3) to eq. (6).

The estimated bottom shear stress in eq. (8) is studied against the rate of deposition at different time steps (Fig. 7 for $c_0 = 3.5$ g/l). Measured dynamic wave pressure on the bottom is studied also against the rate of deposition. It is found that the relation has a more understandable trend, when pressure is expressed as a steepness ($P' = P/gT^2$). As shown in Figs. 8 and 9, the rate of deposition is decreased as bottom shear stress and pressure steepness increase.

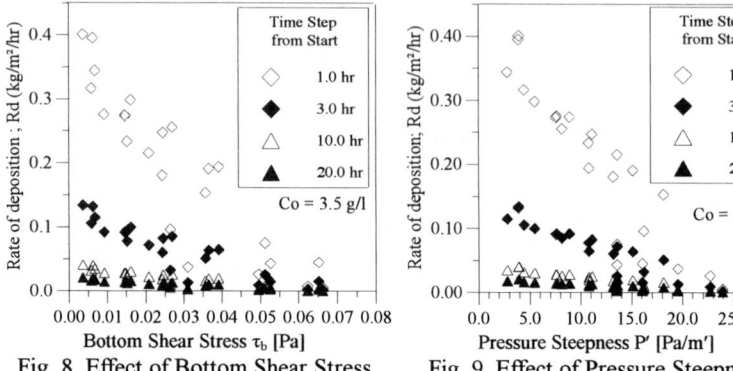

Fig. 8. Effect of Bottom Shear Stress on Deposition

Fig. 9. Effect of Pressure Steepness on Deposition

Looking through all the bottom shear stress values and selecting some very close values, the rate of deposition is studied against these nearly identical shear stress values (Fig. 7). It is notable, that rate of deposition decrease as the pressure steepness increases.

Conclusion

Flocculation and sorting processes control deposition of cohesive sediments, which was shown in all test cases.

The sediment deposition decreases with increasing wave period for constant wave steepness, because bottom shear stress and diffusion under these conditions increase.

It was also found, that not only the bottom shear stress controls the deposition of cohesive sediments, but also the pressure steepness has this role. The probability factor in eq. (1) should incorporate also the pressure steepness to express the deposition mechanism under waves. Further research will study ranges of initial concentrations, water depths and wave characteristics.

Appendix I. References

Cox, D.T., Kobayashi, N., and Okayasu, A. (1996). "Bottom shear stress in surf zone." J. Geophys. Res., 101(c6 14), 337-348.

De Wit, P.J. (1995). "Liquefaction of cohesive sediments caused by waves." Rep. No. 95-2, Ph. D. thesis, Delft University of Technology, Delft, Netherlands.

Feng, J. (1992). "Laboratory experiments on cohesive soil bed fluidization by water waves." M.S. thesis, University of Florida, Gainesville, Florida.

Hawang, K. (1989). "Erodibility of fine sediment in wave-dominated environments." M.S. thesis, University of Florida, Gainesville, Florida.

Horikawa, K. and Shibayma T. (1986). "Mud mass transport due to waves." Coastal Engineering in Japan, Vol. 29, 151-161.

Jonsson, I.G., and Carlsen, N.A. (1976). "Experimental and theoretical investigations in an oscillatory turbulent boundary layer." J. Hydraul. Res., Vol. 14, 45-60.

Krone, R.B. (1962). "Flume studies of the transport of sediments in estuarial shoaling processes." Final Rep., Hydraulic Engineering Laboratory and Sanitary Engineering Research Laboratory, University of California, Berkeley.

Mehta, A.J. and Jiang, F. (1993). "Some observations on bottom mud motion due to waves." Report UFL/coel/mp-93/01, Coastal and Oceanographic Eng. Dept., University of Florida, Gainesville, Florida.

Mehta, A. (1996). "Interaction between fluid mud and water waves." Env. Hydr., Kluwer Academic Publishers, 153-187.

Mimura, N. (1993). "Rates of erosion and deposition of cohesive sediments under wave action." In: Near shore and estuarine cohesive sediment transport, Mehta, A.J. (Ed.), 247-264.

Otsubo, K. and Muraoka, K. (1988). "Critical shear stress of cohesive bottom sediments."J. Hydr. Eng., 114 (10), 1214-1256.

Teisson, C., Ockenden, M., Kranburg, C. and Hamm, L. (1993). "Cohesive sediment transport processes." Coast. Eng., Vol. 21, 129-162.

van Rijn, L.C. (1993). "Principles of sediment transport in rivers, estuaries and coastal seas." Aqua Publications, Netherlands.

van Kessel, T. (1997). "Generation and transport of subaqueous fluid mud layers." Rep. No. 97-5, Dept. Civ. Eng., Delft University of Technology, Delft, Netherlands.

Appendix II. Notation

The following symbols are used in this paper:

τ_b	=	bottom shear stress [N/m^2]
τ_d	=	critical shear stress for deposition [N/m^2]
w_s	=	settling velocity [m/s]
c	=	concentration [kg/m^3]
h	=	still water depth [m]
a, b, n, m	=	sediment dependent empirical coefficients
		b = 1 to 10, n = 0.8 to 2.5, m = 1 to 3
u_a	=	phase-averaged horizontal velocity [m/s]
$u_{a,\,max}$	=	maximum phase-averaged horizontal velocity [m/s]
w_a	=	phase-averaged vertical velocity [m/s]
Z_b	=	vertical coordinate (upward) with $Z_b = 0$ at the bottom [m]
u_*	=	shear velocity [m/s]
$u_{*,\,max}$	=	maximum shear velocity [m/s]
i	=	index for measuring point elevation [-]
k	=	van Karman constant 0.4 [-]
d_*	=	displacement distance [m]
Z_0	=	bottom roughness [m]
ρ	=	fluid density [kg/m^3]

Risk Analysis for Wave-Suspended Sediments

Katherine A. Larm[1] and Billy L. Edge[2]

Abstract

To determine the risks of resuspension of contaminated sediments in Lavaca and Matagorda Bays, Texas, it is necessary to determine the scour or resuspension potential of the conditions in the Bays. Typically, conditions in the Bays are driven by low-energy winds that do not affect the bottom sediments through the wind-generated waves and currents. A layer of clean sediments typically overlies the contaminated sediments. Therefore, the objective of this study is to determine the expected event that would cause a removal of the overlying clean sediments and the amount of contaminated sediments that would be removed and suspended. ADCIRC is used to develop a method to determine the dynamic water levels in the Bays from hurricanes moving across the Gulf of Mexico and causing a large storm surge with accompanying large currents. The dynamic water levels and winds from a planetary boundary layer model are then used with SWAN, a regional wave prediction model, to determine the dynamic wave conditions in Lavaca Bay. The current from the storm surge and the wave generated bottom currents are combined to determine the amount of sediment removal that may occur.

Introduction

Matagorda and Lavaca Bays, in Texas, are areas typical of a low energy environment and thus are not subject to frequent erosion or resuspension of bottom sediments. On the other hand, because the Bays are connected to the Gulf of Mexico (Fig. 1), the storm (hurricane) conditions produce high-energy events with elevated storm surges and large waves. These events contribute to resuspension of some bottom sediments and erosion of the banks. It is the resuspension of the bottom sediments that are the primary concern in this paper. As a result of historical activities in Lavaca Bay, localized areas of contaminated sediments exist in the Bays with higher concentrations in Lavaca Bay. Most of the contaminated sediments are covered with a thin layer of recently deposited clean sediments, from natural processes, since the sources of contamination have been contained. Thus ignoring bioturbation and uptake through plants attached to the bottom, the contaminated sediments, in most locations, would present a concern only if they were suspended and redistributed in the Bay.

The objective of this study is to develop a method to determine the risk of exposing the contaminated sediments by eroding the thin layer of clean sediments. Events which can cause significant resuspension and erosion of the contaminated sediments are expected to be hurricanes. The procedure in this study uses the parameters of previous storms that affected the area with both storm surge and wind waves to model the probable sediment suspension and removal that results from those specific historical conditions. From the results of the hindcasted events, an empirical simulation approach can be used to determine the probabilities of exposure or the risk associated with a storm having a particular return frequency. Following such a procedure requires an analysis of all tropical storms which have affected the area and possibly caused sediment resuspension.

---

1 Research Assistant, Ocean Engineering Program, Department of Civil Engineering, Texas A&M University, College Station, TX 77843-3136.

2 W. H. Bauer Professor of Dredging Engineering, Ocean Engineering Program, Department of Civil Engineering, Texas A&M University, College Station, TX 77843-3136.

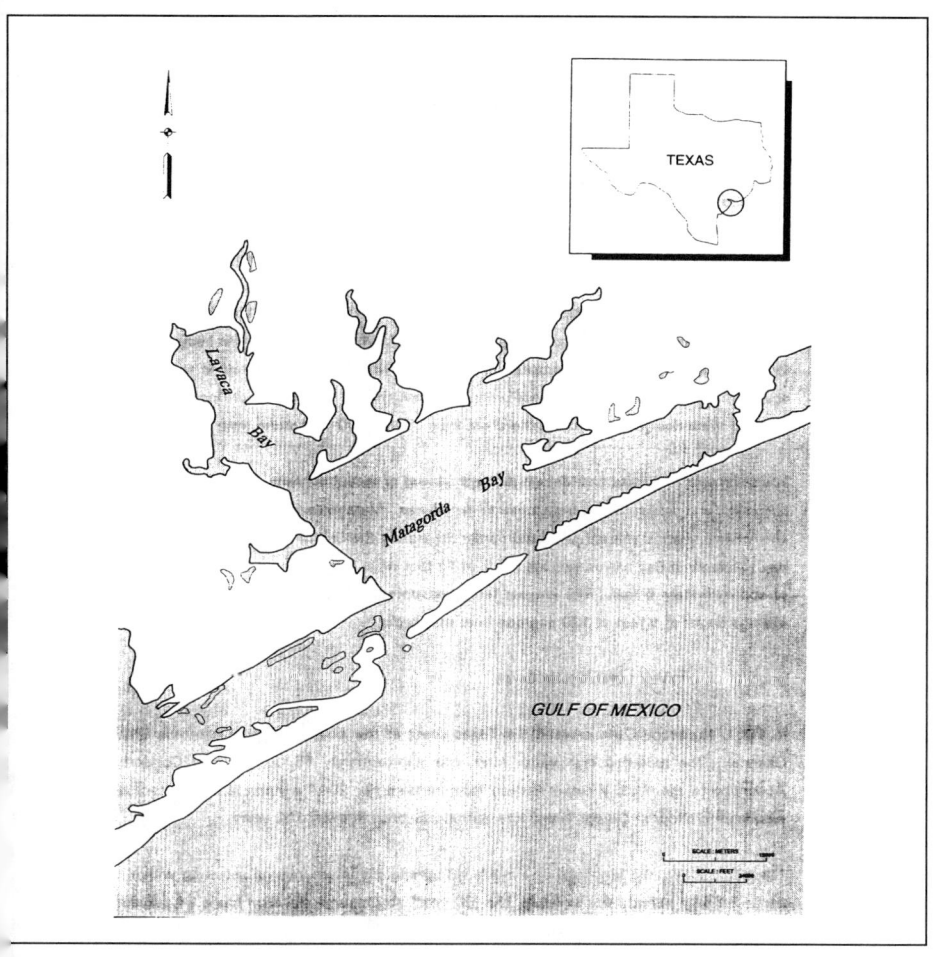

Figure 1 - Lavaca and Matagorda Bays

Hurricane Related Storm Surges

Infrequent events, such as hurricanes, can significantly alter the coastal bathymetry and move large quantities of sediment. Wind velocities, wave steepness, storm duration, tidal stage, and storm surge are factors of large storm events (such as hurricanes) that influence sediment transport. Often, extreme storm events produce a significant affect on the shoreline. This is especially the case in regions that exhibit a typically low-energy wave climate, such as areas adjacent to the Gulf of Mexico. Moreover, sediment transport is experienced not only along the coastline but also within interior bays such as along the shorelines and the bay bottoms.

To determine the storm surges for each storm listed in Table 1, the ADCIRC (Luettich et al. 1992) model is used. Historical hurricane parameters are obtained from the Hurricane Database (HURDAT) maintained by the National Hurricane Center. A planetary boundary layer (PBL) model (Cardone et al. 1992) is used to determine the wind fields that result from historical hurricanes. ADCIRC is a two-dimensional, depth-integrated hydrodynamic model, (Luettich et al. 1992) which is forced by the winds from the PBL model, pressure fields, natural runoff and tidal conditions. ADCIRC solves the generalized wave continuity equation to determine the storm surge and depth-averaged velocities.

Table 1. Historical Hurricanes Affecting Lavaca and Matagorda Bays

Storm Name	Year	Category
Not Named	1886	2
Not Named	1888	1
Not Named	1891	1
Not Named	1900	3
Not Named	1902	1
Not Named	1909	1
Not Named	1915	1
Not Named	1921	1
Not Named	1929	1
Not Named	1932	1
Not Named	1934	1
Not Named	1934	1
Not Named	1941	1
Not Named	1942	1
Not Named	1945	5
Not Named	1947	1
Not Named	1949	3
Debra	1959	1
Carla	1961	3
Celia	1970	3
Edith	1971	1
Fern	1971	1
Alicia	1983	3
Jerry	1989	1

The ADCIRC model is set up with an irregular grid. For this analysis, the entire Gulf of Mexico was used for the grid and area of study as shown in Figure 2. The grid uses approximately 55,000 elements and 30,000 nodes. The output from the model is shown for the case of Hurricane Carla in Figure 3 in which the storm surge conditions are given for immediately offshore and at two typical locations in Matagorda and Lavaca Bays. Figure 4 illustrates the storm surge and tide setup in the Bays and the Gulf at the time of maximum setup in the most northerly portion of the Lavaca Bay. Note that the Eastern Matagorda Bay has become dry due to the wind stress. From this figure, the influence of the winds and the storm surge is clear. ADCIRC also provides currents from the combined tide, storm surge and wind as shown in Figure 5. Currents flowing the channel out to the Gulf are about 1.5 m/s and currents in the flats are typically 0.5 m/s.

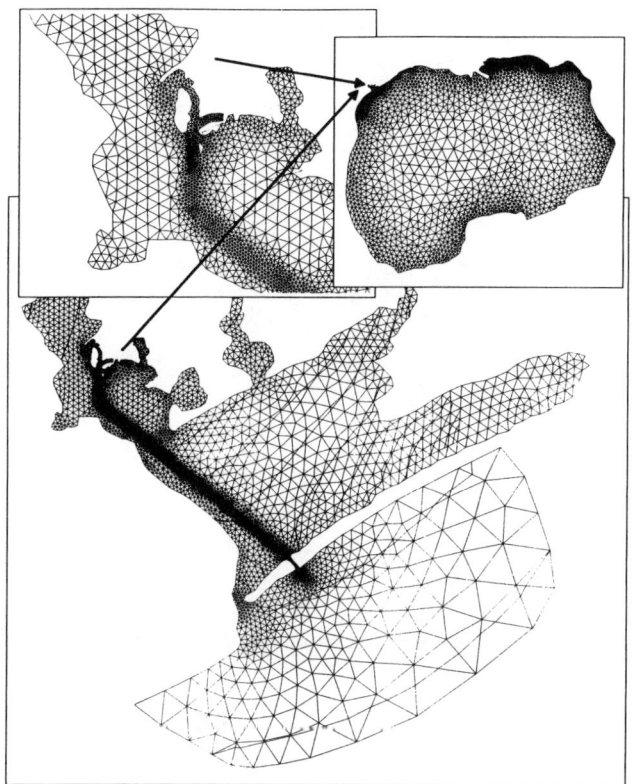

Figure 2 - ADCIRC Grid for the Gulf of Mexico and the Detail for Lavaca Bay

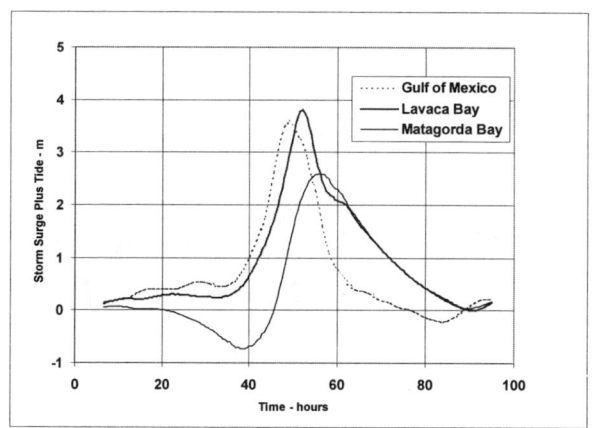

Figure 3 - Selected Water Surface Elevations for the Gulf - Bay System During Hurricane Carla

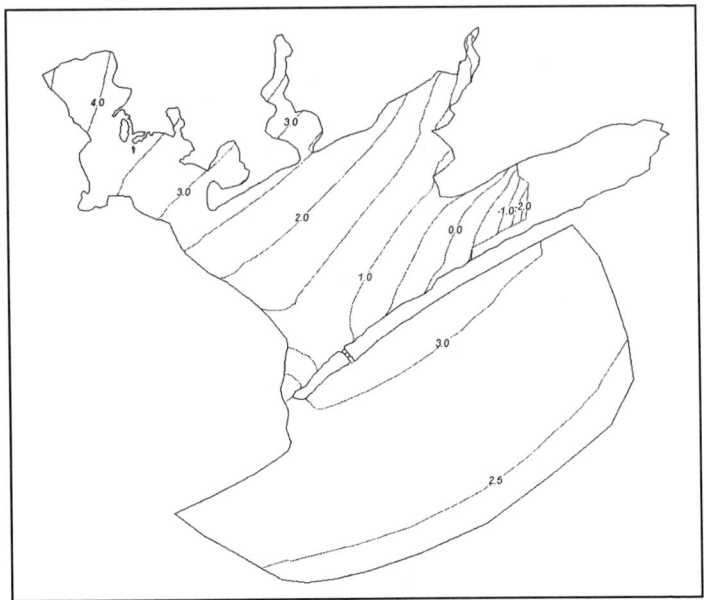

Figure 4 - Contours of Water Surface Elevation at the Peak of Hurricane Carla

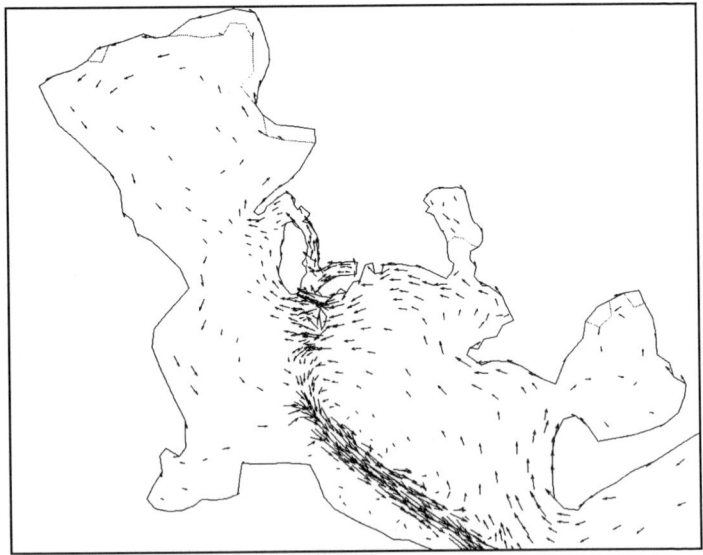

Figure 5 - Currents from ADCIRC at Maximum Surge Level in Lavaca Bay

The Wave Model

Coupling the results of ADCIRC with the SWAN (Simulating WAves in the Nearshore) model (Ris et al. 1996) provides a means to determine both wave and current interactions and the resulting sediment suspension and transport. The model SWAN is a finite-difference spectral model that solves the action balance equation to estimate the wave conditions. SWAN was run in 2nd generation mode, allowing for wave energy to be dissipated due to wave breaking, currents, bottom friction, and triad-triad interactions. The winds from the Planetary Boundary Layer model provided the source of energy for model runs; the water levels and currents from the ADCIRC results were included to simulate hurricane conditions. SWAN was run in a series of static simulations at 0.5 hour time intervals over a period of 8 hours for Hurricane Carla. SWAN accounts for refraction of waves due to changes in bathymetry and currents. Diffraction, however, is not included in the model.

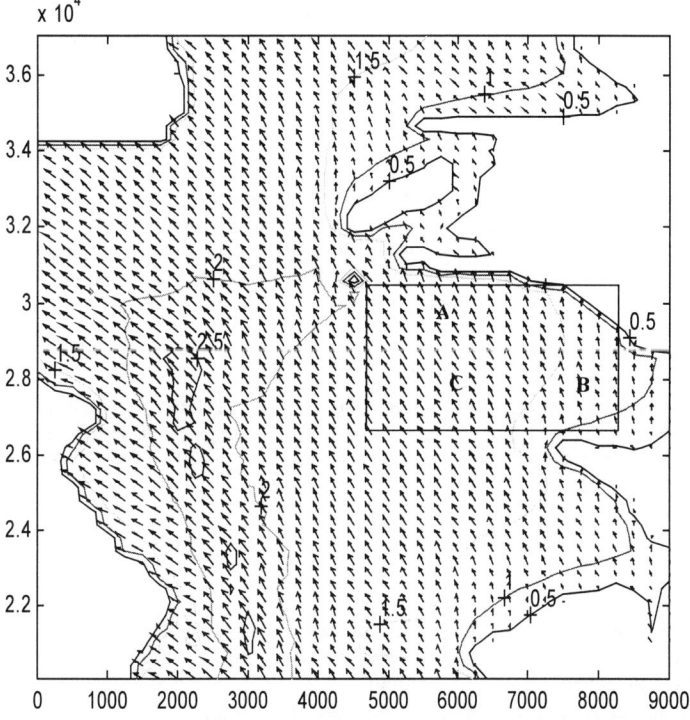

Figure 6. H_s (m) and mean direction as indicated by the vectors. Distances are in meters.

Figure 6 shows the contours of the significant wave height, H_s, for Lavaca Bay. The time of the simulation was chosen around the point of maximum surge in Cox Bay (see Fig. 6 for an outline of the area of interest in Cox Bay). The geographical location of Cox Bay and the direction of the winds during the simulation are such that waves are fetch limited. The vectors indicate the mean direction of the wave at the point of maximum surge and the length of the vector indicates wave height. Wave heights of about 1.5 to 1.75 m are travelling in roughly the same direction as the wind.

Sediment Resuspension

Sediments are suspended and mobilized by wave-induced fluid motion as well as tidal and riverine currents in the coastal zone. The wind field and associated wave climate provide the energy and momentum to resuspend and transport sediments in this area. On a daily basis, a certain amount of sediments are suspended and transported due to these regular processes.

For non-cohesive sediments, the relationships of Akers and White (1973) are used to determine the rate of resuspension. These relationships apply to non-cohesive sediments with grain sizes ranging from 0.04 to 4.0 mm in diameter. The sediment resuspension is predicted by the grain size, depth of the water column and depth-averaged velocity.

$$Q = C\left[\frac{F_{gr}}{A} - 1\right]^m S \frac{D}{(1-e)} \left(\frac{V}{v_*}\right)^n V \tag{1}$$

where Q respresents the sediment transport in volume of sediment mixture per second per unit width. C, A, n, and m are variables that depend on the ratio of the particle weight and the viscous forces acting on the particle, S is the specific gravity of the sediment mixture, D is the diameter, e is the porosity, V is the depth-averaged velocity, and v_* is the shear velocity. F_{gr} represents particle mobility:

$$F_{gr} = \frac{v_*^n}{\sqrt{gD(S-1)}} \left[\frac{V}{\sqrt{32}\log(10d/D)}\right]^{1-n} \tag{2}$$

where d is the mean depth of the water column. The concepts of Bijker (1967) and Swart (1976) are used to incorporate surface waves and currents in the process of sediment transport. The effect of the waves are included as an increase in the depth-averaged velocity, V_{wc}, based on the velocity in the presence of currents only, V_c.

$$V_{wc} = V_c\left[1 + \frac{1}{2}\left(\frac{\xi u_o}{V_c}\right)^2\right]^{\frac{1}{2}} \tag{3}$$

where ξ is a function of the hydraulic bed roughness and orbital excursion at the bed, and u_o is the orbital velocity at the bed based on linear wave theory. A full description of the sediment transport equation and parameters can be found in Scheffner (1996).

The results of the sediment model for Cox Bay are given in Table 2. These results are the cumulative effect of sediment scour for the 8 hour duration around the time of maximum surge. The scour depths are not coupled with the hydrodynamic or the wave model, and the scoured sediment is assumed to be clear water scour. The sediment model is applied to a time frame that occurs at the height of the storm; the assumption of sediments being suspended and not resettling seems reasonable for this instance. interest is the greater depth of scour at site C as compared to site B. This can be attributed to the fast currents and wave heights at C, whereas site B is somewhat more protected from the effects of currents and waves.

Table 2. Sediment Erosion and Removal Rate			
Location	Diameter (mm)	Depth of Scour (cm)	Sediment Removal Rate (cm/hr)
A	0.1	3.5	0.44
B	0.1	0.7	0.09
C	0.1	2.6	0.33

The degree of sediment removal is linked with the category of storm, and an empirical simulation technique (EST) (Scheffner et al. 1996) is applied to develop an extreme probability distribution for the limits of scour and the amount of contaminated sediment that may be removed. With the historical database, a limited set of storms can be applied to run the hydrodynamic, wave, and sediment model components.

The potential and extent of sediment scour depends on the severity of the storm. Assessing the likelihood that the contaminated sediments will be scoured will assist in determining areas that may require special consideration or protection from certain storm events. The results for Hurricane Carla indicate that sediment erosion below 5 cm in certain areas of Cox Bay is possible. Hurricane Carla is an extreme event, listed as a Category 5 storm on the Saffir Simpson scale when offshore, and it is a Category 3 storm as it approaches land. With the complete set of storms and the EST analysis, the potential of sediment scour due to more severe and less severe events can be examined.

References

Cardone, V. J., C. V. Greenwood, and J. A. Greenwood (1992). "Unified Program for the Specification of Hurricane Boundary Layer Winds over Surfaces of Specified Roughness." Contract Rep. CERC-92-1, U.S. Army Eng. Waterways Experiment Station, Vicksburg, Miss.

Krone, R. B. (1962). "Flume Studies of the Transport of Sediment in Estuarial Shoaling Processes." Final Report, Hydraulic Engineering Laboratory and Sanitary Engineering Research Laboratory, Univ. of Calif. Berkeley.

Ris, R. C., N. Booij, L. H. Holthuijsen, R. Padilla-Hernandez (1996). "SWAN Cycle 2 User Manual (Draft)." Delft University of Tech., the Netherlands.

Scheffner, N. W., L. E. Borgman, D. J. Mark (1996). "Empirical Simulation Technique Based Storm Surge Frequency Analyses." J. Wtrwy., Port, Coastal, and Ocean Engrg., ASCE, 122(2), 93-101.

Swart, D. H. (1976). "Predictive Equations Regarding Coastal Transports." Proc. 15th Coast. Engrg. Conf., ASCE.

Westerlink, J. J., Luettich, R. A., and Scheffner, N. W. (1994). "ADCIRC: An Advanced Three-Dimensional Circulation Model for Shelves, Coasts, and Estuaries." Tech. Rep. DRP-92-6, U.S. Army Eng. Waterways Experiment Station, Vicksburg, Miss.

A Risk Analysis of a Submerged Breakwater

Chai-Cheng Huang[1]   Jia-Shiang Lin[2]

Abstract

Submerged breakwaters have been frequently applied in shoreline protection in recent years to avoid unsightliness. Waves are the most dominant and influential environmental factor to be considered in the design of marine structures. How to estimate the risk of structures throughout their designed life is often concerned by engineers. A risk direct integration method was adopted in this study, which is based on the probability distribution functions of wave loading and the results of the damaged physical model of a submerged breakwater. It is shown that both Gumble's and Weibull's distribution functions are good for significant wave heights of local typhoons, and that if the armor weight is obtained from the design curves of stability, the risk of damage will be less than 2% in 50 years.

Introduction

Submerged breakwaters have been frequently applied in shoreline protection in recent years. These offshore structures are subject to hostile environmental conditions, and it is critical that they maintain their functions appropriately throughout their designed life. The major factors of the safety of submerged breakwaters are whether the stability of the armor layer has been designed properly and the reliability of information on wave conditions at the site of interest. Underestimation of wave severity leads to the failure of a submerged breakwater, whereas overestimation results in unnecessary costs. A risk direct integration method used by Yen (1987) was adopted in this study, which is based on the probability distribution functions of wave loading and the results of the damaged physical model of a submerged breakwater.

---

[1] Associate Professor, Dept. of Marine Environment, Sun Yat-sen University, 70 Lian-Hae Road, Kaohsiung 804 Taiwan

[2] Technician, Ocean Research III, Sun Yat-sen University, 70 Lian-Hae Road, Kaohsiung 804 Taiwan

Basic Theory

The direct integration method was adopted in analyzing the risk of a submerged breakwater. In this method the risk is evaluated through a direct, analytical or numerical integration of the probability density function of wave loading and the submerged breakwater resistance. In general, the risk can be expressed as

$$P_f = P(L > C) \tag{1}$$

in which P (L > C) is the joint probability density function of C(capacity) and L(loading). For statistically independent C and L, eq. 1 can be simplified as

$$P_f = \int_0^\infty F_c(H) \, dF_L(H) \tag{2}$$

in which $F_C(H)$ and $F_L(H)$ is the cumulative distribution function of C and L, Figure 1.

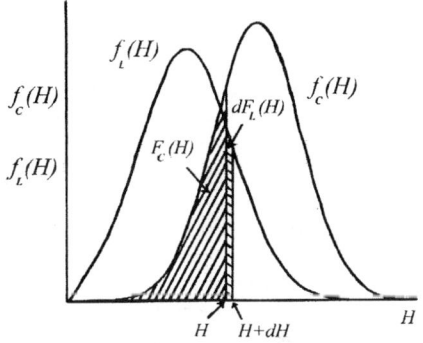

Figure 1. Probability density functions for loading and resistance

To facilitate the direct integration, the exact cumulative distribution functions of wave loading, $F_L(H)$, and the submerged breakwater resistance, $F_C(H)$, should be analyzable and definable.

In principle, $F_L(H)$ may be obtained through the analysis of the historical typhoon's data, while $F_C(H)$ may be determined by hydrodynamic model test.

Cumulative Distribution Function of Resistance

The wave flume with the scale (1:20) submerged breakwater model is shown in Figure 2. The flume is 1m wide, 35m long, and 1.2m high. The submerged breakwater model is made of boulders in the core and two layers of tetrapod armor units on the top. Both of them have similar porosity, 48% - 55%. The ratios of the height of submerged breakwater to the water depth, h/d, are 0.7, 0.8, 0.9, and 1.0 etc.

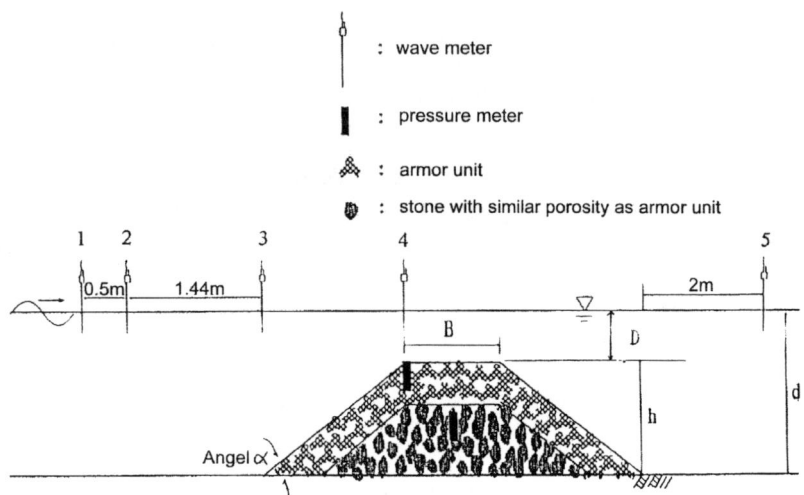

Figure 2. Submerge breakwater physical model arrangement

Based on Taiwan Harbor Institute report (1995), it shows that the wave periods at southwest part of Taiwan concentrate around 5-6 seconds in regular sea states, while around 8-9 seconds in rough sea states, especially during typhoon attacking. The report also indicates that JONSWAP wave spectrum mode is appropriate for the site of interest, and thus was adopted in the model test to simulate the wave loading.

Statistic Extreme Values for Wave Height

It is assumed that the cumulative distribution function of typhoon is Rayleigh distribution. If an attacking typhoon has total n waves acting on the submerged breakwater, then based on the type I of Gumble distribution (1958), the expected extreme value and its standard deviation are as follows:

$$E[H_{max}] \cong \frac{H_S}{2}\sqrt{2\ln(n)}\left[1 + \frac{0.2886}{\ln(n)}\right]$$
$$\sigma_{H\,max} \cong \frac{\pi H_S}{4\sqrt{3\ln(n)}} \quad (3)$$

The waves generated by wave maker in the flume is according to Rayleigh distribution. The maximum number of digital water level is $2^{11} = 2048$. If the output rate of digital water level is 10 Hz, then the period of one cycle of wave series is 204.8 seconds, and divided by setting wave period, 1.79 seconds, then we will have n=115 waves in a typical test. Substitute n=115 into eq.3 . We have

$$E[H_{max}] \cong 1.63 H_S$$

$$\sigma_{H max} \cong 0.21 H_S \qquad (4)$$

In fact, the rough sea state in a attacking typhoon may contain thousands of waves, thus for conservative estimation of H_{max} the following equation

$$E[H_{max}] \cong 2 H_S \qquad (5)$$

is usually adopted for the maximum nonbreaking wave.

Definition of Damage probability, and Test Results

According to Huang's (1996) stability study of submerged breakwaters, the damages of submerged breakwaters frequently occur at the joint of top level and frontward slope. The total amount of armor unit at that conjunction is about 300 units. This number is chosen as statistical population. The types of movement of armor units can also be identified as rocking or rolling. Thus based on Franco (1986), the damage rating may be classified as A, B, and C, i.e.

A: the number of armor units that are rocking at least once
B: the number of armor units that are rocking at least four times continuously
C: the number of armor units that are rolling at least once

For the probability of each damage rating may be obtained by the ratio of rocking (or rolling) units to the total population (300).

Each run of testing wave height, H_s, is set from 5 to 20cm, and the wave period is set to 1.79 seconds which is equivalent to 8 seconds of prototype wave period for JON-SWAP wave spectrum mode. If the damages of physical model occurred, we immediately record the damage phenomenon including the types of movement and the number of moving armor units. Following the same procedure, we complete 4 different water depths. The results are shown in Table 1. Its prototype data is also plotted as Figure 3.

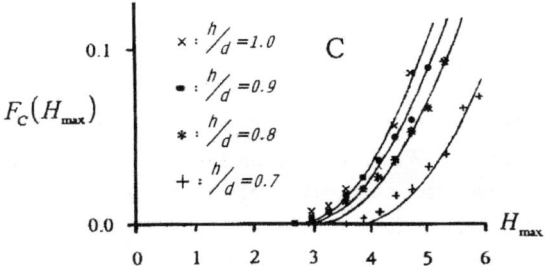

Figure 3. Regression curves (C rating) for each different water depth ratio

Table1. The results of damaged physical model

$H_z(cm)$	$H_{max}(cm)$	H_{max}/D_n	water depth53.6cm(h/d=0.7)			water depth46.9cm(h/d=0.8)			water depth41.7cm(h/d=0.9)			water depth37.5cm(h/d=1.0)		
			A	B	C	A	B	C	A	B	C	A	B	C
5	7.43	2.03	0	0	0	0	0	0	0	0	0	0	0	0
6	8.92	2.44	0	0	0	0	0	0	0.0033	0	0	0.0067	0.0033	0
7	10.41	2.84	0	0	0	0.0067	0.0033	0	0.0133	0.0067	0	0.0167	0.0100	0
8	11.89	3.25	0	0	0	0.0167	0.0100	0	0.0300	0.0133	0	0.0433	0.0233	0
9	13.38	3.66	0.0067	0.0033	0	0.0333	0.0167	0	0.0533	0.0300	0	0.0600	0.0367	0
10	14.87	4.06	0.0133	0.0067	0	0.0500	0.0300	0	0.0733	0.0433	0.0033	0.0933	0.0567	0.0067
11	16.35	4.47	0.0167	0.0067	0	0.0733	0.0433	0.0067		0.0500	0.0067		0.0700	0.0100
12	17.84	4.87	0.0300	0.0133	0		0.0633	0.0133		0.0667	0.0167			0.0200
13	19.33	5.28	0.0733	0.0367	0.0033			0.0200			0.0267			0.0267
14	20.81	5.69		0.0667	0.0067			0.0267			0.0367			0.0333
15	22.30	6.09		0.0767	0.0167			0.0367			0.0500			0.0567
16	23.79	6.50			0.0200			0.0533			0.0600			0.0867
17	25.27	6.91			0.0333			0.0667			0.0900			
18	26.76	7.31			0.0400			0.0933						
19	28.25	7.72			0.0667									
20	29.73	8.12			0.0733									

A: the number of armor units that are rocking at least once
B: the number of armor units that are rocking at least four times continuously
C: the number of armor units that are rolling at least once

Hmax=0.91 × 1.63 × HS
0.91=adjusted factor due to wave maker capability

Cumulative Distribution Function of Wave Loading

The cumulative distribution function of significant wave height may be obtained according to the long term record of typhoons passing through the island of Taiwan.

Two kinds of distribution function, Gumble's and Weibull's, are mostly used as the extreme value distribution functions. From 1940 to 1989, in this 50-year period, there are 75 typhoons passing through the vicinity of Kaohsiung Harbor. The maximum wave height is 9.83 m. Plotting these data in Gumble's and Weibull's probability papers, straight lines are obtained, which basically means that these two kinds of probability distribution function are good for typhoon extreme value distribution.
Gumble's cumulative distribution function is

$$F_{HS}(H) = \exp\{-\exp[-0.613(H - 2.6335)]\} \quad (6)$$

Weibull's cumulative distribution function is

$$F_{HS}(H) = 1 - \exp\left[-\left(\frac{H-1.2}{2.5}\right)^{1.13}\right] \quad (7)$$

χ^2 goodness-of-fit test with significance level 5% may be used to verify Gumble's and Weibull's distribution functions. The results are shown that

χ^2 goodness-of-fit test for Gumble distribution is 8.94<($\chi^2_{0.95,6}$=12.59), and

χ^2 goodness-of-fit test for Weibull distribution is 4.00<($\chi^2_{0.95,5}$=11.07).

In both cases, the sum of the difference between observation and calculation values is less than the specific value χ^2, which means that the suggested distributions are valid or can't reject for typhoon wave loading distribution.

Risk Evaluation of a Submerged Breakwater

To evaluate the risk of a submerged breakwater, we have to make several assumptions:
1. each attacking typhoon is regarded as independent event;
2. the number of occurred typhoons in a T years return period is a known value, n;
3. the distribution of submerged breakwater and the distribution of typhoon wave loading are independent.

The risk of a submerged breakwater may be obtained by the following integration equation

$$P_F = \int_0^\infty F_c(H_{max})d[F_{Hs}{}^n(H_{max})] \qquad (8)$$

For simplification, eq. 8 can be written in a summation form

$$P_F \cong \sum_{j=1}^K F_c(H_{max})\Delta[F_{Hs}{}^n(H_{max})] \qquad (9)$$

Where the values of wave loading, H_{max}, are restricted by two factors: one is the possibility of extreme value $H_{max} = 2H_s$, and the other is the sustainable water depth(d), $H_{max} = 0.85d$, based on Chen (1985) suggestion. Therefore the possible maximun values of wave loading may be expressed as

$$H_{max} = \min\{2H_S, 0.85d\} \qquad (10)$$

An Example of Risk Evaluation

Based on the results of experimental study, Huang (1996), conducted in the flume. The rolling damage (C rating) cumulative distribution of a typical submerged breakwater (h/d=0.7) is

$$F_c(H_{max}) = 0.0085[\frac{H_{max}}{D_n} - 5.06]^2 \qquad \frac{H_{max}}{D_n} > 5.06 \qquad (11)$$

while Gumble cumulative distribution of wave loading is as eq.6. A failure risk result is obtained by substituting eqs.11 and 6 into eq.9, and setting n=75 for 50-year return period, see Table 2. The column C values seem too high to be believable. In fact, the reason for this high value is due to the designed weight of armor unit that intended to withstand 3m wave height according to the Hudson formula. Under this arrangement, we may investigate the damage phenomenon in the flume relatively easier. In the real

Table2. The results of risk analyses by model test

h/d	Model D_n/Design D_n		Risk P_f	
		A	B	C
0.7	0.51	1	0.9427	0.4622
0.8	0.49	1	0.9084	0.4179
0.3	0.52	1	0.7249	0.3234
1.0	0.55	1	0.6594	0.2607

world, typhoon waves are frequently higher than 3m such that the high risk values are acceptable.

Risk Evaluation Combined with Stability Formula

For the same model setting, Huang (1996) has developed a stability formula for estimating the weight of armor unit, i.e.

$$\frac{D_n}{H} = 0.00228\xi + 0.1062 \tag{12}$$

where $\xi = \dfrac{\tan \alpha}{\sqrt{H/L_0}}$, surf similarity parameter

The H is chosen according to $H_{max} = \min\{2H_s, 0.85d\}$. For a 50-year return period typhoon, the maximum wave height at the vicinity of Kaohsiung Harbor is H=9.11m and the wave period is assumed 8 seconds. Substituting above values into eq. 12, we got D_n=1.427m for the practical design case. Then we may put D_n=1.427m into eq. 11 and recalculate eq. 9. The rolling damage risk (C rating) becomes 1.49%, see table 3, which is quite reasonable situation for an engineering practice.

Table3. The results of risk analyses by stability formula

		Risk P_f		
h/d	Design D_n	A	B	C
0.7	1.427	0.1555	0.0993	0.0149
0.8	1.717	0.1614	0.0983	0.0145
0.3	1.413	0.1779	0.084	0.0132
1.0	1.084	0.1815	0.1041	0.0131

Conclusions

(1) Under the condition of 5% significance level, Gumble and Weibull distribution functions are appropriate for historical typhoon significant wave height data.
(2) If the weight of armor units are estimated by the stability curves, the risk values may be less than 2%, which means that the method of risk analysis is acceptable.
(3) The determination of failure probability for a submerged breakwater is based on the ratio of the number of rolling units to the total population, which may involve certain amount of uncertainty.

Acknowledgment

This study has been supported by the National Science Council, R.O.C. under contract No. NSC85-2611-E-110-002. This support is gratefully acknowledged.

References

1. Ahrens, J. P.(1989). "Stablity of Reef Breakwaters." J. Watrway, Port,Coast., and Oc. Engrg., ASCE, 115(2), PP.221-234.
2. Ang, A. H-S. and Tang, W. H. (1984). Probability Concepts in Engineering Planning and Design, Volume II Decision, Risk, and Reliability.
3. Chang, C.K , etal. (1995). "The Investigation of Waves around the Island of Taiwan.", the Report of Institute of Harbor & Marine Technology, Taichung, Taiwan.
4. Chen, Y.Y. (1983). "The Wave General Formula and the Maximum Wave Height Analysis.", the dissertation of National Cheng Kung University, Taiwan.
5. Franco, L., Lamberti, A., Noli, A. and Tomasicchio, U.(1986). "Evaluation of Risk Applied to the Designed Breakwater of Punta Riso at Brindisi, Italy." Coastal Eng., Vol. 10, PP.169-191.
6. Goda, Y. and Suzuki, Y. (1976). "Estimation of Incident and Reflected Wave in Random Experiments. ", Proc. 15th Coastal Eng. Conf., PP.828-845.
7. Goda, Y. (1985). Random Seas and Design of Maritime Structures, University of Tokyo Press, P.84.
8. Huang, C.C. (1996). "An Experimental Study of a Submerged Breakwater Stability.", the Report of National Science Council, Taiwan,NSC85-2611-E-110-002, pp.72-73.
7. Hudson, R.Y. (1959). "Laboratory Investigation of Rubble Mound Breakwaters.", ASCE, J. Waterw. Harbors Div., 85(WW3) PP.93-121.
8. Tanimoto, K. et al.(1982). "Stablity of Armour Unit for Foundation Mounds of Composite Breakwaters by Irregular Wave Tests." Rept. Port and Harbour Res. Inst., Vol. 21, No. 3, PP.3-42.
9. Van der Meer, J. W.(1987). "Stability of Breakwater Armor Layers-Design Formula." J. of Coastal Engineering,Vol.11,No.3,Sept., PP.219-239, Amsterdam, The Netherlands.

Keywords : submerged breakwater, stability, risk analysis, model test

Effects of Onshore Winds on Velocity of Wave Runup

Donald L. Ward[1], Assoc. Member, ASCE; Jun Zhang[2], Assoc. Member, ASCE; Christopher G. Wibner[3]

Abstract

A physical model study investigated effects on runup on coastal structures of the strong onshore winds usually associated with design storm conditions. This paper examines runup velocities at the beginning of runup and midway between elevation of maximum rundown and maximum runup. It is found that runup velocities at the onset of runup in the presence of wind are higher than without wind for similar incident wave conditions, indicating that a stronger surface current is created after wave breaking under the influence of wind. Wind sheer effects directly on the runup bore appear to have little effect on runup velocities.

Introduction

It previously has been shown that the addition of a strong onshore wind will increase runup on coastal structures (Zhang et al. 1995). Possible effects of the wind may be broken into the following categories: an increase in incident wave energy due to energy transfer to the wave field, an increase in still water level (swl) due to wind-induced setup, changes in wave kinematics due to a wind-induced horizontal force component at the point of breaking increasing the initial runup velocity on the structure slope, and direct shear effects of the wind acting on the runup bore and/or pressure differentials between the windward and leeward sides of the runup bore.

---

[1] Research Hydraulic Engineer. US Army Engineer Waterways Experiment Station, CN-S, 3909 Halls Ferry Rd., Vicksburg, Mississippi 39180-6199, USA.

[2] Assoc. Professor. Texas A&M University, Mail Stop 3136, Ocean Engineering Program, Department of Civil Engineering, College Station, Texas 77843-3136, USA.

[3] Engineer. Aker Omega, Inc., 11757 Katy Freeway, Suite 1300, Houston, Texas 77079, USA.

After making appropriate adjustments for increased wave energy and wind-induced setup, observed increases in runup during tests with onshore winds appear to be either a direct result of sheer forces acting on the runup bore or caused by changes in wave kinematics not related to changes in wave height. This paper examines the velocities of waves as they run up a structure slope with or without the influence of a following wind.

Test Facility and Equipment

All experiments were conducted in a covered 36-m-long by 0.6-m-wide by 0.9-m-deep glass-walled wave flume in the Hydromechanics Laboratory at Texas A&M University (TAMU). The wave flume was equipped with two dry back flap-type wave makers that spanned the width of the flume and operated in tandem. Wave makers were controlled by a standard IBM personal computer interfaced to the wave makers through an analog output card and using software developed at TAMU.

An elevated extension to the wave flume led to a rheostat-controlled blower that pulled air along the length of the flume. Air input was through a vertically-adjustable air intake manifold located 4-m in front of the wave makers and equipped with horizontal guide vanes to smooth the flow of air entering the flume. Twin-wire resistance-type wave gauges were placed through holes in the flume cover to record changes in water surface elevation. Wind speed was recorded with a three-cup anemometer located mid-way down the flume. Wave gauges and anemometer were sampled at 30 Hz during calibration of incident wave conditions, and at 45 Hz during all test runs.

Test structures were 1:1.5 (V:H) and 1:3 slopes constructed of plywood with varying amounts of freeboard (vertical distance from swl to structure crest). The flume cover was modified to allow the cover to be raised so that all tests had a similar cross-sectional area between structure crest and flume cover. Slopes were tested both as smooth slopes and covered with gravel as a riprap revetment. Capacitance-type runup gauges were suspended just over the riprap on the rough slope or placed on a thin sheet of plastic and mounted directly on the smooth slope. Runup gauges were sampled at 45 Hz during all test runs.

Test Procedures and Conditions

Incident wave conditions were determined by placing a wave-absorbing beach at the location where the test structures would later be placed and measuring incident conditions with a single wave gauge placed at the toe of the slope. Wave absorption was not perfect, but measurements made in the absence of wind indicated reflection coefficients were of the order of five to seven percent and were disregarded. All tests were conducted for only two minutes to insure the reflected wave train could not reach the structure after being re-reflected off the wave board. After the flume had fully stilled between tests, wind was blown along the flume for two minutes prior to initiating wave generation to insure the wind-wave field was established (Zhang et al. 1995).

All tests were conducted at swl of 0.5 m. Monochromatic wave conditions tested included 1.0 sec wave periods with wave heights of 5.0, 7.0, and 10.0 cm; 1.75 sec wave period with wave heights of 3.2 and 5.4 cm; and 2.5 sec wave periods with wave heights of 2.2 and 3.8 cm. Each wave period/wave height combination was tested with wind speeds of 0 m/s, 6.5 m/s, 12 m/s, and 16 m/s.

The addition of wind to generated wave periods of 1.75 and 2.5 sec resulted in dual-peaked wave energy spectra, with the wind energy adding a second spectral peak at a frequency of about 1.5 Hz. Although dual-peaked spectra are commonly found along our coastlines, this paper will be limited to the single-peaked spectra from either the 1-sec period waves with and without wind, and the 1.75 and 2.5-sec waves without wind.

With the addition of wind, monochromatic generated waves increased in height and gained some spectral width, but remained as a very narrow banded spectrum. The average of the one-third highest wave heights (significant wave height, H_s) and the average wave period associated with the one-third highest wave heights (significant wave period, T_s) were determined for each test run. Deepwater wave celerity (C_o) was determined with linear wave theory based on significant wave period, where $C_o = g\,T_s/(2\pi)$ and g is gravitational acceleration.

Records from runup gauges were analyzed by determining the mean water level of the runup on the slope and locating zero up-crossings of the mean water level. From each up-crossing, the runup record was followed back to the first trough and followed forward from the mean water level to the first crest. This was considered the runup portion of a single wave, and runup velocity was determined for each time step during the runup (distance runup traveled up the slope in each time step, multiplied by sampling rate). For each wave, the maximum runup velocity was determined, and also the velocity occurring at 10 percent and at 50 percent of the runup elevation range for that wave. Velocities discussed herein are the average of the one-third highest maximum velocities in a test run (significant velocity, V_s), and the average of the velocities recorded at the 10 percent and 50 percent elevations (V_{ave}).

The test program at TAMU included both runup and overtopping by testing three different freeboards with each incident wave condition. Because overtopping reduces the amount of rundown which must be overcome by the following wave's runup, the data presented herein is limited to those test runs with sufficient freeboard that there was no overtopping.

Results and Discussion

The role of wind in controlling the form of wave breaking as waves shoal into the nearshore has been investigated both in the field and in the wind/wave flume. Papers by Galloway et al. (1989) and Douglass (1990) both reported that waves breaking under the influence of onshore winds tend to be spilling breakers, while offshore winds tend to form plunging breakers. Galloway et al. also reported that very strong onshore winds may reintroduce a plunging breaker under conditions where waves would be spilling under a lighter onshore wind. In the tests reported herein, the wave flume had a flat bottom at an intermediate water depth and there was no shoaling until the waves reached the revetment. However, wave growth was apparent due to the wind, and it is possible that onshore winds initiated breaking earlier in the growing waves.

Van der Meer and Breteler (1990) reported on runup velocities recorded in a series of large-scale runup tests conducted with monochromatic waves on a smooth 1:3 slope. Runup velocities were recorded just above the structure slope in the vicinity of swl with an electro-magnetic velocity meter. They report that velocities recorded from swl to a depth of 40 percent of runup elevation were not dependent on location but varied with incident wave steepness. Because the runup velocity must go to zero at the point of maximum runup, van der Meer and Breteler assumed a decay function of $(1-z/RU)^{1/2}$, where z is elevation and RU is elevation of maximum runup, to estimate runup velocities at various elevations on the runup slope.

Figure 1 plots data from smooth 1:3 slopes in the present test series using the same nondimensional forms and same range of conditions used in van der Meer and Breteler (1990). Solid symbols in Figure 2 show average velocities taken at 10 percent of the distance between maximum rundown and maximum runup for each wave, and open symbols are average velocities at 50 percent of the runup distance. The straight line (labeled "vdM and B") shows the line of best fit to the data of van der Meer and Breteler. Results of the small-scale tests presented herein are generally within the scatter from the large-scale tests in van der Meer and Breteler and indicate that data presented herein is reasonable.

Figure 1 also plots data from tests with onshore wind speeds of 6.5 and 12 m/sec. It is seen in Figure 1 that an onshore wind velocity of 6.5 m/sec does not have any noticeable effect on runup velocities. An onshore wind velocity of 12 m/sec, while within the scatter of the data, appears to slightly lower average velocities. The slight reduction in velocities observed with the 12 m/sec wind may be due to incident waves having to overcome greater rundown velocities due to the higher runup observed

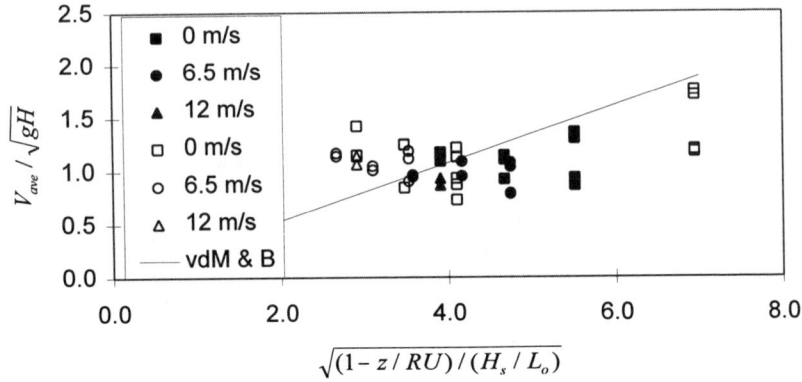

Figure 1. Average wave velocities at 10 percent runup elevation (solid symbols) and 50 percent runup elevation (open symbols) for smooth 1:3 slope.

with higher wind speeds.

The surf similarity parameter (ξ) has been shown to be an excellent indicator of the type of wave breaking that may be expected (Battjes 1974). The surf parameter is defined as the tangent of structure slope to the horizontal divided by square root of wave steepness, where wave steepness is significant wave height divided by deepwater wavelength (determined here by linear wave theory and significant wave period). In general, surf parameters on the order of 1.5 tend to have plunging breakers, with surging breakers occurring for surf parameters over 3.0. The type of wave breaking has a significant influence on the amount of energy dissipation during breaking, which affects the amount of energy that remains for runup. The following figures will therefore present runup velocities as a function of surf parameter.

Instead of the average velocities from different elevations presented in Figure 1, Figure 2 examines the average of the one-third highest maximum velocities (significant velocity) from each test run. Significant velocity is nondimensionalized by the deepwater wave celerity and plotted against the surf parameter. Solid symbols in Figure 2 are from tests on a smooth 1:1.5 slope, and open symbols are from tests on a smooth 1:3 slope. The data set is limited because many of the tests conducted on the smooth slopes produced overtopping and are not included. For the data reported herein, no effect of wind speeds up to 12 m/sec is noted.

Figure 3 is similar to Figure 2 but presents data from tests with 1:1.5 (solid symbols) and 1:3 (open symbols) rough slopes. Because energy dissipation in the

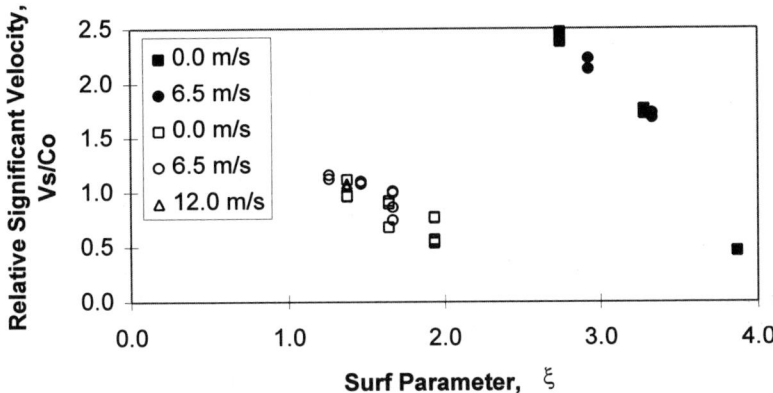

Figure 2. Relative significant velocity as a function of surf parameter for smooth 1:1.5 (solid symbols) and 1:3 (open symbols) slopes.

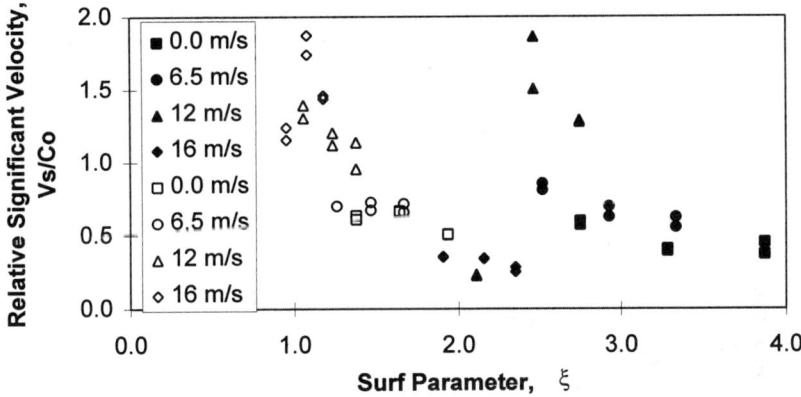

Figure 3. Relative significant velocity as a function of surf parameter for rough 1:1.5 (solid symbols) and 1:3 (open symbols) slopes.

riprap limited maximum runup levels, fewer tests had overtopping and more results may be presented. Surprisingly, although no wind effects were observed for the smooth slopes in Figure 2, there is a noticeable increase in runup velocities at higher wind speeds on the rough slope. It is also noted that a few tests with higher wind speeds on the 1:1.5 slope show very low relative significant velocities (≈ 0.2) due to early breaking of the waves.

It should be remembered that runup velocities on smooth slopes were determined from a runup gauge located almost directly on the slope, whereas the runup gauge for rough slopes was suspended above the riprap. It is reasonable to assume that the wind is causing a higher surface current velocity on the runup, but has less effect near the structure slope.

As mentioned earlier, higher surface current velocities under the influence of wind may be due to stronger onshore currents after breaking and/or sheer effects acting on the runup bore. To examine these possibilities, average runup velocities at the beginning of wave runup before there can be much effect from wind sheer on the runup bore (10 percent elevation), and runup velocities higher in the runup cycle (50 percent elevation) are plotted in Figure 4 for the rough 1:3 slope.

Results from the 10 percent and 50 percent elevations (solid and open symbols, respectively) are similar but there is less scatter at the 50 percent elevation. It is clear that the 12 m/sec wind has a significant effect on average velocities, with a minor effect noted for the 6.5 m/sec wind. Average velocities are greater at the 50 percent elevation than at the 10 percent elevation, but the increases in runup velocities with wind are not greater than the increases in velocity without wind. No effects are observed for the 6.5 m/sec wind at the 50 percent elevation. It appears that under the influence of a strong onshore wind, runup begins at a higher velocity than for similar waves in the absence of wind. Wind sheer acting on the runup bore does not appear to play a significant role in the runup velocity on the rough slope.

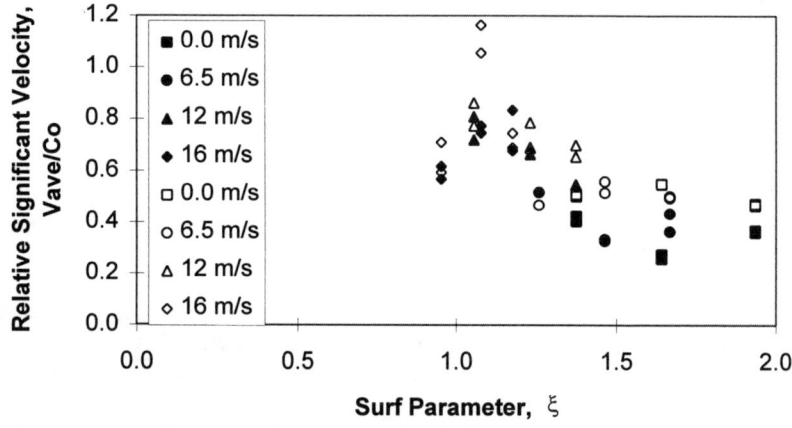

Figure 4. Average velocities at the 10 percent elevation (solid symbols) and 50 percent elevation (open symbols) for the 1:3 rough slope.

Conclusions

Strong onshore winds appear to increase the surface current velocity of waves breaking on sloping coastal structures with structure toes in reasonably deep water. The increase in surface current velocity results in higher runup elevations on the structure. Direct sheer effects from wind acting on the runup bore appear to have little effect on runup velocities on the structure.

Acknowledgment

The writers wish to acknowledge the Office, Chief of Engineers, U.S. Army Corps of Engineers, for authorizing publication of this paper. It was prepared as part of the Coastal Program of the Civil Works Research and Development Program.

References

Battjes, J.A. 1974. "Surf similarity," Proc. 14th Coastal Engr. Conf., Copenhagen, Denmark, ASCE. Pp. 465-480.

Douglass, S.L. (1990). "Influence of wind on breaking waves," *J. Waterway, Port, Coastal, and Oc. Engr.*, Vol 116(6), ASCE. Pp. 651-663.

Galloway, J.S., Collins, M.B., and Moran, A.D. (1989). "Onshore/offshore wind influence on breaking waves: an empirical study," *Coastal Engineering*, vol. 13, Elsevier Science Publ. Pp. 305-323.

van der Meer, J.A., and Breteler, M.K. (1990). "Measurement and computation of wave induced velocities on a smooth slope," Proc 22nd Coastal Engr Conf., ASCE. Pp. 191-204.

Zhang, J., Wibner, C.G., and Cinotto, C.M. (1995). "Experimental Studies of Wind Effects on Wave Runup and Overtopping of Revetments," COE Report No. 347, Texas Engineering Experiment Station, College Station, Texas.

On the Radiation Boundary Conditions for Time-Dependent
Parabolic Wave Calculations

Tai-Wen Hsu[1] and Chih-Chung Wen[2]

Abstract

This paper presents a wide-angle parabolic equation for describing wave refraction, diffraction, reflection, breaking and energy dissipation. Wave breaking and energy losses due to the influence of turbulence are accounted for in a relatively straightforward manner. The radiation boundary condition is solved directly in the parabolic equation without using approximations of a certain order. This method maintains the exact relationship of a circle between the wave number components. The test of wave height pattern around a circular shoal with large-angle wave incidence show that the present model gives the reasonable results even for large incident angles up to $80°$. The applicability of the present model is verified by comparing with experiments conducted in a wave basin.

Introduction

The mild-slope equation which was derived by Berkhoff (1972) is widely employed in the calculations of wave transformation in coastal engineering. The applicability of the mild-slope equation to the problem of wave scattering in slowly-varying domains has been well established. The transient mild-slope equation which is given by Booij (1981) is written as

$$\frac{\partial^2 \Phi}{\partial t^2} - \nabla \cdot (CC_g \nabla \Phi) + (\omega^2 - k^2 CC_g)\Phi = 0 \qquad (1)$$

---

[1] Professor, Department of Hydraulics and Ocean Engineering, National Cheng Kung University, Tainan 70101, Taiwan

[2] Ph. D. student, Department of Hydraulics and Ocean Engineering, National Cheng Kung University, Tainan 70101, Taiwan

where Φ = the velocity potential; t = the time; $\nabla = (\partial/\partial x, \partial/\partial y)$ = the horizontal gradient operator; x and y = the coordinates describing the horizontal plane, x = direction, y = alongshore direction; C = the wave celerity; C_g = the group velocity; $\omega = 2\pi/T$ = the angular frequency; T = the period and k = the wave number.

In (1), a hyperbolic or time-dependent parabolic equation which march in time domain can be formulated as an initial value problem and solved by numerical schemes. In addition, an elliptic equation or time-independent parabolic equation marching in space domain can also be obtained from (1) and solved as a boundary value problem in the case of a steady-state. The time-independent parabolic approximation, which has been used by many researchers (Radder, 1979; Kirby, 1989; Li, 1997) is inadequate in determining the wave field in the presence of a strong reflection. Several numerical models which retain the reflected wave have been proposed to solve the mild-slope equation. Among them, the amount of computer memory and calculation required render the elliptic equation inefficient for application to a large coastal areas. The computer time required by the hyperbolic equation is one order of magnitude smaller than that required for the solution of the elliptic equation solved by the matrix equation (Booij, 1981; Li, 1994). A time-dependent parabolic equation which retains the reflected waves was developed by Li (1994) on the basis of the perturbation method. Models of the time-dependent parabolic equation are usually expanded in the tridiagonal matrix form which can be easily solved by the Alternating Direction Implicit (ADI) method. The requirements of this equation in terms of computer storage and CPU time are relatively small when compared to those for other models. Therefore, this method is inadequate to determine the wave field in the presence of a strongly reflecting structure. The second limitation restricts confines the model to simulate wave field in the entire surf zone. In engineering practice, it is essential to develop a model describing wave transformation in the surf zone to provide reasonable information of nearshore currents and sediment transports.

An extensive study of the time-dependent parabolic approximation method for waves subject to combined refraction, diffraction, reflection, breaking and energy dissipation is performed in the present paper. Wave breaking and energy dissipation are accounted for in a relatively straightforward manner based on the model of Isobe (1987). For large-angle wave incidence, as opposed to other existing models using approximations of radiation boundary to a certain order, the present parabolic model solves the radiation boundary condition directly without using series expansions. This method maintains the exact relationship of a circle between two wave components in the x and y direction. The angle limitation of the time-dependent parabolic model can be improved to reduce the reflected waves back into the computational domain. Several computational cases involving incident wave angles up to 80° over a circular shoal is adopted to test the model. The accuracy of the wave breaking and energy model is verified by a comparison with well-documented experiments conducted in a wave basin.

Mild-slope equation with wave energy dissipation

Waves approaching the nearshore zone tend to steepen and eventually break because of decreasing water depths. The energy dissipation due to the turbulence shoreward of this breaking point influence the wave height. Approximate models combined with a parabolic equation are required to describe the wave breaking point and the energy losses in the dissipation process. To incorporate the energy dissipation due to wave breaking into mild-slope equation, an empirical energy dissipation presented by Isobe (1987) is straightforwardly added to (1):

$$\frac{\partial^2 \Phi}{\partial t^2} - \nabla \cdot (CC_g \nabla \Phi) + [\omega^2 - k^2 CC_g (1 + if_d)]\Phi = 0 \tag{2}$$

where $i = \sqrt{-1}$ is a unit complex number, f_d is the energy dissipation coefficient

Time-dependent parabolic equation

Following Mei (1983), we introduce the slow coordinate for the time variable

$$\bar{t} = \varepsilon t \tag{3}$$

where ε is a perturbation coefficient and is much smaller than unity. It is thus reasonable to assume

$$\Phi(x,y,t) = \psi(x,y,\bar{t})e^{-i\omega t} \tag{4}$$

By substituting (3) and (4) into (2) and taking only the leading term, we obtain a new parabolic equation for the mild-slope equation:

$$-2\omega i \frac{\partial \psi}{\partial t} = \nabla \cdot (CC_g \nabla \psi) + CC_g k^2 (1 + if_d)\psi \tag{5}$$

Using the transformation introduction by Radder (1979):

$$\psi = \frac{\phi}{\sqrt{CC_g}} \tag{6}$$

Eq. (5) can be rewritten as

$$-\frac{2\omega i}{CC_g} \frac{\partial \phi}{\partial t} = \nabla^2 \phi + k_c^2 \phi \tag{7}$$

where

$$k_c^2 = k^2(1 + if_d) - \frac{\nabla^2 \sqrt{CC_g}}{\sqrt{CC_g}} \tag{8}$$

is the pseudo wave number. In the present approach, the ADI method is used to solve (7). In both x and y direction, a tridiagonal algebraic system is generated and can be solved by the Gauss elimination method.

Radiation boundary conditions

Partial reflection boundary conditions

The radiation boundary condition specifies that there is no change of a velocity potential Φ when a wave train cross the boundary. This boundary condition reduces the reflection of waves back to the computational domain. The expression of Φ satisfying a partial dissipation radiation boundary condition is given by Behrendt (1985) as

$$\frac{d\Phi}{dt} = \alpha \frac{\partial \Phi}{\partial t} + C \frac{\partial \Phi}{\partial r} = 0 \tag{9}$$

where $\alpha = (1 - R) / (1 + R)$ = the absorption coefficient; R = the reflection coefficient; $r = |r| = x \cos\theta + y \sin\theta$ = the position coordinate. When $\alpha = 0$ and $\alpha = 1$, the above boundary condition represents total reflection and total passing through boundary conditions, respectively. Using $\partial r / \partial x = \cos\theta$, $\partial r / \partial y = \sin\theta$ and (4) and (6), then (9) in x direction becomes

$$\frac{\partial \phi}{\partial x} \mp i\alpha k_x \phi = \frac{\partial \phi}{\partial x} \mp i\alpha k \sqrt{1 - (\frac{k_y}{k})^2} \ \phi = 0 \qquad \text{,on } \partial B_{x^+} \text{ or } \partial B_{x^-} \tag{10}$$

where $k_x = k \cos\theta$ and $k_y = k \sin\theta$ are wave number components in x and y directions, respectively, θ is the approaching wave angle, and $\partial B_x (\partial B_y)$ is the boundary perpendicular to x (y) axis.

Given boundary conditions

The given boundary condition is used at the driving boundaries where input wave information is specified. There are two velocity potential values at a given and the outgoing scatter waves should have passed through this boundary and remain unaffected. The given boundary condition in x direction is then written as:

$$\frac{\partial \phi}{\partial x} = \mp k_x \phi + 2 i k_{xi} \phi_i \qquad \text{, on } \partial B_{x^+} \text{ or } \partial B_{x^-} \tag{11}$$

where ϕ_i is an incident wave potential. For a given wave with wave height, H_0, period, T, and direction, θ_0, the give wave potential can be calculated and

$$\phi_i = \frac{i g H_0}{2\omega} e^{iS} \tag{12}$$

where $S = k \cos\theta_0 x + k \sin\theta_0 y - \omega t$ is the phase function.

In the numerical calculation, it is impossible to know k_x (k_y) or the approaching wave angle in advance. Therefore, there are various ways to calculate k_x and k_y by expanding the term

$$k_x = \sqrt{1-(k_y/k)^2} \quad , \quad k_y = \sqrt{1-(k_x/k)^2} \tag{13}$$

as different approximations. Without losing generality, we don't expand the term $k_x = [1-(k_y/k)^2]^{1/2}$ and $k_y = [1-(k_x/k)^2]^{1/2}$ to different approximations. This method has an advantage to keep the exact relationship between the wave number components. The wave number components, k_x and k_y in (11) can be determined by the phase function, S, from the complex velocity potential ϕ:

$$\phi = \sqrt{CC_g}\, Ae^{iS}, \quad S = tan^{-1}\frac{Im(\phi)}{Re(\phi)} \tag{14}$$

where A is the amplitude of velocity potential, S is the wave phase, Re and Im are real and imaginary part of a complex value, respectively. From the definition of wave number vector

$$\vec{k} = \nabla S \tag{15}$$

we can calculate the approaching wave angle θ and $k_x = k\cos\theta$ and $k_y = k\sin\theta$ by the equation:

$$\theta = tan^{-1}(\frac{\partial S/\partial y}{\partial S/\partial x}) \tag{16}$$

Model verification

A well-documented experiments were used to verify the applicability of the numerical model. The cases concerning wave refraction, diffraction, reflection and breaking around parallel and perpendicular breakwaters attacked by oblique incident waves. In experiment of Watanabe and Maruyama (1986), the study domain was a rectangular basin with an area of $5m \times 8m$. The input parameters of model performance for each case are shown in Table 2.

Table 2. Input parameters of the study cases

grid number	dx (m)	dy (m)	wave period (sec)	incident wave height (cm)	incident wave angle (degree)	iterative numbers
51×81	0.1	0.1	1.2	2.0	0°	200

A detached breakwater model made of a steel plate of 2.67 m long was placed at the water depth of 6.0 cm in parallel to the shoreline on a plane beach made of mortar with a uniform slope of 1/50. The case where the breakwater is perpendicular to the shoreline, experimental data obtained by Watanabe and Maruyama (1986) are

patterns around the breakwater is shown in Fig. 1.

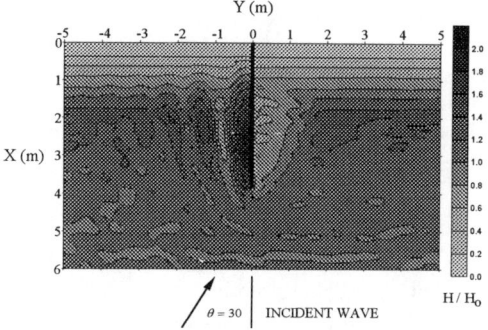

Fig. 1. Wave pattern around a perpendicular breakwater for $30°$ angle of incidence

Comparisons of wave distribution along the cross sections $y = -0.2m$ (upwave region); $0.2m$ and $2.0m$ (downwave region) are presented in Fig. 2. It is seen that the distribution of wave heights on the reflection side is quite different from that of the parallel detached breakwater (Fig. 2).

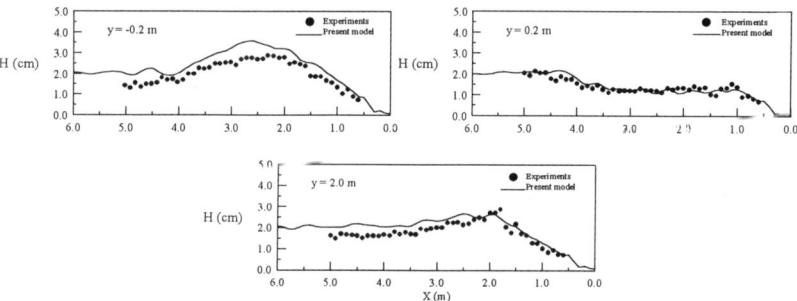

Fig. 2. Comparison of cross-shore wave distribution around a perpendicular breakwater for $30°$ angle of incidence

It is noted that the overall agreement between numerical results and experimental data is quite good. We find the present model predicts reasonable locations of breaking points and wave transformations across the surf zone. This result confirms that the present model is capable of representing wave breaking and energy dissipation in the entire surf zone.

In order to examine the accuracy of large angle incidence by the present model, waves propagating over a circular shoal resting on a flat bottom are simulated. This test was carried out by some researchers [Kiry (1986), Dalrymple et al. (1989) and Li (1994)] for their large-angle models. Owing to the axisymmetry of the

In order to examine the accuracy of large angle incidence by the present model, waves propagating over a circular shoal resting on a flat bottom are simulated. This test was carried out by some researchers [Kiry (1986), Dalrymple et al. (1989) and Li (1994)] for their large-angle models. Owing to the axisymmetry of the circular shoal, predictions of the wave focusing pattern behind the shoal should be independent of the incident wave angle. The depth used here is the same as that adopted by Dalrymple et al. (1989), which is expressed as

$$h = \begin{cases} h_0 & , r_1 > R_1 \\ h_0 + h_1 - h_2 [1-(0.2X')^2 - (0.2Y')^2]^{1/2}, & r_1 < R_1 \end{cases} \quad (17)$$

where $h_0 = 0.336m$; $h_1 = 0.12m$; $h_2 = 0.2m$; $R_1 = 4m$ is the radius of the shoal; $r_1 = (X'^2 + Y'^2)^{1/2}$; and (X',Y') = coordinates with the origin at the crest of the shoal. The wave period is 1 sec and the incident wave angles are $0°, 45°$, and $80°$, respectively. Wave height patterns of different incident wave angles are shown in Fig. 3. In the computational domain the grid size $\Delta x = \Delta y = 0.1m$ is used, which corresponds to a grid of 251×601 points.

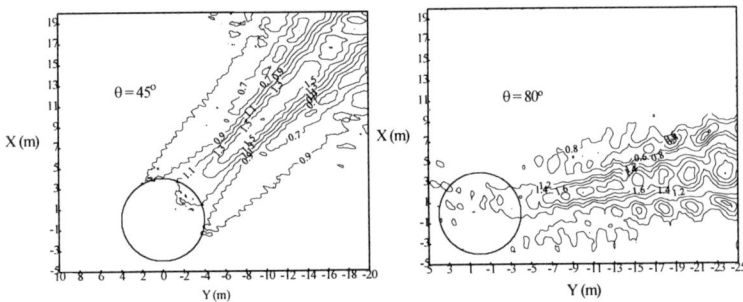

Fig. 3 Wave height pattern for circular shoal cases

It is interesting to note that the wave pattern corresponding to $0°$ and $45°$ preserves the axisymmetry very well without any angle distortion. Although the axisymmetry of the wave pattern is distorted in the case of $\theta = 80°$, the angle along the wave focusing crest is still reasonable.

Conclusions

The present method solves the radiation boundary condition in a straightforward manner without using approximations of a certain order. The approaching wave angle and wave number components appearing in the radiation boundary conditions are obtained through iteration in each time step. This method keeps the exact relationship of a circle between two wave number components. It has the advantage to improve limitation of oblique incident waves of the existing parabolic models. The accuracy of the model has been shown by a comparison with experiments conducted in wave basins. Computational cases involving incident wave angles up to $80°$ over a circular shoal was to test the model. Predictions of the axisymmetry of wave focusing pattern behind the shoal show that the present model can give reasonable results for large incident wave angles. The chosen grid spacing in the present model should be less than 1/12 of a wave length to achieve a good resolution of the computational domain. It is recommended that the chosen grid spacing should be less than one-length of wavelength to archive a good resolution of wave patterns.

References

1. Behrendt, L., (1985). "A finite element model for water wave diffraction including boundary absorption and bottom friction." *Series Paper 37, Institute of Hydrodynamics and Hydraulic Engineering*, Tech. Rept, University of Denmark.
2. Berkhoff, J. C. W., (1972). "Computation of Combined Refraction-Diffraction." *Proc. 13th Conf. on Coastal Eng.*, Canada, ASCE, 471-490.
3. Booij, N., (1981). "Gravity waves on water with non-uniform depth and current." *Rep. No.* 81-1, Dept. Civil Engrg., Delft Univ. of Tech., Delft, The Netherlands.
4. Dalrymple, R. A., Suh, K. D., Kirby, J. T., and Chae, J. W., (1989). "Models for very wide-angle water waves and wave diffraction. 2: Irregular bathymetry." *J. Fluid Mech.*, U.K., 201, 299-322.
5. Isobe, M. (1987). "A parabolic equation model for transformation of irregular waves due to refraction, diffraction and breaking." *Coastal Eng. in Japan*, Vol. 30, 33-47.
6. Kirby, J. T., (1986). "Rational approximations in the parabolic equation method for water waves." *Coastal Eng.*, 10, 355-378.
7. Kirby, J. T., (1989). "A note on parabolic radiation boundary conditions for elliptic wave calculations." *Coastal Eng.*, 13, 211-218.
8. Li, B., (1994). "An evolution equation for water waves." *Coastal Eng.*, 23 227-242.
9. Li, B., (1997). "Parabolic model for water waves." *J. Waterway, Port, Coastal and Ocean Eng.*, Vol. 123, No. 4, 192-199.
10. Mei, C. C., (1983). *"The applied dynamics of ocean surface waves,"* Wiley-interscience, New York.
11. Radder, A. C., (1979). "On the parabolic equation method for water wave propagation." *J. Fluid Mech.*, Vol. 95, No. 1, 159-176.
12. Watanabe, A. and Maruyama, K., (1986). "Numerical modeling of nearshore wave field under combined refraction, diffraction and breaking." *Coastal Eng. in Japan*, Vol. 29, 19-39.

A Response Based Method for Developing Joint Metocean Criteria for Seabed Pipeline Design

Kevin C. Ewans[1]

Abstract

A response-based method is proposed for deriving joint metocean criteria for the on-bottom stability design of submarine pipelines. The method consists of applying a response function to convert a long time history of oceanographic variables into a corresponding response time history, from which extreme values and the joint metocean criteria can be obtained. Response functions are developed for both the traditional static design approach and the state-of-the-art dynamic simulation design approach. In the case of the latter, simple response functions are developed from simulations of a pipeline's dynamic response for a test location with a water depth of 43 m. The joint wave and current criteria obtained from the different response functions are compared and evaluated against the independent wave and current criteria. The method gives a consistent reduction of the steady current speed associated with the independent 100-year return period wave-induced water particle velocity for all the response functions. However, inconsistencies occur in the wave-induced water particle velocity when it is estimated from the independent 100-year return period steady current. Good agreement between the normalised response curves of the respective response function indicates that they can be used to give consistent reliability estimates of the pipelines lateral stability.

---

[1] Shell International Deepwater Services B.V., P.O. Box 3006, 2280 MH Rijswijk, The Netherlands, k.ewans@sidsbv.shell.com

Introduction

Traditional techniques for on-bottom stability design of submarine pipelines require the specification of an extreme sea state and steady current. These have often involved taking the n-year return period wave conditions combined with the n-year return period currents, but it has long been recognised that this approach may be conservative when the desired level of structural reliability is meant to correspond to the n-year return period load. The reason is that the highest waves and currents associated with extreme events do not occur simultaneously at a given location. Additional conservatism is often introduced by the assumption that the wave induced currents and the steady currents will act simultaneously from the worst possible direction.

Recent advances in the calculation of the on-bottom stabilisation of submarine pipelines, largely due to research sponsored by the American Gas Association (AGA, 1993), now include more sophisticated load analyses, but they have not altered the way in which the metocean conditions are specified; specifically, they do not account in a rationale way for the joint occurrence of wave and current conditions. Rational methods do however exist for developing joint metocean criteria for the design of offshore structures (for example, Forristall et al., 1991; Tromans and Vanderschuren, 1993). The methods are response based, in which the joint metocean criteria are established from the design load which itself has been derived from a population of loads associated with extreme environmental events. This paper describes a rational approach to deriving joint wave and current criteria for calculating the stability of pipelines on the seabed. The approach follows established response based methods for fixed structures.

Response Based Design Criteria Methodology

General Approach

The starting point of the method is a long time history of metocean variables, specifically waves and steady currents from which near-bottom water velocities can be calculated. This data base should extend for many years and include all the meteorological events responsible for the extreme wave and current events along the pipeline route. Such time histories are available from hindcast studies, which typically provide a data base of metocean variables specified at regular time intervals (usually 1 to 3 hours) for the largest extreme meteorological events over many years (often longer than 25 years).

The time history of metocean variables is transformed into a time history of near-bottom wave-induced water velocities and steady currents which in turn is converted, through a response function, into a time history of a single load or response variable, relevant for the pipeline stabilisation. A response function is used to convert the water velocities into the response variable for each time step of the metocean time series. Extreme value analysis of the response time history then provides the long-term statistics necessary both for estimating the n-year return period response and for making reliability estimates of the pipeline's lateral stability. The n-year joint wave and current criteria which produce the n-year response are then derived through the application of an inverse form of the response function.

Static Case

The traditional pipeline stability design approach is a static one, involving the selection of a pipe weight which will ensure that the pipeline remains static when subjected to the hydrodynamic forces of the design oceanographic conditions. The horizontal forces acting on the pipeline are the drag forces, F_d, due to the wave-induced and steady current, the inertial force, F_i, produced by the wave-induced current, and the opposing soil resistance force, which in turn is a function of the lift force, F_l, arising from the wave-induced and steady currents, the pipeline weight, wt, and the soil friction, with friction coefficient, μ. Thus stability, with a particular factor of safety, fos, occurs when,

$$wt \geq \frac{fos\,|F_d + F_i|}{\mu} + F_l \quad (1)$$

The hydrodynamic forces are obtained from the component of the steady and wave-induced currents perpendicular to the pipeline with the Morison equations, as follows.

$$\begin{aligned} F_d &= \tfrac{1}{2}\rho_w D C_d \,|U_s \cos\theta + U_c|\,(U_s \cos\theta + U_c) \\ F_i &= \tfrac{\pi D^2}{4}\rho_w C_m A_s \sin\theta \\ F_l &= \tfrac{1}{2}\rho_w D C_l \,(U_s \cos\theta + U_c)^2 \end{aligned} \quad (2)$$

where ρ_w is the density of sea water, D is the pipeline outside diameter, C_d is the drag force coefficient ($C_d = 0.7$), C_i is the inertia force coefficient ($C_i = 3.29$), C_l is the lift force coefficient ($C_l = 0.9$), U_s is the significant near-bottom wave-induced velocity amplitude perpendicular to the pipeline, U_c is the steady current perpendicular to the pipeline, and θ is the phase angle of the wave force that maximises wt in Equation 1.

Equations 1 and 2 transform the oceanographic variables U_s and U_c into the single pipeline design variable wt, allowing a data base of sea states and steady currents to be converted into a data base of pipeline weights.

Dynamic Case

Detailed analysis of pipeline stability involves a full dynamic simulation of a pipeline resting on the sea bed, taking account of wave wake effects, through the use of Fourier force coefficients, and pipeline embedment. The analysis involves the simulation of the pipeline response at various nodes spaced along a length of pipe. The oceanographic input required is the specification of a random sea state and steady current velocity. The sea state specified is used to simulate a time series of near-bottom, wave-induced water particle velocities at each of the pipe nodes. These velocity time series, together with the steady current are converted into a time series of hydrodynamic drag and lift forces, based on a Fourier summation of the wave forces. The forces are then used to simulate the pipeline response.

The American Gas Association (AGA, 1993) dynamic simulation program was used to develop two response functions - one for pipeline displacement and one for pipeline stress. The AGA program consists of three parts - L3WSIMQ, L3FORCE, and L3PIPDYN. These three parts are collectively referred to as the Level 3 program.

The current velocity simulations are produced with L3WSIMQ, which calculates near-bottom water particle velocities based on Airy wave theory and a set of randomly phased waves with different wave frequencies and directions. The sea state is specified by an Ochi-Hubble variance density spectrum and a wrapped normal spreading distribution. The Ochi-Hubble spectrum is a function of three parameters the peak frequency, f_p, the significant wave height, H_s, and the spectral peakedness parameter, λ. A Pierson-Moskowitz spectrum results if $\lambda = 1.0$. The spreading distribution is a function of two parameters, the mean wave direction, θ_0, and the spreading parameter, σ.

Program L3FORCE produces a time series of hydrodynamic drag and lift force at each pipe node, for a stationary, unburied pipe. The program decomposes the irregular wave time series into half-wave regular waves for which the hydrodynamic force is calculated from a set of Fourier coefficients and phases defined by the half-wave Keulegan-Carpenter number and the ratio of the wave induced particle velocity amplitude to the steady current speed. The inertial force as defined in Equation 2 is added to give the total in-line force.

Program L3PIPDYN performs a finite-element analysis of the pipe string, using the L3FORCE output as the forcing function, and estimates the pipe displacement and stress at each node for each time step, allowing for possible pipe embedment. The pipe parameters, such as the outside diameter and weight, and soil properties (for each node) are specified before running.

Together these programs could be used directly to transform a long time series of wave and current data into a pipeline response and stress time series, but this is impractical on account of the number of simulations that would be necessary to produce a statistical reliable estimate of the design conditions; and it would still leave the problem of inverting the design response to obtain the joint wave and current conditions. Accordingly, a simple response function was developed.

The Dynamic Response Functions

This section describes the development of simple response functions for the pipeline's displacement and stress, in which the response, R, be it displacement or stress, is described as a polynomial equation of the wave parameters - H_s and $T_p(=1/f_p)$ - and the steady current perpendicular to the pipeline, U_c. The objective then is to derive coefficients, A_i, in the following equation, for both the displacement and stress responses.

$$R = A_1 + A_2 H_s + A_3 T_p + A_4 U_c \tag{3}$$

The derivation involved running the AGA Level 3 software on a set of 26 'sea states', each specified by H_s, T_p, and U_c. Within the 26 sea states, H_s ranged from 5 m to 15 m, T_p ranged from 10 s to 25 s, and U_c ranged from 0.5 m/s to 2.0 m/s. For each 'sea state' 30 simulations were made to obtain statistical reliability. The wave parameters, θ_0, λ, and σ were fixed at $\theta_0 = 90°$ (perpendicular to the pipeline), $\lambda = 1$, and $\sigma = 30°$ for all simulations.

The simulations were made for 5 pipe nodes spaced evenly on a pipe segment of diameter 1 m and length of 500 m, in 43 m water depth. At this configuration, the near-bottom wave particle velocity time series were essentially uncorrelated. The simulations produced half-hour long time series of pipe displacement and stress at each of the pipe nodes. From these the maximum displacement and stress of any node was obtained, resulting in a data set of 780 maximum displacements and stresses - 30 pairs for each sea state. The 30 maximum responses for each sea state were averaged to obtain a 'best estimate' of the maximum displacement and stress for each combination of H_s, T_p, and U_c. Accordingly, the response function coefficients followed from a solution in the least squares sense to the over-determined system of equations $\mathbf{XA} = \mathbf{R}$, where $\mathbf{X}$ is the 26 by 4 matrix of parameters in equation 3, and $\mathbf{R}$ is the 26-element vector of responses (displacement or stress).

The analysis resulted in the following response functions for the displacement, d, and stress, s.

$$\begin{aligned} d &= 176.0 - 13.14 H_s - 4.071 T_p + 727.8 U_c \\ s &= -198.8 + 11.16 H_s + 5.936 T_p + 830.3 U_c \end{aligned} \tag{4}$$

Figure 1 gives a comparison between these formulae and the simulations from which they are derived. The plots shows that the response functions, which are in terms of linear terms only, give a good approximation to the simulated responses, over the range of parameters considered.

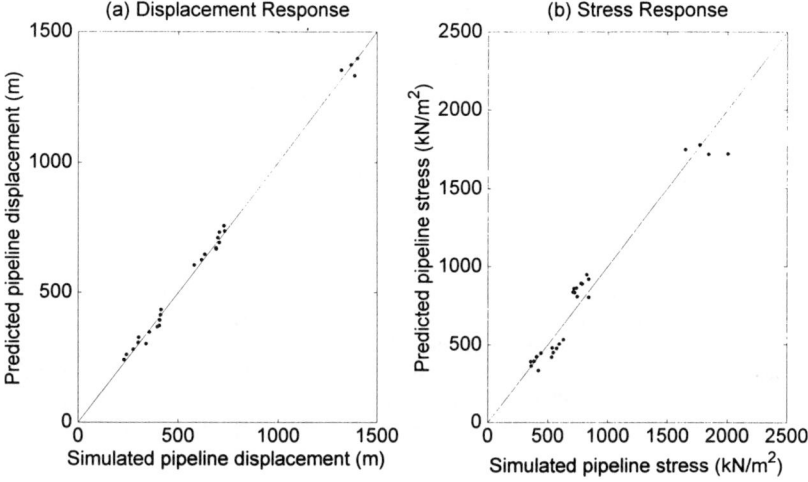

Figure 1: Scatter plot of simulated and predicted responses.

Application to a Test Location

In this section, the method and response functions are evaluated and compared for a site with a water depth of 43 m offshore Philippines in the South China Sea.

The Metocean Data

The wave and current data used for the evaluations were obtained from hindcasts of the most important tropical cyclones for the region over a 47 year period. The data available for this study consisted of time series of the wave and steady currents for the 9 most important tropical cyclones during the period. In practice, a larger data set should be used for establishing criteria, but the 9 tropical cyclone data set is sufficient for the purposes of the comparisons given here.

The data consisted of concurrent hourly wave frequency-direction spectra (produced with a 3G wave model) and current profiles (produced with a 17-layer implementation of the Princeton Ocean Model).

Static Case

Equations 1 and 2 were used to produce a data base of pipeline weights from the hindcast data base of waves and currents. The weights produced were based on a

$fos = 1$ and a pipe outside diameter of 1 m.

The perpendicular wave-induced significant water particle velocities were obtained by transforming the surface spectra to a height of 1 m above the bottom, using Airy theory. The perpendicular steady current speeds at 1 m above the sea bed were obtained by fitting a logarithmic boundary layer profile to the near-bottom levels of the current profile. A fixed pipeline orientation was used to obtain the perpendicular components for both the wave-induced and steady currents.

The data set of pipeline weights was then subjected to a peaks-over-threshold extreme value analysis, in which a 3-parameter Weibull distribution was fitted to the distribution of 9 maximum weights associated with the 9 cyclones.

The response based design conditions, for a particular return period can be obtained by inverting Equations 1 and 2. This requires the selection of an appropriate period for the wave induced water particle velocities; this period was taken as the mean of the 2nd moment periods of the 9 maxima used in the extreme value analysis. Figure 2 is a plot of the 100-year return period weights, wt_{100} as a function of the perpendicular significant water particle velocity and steady current velocity.

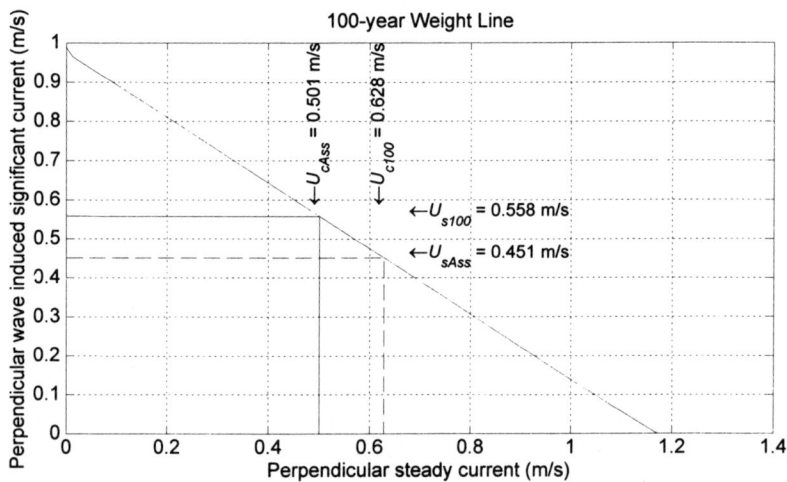

Figure 2: 100-year return period pipeline weight as a function of perpendicular significant wave induced current and perpendicular steady current

Any combination of wave induced and steady currents defined by the line will give wt_{100}. One approach to selecting an appropriate pair of the wave and current velocities is to select the independent values of either one - i.e. U_{s100} or U_{c100} - and determine the associated value of the other parameter. The pairs of numbers found in this way are given in Table 1 and indicated in Figure 2. Table 1 also gives the return periods of the estimates, if they had been derived from extrapolation of the independent probability distribution for the parameter concerned. Thus, the associated steady current, U_{cass}, corresponds to an independent current with a return period of 52 years, while the associated significant wave induced current, U_{sass}, corresponds to a return period of 29 years. By comparison the DNV RP305 (1990) recommended practice for on-bottom stability design of submarine pipelines suggests 10-year return period values for the associated variable, the associated variable being steady current or wave-induced current depending on whether respectively wave or current forces dominate. The results suggest that the use of these return periods would be optimistic for the test location.

Response function	U_s(m/s)	Return Period (years)	U_c(m/s)	Return Period (years)
Static	0.558	100	0.501	52
Static	0.451	29	0.628	100
Displacement	0.559	100	0.567	72
Displacement	1.120	>1000	0.629	100
Stress	0.559	100	0.579	77
Stress	0.190	<5	0.629	100

Table 1: Joint wave-induced and steady current estimates for the three response functions.

Dynamic Case

The equations 4 give the dynamic displacement and stress responses of the pipeline as function of the H_s and T_p of the surface wave field and U_c the perpendicular current speed. The surface wave field has a Pierson-Moskowitz spectrum and a wrapped normal direction distribution, with a mean wave direction which is perpendicular to the pipeline and a spreading parameter of 30°. The frequency-direction spectrum of such a wave field is of course different from those hindcast for the test location. Therefore, to enable comparison of response analyses with the static case for the test location, it was necessary to obtain a consistent set of surface wave field parameters that when used to describe the appropriate Pierson-Moskowitz spectrum and the wrapped normal direction distribution will give the same perpendicular significant wave induced water particle velocities as derived from the hindcast data base for the static case. Accordingly, the U_s values and their associated 2nd moment periods, derived in the static case for the test location were back-transformed to an equivalent surface set of H_s and T_p of a wave field that has a Pierson-Moskowitz frequency spectrum and a wrapped normal direction distribution, with a mean wave direction which is perpendicular to the pipeline and a spreading parameter of 30°. The new values of H_s and T_p will in general be different to the hindcast values, but they will give the same values of U_s.

The new set of parameters H_s, T_p, and U_c were then transformed into displacement and stress responses, using Equations 4, and following the static case an extremal analysis performed to obtain the response extremes.

For comparison with the joint criteria derived for the static case, the equations 4 were inverted to obtain the values of U_{cass} and U_{sass} associated with the U_{s100} = 0.558 m/s and U_{c100} = 0.628 m/s respectively. These values are given in Table 1. There is good agreement between the associated currents of the two dynamic response functions - 0.567 m/s and 0.579 m/s, but both values are approximately 15% larger than 0.501 m/s predicted by the static response function. However, there is little agreement between the estimated associated wave-induced currents. Clearly, with negative coefficients for H_s and T_p the displacement response function given in equation 4 will give increasing wave-induced currents for increasing steady current speeds, explaining the anomalously large associated wave-induced current speed of 1.12 m/s. By comparison, the associated wave-induced current speed for the stress response function is substantially lower, even when compared with the static estimate, and with a return-period less than 5 years, this value would be intuitively unacceptable for design purposes. Apparently, the dynamic response functions are able to give acceptable estimates of the associated steady current speeds, but this is not so when they are used to estimate the associated wave-induced currents.

A further comparison of the respective response functions is given in Figure 3, which is a plot of the normalised extreme response against return period. Given the small size of the hindcast data set on which the extreme value analysis was based, the figure indicates that there is good agreement between the normalised response curves of the three responses, indicating that these functions could be used to give consistent reliability estimates of the pipelines lateral stability.

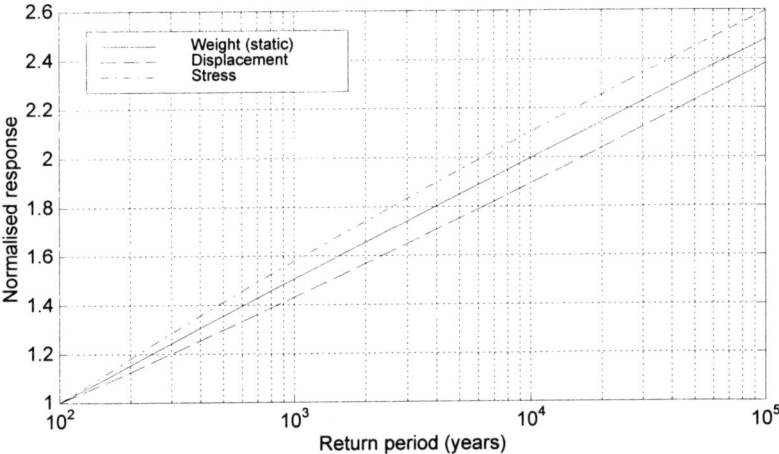

Figure 3: Normalised responses as a function of return period for the test location.

Conclusions

The values of the joint wave and current criteria derived with the response based methods are dependent on the response function used. All the response functions give a consistent reduction of the steady current speed associated with the independent 100-year return period wave induced water particle velocity; but inconsistencies in the reverse case, of estimating the associated wave induced water particle velocity from the independent 100-year return period steady current, highlights inadequacies in the simple response functions derived. As the static response function was a full implementation of the static design method and it produced realistic values of the associated wave induced current, the results suggest the need for further development of the dynamic pipeline response functions, so that they properly account for the relative contributions of the wave-induced and steady currents to the pipeline's dynamic response.

Good agreement between the normalised response curves of the respective response functions indicate that the simple functions derived give reliable estimates of the long-term behaviour of the pipeline responses and can be used to give consistent reliability estimates of the pipelines lateral stability.

References

[1] American Gas Association, 1993: Submarine Pipeline On-Bottom Stability. AGA Project Report PR-178-9333, September, 1993.
[2] Forristall, G.Z., Larrabee, R.D., and Mercier, R.S., 1991: Combined Oceanographic Criteria for Deepwater Structures in the Gulf of Mexico. OTC paper 6541, 377-390. May, 1991.
[3] Tromans, P.S. and Vanderschuren, L, 1995: Response Based Design Conditions in the North Sea: Application of a New Method. OTC paper 7683, 387-397. May, 1995.
[4] Veritec Offshore Technology and Services, 1988: On-Bottom Stability Design of Submarine Pipelines. Recommended Practice RP E305.

Wave Forces on Partially Submerged Pipe Breakwater

J.S. Mani[1]

Abstract

Partially submerged pipe breakwater has been designed exclusively for fishery and pleasure - craft harbours, wherein moderate wave agitations are permissible. The breakwater consists of a row of closely spaced pipes mounted on a frame and suspended between the support piles spaced far apart. Experimental studies conducted to determine wave forces indicate that the maximum force would reduce by about 60% when compared with the force that an equivalent solid plate would experience. Further studies suggest that by suspending a row of closely spaced pipes (with a gap to diameter ratio of 0.22 and draft to water depth ratio of 0.46), 50% reduction in incident wave height can be achieved. Based on Kriebel (1992), modified expressions for wave transmission and forces were obtained and results compared with experimental values.

Introduction

Recreational harbours, marinas and fishing harbours are constructed for varied purposes and depending on the need, the required degree of tranquillity for each facility differ. For example, in case of recreational harbours, coastal swimmers and surfers prefer to have acceptable wave conditions to suit their sporting activities and for fishing harbours, creation of still water condition is not necessary. In such circumstances, expensive rubble mound breakwaters may not be the right choice as they are meant for providing very calm waters. Alternately to control the wave disturbance, floating, tethered, submerged and pile breakwaters are adopted. These breakwaters can be considered if the water depths are relatively shallow.

---

[1] Associate Professor, Ocean Engineering Centre, Indian Institute of Technology, Madras, Chennai - 600 036. [INDIA]

In the present investigations, attention has been focussed on pile breakwaters. As the performance and cost of a closely spaced pile breakwater is directly related to the number of piles, an attempt is made to develop partially submerged pipe breakwater which is cost effective, easy to install and capable of arresting the incident wave height to a desirable limit. Results of the studies conducted to (a) determine the forces on the pipe breakwater and (b) evaluate its performance capabilities are discussed in this paper. The experimental results are compared with those of theoretical estimates obtained based on expressions derived for partially submerged breakwater. The modified expressions for transmission coefficient and force ratios derived herein are according to the procedure suggested by Kriebel (1992).

Theoretical solution for wave transmission and wave force

Expressions for wave transmission and wave force were derived for the submerged pipe beakwater in line with the procedure detailed by Kriebel (1992) for vertical barriers. Appendix I presents the expressions for wave wave transmission and wave forces as obtained by Kriebel.

Modified expressions for the submerged pipe breakwater are as follows.

For wave transmission

$$K_t = \frac{-1 + \sqrt{1+4T_t}}{2T_t} \qquad (1)$$

where

$$T_t = \frac{H_i}{6L \sinh^2 kh} [K_{loss}(S_a)]$$

$$S_a = 2ky + \sinh 2kh - \sinh 2k(h-y)$$

$$K_{loss} = \left[\frac{1}{C_c P} - 1\right]^2$$

C_c : coefficient of contraction $= 0.48 + 0.4P^3$

P : $b/b+d$

For wave force

$$\frac{F}{F_o} = \frac{(b+d)/m\exp(K_t)}{d} \quad (2)$$

where $m = \log(H_i/gT^2) + C\,e^{(1-(y/h))+\frac{2\pi H_i}{gT^2}}$

C : a constant = 2.5

Experimental Setup

Experiments were conducted in a 30m long, 2m wide and 1.5m deep wave flume. The wave maker machine installed in the flume is capable of generating monochromatic incident waves of height H_i varying between 6cm and 24cm with period ranging between 0.80 and 2 sec. The dimensions of the flume facility and the details of submerged pipe breakwater are shown in figure 1. The support piles, 16cm in diameter, (representing a pile in nature) were positioned close to the flume walls and a frame containing a row of 4cm diameter pipes spaced with a b/d ratio (b = clear gap; d = pipe diameter) was suspended in between the pipes (section A-A, figure 1). The frame was made to slide through vertical slot provided in support pipes, so as to facilitate testing the breakwater with different y/h ratios (y = draft of the pipe; h = water depth). Experiments were conducted for b/d ranging from 0.11 to 1.0 and y/h ranging from 0.26 to 0.56. With strain gauges positioned at different elevations along the pipe, forces were determined and the performance characteristics evaluated by analyzing the wave profiles recorded through wave gauges.

Experimental results

Wave transmission

Experimental studies conducted on the wave transmission characteristics of the partially submerged pipe breakwater (Mani, 1995) indicate that the wave transmission can be effectively controlled by adopting a gap to diameter ratio (b/d) of 0.22 and a depth of submergence to water depth ratio (y/h) of 0.46. Figure 2 shows the results on wave transmission characteristics. Figure 3 shows a comparison of experimental results and those obtained based on equation 1. Good correlation between the theory and the experimental results validate the theory and supports the expression for coefficient of contraction.

Wave force

Figure 4 shows the comparison of experimental results with those obtained based on modified expression (eqn.2). For the given breakwater configuration (y/h = 0.46 & b/d = 0.22), reasonable agreement between the experimental and theoretical results exists and it is inferred that

(a) the non-dimensional force decreases exponentially with increase in wave steepness parameter

(b) the force on the structure depends largely on (i) wave steepness (ii) depth of submergence and (iii) gap to diameter ratio.

The results further indicate that the submerged pipe breakwater would experience a reduction in the force varying between 60 and 80% when compared with the force that would be experienced by an equivalent solid plate.

Conclusion

(1) The suspended pipe breakwater would experience a reduction in the force (varying between 60 & 80%) when compared with the force on an equivalent solid plate.

(2) The modified theoretical expressions for determination of forces and transmission characteristics are capable of predicting the respective phenomenon.

(3) The structure is capable of attenuating the waves effectively which can be adopted for protecting recreational harbours.

Acknowledgement

The writer would like to express his sincere thanks to the authorities of the Indian Institute of Technology Madras, Chennai, India for providing all the facilities.

The author expresses his thanks to Ms.V.Saradha and K.Lakshmi for the support extended by her during various stages of this research work.

References

Kriebel. D.L., (1992) "Vertical wave barriers : Wave transmission and wave forces." 23rd Int. Conf. on Coastal Engrg., ASCE, Vol.2, pp.1313 - 1326.

Mani, J.S. and Jayakumar, S. (1995). "Wave transmission by suspended pipe breakwater." J. Wtrwy., Port Coast. and Ocean Engrg., ASCE, Vol.121, No.6, pp.335-338.

Appendix I

Solution for wave transmission and wave force.

Following expression are given by Kriebel (1992) for determination of wave transmission and wave force.

Wave transmission

$$K_t = \frac{-1+\sqrt{(1+4T_t)}}{2T_t} \quad (A1)$$

$$T_t = K_{loss}\frac{1}{6}\frac{H_i}{L}\frac{\sinh 2kh + 2kh}{\sinh^2 kh} \quad (A2)$$

where

$$K_{loss} = \left[\left(\frac{1}{C_c P}\right)-1\right]^2$$

C_c = coefficient of contraction = $0.6 + 0.4P^3$

$P = b / (b+d)$

$$F = \rho g \frac{H_i}{k} \tanh(kh)(1-K_t)(b+d) \quad (A3)$$

$$F_o = \rho g \frac{H_i}{k} \tanh(kh) d \quad (A4)$$

Appendix II

Notation

The following symbols are used in this paper :

b	=	gap width
C	=	constant
C_c	=	coefficient of contraction
d	=	diameter of pipe
F	=	force on partially submerged pipes
F_o	=	force on partially submerged solid plate
g	=	acceleration due to gravity
H_i	=	incident wave height
H_t	=	transmitted wave height
h	=	water depth
K_t	=	transmission coefficient = (H_t/H_i)
k	=	wave number
L	=	wave length
l	=	length of the pipe
P	=	b / b + d
T	=	wave period
y	=	draft
γ	=	specific weight of water

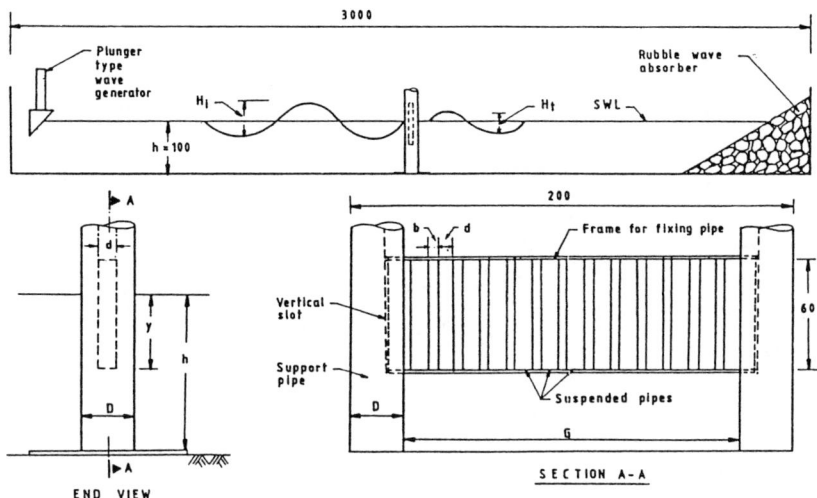

Figure 1. Partially Submerged Pipe Breakwater

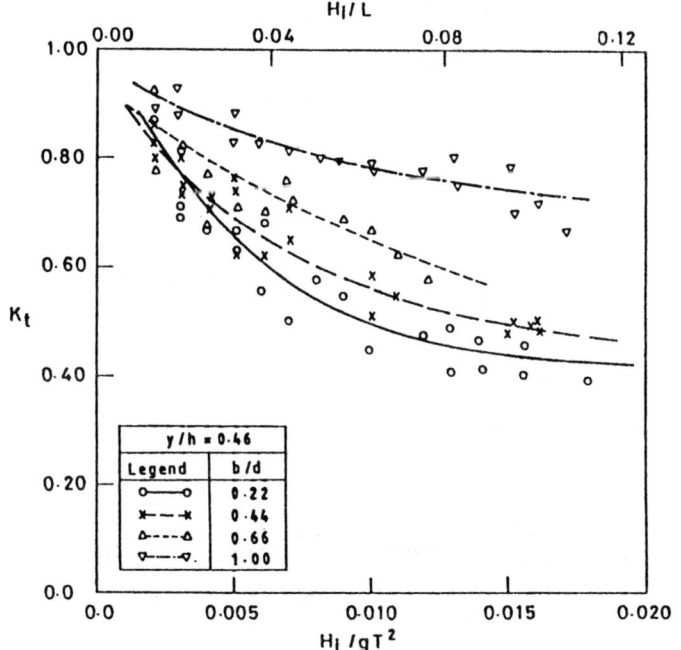

Figure 2. Wave Transmission Characteristics

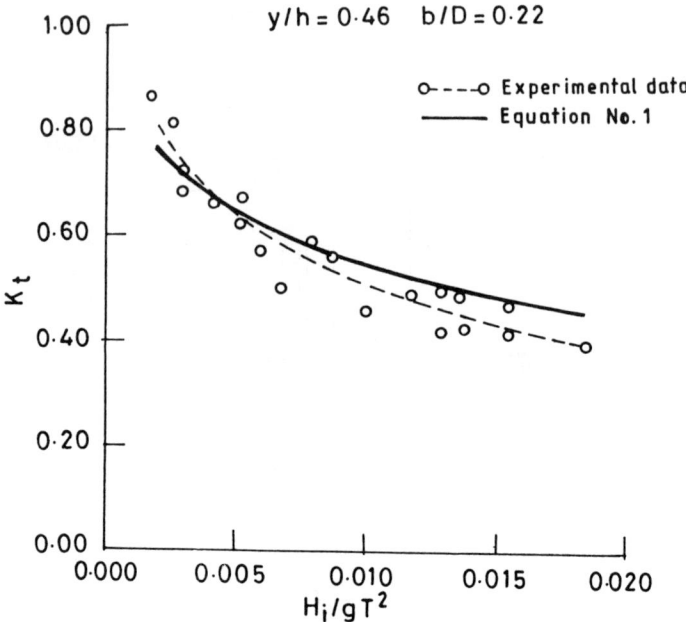

Figure 3. Observed and Predicted Wave Transmission - A Comparison

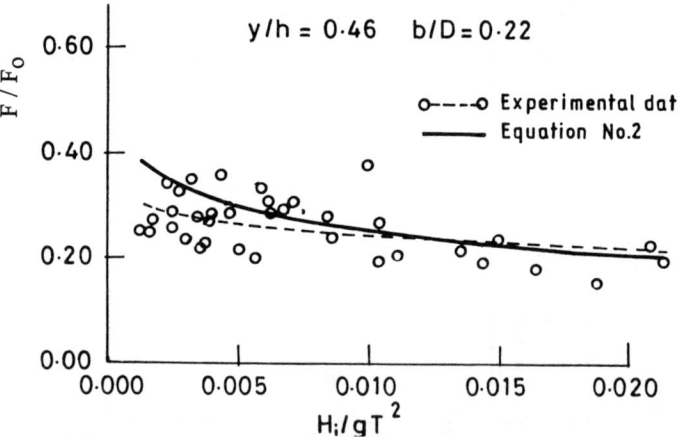

Figure 4. Comparison of Observed and Predicted Non-dimensional Force

ON THE AVERAGE WAVE STEEPNESS

Elzbieta M. Bitner-Gregersen [1]   Carlos Guedes Soares [2] and Alexandra Silvestre [3]

Abstract

In the present study the mean as well as extreme values of average wave steepness are evaluated for different locations using wave buoy, satellite, hindcast as well as visual wave data. Uncertainties related to the wave steepness estimates are discussed. The obtained values are compared with the values specified in design guidelines. Finally, recommendations concerning the estimates of extreme mean wave steepness are provided.

Introduction

The "average" (or "significant") wave steepness defined in terms of the significant wave height and the zero-crossing wave period (or the spectral peak period) represents one of the sea state characteristics of interest for the structural design of offshore structures. It is related to wave breaking and it affects the wave forces on structures, influencing structural response analysis.

Battjes (1972) used Shipborne Wave Recorder measurements from the UK to show that average wave steepness appeared to have an upper limit of around 1/16 – 1/20. These results were later confirmed by Draper (1976), who documented that the higher values of the average wave steepness were generally lying within the range 1/16 –1/20. The lower limit of the steepness range is close to 1/19.7, which is the wave steepness calculated from a Pierson-Moskowitz spectrum, see Carter et al. (1986).

Carter et al. (1986) calculated the average wave steepness based on a formula from the JONSWAP measurements for significant wave height and zero-crossing wave period in terms of fetch and wind speed. For a fully developed sea

---

[1] Principal Research Eng., Det Norske Veritas AS, 1322 Høvik, Norway

[2] Prof. Unit of Marine Techn.& Eng., Instituto Superior Tecnico, 1096 Lisboa, Portugal

[3] Res. Assistant, UMTE, Instituto Superior Tecnico, 1096 Lisboa, Portugal

the steepness was 1/20.2, thus close to the value for the Pierson-Moskowitz spectrum. For a growing sea the steepness values were provided for various fetches and two wind speeds: 30m/s and 40m/s. These values covered the range of 50-year return value of wind speed in UK waters. The authors showed that in open waters, for a growing sea the average steepness was generally less than about 1/16. For limited fetches the waves were considerably steeper, in the range of 1/11.1 –1/14.8.

Recent investigations carried out by Bjerke et al. (1990), Mathiesen and Torsethaugen (1992), and Torsethaugen (1993) have shown that for high sea states the extreme steepness for the Northern North Sea and Haltenbanken is in the range of $s_s = 0.067 - 0.070$. However, investigations of Bitner-Gregersen et al. (1996) bring some doubt concerning this steepness limit.

The number of reliable simultaneous data of significant wave height and zero-crossing wave period collected by different instruments has increased significantly in the last years, particularly for the North and Norwegian Sea. This allows the verification of the commonly used values of wave steepness. In the present study the mean as well as the extreme values of average wave steepness are evaluated for different locations using wave buoy, satellite, hindcast as well as visual wave data. Uncertainties related to the wave steepness estimates are discussed. Furthermore, the obtained values are compared with the values given by the design guidelines. Finally, recommendations are provided of the extreme values of average wave steepness to adopt in design.

Design Guidelines

The average wave steepness may be defined as

$$S_s = 2\pi H_s / gT_z^2 \qquad (1)$$

where H_s denotes significant wave height and T_z zero-crossing wave period (H_{mo}, T_{mo2} respectively if wave height and period are evaluated from the wave spectrum).

The DNV Classification Notes No.31.4 (DNV, 1987) recommends that the sea steepness need not to be taken greater than the 100-year steepness for unrestricted service, which normally may be adopted as:

$$h_s = \begin{cases} 0.156 t_z^2 & \text{for } t_z \leq 6s \\ 0.206 t_z^2 - 0.0086 t_z^3 & \text{for } 6 < t_z < 12s \\ 0.104 t_z^2 & \text{for } t_z \geq 12s \end{cases} \qquad (2)$$

The UK Department of Energy (1990) gives the following limits on wave steepness:

$$3.2h_s^{1/2} < t_z < 3.6h_s^{1/2} \qquad (3)$$

which reflect the maximum and minimum wave steepness observed around the British Isles.

Thus for H_s=15.0m the DNV recommendations give $s_s = 1/15 \cong 0.067$ while the UK Department of Energy $s_s = 1/16 - 1/20$.

Average Wave Steepness for Different Locations

The average wave steepness has been calculated based on (1) for different data types and for different locations. The following locations are considered in the study: Sines and Figueira da Foz in the Portuguese waters, the Southern North Sea (SNS), the Northern North Sea (NNS), the Norwegian Sea (NS), Haltenbanken, Tromsøflaket and the North Atlantic Ocean (NAO).

The Sines, Figueira da Foz, SNS, NNS, NS, Haltenbanken and Tromsøflaket data have been collected by waverider buoys. The measured data records cover a period of 10 years for Sines and Figueira da Foz, 16 years for SNS, 14 years for NNS, 2.5 years for NS, 10 years for Haltenbanken and 12 years for Tromsøflaket. Additionally, for the NS location the 42-year hindcast data generated by the Norwegian hindcast model are analysed. The wave buoy data are compared with the Global Wave Statistics data published by Hogben et al. (1986), from the worldwide (GWS) and European (EGWS) database. Two data types are considered for the NAO route, which includes the GWS zones 11, 9, 16, 24, and 3: 37-year GWS data and 4-year satellite data originated from CLIOSat, (Lasnier et al. 1995). For more details concerning the satellite data the reader is referred to Bitner-Gregersen et al. (1996). Only omnidirectional data are considered. All data sets, apart from the Global Wave Statistics one, which is long enough, are affected by the climatic uncertainty.

Based on the data, the mean values of average wave steepness have been calculated and are presented in Figures 1, to 4. In order to be able to extrapolate the data to extreme conditions the steepness data has been fitted, using the nonlinear regression model:

$$s_s = c_1 h_{mo}^{c_2} \qquad (4)$$

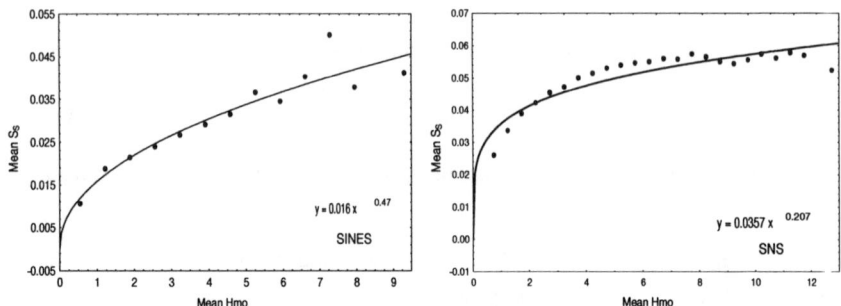

Figure 1. Mean value of average wave steepness of Sines and SNS

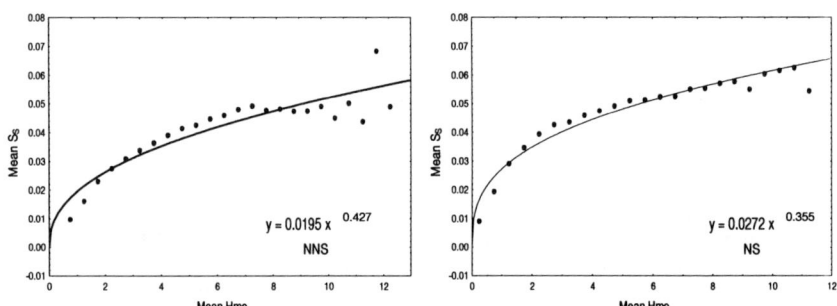

Figure 2. Mean value of average wave steepness of NNS and NS

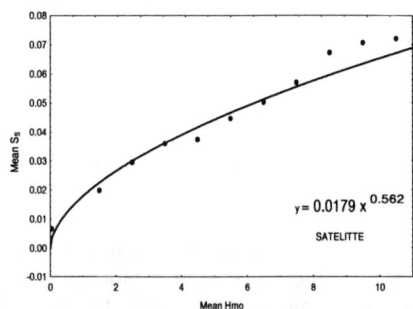

Figure 3. Mean value of average wave steepness for NAO (satellite data).

Figure 4 shows that the Global Wave Statistics data leads to larger steepness.

The extreme wave steepness has been calculated by fitting a joint 3-parameter Weibull distribution for significant wave height and a lognormal conditional distribution for the zero crossing wave period given the significant wave height:

$$f_{H_{mo}}(h_{mo}) = \frac{\beta}{\alpha}\left(\frac{h_{mo}-\gamma}{\alpha}\right)^{\beta-1} \exp\left[-\left(\frac{h_{mo}-\gamma}{\alpha}\right)^{\beta}\right] \quad (5)$$

where α is the scale parameter, β the slope parameter, γ the location parameter, and

$$f_{T_{mo2}|H_{mo}}(t_{mo2}|h_{mo}) = \frac{1}{\sqrt{2\pi}\sigma t_{mo2}} \exp\left[-\frac{(\ln t_{mo2}-\mu)^2}{2\sigma^2}\right] \quad (6)$$

where:

$$\mu = E(\ln T_{mo2}) = a_1 + a_2 h_{mo}^{a_3}$$
$$\sigma = Std(\ln T_{mo2}) = b_1 + b_2 e^{b_3 h_{mo}}$$

The parameters α, β, γ, a_1, a_2, a_3, b_1, b_2 and b_3 have been determined from the data by the least squares technique.

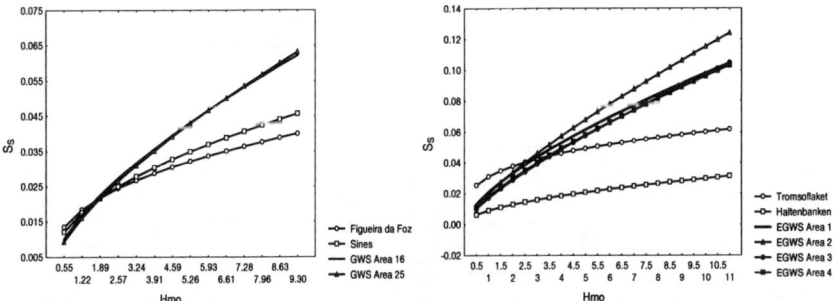

Figure 4. Average wave steepness; comparison of GWS, EGWS and wave buoy data.

Furthermore, the 100-year characteristic largest value of the significant wave height has been evaluated according to the expression

$$h_{100} = \gamma + \alpha(\ln N)^{1/\beta} \quad (7)$$

where N is the number of observations during the 100-year period (N=292 000 for the wave buoy data, N=219 000 for GW visual observations, N=146 000 for hindcast data and N=36 500 for the satellite data).

In Table 1 the extreme significant wave heights and related wave steepness calculated by formula (4) as well as from the joint distribution of wave height and wave period ((5) and (6)) are shown. The presented values cover the range of 100-year return values.

As seen, the mean average wave steepness increases for higher sea states for all locations and data types considered. In the European waters the evaluated mean wave steepness is s_s=1/13.18-1/22.27, for the wave buoy data, and s_s=1/6.85-1/9.2 for the EGWS data, while it is s_s=1/11 -1/12.67 for the North Atlantic.

Table 1 Average wave steepness given significant wave height

Locat.	hmo/hs (m)	Ss Acc. To (5)	Ss From joint model
Sines	9.0	0.0449	0.0445
Sines	10.0	0.0472	0.0466
Sines	11.0	0.0494	0.0485
Sines	12.0	0.0514	0.0503
SNS	14.0	0.0617 (0.123)#	0.0555
SNS	15.0	0.0626 (0.129)#	0.0548
SNS	16.0	0.0635 (0.134)#	0.0541
SNS	17.0	0.0643 (0.140)#	0.0532
NNS	14.0	0.0601 (0.109)#	0.0486
NNS	15.0	0.0619 (0.114)#	0.0478
NNS	16.0	0.0637 (0.120)#	0.0467
NNS	17.0	0.0653 (0.125)#	0.0456
NNS	18.0	0.0670 (0.130)#	0.0444
NS	14.0	0.0694 (0.123)#	0.0642 (0.0623)*
NS	15.0	0.0711 (0.129)#	0.0629 (0.0628)*
NS	16.0	0.0728 (0.135)#	0.0612 (0.0632)*
NS	17.0	0.0743 (0.140)#	0.0593 (0.0636)*
NS	18.0	0.0759 (0.146)#	0.0574 (0.0639)*
NAO	14.0	0.0789	0.0780 (0.0739)"
NAO	15.0	0.0820	0.0800 (0.0771)"
NAO	16.0	0.0850	0.0819 (0.0803)"
NAO	17.0	0.0880	0.0837 (0.0833)"
NAO	18.0	0.0908	0.0853 (0.0863)"

*hindcast data, #EGWS data, "GWS data

For the wave buoy data in the European waters the extreme values of average wave steepness show an opposite trend to the mean values for all locations apart from Sines. They decrease with increase of the wave height. A slight increase of the average wave steepness is obtained also for the hindcast data for higher sea states, but these data are affected by the model uncertainty, while the instrumental uncertainty of the wave buoy data can be regarded as negligible for the wave parameters considered. The calculated steepness limits lie within a range $s_s=1/15.58$ -$1/22.52$, thus well within the design recommendations. The obtained results are in good agreement with the findings of Battjes (1972), Draper (1976) and Carter et al. (1986).

The North Atlantic data (the GWS and the satellite data) show an increase of the extreme average steepness for the higher sea states. Furthermore, the calculated extreme wave steepness values are outside the limits provided by the industry ($s_s=1/11.59$ -$1/13.53$), indicating that in this region waves are steeper than in the European waters. However, some doubt still remains concerning the GWS data accuracy, especially wave period, confer e.g. Guedes Soares and Moan (1991), and Bitner-Gregersen et al. (1994, 1995). The evaluated mean wave steepness for the North and Norwegian Sea supports this doubt.

Furthermore, an increase of the wave steepness can be observed from Sines to the North Atlantic, i.e., when going to more open waters.

The calculated mean average wave steepness is systematically higher than the extreme average steepness evaluated from the joint model for all locations and data types considered. One should expect that for high significant wave heights the extreme wave steepness converge against the mean average wave steepness. However, the presented results are valid within the model applied.

Conclusions

The mean as well as the extreme values of average wave steepness are evaluated for different locations using waverider buoy, satellite, hindcast as well as visual wave data. The European waters as well as the North Atlantic region have been studied.

For high sea states an increase of the wave steepness is observed when going from coastal to more open waters of the North Atlantic.

In the European waters the extreme average wave steepness evaluated based on the wave buoy data decreases with increasing significant wave height for all locations apart from Sines. There is some indication that the higher sea states may be characterised by lower average wave steepness. The calculated steepness limits $s_s=1/15.58$ -$1/22.52$ confirm the findings of Battjes (1972), Draper (1976) and Carter et al.(1986). The North Atlantic data (the GWS and the satellite data) show that the extreme average steepness increases for the higher sea states. It should be noticed that the presented results are valid within the models applied.

For the European waters the evaluated extreme wave steepness based on the wave buoy data is within the limits given in the guidelines, while the calculated North Atlantic wave steepness is outside the recommendations.

The study strongly indicates that the accuracy of the Global Wave Statistics worldwide and European databases should be investigated.

References

Bitner-Gregersen, E.M., and Cramer, E., H., "Accuracy of the Global Wave Statistics Data", *Proc. of the ISOPE-94 conference*, Osaka, 1994.

Bitner-Gregersen, E.M., Cramer, E.H. and Korbijn, F., "Environmental Description for Long Term Response of Ship Structures ", *Proc. of the ISOPE-95 conference*, the Hauge 1995.

Bitner-Gregersen,E., Bonicel, D., Hajji, H., Olagnon, M., and Parmentier, G., " World-wide Characteristics of H_s and T_z for Long-term Load responses of Ships and Offshore Structures", *Proc. ISOPE-96 conference*, Los Angeles, 1996.

Bjerke, P.L., Mathiesen, M., Torsethaugen, K., "Haltenbanken Area Metocean Study", *Main Report SINTEF NHL 1990*, STF60 A90055.

Hogben, N., Da Cunha, L.F. and Olliver, H.N. "Global Wave Statistics", *Unwin Brothers Ltd, London* 1986.

DNV Classification Note No.31.4 "Column Stabilized Units (Semisubmersible Platforms"), September 1987.

Guedes Soares, C., and Moan, T., "Model Uncertainty in the Long-term Distribution of Wave-induced Bending Moments for Fatigue Design of Ship Structures", *J. Marine Structures*, Vol.4, 1991.

Lasnier, P., Loeul, S., Hajji, H., Bonicel, D., and Charriez, P., "Contribution of ERS data to CLIOSat Project, the Satellite Metocean Atlas", *Proc. of the 2nd ERS-1 International Workshop* -London, Dec. 6-8, 1995.

Mathiesen, M, and Torsethaugen, K., "Some Aspects of the Wave Climate at the Draugen Field", *SINTEF NHL, Report No. SFT60, F92087*, October 1992.

Torsethaugen, K., "Wave Steepness for the Draugen and Troll Fields", *SINTEF NHL Report, No.STF60 F93029*, 8th March 1993.

Subject Index

Page number refers to the first page of paper

Acceleration, 126
Acoustic measurement, 297
Algorithms, 40, 273
Altimeters, 354
Anemometers, 428
Armor units, 289

Barbados, 215
Bathymetry, 412, 432
Beaches, 432
Bed load movement, 446
Boundary conditions, 40, 48, 412, 486
Boundary element method, 108
Boundary value problems, 9
Breaking waves, 17, 273, 281, 289, 297, 305, 486
Breakwaters, 116, 470, 502
Bubbles, 297
Buoys, 513

Caribbean, 215, 257
Cnoidal waves, 100
Coastal environment, 454
Coastal structures, 289, 470, 478, 502
Coefficients, 147
Cohesive sediment, 454
Columns, 203
Computation, 1, 76, 84, 155, 486
Computer aided instruction, 64
Computer networks, 64
Computer programs, 155
Contaminants, 462
Currents, 462
Cylinders, 108, 126, 179

Damping, 147
Data analysis, 513
Databases, 354
Decks, 195
Deep water, 265, 305, 420
Density measurement, 297
Design, 215, 257, 265, 362, 470
Design criteria, 494
Design data, 240
Design waves, 171
Diffraction, 203
Distribution functions, 372
Drag, 134
Dynamic response, 134, 171, 240, 494
Dynamics, 1

Energy dissipation, 486
Experimental data, 76

Field tests, 313, 350
Finite differences, 321
Fixed structures, 240
Floating structures, 116, 147, 179, 240
Flow visualization, 64
Fluid-structure interaction, 249
Flumes, 56, 281
Forced vibration, 179
Forecasting, 225, 404
Fourier analysis, 34
Free surfaces, 281
Frequency analysis, 147

Geometry, 281
Gravity waves, 404
Greens function, 412

Guidelines, 195
Gulf of Mexico, 334, 342

Harbors, 116, 404, 454, 502
Heaving, 225
Hurricanes, 195, 257, 321, 342, 362
Hybrid simulation, 25
Hydrodynamics, 1, 179, 215, 233, 446

Image analysis, 281
Impact, 187
Interactions, 34, 126, 155
Interactive graphics, 64
Internal waves, 34

Kinematics, 1, 17, 25, 34, 40, 48, 64, 76, 203, 350, 420

Laboratory tests, 203, 305, 439
Lagrange's equations, 56
Laplace transform, 96, 116
Linear systems, 203
Littoral currents, 446
Loads, 1, 108, 126, 134, 155, 195, 203, 233, 240, 273, 470

Mass transport, 9
Measurement, 428, 513
Mexico, 257
Model studies, 350, 454, 478
Model tests, 249, 470
Modeling, 249, 439
Models, 17, 25, 203, 305, 334
Mooring, 147
Motion effects, 225

Nonlinear analysis, 96
Nonlinear response, 84, 249
Nonlinear systems, 100, 126, 134, 396

North Sea, 372
Numerical models, 179, 289, 321, 404, 446

Ocean waves, 1, 9, 17, 25, 34, 40, 48, 96, 171, 273, 281, 342
Offshore drilling, 225
Offshore pipeline, 494
Offshore platforms, 134, 155, 195, 233, 257
Offshore structures, 1, 265, 273, 396, 428
Oscillations, 147, 179

Parameters, 362
Parametric hydrology, 350
Piers, 215
Piles, 502
Pipeline design, 494
Plunging flow, 187
Pneumatic systems, 179
Porous media, 289, 502
Predictions, 25, 84, 240, 313, 321, 334
Pressure measurement, 187
Public opinion, 265

Random waves, 155, 233, 249, 305
Reliability, 171, 265
Revetments, 187, 478
Risk, 265
Risk analysis, 462

Satellites, 354, 396, 513
Sediment deposits, 454
Sediment transport, 432, 446
Sensitivity analysis, 362
Shallow water, 48, 100, 420, 432
Ships, 155, 225
Shoaling, 404
Shore protection, 470

Simulation, 84, 96, 100, 116, 273, 396, 446
Spectral analysis, 171
Stability, 494
Standards, 265
Statistics, 297, 342, 350, 354, 362, 372
Stoke's law, 9
Storms, 233, 257, 321, 342
Structural failures, 265
Structures, 108
Submarine pipelines, 494
Surf zone, 297
Surface waves, 34, 40, 350
Surge, 249
Suspended sediments, 462

Texas, 462
Theories, 9, 64, 96, 372
Three-dimensional analysis, 108
Three-dimensional models, 34
Time series analysis, 48
Time studies, 96
Towing, 428
Transient flow, 17
Tropical cyclones, 342
Turbulence, 428
Turbulent boundary layers, 432
Two-dimensional models, 126

Undertow, 439
Unsteady flow, 116
Uplift, 187, 215

Velocity, 76

Water depth, 56, 76, 420
Water levels, 362
Water surface profiles, 56
Water waves, 56, 64, 76, 100, 126, 187, 412, 432, 446, 470
Wave climatology, 354, 396
Wave crest, 372, 396
Wave diffraction, 84, 486
Wave energy, 17
Wave forces, 108, 116, 134, 195, 215, 233, 257, 273, 289, 454, 462, 494, 502
Wave generation, 9
Wave groups, 17, 420
Wave height, 313, 334, 342, 354, 396, 432, 439, 470, 502, 513
Wave measurement, 25, 40, 48, 56, 342, 372, 513
Wave propagation, 100, 412
Wave reflection, 486
Wave refraction, 404, 486
Wave runup, 84, 478
Wave spectra, 313, 334, 404, 420
Wave tanks, 84, 428
Wave velocity, 478
Waves, 203, 362, 513
Wind forces, 321, 439, 478
Wind velocity, 354
Wind waves, 313, 439

Author Index

Page number refers to the first page of paper

Abdel-Mawla, S., 454
Arney, Cyril E., 265
Arntsen, Øivind A., 56
Athanassoulis, Gerassimos A., 412

Bashir, T., 126
Bea, Robert, 195, 257
Belibassakis, Konstadinos A., 412
Bender, L. C., III, 334
Bingham, Harry B., 108
Bitner-Gregersen, Elzbieta M., 513
Büchmann, Bjarne, 108
Burge, R. E., 297

Carter, Joshua, 439
Celebi, M. S., 84
Chan, Andy T., 34
Chan, Eng-Soon, 187, 273
Chang, Kuang-An, 289
Chen, Hamn-Ching, 116
Chen, Xiaohong, 147
Cheong, Hin-Fatt, 187
Cheung, Kwok Fai, 108, 179
Chow, Kwok W., 34
Cornett, Andrew, P.E., 215
Cornut, Stéphane, 225
Cox, Daniel, 439

Di Natale, M., 446
DiMarco, S. F., 334
Dyer, Roger, 225
Dysthe, Kristian B., 76

Edge, Billy L., 462
Ewans, Kevin C., 494

Fan, Ju, 147
Forristall, George Z., 372

Gjøsund, Svein H., 56
Gopinath, A. L., 187
Gran, Tone M., 240
Grant, Colin, 225
Grytøyr, Guttorm, 281
Gudmestad, O. T., 126
Gudmestad, Ove T., 1, 76, 134

Halkyard, John, 40
Hardjanto, F. A., 64
Harland, Léon A., 233
Hart, H., 64
Haver, Sverre, 240
Hobensack, William, 439
Holt, Martin, 225
Hsu, Tai-Wen, 486
Huang, Chai-Cheng, 470
Huang, Xianglu, 147
Hughes, Steven A., 48
Hurdle, David P., 362

Im, Sungbin, 249

Jin, Zhaohui, 257
Johannessen, T. B., 17

Karunakaran, Daniel, 134
Kennedy, Douglas, 439
Kim, C. H., 203
Kim, M. H., 64, 84
Kim, S., 155
Kim, Y., 155
Kinnas, S. A., 64
Kring, D. C., 155

Lader, Pål F., 281
Lai, Derek W. C., 34
Lan, Yuan-Jyh, 9

Larm, Katherine A., 462
Lee, Jaw-Fang, 9
Lee, Sing Kwan, 179
Li, Weimin, 100
Liaw, Chih-Young, 273
Lin, Jia-Shiang, 470
Lin, Lihwa, 321
Lin, Pengzhi, 289
Lin, Woei-Min, 116
Liu, Cheng-Chi, 9
Liu, Philip L.-F., 289

Mani, J.S., 502
Matheja, A., 454
Mattila, Mark, P.E., 215
Mehlum, Even, 96
Mitchell, John, 225
Moe, Geir, 56
Müller, Volker, 432
Myrhaug, Dag, 281

Nerzic, Raymond, 396
Niedzwecki, J. M., 350

Park, In-Seung, 249
Park, J. C., 84
Pettersen, Bjørnar, 281
Phadke, Amal, 179
Piccirillo, G., 446
Powers, Edward J., 249
Prevesto, Marc, 396
Prislin, Igor, 40

Ramos, Rafael, 195, 257

Sagli, Gro, 240
Sakakiyama, Tsutomu, 289
Sclavounos, P. D., 155
Seidl, Ludwig H., 179
Seymour, Richard, 305
Sheremet, Alexandru, 404
Silvestre, Alexandra, 513
Skourup, Jesper, 108

Soares, Carlos Guedes, 513
Sobey, Rodney J., 48
Spidsøe, Nils, 134
Stear, James, 195
Su, Ming-Yang, 297
Suastika, I. Ketut, 171
Suhayda, Joseph, 257
Swan, C., 17, 126

Tarbotton, Michael, P.E., 215
Taylor, Paul H., 233, 420
Teague, W. J., 297
Temmos, Jean-Marc, 76
Teng, Chung-Chu, 342
Tromans, Peter S., 171
Trulsen, Karsten, 76

van de Lindt, J. W., 350
van Vledder, Gerbrant Ph., 362
Velarde, Manuel G., 76
Vijfvinkel, Erwin, 420

Wang, Keh-Han, 100
Ward, Donald L., 478
Wen, Chih-Chung, 486
Wen, Jiangang, 25
Wesson, J. C., 297
Wibner, Christopher G., 478
Wisch, David J., 265
Wu, Shukai, 428

Xu, Tao, 195

Yang, Jun, 25
Young, I. R., 313, 354

Zhang, Jun, 25, 305, 478
Zimmermann, C., 454
Zimmermann, Charles-Alexandre, 305
Zong, Zhi, 273
Zou, J., 203